PT2920

Process Engineering of Size Reduction: Ball Milling

by

L. G. Austin
The Pennsylvania State University
University Park, PA

R. R. Klimpel
Dow Chemical Company
Midland, MI

P. T. Luckie
The Pennsylvania State University
University Park, PA

Published by
Society of Mining Engineers
of the
American Institute of Mining, Metallurgical, and Petroleum Engineers, Inc.
New York, New York ● 1984

Library of Congress Catalog Card Number 83-73512
ISBN 0-89520-421-5

*This book is dedicated with thanks
to the students who have performed
much of the work reported here
and to
the National Science Foundation
for financial support over many years.*

Acknowledgments

We wish to thank the many companies and individuals who have supplied data and helpful discussions, especially Dr. H. M. von Seebach of the Polysius Corp., Atlanta, GA; Dr. R. S. C. Rogers of the Kennedy Van Saun Corp., Danville, PA; and Mr. C. Rowland of the Allis Chalmers Co., West Allis, WI.

Thanks are extended to B. G. Ateya, P. S. Bagga, A. Beattie, V. K. Bhatia, K. A. Brame, M. Celik, R. P. Gardner, S. Lohrasb, T. Miles, I. Shah, K. Shoji, F. M. Smaila, C. Tangsathitkulchai, K. Suryanarayanan, T. J. Trimarchi, and N. P. Weymont.

Special thanks are due to Miss Sharon Rishel for her diligence and patience in typing the manuscript and its many revisions. We also aknowledge with gratitude the assistance of Carmel A. Huestis of the Society of Mining Engineers in the production of this volume.

Table of Contents

Introduction: Formulation of the Problems Facing the Designer of Grinding Circuits

Grinding as a Unit Operation

Size reduction by mechanical crushing and grinding is an important operation in the mineral, metallurgical, power, and chemical industries. The quantity of brittle materials such as rocks, ores, coal, and cement products ground annually in the United States is at least 1 Gt (one billion tons) (Rumpf, 1973) with large associated energy consumption (Sheridan, 1977). Single plants handling 9 Mt (10 million tons) or more per year are quite common. Surprisingly, for a unit operation of such basic importance to industrial technology, there have been, till recently, no up-to-date standard texts dealing with process design principles for mills and grinding circuits. Several books describing various features of milling have become available in the last few years (Rowland and Kjos, 1980; Lynch, 1977; Marshall, 1975); the chapter by Rowland and Kjos especially is a good condensed guide to conventional design practice using the Bond method. However, much of the literature presents other design methods either as completely empirical or as *semi-theoretical* based on reasoning which can be shown to be unsound. It is our contention that conventional design procedures can be clearly compared and explained, and one objective of this book (see Chapters 2 and 3) is to collect and outline these empirical procedures for ball mills and demonstrate the errors that can arise when the procedures are applied without a clear understanding of the underlying assumptions and approximations involved.

In addition, the unit operation of grinding has a more elaborate theoretical basis which has developed in the last two decades. Although by no means complete as yet, it will undoubtedly be used more and more in the future. This theoretical basis is comparable to that which exists for heat transfer, distillation, etc. and in particular it has many close similarities to the theory of chemical reactor design and uses many concepts and terminology in common with reactor design. Major objectives of this book are to present this more elaborate treatment in depth and to show the correlations and divergences of its results from the older treatments.

As there are many different types of grinding machines and the physical properties of the types of materials ground in large quantities vary over a wide range, no attempt will be made here to prepare an exhaustive handbook of grinding. This volume will deal almost exclusively with tumbling ball mills performing size reduction on brittle materials, although many of the concepts will be applicable to other types of mills. We will not give an extensive literature review or quote in detail the history and source of the concepts discussed. This book is intended as a compact introduction to the mathematical treatment of the *unit operation* of size reduction by mechanical means, that is, the sizing, behavior, and performance of grinding circuits using ball mills, and mechanical engineering aspects of mill design are mentioned only as they relate to the process design. It is hoped that the book will be suitable as an advanced text for mineral and chemical engineering curricula, since it stresses the fundamental concepts and calculational procedures for size reduction in mills rather than equipment selection or mechanical design.

We would be delighted if teachers of chemical and process engineering could be convinced that size reduction by grinding is not a *black art* which can only be learned by years of experience, but that it forms a developing body of knowledge which makes sense to students and which can be taught with as much intellectual stimulation as the concepts of heat transfer, mass and momentum transfer, and chemical kinetics.

Formulation of the Problems Facing the Designer of Grinding Circuits

In designing any type of reactor a primary task of the process engineer is to size the reactor for a desired output rate of a desired quality of product, using rate constants, mass and thermal balances, heat transfer coefficients, etc. He must allow for the input or extraction of sufficient energy to accomplish the desired reactions and he must design to minimize undesired reactions. The operating system must be stable and controllable, to meet a range of product specifications if necessary. It should produce the specified quantity of output in as efficient a manner as possible, with a minimum of capital expenditure, energy costs, and maintenance/labor costs.

Similar considerations apply to mill design. Consider, for example, the most widely used type of mill, the tumbling ball mill, shown in Fig. 1. Coarse powder feeds into one end of the mill, passes down the mill receiving breakage actions due to the heavy tumbling balls, and leaves as product with a finer size distribution. This can be considered as a continuous *reactor* where energy input is converted to mechanical breakage action, and the *reaction* accomplished is size reduction. All the requirements mentioned above must be met. A basic step in mill circuit design is to *size a mill to produce a desired tonnage per hour of a required product from a specified feed*. The capital expenditure per unit of mill capacity has to be minimized, which involves the correct choice of mill conditions such as rotational speed, ball load, and ball sizes.

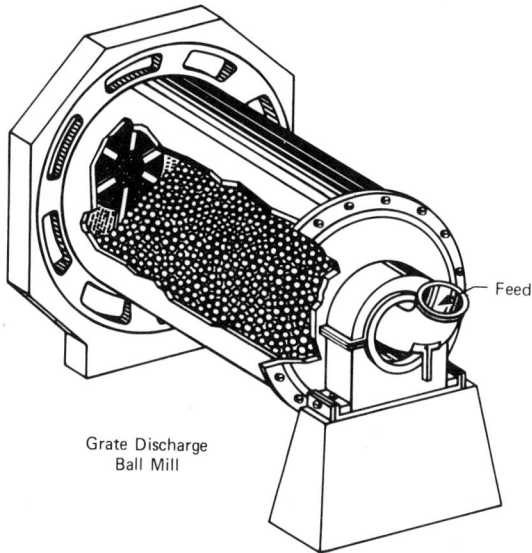

Feed

Grate Discharge
Ball Mill

Fig. 1. Illustration of a simple tumbling ball mill at rest.

Associated with the basic step of mill size is the specification of the power necessary to run the mill and the expected use of energy per ton of product. Obviously the designer wishes to be able to specify the grinding conditions which produce a minimum use of energy per ton of product. However, it must be remembered that the conditions for minimum energy are not necessarily those for maximum capacity. In general, the mill should be designed to operate with as efficient grinding as possible, defined by high specific mill capacity and low specific energy consumption, subject to constraints of wear, maintenance costs, contamination of product, etc. Further, it is usually very desirable to know how the circuit will react to changes in operating conditions, so that advice can be given to the operator who has to run the circuit to meet different specifications.

As in many reactor systems, the use of several stages of grinding combined with recycle can be advantageous. It is common practice to pass the material coming out of a mill through a size *classifier*, which splits the mill product into two streams, one containing coarser (oversize) particles and the other finer particles. The coarse stream is recycled back to the mill feed (see Fig. 2). The process of selective size separation is known as *classification*, and there are several types of devices which perform this classification action: continuous screens, spiral and rake classifiers, hydro-cyclones, air separators. The design must include a specification of the optimum amount of recycle and how to obtain it.

There may be two mills in series, with appropriate classifiers and recycle or there may be recycle and regrinding of material from some later stage in the process, e.g., from flotation cells. It is thus often necessary to choose

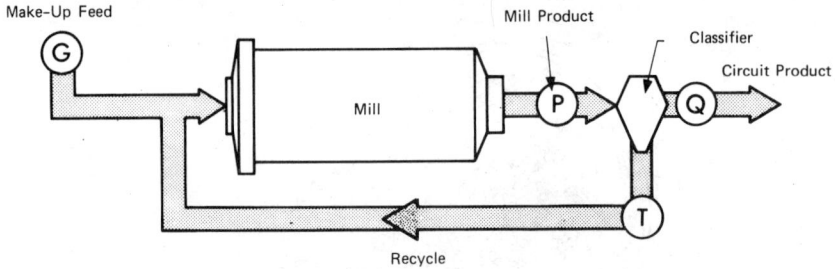

Fig. 2. A grinding circuit with recycle of the coarse stream to the mill feed. The symbols represent mass flow rates of solid.

between several alternative milling circuits and define the sizes of a number of components to obtain the most efficient system for a particular task. For example, the designer can be faced with the choice between a circuit containing primary crusher-secondary crusher-tertiary crusher-rod mill-ball mill and one consisting of a primary crusher-autogenous mill sequence. Both circuits may be technically feasible and the choice then becomes a matter of overall economics.

It is not often possible to design the grinding components optimally without knowing the rest of the system and the system engineering of the whole becomes important. A complete design involves the rest of the milling and treatment circuit, with descriptions of classifiers, flotation cells, thickeners, etc., being necessary for total systems analysis. Optimal design involves making a desired product or products with a balanced minimum of capital investment, power and maintenance costs, and a maximum of reliability, ease of control, ease of future expansion and so on. However, as in any system analysis, the total system can be split into subsystems amenable to individual analysis and this book is a detailed treatment of ball milling as a separate subsystem.

To summarize then, the following factors must be considered:
1) Mill size.
2) Mill power, specific grinding energy.
3) Efficient grinding conditions.
4) Recycle, classification efficiency.
5) Mill circuit behavior under varying conditions.
6) Mill selection for complex circuits.
7) Economic optimization.

Definition of Terms and Concepts

A grinding mill is essentially a *reactor* which is *reacting* large particles to smaller particles. There are, of course, many ways of applying force to particles to cause breakage but mineral engineers are primarily interested in large, continuous devices operating on brittle materials at high throughput rates, often with approximately steady output throughout the twenty-four hours of the day. Widely used mills under these circumstances are rod mills and ball mills, illustrated in Figs. 3 and 1. These mills are simple, relatively cheap to construct, reliable, easy to control and maintain, and have low energy requirements per ton of product compared to many other types of grinding machines.

The *reactant* in the mill is the feed going into the mill; rarely is this a single size, and the feed is normally a complete distribution of sizes, so that a set of reactants must be considered. This size distribution can be represented by a continuous curve, or a set of numbers $P(x)$, representing the cumulative weight fraction below size x vs. x; it is normally convenient to use log-log scales for this plot, as in Fig. 4. The simplest and most reliable method of size analysis is sieving, and size frequently refers to sieve size (see Table 1). The weight fraction, w, in the various sieve size intervals contains the same information as Fig. 4, so that a set of w numbers also represents the size distribution. It is usually convenient to use size intervals in geometric progression corresponding to the standard sieve progression. We will use the arbitrary convention of numbering the top size of interest as 1, 2 for the next smaller, etc. (see Fig. 4). Considering any general size interval, the ith size interval, the fraction of the material in the ith interval is w_i. It is not easy to extend size distributions to very fine sizes, beyond 38 μm (400 mesh) because of the experimental difficulty of accurately measuring small sizes. The final size interval containing the weights of the smallest material is defined by the weight fraction, w_n, of sizes less than or equal to the smallest measuring size used. This is referred to as the *sink* interval, since it is

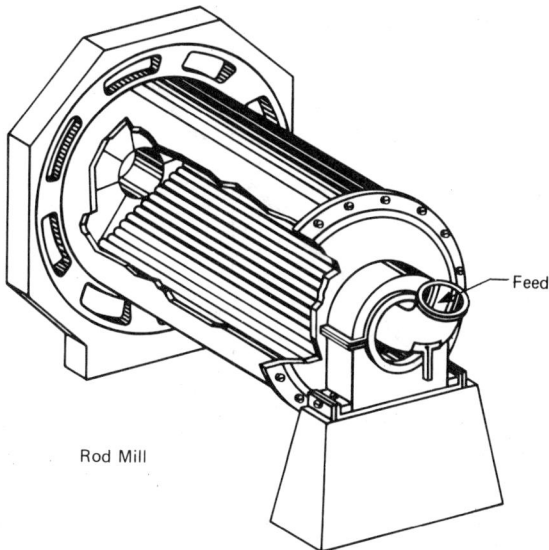

Fig. 3. Illustration of a simple tumbling rod mill at rest.

receiving material from breakage of all larger sizes but material cannot be broken out of the interval.

The *product* from the mill is the size distribution of the material coming out of the mill. Again, this is never a single size and a curve or set of numbers is used to characterize the size distribution in the same way as for the feed material. To define a grinding system, the desired product must be clearly specified. It is not generally possible to specify the complete size distribution and it is usually specified by: (1) a single point, e.g., 80%-by-weight passes 75 μm (200 mesh); (2) two points, e.g., 50% less than 38 μm (400 mesh), no more than 5% greater than 208 μm (65 mesh); or (3) an equivalent specific surface area of the distribution. These specifications are a function of the process following the milling. For example, lumps of fired cement clinker are ground to a powder fine enough to give suitable properties to concrete, and the concrete industry has found a number of empirical

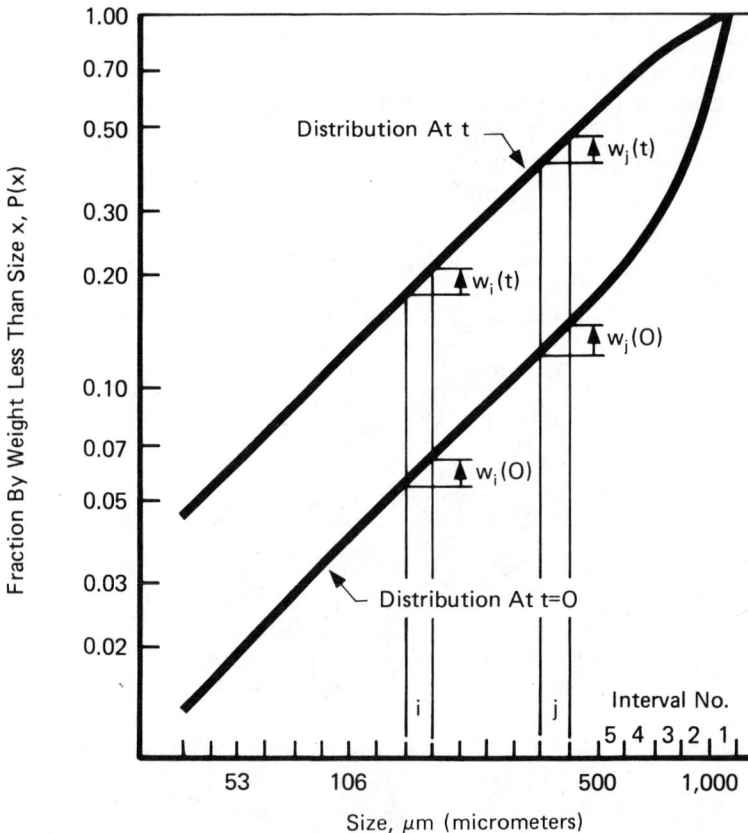

Fig. 4. A log-log plot of cumulative size distribution data: *t* is the time of grinding.

Table 1. The International Standard Sieve Series

Standard	US Sieve Designation	Standard	US Sieve Designation
125 mm	5 in.	850 μm	No. 20
106 mm	4.24 in.	710 μm	No. 25
100 mm	4 in.	600 μm	No. 30
90 mm	3 1/2 in.	500 μm	No. 35
75 mm	3 in.	425 μm	No. 40
63 mm	2 1/2 in.	355 μm	No. 45
53 mm	2.12 in.	300 μm	No. 50
50 mm	2 in.	250 μm	No. 60
45 mm	1 3/4 in.	212 μm	No. 70
37.5 mm	1 1/2 in.	180 μm	No. 80
31.5 mm	1 1/4 in.	150 μm	No. 100
26.5 mm	1.06 in.	125 μm	No. 120
25.0 mm	1 in.	106 μm	No. 140
22.4 mm	7/8 in.	90 μm	No. 170
19.0 mm	3/4 in.	75 μm	No. 200
16.0 mm	5/8 in.	63 μm	No. 230
13.2 mm	0.530 in.	53 μm	No. 270
12.5 mm	1/2 in.	45 μm	No. 325
11.2 mm	7/16 in.	38 μm	No. 400
9.5 mm	3/8 in.		
8.0 mm	5/16 in.		
6.7 mm	0.265 in.		
6.3 mm	1/4 in.		
5.6 mm	No. 3 1/2		
4.75 mm	No. 4		
4.00 mm	No. 5		
3.35 mm	No. 6		
2.80 mm	No. 7		
2.36 mm	No. 8		
2.00 mm	No. 10		
1.70 mm	No. 12		
1.40 mm	No. 14		
1.18 mm	No. 16		
1.00 mm	No. 18		

correlations between the Blaine air-permeability surface area* of the powder and the performance of the concrete. Again, it is found that pulverized coal fired into large steam generators burns satisfactorily when it is mainly less than 75 μm (200 US mesh) and contains no more than 5% by weight greater than 300 μm (50 mesh).

Another very important application of a product size specification occurs in the freeing (*liberation*) of a valuable material from a mass of rock in extractive metallurgical operations. By trial-and-error laboratory tests,

---

*ASTM Designation C204-73.

the metallurgist arrives at a desired *fineness of grind* to give sufficient liberation, which he then specifies to the mill designer. In concentration of the valuable component by flotation, it is known that very fine particles (e.g., minus 5 μm) float poorly, as do large particles (e.g., plus 300 μm). This is an example of a specification where the product should be mainly less than a specified size, but should also have the minimum of fines.

As will be shown later, and as might be expected from common-sense reasoning, the rates at which particles break in a grinding device depend on the size of the particle. Unlike a chemical reactor which simply converts A to B, a mill operates on a whole set of feed sizes and produces a whole set of final sizes. The statement that a mill produces 10 tons per hour with an energy expenditure of 20 kWh/t is meaningless without defining the feed and product size distributions. Obviously, more breakage actions are required to grind a coarser feed to a desired product size than to grind a finer feed to the same product size and hence, the energy per ton is higher and the mill output lower for the coarser feed under normal conditions. Similarly, the energy to grind a given feed to 80%-by-weight less than 500 μm is clearly much lower than taking the same feed to 80% less than 50 μm, for example, and a primary function of laws or rules of grinding is to develop methods for accurately calculating the difference.

Analogous to a chemical reactor, knowledge of rates at which materials break enables the prediction of how rapidly these particle sizes disappear from the mill contents. However, unlike a simple chemical reaction $A \rightarrow B$, the fragmentation of even a fixed size of particle produces a complete range of product sizes. If the size range is split into a number of intervals, the fraction of material broken from a fixed size which falls into a smaller size interval can be considered a *product*, as illustrated in Fig. 5. It is clear that a reasonably detailed understanding of the mill must involve a knowledge of these *primary progeny fragment distributions* or the *primary breakage distribution function*. Knowing how quickly a size breaks, and in what sizes its products appear, is the elemental description of the *size-mass balance* or *population balance* of the reactor.

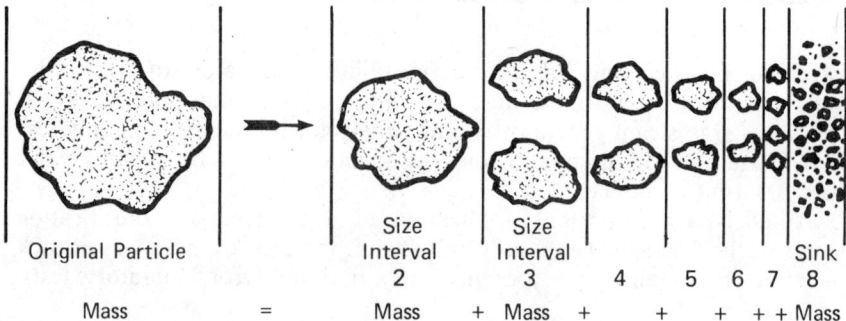

| Original Particle | | Size Interval 2 | Size Interval 3 | 4 | 5 | 6 | 7 | Sink 8 |
| Mass | = | Mass | + Mass + | | + | + | + | + Mass |

Fig. 5. Illustration of the fraction of material broken from a fixed size which falls into smaller size intervals

To define the various rates of breakage in a mill, consider the mill as a *black box* of volume V containing a mass of powder W. Looking at a particular size interval i, the fraction of W which is of size i is w_i, therefore the mass of size i is w_iW. The *specific rate of breakage* S_i, of this size is the *fractional* rate of breakage, e.g., kg of size i broken per unit time per kg of size i present $= $ kg time^{-1}/kg $=$ time^{-1}. Thus S_i is defined by

$$\text{rate of breakage of size } i = S_iw_iW \tag{1}$$

This is the equivalent of a first-order chemical rate constant: note, however, that a chemical rate is usually defined per unit reactor volume whereas the rate of breakage is defined per unit mass in the mill. We will also use the term *absolute rate of breakage* of size i, to denote the product S_iW; this would be the total rate of breakage if all of W were of size i, and it has units of kg min^{-1}, for example. Another term which is widely used is the term *mill capacity*, often expressed in ton per hour, tph. This means the rate of production of a desired product from a specified feed for a given mill. Obviously this lumps all the specific rates of breakage, progeny fragment distributions, feed-product size specifications and mill size into a single number which can only be constant for precisely constant conditions. The term *specific mill capacity* is used for mill capacity per unit mill volume, with units of kg m^{-3} min^{-1}, for example, but is also used for mill capacity per unit powder mass in the mill, with units of min^{-1}, for example.

Again, considering the mill as a reactor, another fundamental concept emerges. If the feed rate to a given size of ball mill is decreased, the material spends longer in the mill, receives more breakage actions and is hence ground finer. Thus the *residence time* (also called *retention time*) is one fundamental descriptor of the mill. On the other hand, if the mill geometry or the ball loading conditions in the mill are changed, the intensity and statistics of breakage actions per unit mill volume are also changed, so a specification of residence time is a necessary but not a sufficient design statement (changing mill conditions is analogous to changing temperature in a chemical reactor). As in any reactor system, the above concept also leads to the concept of a *residence time distribution* (Levenspiel, 1962). From a traced pulse of feed admitted to the mill over a very short time, some traced material may leave the mill almost immediately, while some may hold up in the mill for longer periods of time, and a complete distribution of residence time exists. This is illustrated in Fig. 6. *Plug flow* is here defined as the sudden emergence of all traced feed after the mean residence time τ, and implies no forward or backward mixing as the material moves along the mill:

$$\left.\begin{array}{ll} \Phi(t) = 0, & t < \tau \\ \Phi(t) = 1, & t > \tau \end{array}\right\} \tag{2}$$

where $\Phi(t)$ is the cumulative fraction of feed which has emerged at time t after admission. At the other extreme, a *fully mixed* system immediately mixes the traced material into the bulk of the charge, and in the usual way

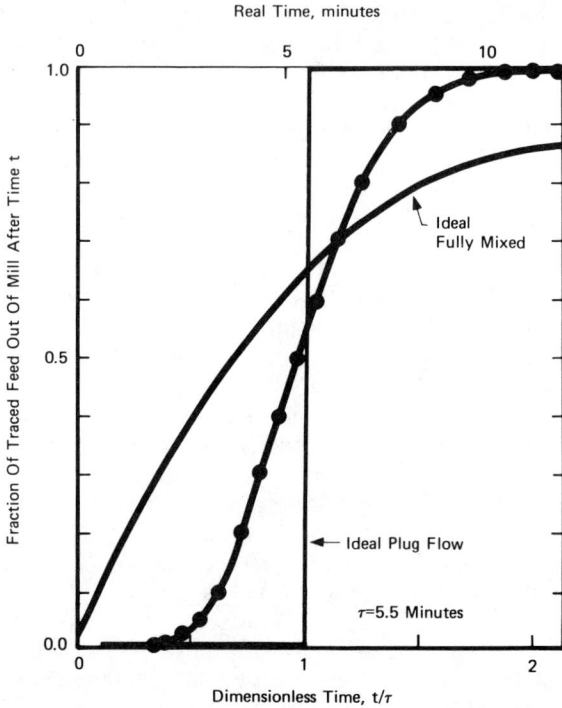

Fig. 6. Cumulative residence time distribution of a tumbling ball mill dry grinding cement clinker.

(Levenspiel, 1962),

$$\Phi(t) = 1 - \exp(-t/\tau) \tag{3}$$

The *mean residence time* τ is defined by W/F, W being the mass of powder material in the mill (e.g., kg) and F the feed rate (e.g., kg/min). W is commonly called the *hold-up* in the mill.

The *shape* of the product size distribution can be altered by the manner in which the milling circuit is designed and operated. By shape, we mean the steepness of the size distributions shown in Fig. 4, that is, the relative proportions of fines to medium to coarse material. In many industries, the milled product has to be less than a certain size, but the presence of *excess fines* is undesirable. Less relative amounts of fines appear as a *steeper* size distribution, as shown in Fig. 7. The production of excess fines can be considered as analogous to an undesirable chemical side reaction which has to be minimized for efficient operation. In wet grinding the excess fine material is sometimes referred to as *slimes*.

An important general principle is that to avoid the production of excess fines it is necessary to remove material which is already fine enough from the mill as quickly as possible, thus avoiding *overgrinding*. Fig. 7 shows a

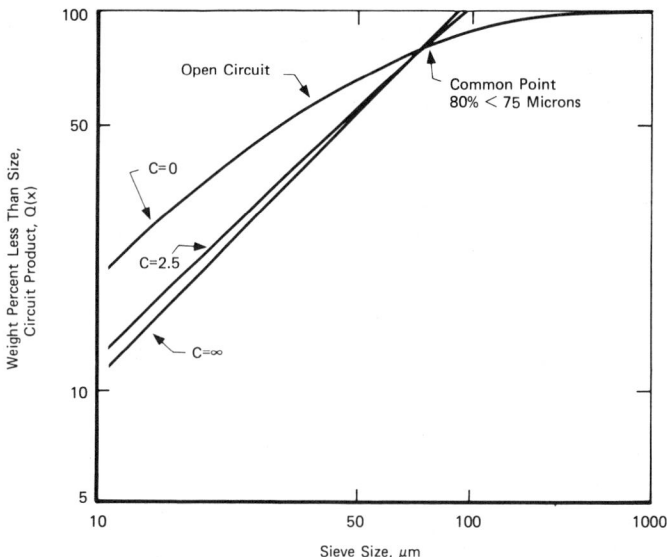

Fig. 7. Illustration of the range of product size distributions with a common point of 80% < 75 μm obtained by varying the amount of recycle.

theoretical result (deduced, as will be described in Chapter 7) from a mill circuit operated to produce a size distribution through the point 80% less than 75 μm. Under open circuit conditions (no classification or recycle), material already fine enough still passes naturally along the mill and is ground finer at the same time as coarse material is being reduced below the control size. Incorporating a classifier and closing the circuit means that the mill is run at higher mass flow rates and shorter residence times. If the steady rates of raw feed and final product are Q tph and if the amount recycled is T tph, the actual flow rate through the mill is $Q + T$. This higher flow rate removes material faster, the fines are separated in the classifier, and the coarser particles are returned to the mill feed. The net effect is that the particle size distribution being acted on in the mill is weighted toward the coarser particles and less fines are present. If fines are not present they are not rebroken. The ratio $(Q + T)/Q$ is called the *circulating load* and is often expressed as a percentage. The ratio T/Q is called the *circulation ratio C*.

Two types of mill inefficiency can be defined. The first type, which we will call *indirect* inefficiency, is that discussed in the previous paragraph. The mill can be breaking efficiently, but energy is wasted in overgrinding material which is already fine enough. The second type, which we will call *direct* inefficiency, occurs when the mill conditions cause poor breakage actions. Examples are: (1) underfilling of the mill by powder, so that the energy of tumbling balls is used in steel-to-steel contact without causing particle breakage; (2) overfilling of the mill so that the ball-powder-ball action is cushioned by excessive powder; and (3) too high slurry density in wet grinding, which gives a thick slurry that can absorb impact without giving breakage (see Chapter 5).

It is sometimes desirable to give an approximate indication of the extent of size reduction accomplished or desired. A term which is often used is the *reduction ratio*. This might be defined as follows: If a feed of maximum particle size x_F is ground to a product of maximum particle size x_P, the reduction ratio is x_F/x_P. Thus to calculate a reduction ratio it is necessary to characterize the size of the feed and the size of the product by single numbers, which is obviously only a crude description. It is found that the maximum size is not a reliable single characterization number because the top sizes may be only a small percentage of the total material. It might be more logical to use the 50% weight passing size (median) as the single size to characterize the size distribution, but this in turn is not descriptive of the largest size present. In practice, size distributions from industrial mills have size distributions which are conveniently characterized by the size for which 80%-by-weight is finer, 20% is coarser. This *80%-passing size* gives a rough indication of the largest sizes present in any significant quantity and yet it is still reasonably characteristic of the size distribution as a whole. Thus the reduction ratio is commonly taken to be the ratio of the 80%-passing sizes of feed to product.

Mill Conditions in Tumbling Ball Mills: Definitions

The tumbling ball mill is the most important type of industrial mill, and much of the work in the following pages is specifically concerned with this type of mill, so the terms describing mill conditions in ball mills are now defined. The tumbling ball mill contains a reservoir of powder being acted on by the breakage action and the fineness of grind depends on how long the material is retained, that is, the product becomes coarser as the feed rate increases, as discussed above. This type of machine is a *retention* device.

The *critical speed* of the mill is defined as the rotational speed at which balls just start to centrifuge on the mill case and not tumble. By balancing the force of gravity on a ball at the top of a mill against the radial centrifugal force for a ball on the case,

$$\text{critical speed, rpm} = 76.6/\sqrt{D - d}\,(D, d \text{ in ft}) \qquad (4a)$$

$$= 42.2/\sqrt{D - d}\,(D, d \text{ in m}) \qquad (4b)$$

where D is the internal mill diameter and d the maximum ball diameter. It is reasonable to expect that the tumbling action in a mill will depend on the fraction of the critical speed at which the mill is run, thus the rotational speed condition of the mill is normally specified by ϕ_c, the *fraction of critical speed*.

The tumbling action and the rates of breakage will clearly depend on how much of the mill volume is filled with balls. The most precise measure of this is the fractional volume filled by ball volume. However, in tests on large scale mills it is often not possible to determine the weight of balls in a mill and therefore not possible to determine their volume, but it is possible to stop the mill and measure the height from the ball bed to the top of the mill. This enables the estimation of the fraction of mill volume filled by the ball bed, Fig. 8. Thus the *fractional ball filling, J*, is conventionally expressed as the fraction of the mill filled by the ball bed at rest. To convert the bed volume to mass of balls present, or vice versa, it is necessary to know the bulk density of the ball bed. The bed porosity varies slightly depending on the ball size mix, powder filling, etc., but it is conventional to define a constant *formal* bed porosity for all calculations. Different industries and different manufacturers use slightly different values of porosity. We will use

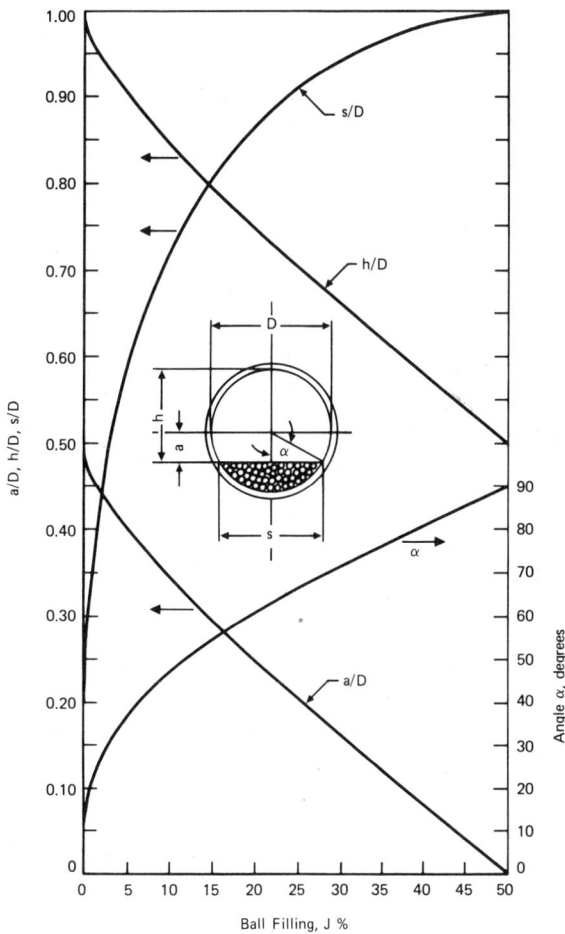

Fig. 8. Geometry of ball charge filling.

a formal bed porosity of 0.4 which gives

$$J = \left(\frac{\text{mass of balls/ball density}}{\text{mill volume}} \right) \times \frac{1.0}{0.6} \qquad (5)$$

since a true steel volume of 0.6 gives a steel-plus-porosity volume of 1. For normal forged steel balls, a formal porosity of 0.4 gives a bed bulk density of 4.70 t/m³ (295 lb/ft³).

Similarly, the mill filling by powder is expressed as the *fraction of mill volume filled by powder bed*, f_c, again using a formal bed porosity of 0.4;

$$f_c = \left(\frac{\text{mass of powder/powder density}}{\text{mill volume}} \right) \times \frac{1.0}{0.6} \qquad (6)$$

In order to relate the powder loading to the ball loading, the formal bulk volume of powder is compared to the formal porosity of the ball bed,

$$U = \frac{f_c}{0.4J} \qquad (7)$$

Thus U is the fraction of the spaces between the balls at rest which is filled with powder; it is found empirically that $U = 0.6$ to 1.1 is a good powder ball loading ratio to give efficient breakage in the mill.

If water is present, the *pulp density*, or *slurry density*, is defined by the fraction by weight of solids in the mixture. Actually, the rheological properties of a slurry are better defined by the volume fraction of solids, c:

$$c = \frac{(w_s/\rho_s)}{(w_s/\rho_s) + [(1 - w_s)/\rho_l]} \qquad (8)$$

where w_s is the weight fraction of solid and ρ_s, ρ_l are the specific gravities or densities of the solid and the liquid. The viscosity of a slurry also depends on the size distribution of the particles (see Chapter 5).

Levels of Complexity: The Different Approaches to Mill Sizing

In describing even a simple grinding system, there are a number of levels of complexity which can be used. These can be categorized as follows, in increasing order of complexity:

1) Global rate-of-production method.
2) Global specific energy method.
3) Global Bond/Charles methods.
4) Size-mass balance method.

The essence of Method 1 is to determine experimentally the rate of production per unit of mill volume (kg/m³s or lb/ft³h) of a desired product from a given feed in a laboratory or pilot-plant test mill, where the conditions in the test mill are chosen to be as close as possible to those in the production mill, and the grinding time is adjusted to produce the desired product size. It is then assumed that the rate of production in the big mill is the test rate multiplied by the volume of the production mill multiplied by a *scale-up factor* based on experience. This technique is quite widely used, but it gives no information on the effect of recycling with classifiers of various efficiencies, the effect of changes in feed size or product specifications, etc.

Method 2 involves a laboratory or pilot-plant test procedure similar to that of Method 1 but with the addition of the measurement of the power to the test mill. The power is used to calculate the *specific energy of grinding* (kWh/t) to go from the given feed to the desired product size. It is then assumed that the specific energy of grinding to go from the given feed to the desired product is *independent* of the mill design or operation. Thus, by measuring the power (m_{p_1}) into a laboratory or pilot-plant mill while it is operating at a steady output rate Q_1 of desired product from given feed, the specific energy is obtained from

$$\text{specific energy } E = m_{p_1}/Q_1 \qquad (9)$$

Then if a production rate of Q_2 is desired from any other mill, its power will be

$$m_{p_2} = EQ_2 = (Q_2/Q_1)m_{p_1} \qquad (10)$$

if a constant specific energy is assumed. Since the power m_{p_2} required to drive a mill at a desired speed can be calculated from empirical equations using mill size and ball loading, an appropriate mill size is selected to give m_{p_2}.

This approach is often surprisingly successful, but its application without prior experience is fraught with danger. There is no fundamental reason why the specific energy of grinding should be constant since it is not a thermodynamic quantity and it is easy to devise systems where it cannot possibly be constant, especially if the production system chosen is more, or less, efficient than the test system. The approach does not address the problems of limited mass flow through the production mill, the correct choice of recirculation, the optimum operating conditions, etc.

The third category uses elements of methods 1 or 2 plus empirical relations, such as *Bond's Law* (Bond, 1960) or *Charles' Law* (Charles, 1957), which describe how the rates of production or specific energies vary with changes in the feed size or product size. We will show that these empirical relations are approximate, limited solutions of the size-mass balance equations of grinding. Again, scale-up factors are used and it is often necessary to perform a series of empirical corrections based on prior experience in order to get accurate results.

All of the above methods are called *global* methods because they are usually applied to the feed entering a circuit and the product leaving the circuit and not to the real feed distribution and real product distribution of the mill (see Fig. 1 in Chapter 3, for example) and they lump all kinetic factors into a single descriptive number, e.g., the Bond Work Index. They are discussed in more detail in Chapters 2 and 3.

Method 4, the highest level of complexity, is to perform a complete size-mass balance on all sizes in the mill, using the concepts of specific rates of breakage, residence time distributions, and mathematical descriptions of classifier action. The scale-up from test results to production conditions, or varying mill conditions, is accomplished via a set of relations which describe how each component in the size-mass balance varies with mill conditions and mill size. It leads to reasonably exact circuit simulations suitable for optimization and system analysis. The advantage of this technique is that alternative circuits can be compared on paper before a final design is adopted. This advanced level of complexity in itself can be treated with various degrees of sophistication (see Chapters 4, 6, and 17). This modern approach, which requires the use of digital computers to perform the calculations for a reasonable number of size intervals (say 20 or 30), is the major topic of this book and is taken up again in Chapter 4.

Summary

Unlike many unit operations in the process oriented industries, grinding has involved the use of design and operating procedures which are almost completely based on empirical correlations. However, because of the inherent similarities of grinding to other process unit operations, very logical and consistent design procedures are evolving. This book presents a mathematical treatment, using concepts very similar to those in chemical reactor design, for assisting in the sizing and performance analysis of grinding circuits. A grinding circuit should produce in a stable, flexible, and safe manner the desired quantity of output (tph) within quality specifications, for specified feed material. This should be accomplished with a minimum of capital, energy, labor, and maintenance costs. The final choice of a design that represents an optimal balance of the above factors requires the availability of sound process design procedures.

The engineer designing a grinding circuit is faced with a number of choices: mill type, mill size, mill power, selective size separation (*classification*), the amount of coarse product recycle, mode of operation (batch or continuous), circuit configuration, the selection of efficient and stable operating conditions, and the meeting of overall cost limitations. This book concentrates on the most common type of mill for large tonnage grinding, the tumbling ball mill. A ball mill is a *retention mill* in which a reservoir of particles is acted on repeatedly, while *once-through* size reduction devices essentially involve one breakage action in one pass, for example, a roll crusher.

To continue the chemical reactor design analogy, the mill is the reactor and the reactions accomplished are breakages of large feed particles into smaller product particles. The feed reactants and reaction products from the mill both consist of a complete distribution (set) of sizes. Each set of sizes can be described by either a continuous curve, $P(x)$, representing the cumulative weight fractions below size x vs. x (see Fig. 4) or by a series of weight fractions occurring in specified size intervals. In the latter description, each ith size interval (usually defined by standard screen intervals, see Table 1) has a corresponding weight fraction w_i.

The quality of the product from the grinding circuit is often specified by one or more of the following measurements: the weight fraction (or %) of product passing one or two preselected sieve sizes; the specific surface area; or the achievement of a preselected degree of liberation of valuable material from a larger mass of rock. The product specifications chosen in a given situation depend on which characteristic is most easily correlated to the desired performance of the final end product (e.g., concrete using cement powder, combustion using pulverized coal, copper recovery by flotation of ground copper ore).

The rates at which particles break depend on a number of factors: particle size, mill diameter, size of grinding media, etc. However, unlike a simple chemical reaction, the single breakage of just one specified particle size (or sieve interval) produces a complete range of finer breakage product sizes (see Fig. 5) which is called the *primary progeny fragment distribution* or *primary breakage distribution function*. Knowing how fast each size breaks, and in what sizes the primary breakage products appear, gives rise to the engineering concept of *size-mass balance* or *population balance* of the reactor. The *specific rate of breakage* of size i material, S_i, is defined by

$$\text{rate of breakage of size } i = S_i w_i W \tag{S1}$$

where W is the total weight or *hold-up* of particles present in a mill of volume V. Thus S_i is the equivalent of a fractional first-order chemical rate constant describing the disappearance of size i particles, with units of time^{-1}. The *absolute rate of breakage* of size i material is defined as $S_i W$ (tph). Several related terms are *mill capacity*, which means the rate of production of a desired product from a specified feed for a given mill and *specific mill capacity* as either mill capacity per unit mill volume or per unit mass of powder in the mill.

The length of time that material spends in the mill, the *residence* or *retention time*, controls the final degree of breakage. The additional concept of a *residence time distribution* is required to account for the observation that some fraction of the feed stays in the mill for a short time, other fractions for longer times, etc. The *mean residence time*, τ, is defined by W/F where F is the solid feed rate into the mill. Two extremes of residence time distribution are useful in mill analysis (see Fig. 6). The first is *plug flow* of material where all feed spends the same length of time in the mill and emerges after the residence time τ. Thus

$$\left.\begin{array}{ll} \Phi(t) = 0, & t < \tau \\ \Phi(t) = 1, & t > \tau \end{array}\right\} \tag{S2}$$

where $\Phi(t)$ is the cumulative fraction of feed which has emerged at time t after entering the mill. The other extreme is *fully mixed* flow, where feed immediately mixes into all of the mill contents, leading to

$$\Phi(t) = 1 - \exp(-t/\tau) \qquad (S3)$$

All real systems lie between these two extremes.

The relative size makeup of the product distributions is affected by the manner in which the circuit is designed and operated with recycle from a classifier. While a common goal is to produce ground product to some percent less than a specified size, it is generally undesirable to produce *excess fines* below the specified size (or in wet grinding, *excess slimes*) by *overgrinding*. To prevent excess fines it is necessary to remove material already fine enough from the mill as quickly as possible. The use of a classifier and closed circuit offers better flexibility of size control than simple open-circuit operation alone (see Fig. 7). If the steady-state rates of new makeup feed and final product are Q tph and if the amount recycled is T tph, the actual flow rate through the mill is $Q + T$. The ratio $(Q + T)/Q$ is called the *circulating load* and the ratio T/Q the *circulation ratio*. If energy is used in overgrinding material already fine enough this is an *indirect* inefficiency. If mill conditions such as powder filling or slurry density are wrongly chosen so that the breakage actions are reduced, this is *direct* inefficiency.

Another term that is often used is the *reduction ratio*, (x_F/x_P) where x_F and x_P are feed and product sizes, respectively, corresponding to some specified weight percent-passing in the distributions (commonly 80%-passing). With regard to tumbling ball mills, several other definitions are also useful. The *critical speed* of the mill is the rotational speed at which the balls on the mill case just start to centrifuge. Thus

$$\text{critical speed} = 76.6/\sqrt{D - d} \ (D, d \text{ in ft})$$

$$= 42.2/\sqrt{D - d} \ (D, d \text{ in m}) \qquad (S4)$$

where D is internal mill diameter and d is maximum ball diameter. The fraction of the critical speed at which a mill is run is denoted by ϕ_c and is normally in the range 65 to 85% to give maximum output from the mill. The *fractional ball* filling, J, is the fraction of the mill filled by the ball bed at rest. Assuming a formal bed porosity of 0.4 gives

$$J = \left(\frac{\text{mass of balls/ball density}}{\text{mill volume}} \right) \times \frac{1.0}{0.6} \qquad (S5)$$

In the same manner, the fraction of mill volume filled by powder bed, f_c, is

$$f_c = \left(\frac{\text{mass of powder/powder density}}{\text{mill volume}} \right) \times \frac{1.0}{0.6} \qquad (S6)$$

The powder-ball loading ratio, U, is the fraction of the spaces between the balls at rest which is filled with powder, and is given by

$$U = \frac{f_c}{0.4J} \qquad (S7)$$

$U \approx 1$ is found to be an appropriate value in practice.

Four levels of complexity can be used in describing any grinding system. The least complex approach is the global rate-of-production method which simply involves measuring the production rate of a suitable smaller test mill per unit mill volume and then multiplying this rate by the volume of the similar larger mill under consideration and by a *scale-up factor* based on experience. The next approach is the global specific energy method which is similar to the previous approach except that the power, m_{p1}, to drive the mill is measured for the test mill, at production rate Q_1. It is then assumed that the specific energy (kWh/t) is constant between the test and larger mills for the same reduction ratios so that

$$m_{p2} = (Q_2/Q_1)m_{p1} \qquad (S8)$$

where the subscript 2 refers to the larger mill. Since there are empirical equations which relate mill size to power, a mill can be sized knowing the required power.

The third approach, called the global Bond/Charles method, is a combination of the first two methods plus empirical relationships that describe how production rates or specific energies vary with changes in feed or product size.

The last and most complete approach is the size-mass balance method which is the major analysis tool of this book. This method combines the concepts of specific rates of breakage, residence time distributions, and mathematical description of classification, plus the set of relationships which describe how each factor in the size-mass balance varies with mill conditions and mill size. Using this methodology, circuits can be simulated, compared, and optimized for technical and economic performance on paper before a final design is adopted.

References

Bond, F. C., 1960, "Crushing and Grinding Calculations," *British Chemical Engineering*, Vol. 6, pp. 378–391, 543–548.

Charles, R. J., 1957, "Energy-Size Reduction Relationships in Comminution," *Trans. AIME*, Vol. 208, pp. 80–88.

Levenspiel, O., 1962, *Chemical Reaction Engineering*, Wiley, New York.

Lynch, A. J., et al., 1977, *Mineral Crushing and Grinding Circuits*, Elsevier.

Marshall, V. C., ed., 1975, *Comminution*, Institute of Chemical Engineering, London.

Rowland, C. A., Jr., and Kjos, D. M., 1980, "Rod and Ball Mills," *Mineral Processing Plant Design*, A. L. Mular and R. B. Bhappu, eds., 2nd ed., AIME, New York, pp. 239–278.

Rumpf, H., 1973, "Physical Aspects of Comminution and New Formulation of a Law of Comminution," *Powder Technology*, Vol. 7, pp. 145–159.

Sheridan, D., 1977, "A Second Coal Age Promises to Slow Our Dependence on Imported Oil," *Smithsonian*, Vol. 8, pp. 30–37.

2
A Review of the Older Laws of Grinding

Introduction

The use of milling circuits to perform size reduction is an old art, widely applied since the start of the industrial revolution, and it obviously predates the availability of computers, online size analyzers, etc. Thus, as will be discussed in Chapter 3, an approximate design can be made on the basis of simple empirical relations, grindability tests, and prior experience. However, over the years there also have been several pseudoscientific attempts to develop fundamental laws of grinding or to explain empirical relations on a scientific basis. We will not spend much time on these explanations, but it is important to understand the distinction between the usefulness of the relations for design and their so-called *theoretical* basis, since a proper comprehension will prevent the empirical laws being applied outside of their true range of validity.

Fig. 1 shows a typical set of results obtained by batch grinding a brittle material in a ball mill, where the size distribution is getting finer and finer for longer periods of time of grinding (similar results are obtained for wet grinding in water). This particular set of data is for a narrow size range of the starting feed, but it is found that even a wide range of feed size will eventually produce the same basic pattern of curves after some time of grinding. Obviously, this data enables the calculation of the capacity (kg/min) of the mill to grind from one size distribution in this set of curves to another. If the mill power is known, the specific energy (kWh/kg) also follows. However, it would obviously be useful to have a mathematical representation of the data to enable interpolation and extrapolation. The following sections involve partial descriptions of the data by various mathematical models, which form the empirical relations useful for mill sizing. Chapter 3 will give methods of scaling-up the data from laboratory to production mills, which can be applied to any of the models.

If the power drawn by the mill, m_p, varies with time in a batch test, it is sensible to use specific grinding energy, $E(\tau) = \int_0^\tau m_p \, dt / W$, as an index of the degree of grinding. The power drawn by the mill at any time is a direct index of the lifting action on the balls at that instant of time (see Chapter 11), and the rate of breakage is expected to be directly proportional to the lifting action. Thus the integrated power can be an index of the integrated

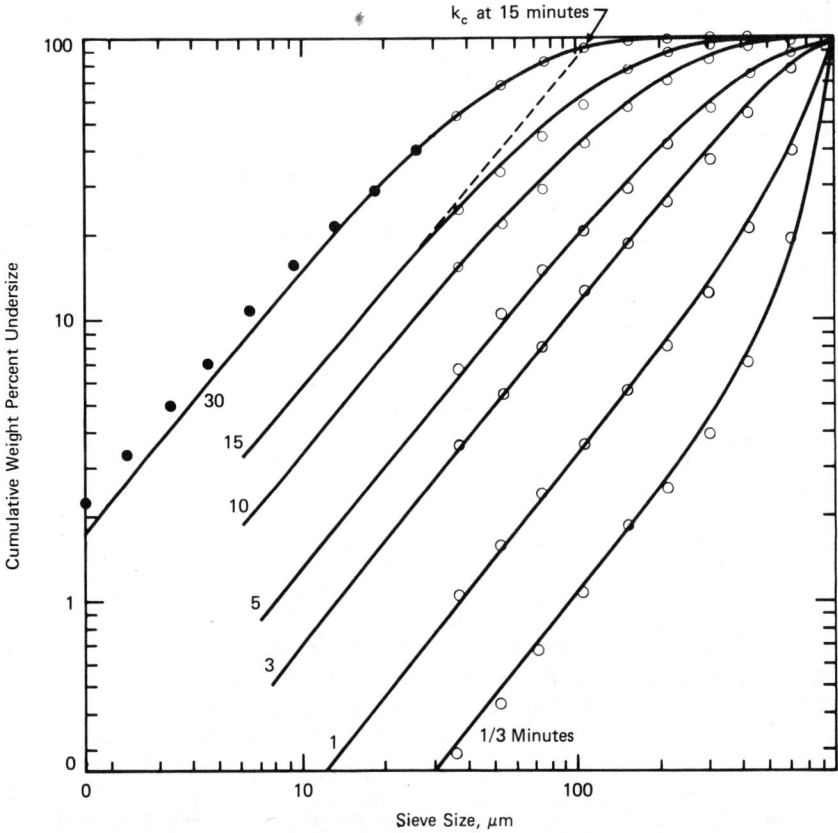

Fig. 1. Experimental and computed size distribution for dry grinding of 850 × 600 μm (20 × 30 mesh) quartz in a 200 mm (8-in.) diam ball mill ($U = 0.5$; $J = 0.2$; $\phi_c = 70\%$ c.s.; 25.4 mm (1 in.) steel balls; $W = 300$ g; $m_p = 0.013$ kW): $-$, computed; $\bigcirc$, experimental sieve sizes; $\bullet$, Sedigraph sizes.

breakage actions. In fact, if m_p is the mean mill power over time τ defined by $m_p\tau = E(\tau)W$, then at any other grinding time t an effective equivalent grinding time t' is defined by $E(t) = m_p t'/W$. However, the data in Fig. 1 is for a test where mill power was constant and $t' = t$, and the following sections make this simplifying assumption.

Kick, Rittinger, and Bond "Laws" of Batch Grinding:
Scientific or Empirical Laws?

Essentially, the laws are all based on the observation that a distinct *pattern* can often be detected in results from batch grinding in a sealed ball mill, as shown in Fig. 1, plus the empirical observation that the Global Specific Energy approach (see Chapter 1, "Levels of Complexity") often works with reasonable accuracy. If a pattern exists, it is natural to try to derive an explanation for it. Thus, these first efforts attempted to relate the degree of size reduction to the energy used for grinding on a *scientific* basis.

Using a highly oversimplified model for the relation of stress and strain energy with particle size and fracture, Kick (1883, 1883a; Austin, 1973) derived the expression

$$E_K = K_K \log(x_F/x_P) \tag{1}$$

where E_K is the strain energy per unit mass to reduce particles of size x_F to particles of size x_P; K_K is the strain energy per unit mass to produce a tenfold reduction in size. As used in practice, the energy supplied to the grinding machine is used in place of E_K, thus apparently assuming that all energy input to the mill goes to strain energy in the solid, which is patently absurd. Also, a *mean* size is used for x_F and x_P, often the 80%-passing size, since real breakage does not occur to a single product size. The correct expression of Kick's Law for a size distribution of feed and the inevitable size distribution of product would be,

$$E_K = K_K \left[\int_{x=0}^{x_{\max}} \log(x/x_u) \, dP(x,0) - \int_{x=0}^{x_{\max}} \log(x/x_u) \, dP(x,t) \right] \tag{2}$$

where $P(x, t)$ is the weight fraction less than size x of the products at time t, $P(x,0)$ the weight fraction less than size x of the feed, and x_u is the unit of size, e.g., one micrometer, one mm, one inch, etc. (It is desirable to use x_u to remind us that the term x/x_u is dimensionless, even though x_u has the value of 1.) The correct *mean* size $\bar{x}$ is therefore,

$$\log(\bar{x}/x_u) = \int_0^{x_{\max}} \log(x/x_u) \, dP(x) \tag{2a}$$

applied to the product or the feed size distribution:

$$E_K = K_K \log(\bar{x}_F/\bar{x}_P) \tag{2b}$$

We can ignore the dubious scientific basis for Kick's Law and use it in the *empirical* form

$$E_K = (\text{power to mill})(\text{time of grind})/(\text{mass of powder in mill})$$

$$= C_K \log(x_F/x_P)$$

where x_F, x_P are arbitrary single sizes characterizing the feed and product distributions, e.g., 80%-passing sizes. Then

$$E_K = m_p t/W = C_K \log(x_F/x_P) \tag{3}$$

is an expression from which the *specific energy E_K can be calculated as a function of the desired degree of size reduction*, if the constant C_K is

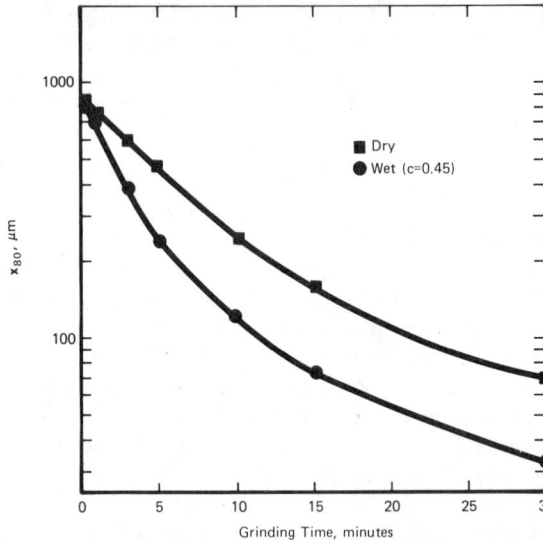

Fig. 2. Test of pseudo-Kick Law for ball milling of 850 × 600 μm (20 × 30 mesh) quartz; see Fig. 1 (c = volume fraction solids).

determined experimentally. We will refer to Eq. 3 as the pseudo-Kick equation. As tested in practice, it is usually assumed that the power input m_P to the test mill is constant with time, so that the value of the 80%-passing size in Fig. 1 should vary with time according to

$$\log(1/x_P) \propto t$$

Fig. 2 shows that this does not satisfactorily describe the pattern of results, although it is sometimes claimed (Andreasen and Jenson, 1955; Hukki, 1961) that it does fit results of coarse grinding.

Rittinger's Law (Rittinger, 1857; Austin, 1973) states that the energy required for breakage is related to the fresh surface produced in a unit amount of mass:

$$\text{specific energy} = (\gamma)(\text{new surface area} - \text{old surface area}) \qquad (4)$$

where γ is the mean surface energy per unit area. A fracture through a plane of unit area produces two unit areas of new surface and requires 2γ energy to break the bonding forces existing before the surfaces are formed (see Fig. 3). It has considerable appeal because it seems a logical physicochemical law; however, the above statement of the law is specious. It should be: the *minimum* energy for breakage is (γ)(surface produced), since the energy input must be *at least* sufficient to break the bonding forces which previously existed across the new surfaces formed. Since it is the mill energy

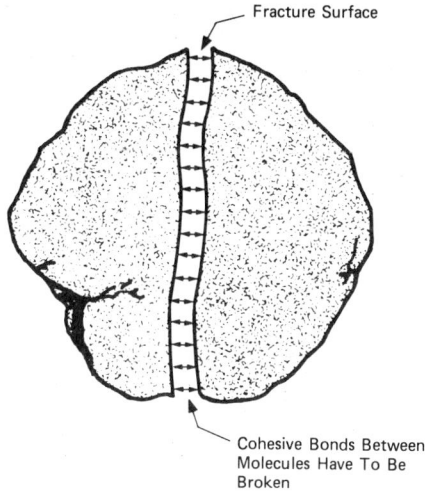

Fig. 3. Illustration of bonding energy between fracture surfaces.

which is our design parameter, it is necessary to find the relation between energy input to the mill and the surface energy.

In a classic series of experiments, Rose (1967) showed, by careful measurement of the energy balance in a mill, that the surface energy is *only a very small* fraction of the energy input to the mill. Within the limits of experimental error, he found that all energy to the mill appears as heat, sound, or the energy of phase transformation. There is no reason to suppose that the energy input to the mill is a large constant multiple of γ and it is difficult to justify the concept that surface energy amounting to, say, one-tenth of one percent of the input energy can control the whole process. The fraction of input energy converted to surface energy in a controlled fracture process varies widely depending on the flaw structure and method of stress application.

As actually used in practice, the law is *empirical*,

$$E_R = K_R(\text{new surface produced per unit mass broken}) \qquad (5)$$

where E_R is the specific grinding energy. Although it is entirely empirical, the value of K_R is often used, especially in the European literature, as an index of the *efficiency* of grinding; the units are J/m^2. A low value means a small energy expenditure per unit of surface produced, which is more efficient grinding than a high value. Similarly, a material which is more difficult to grind will have a high K_R compared to a weaker material.

In principle, the surface area can be obtained by integrating the size distribution to give the total particle area,

$$E_R = K_R\left[\int_{x_{min}}^{x_{max}}(\sigma/x)\,dP(x,t) - \int_{x_{min}}^{x_{max}}(\sigma/x)\,dP(x,0)\right]$$

where σ is a density-shape factor (the area per unit mass for spheres = $4\pi(x/2)^2/(4/3)\pi(x/2)^3\rho = 6/\rho x$, therefore $\sigma = 6/\rho$ for spherical particles of density ρ). Defining a mean size by

$$\frac{1}{\bar{x}} = \int_{x_{min}}^{x_{max}} \frac{1}{x} dP(x) \tag{5a}$$

gives

$$E_R = K_R \left[\frac{\sigma}{\bar{x}_p} - \frac{\sigma}{\bar{x}_F} \right] \tag{5b}$$

One of the problems in attempting to apply the equation is to decide what is the minimum size present, because the integration may give infinity if $x_{min} = 0$ is used.

Again, this *law* is often used in the pseudo-Rittinger form

$$E_R = C_R \left(\frac{1}{x_P} - \frac{1}{x_F} \right) \tag{6}$$

where the *mean* value of size is *not* chosen by the correct definition Eq. 5a, but is taken as the 80%-passing size. Fig. 4 shows the results of Fig. 1 plotted in this way, remembering that E is proportional to the grind time

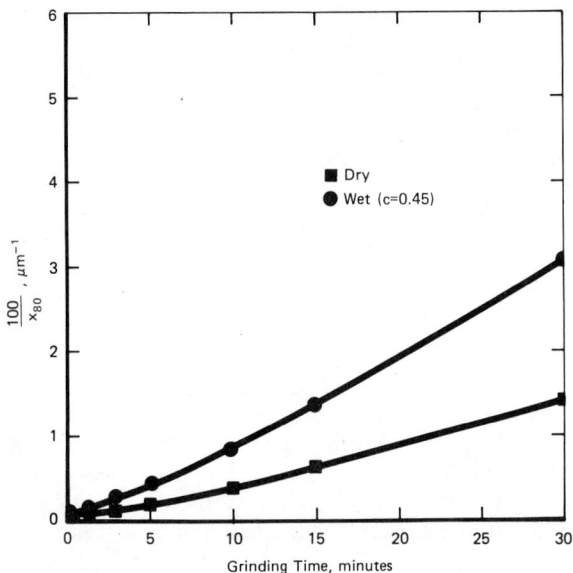

Fig. 4. Test of pseudo-Rittinger Law for ball milling of 850 × 600 μm (20 × 30 mesh) quartz; see Fig. 1.

($E = m_p t/W$). It can be seen that the data fit Eq. 6 only very approximately, over a relatively short grinding time, to a reduction ratio of about 6.

A similar form is Bond's Law of batch grinding (Bond, 1960)

$$E_B = C_B \left[\frac{10}{x_P^{1/2}} - \frac{10}{x_F^{1/2}} \right] \tag{7}$$

where x_P is the 80%-passing size of products and x_F the 80%-passing size of feed. As before, this is an entirely *empirical* equation based presumably on analysis of batch grinding data from laboratory mills. Fig. 5 shows the application of this equation to the results of Fig. 1. It fits the data reasonably well, up to a reduction ratio of 12 in this particular case. Bond (1952) has claimed that the half-power in this expression arises from fracture mechanics because the Griffith crack theory contains critical crack length to the half-power, but his argument is fallacious and cannot be considered seriously by the knowledgeable student of fracture mechanics. This does not, of course, prevent the use of the expression for mill design (see Chapter 3) providing it is recognized that it is not a fundamental law which *must* apply under all conditions.

Walker et al. (1937) claimed that the above relations were specific cases of the *general law*

$$dE = -C_M \frac{dx}{x^n}$$

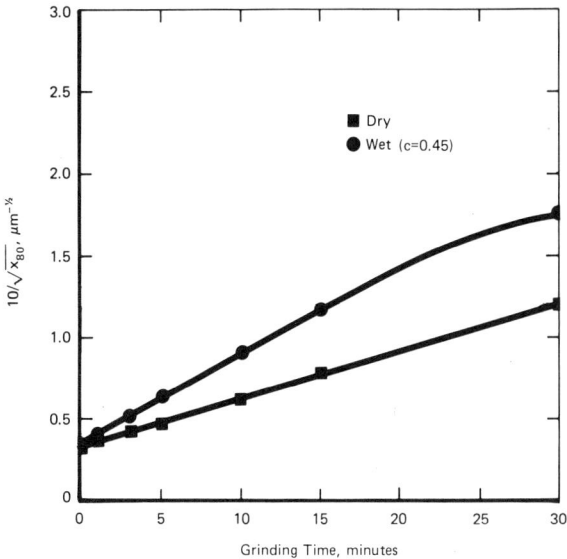

Fig. 5. Test of Bond's Law for ball milling of 850 × 600 μm (20 × 30 mesh) quartz; see Fig. 1.

where x is particle size and n is 1 for Kick's Law, 2 for Rittinger's Law, and 1.5 for Bond's Law. As Austin (1973) has pointed out, however, in this form the equation is absurd as it is not possible to break a particle a differential amount, since the products of breakage must contain small fragments even if the original particle is only slightly broken. The correct expression is

$$dE = -C_M \frac{d\bar{x}}{\bar{x}^n} \tag{8}$$

where $\bar{x}$ is a *mean* particle size; it is perfectly valid to reduce the *mean* size of a size distribution by a differential amount. Integration of the equation gives

$$E = \frac{C_M}{n-1}\left(\frac{1}{\bar{x}_P^{\,n-1}} - \frac{1}{\bar{x}_F^{\,n-1}}\right), \qquad n \neq 1$$

or

$$E = C_M \ln(\bar{x}_F/\bar{x}_p), \qquad n = 1$$

In order for these equations to reduce to the fundamental forms of Kick's or Rittinger's Laws, the *mean size* $\bar{x}$ must be defined to give Eqs. 2a when $n = 1$ and 5a when $n = 2$, that is, the mean size must be defined by

$$1/\bar{x}^{1-n} = \int_{x_{\min}}^{x_{\max}} (1/x^{n-1})\, dP\,(x), \qquad n \neq 1$$

or

$$\log(\bar{x}/x_u) = \int_{x_{\min}}^{x_{\max}} \log(x/x_u)\, dP\,(x), \qquad n = 1$$

The mean size is a function of grinding time, $\bar{x} = \bar{x}(t)$, since $P = P(x,t)$.

Eq. 8 gives Eq. 2, the correct form of Kick's Law, for $n = 1$ and $C_M = K_K$ and it gives Eq. 5, Rittinger's Law, for $n = 2$ and $C_M/(n-1) = \sigma K_R$. However, Eq. 8 does *not* generally integrate to Eq. 7, Bond's Law, when $n = 1.5$. On the other hand, the general law could be *defined* by

$$dE = -C_{80}\frac{dx_P}{x_P^n} \tag{9}$$

where x_P is the 80%-passing size. Again, a differential change in x_P is quite feasible. This equation integrates to Bond's Law for $n = 1.5$, but not the true forms of Kick's or Rittinger's Law for $n = 1$ or 2. Instead, it gives the pseudo-forms of these laws; that is,

$$E = \left(\frac{C_{80}}{n-1}\right)\left(\frac{1}{x_P^{\,n-1}} - \frac{1}{x_F^{\,n-1}}\right), \qquad n \neq 1$$

or

$$E = C_{80}\log(x_F/x_P), \qquad n = 1$$

Conclusions as to whether data fit a particular relation or not can be entirely different depending on the definitions of $\bar{x}$, x_F and x_P and whether the original or pseudo-laws are used.

A Digression: Some Comments on Fracture Mechanics and the Energy of Grinding, as Applied to the Above Laws

Austin and Klimpel (1964) have discussed the fracture process from the point of view of the utilization of energy in creating new surface. In any fracture process the solid must be raised to a state of strain to initiate the propagation of fracture cracks. Creation of this state of strain requires energy, greater than or equal to the stored strain energy. Whether fracture initiates at a low strain energy or at a high strain energy depends on a number of factors in addition to the value of the specific surface energy of the material. These include: (1) the presence of preexisting cracks or flaws; (2) whether plastic flow can occur in the solid or whether it is completely brittle; and (3) the geometry and rate of stress application.

The general treatment of failure of stressed solids is beyond the scope of this book, so let us, as illustration, consider the simple case of a spring under tension as in Fig. 6, with a small flaw as shown. If the spring were perfect, increasing the tensile stress through the spring would eventually separate (fracture) planes of molecules from each other equally at all points, creating a vast number of new surfaces and consuming the strain energy to supply surface energy. It is well known that solids do not fracture like this because fracture initiates at a weak point or flaw, at stress levels which are orders of magnitude below those predicted for perfect solids. Starting from the flaw the crack propagates through the solid, usually at very high speed, with a tree-like structure of branching and secondary cracks, all producing fragments. However, the major part of the stored strain energy is in the two major halves of the spring and it is converted to *heat* by the damped contraction and expansion of each of the separated halves; the pieces get hot, as most of us have experienced. The fraction of total strain energy converted to surface energy along the crack path is extremely small.

This picture applied to compressive or impact fracture in a milling machine immediately explains why the input energy appears primarily as heat. In a tumbling ball mill, for example, the mill energy is used to raise balls against the force of gravity, and when the balls fall the energy is converted to their kinetic energy. The balls strike particles, nipping them between balls, and the kinetic energy is converted to strain energy in the particles. If the impact force is sufficient, a particle is rapidly stressed to the fracture point and it breaks via propagating branching fractures, giving rise to a suite of fragments. Each fragment converts the remainder of its stored strain energy to heat after fracture, just like the two halves of the spring.

The fraction of mill energy not converted to heat is very small. As additional confirmation, it is known that the heat rise of a material flowing through a continuous mill can be calculated quite accurately by direct conversion of the energy input rate to the mill to sensible heat content of the material.

It also follows that the utilization of energy in a mill is extremely inefficient. This has led to the equivalent in grinding of the search for the Philosopher's Stone; that is, a search for ways of breaking material which convert much more of the input energy to surface energy. Unfortunately, an essential feature of a practical machine for size reduction is that it must break materials at high rates, to produce high tons per hour for a given capital investment. One might imagine highly efficient fracture processes induced by ultrafine diamond wedges which initiate crack propagation at low force applications, and hence, low strain energies. However, no one has devised a practical machine to use such a principle, and all industrial milling machines rely on a rapid massive straining of particles, followed by explosive fracture to a variety of fragments, with energy eventually released primarily as heat and sound, with sometimes a certain amount absorbed by endothermic phase change or residual stresses in the fragments. Fig. 7 is an example of a hammer mill, which is used for breaking relatively nonabrasive materials. It is difficult to imagine any kind of controlled, gentle breakage occurring in this type of machine, or in rod or ball mills.

These concepts can be extended to give an important general principle—the process of grinding is a mechanical process and efficiency is controlled by the mechanics of stressing particles to the failure point at high

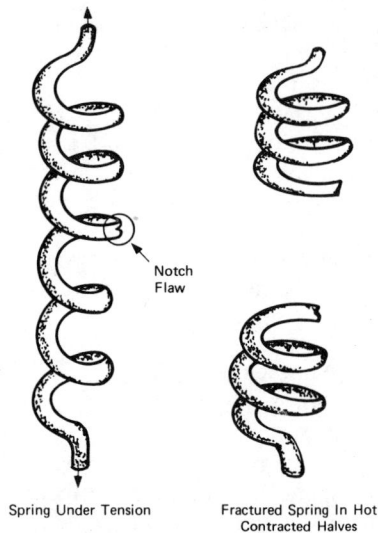

Spring Under Tension   Fractured Spring In Hot
                             Contracted Halves

Fig. 6. Illustration of strain energy in a simple stressed solid, converted to heat after fracture.

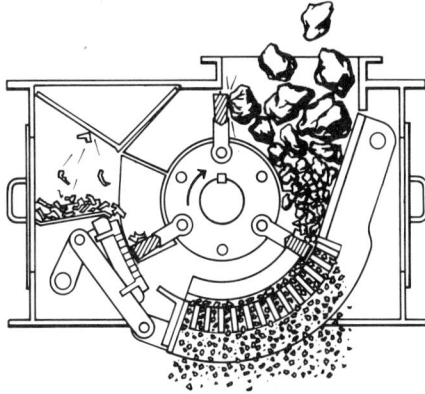

Fig. 7. Nonreversible hammer mill for coal.

rates (high capacity). The power to the mill is used to produce the mechanical action, so the total efficiency involves:

1) Efficient conversion of input energy to mechanical action.
2) Efficient transfer of the mechanical action to the particles.
3) Matching the stress produced by the mechanical action to the failure stress of the particles.

For example, if a tumbling ball mill is operated with too small a loading of particles, the impacts will be of steel ball on steel ball and part 2 will clearly be very inefficient. Again, if the ball density, ball size, and mill diameter are all low and if the particles are large and strong, the mill might not break the particles at all (except by a slow abrasion) and part 3 will be inefficient. These important mechanical concepts do not appear in the laws discussed in the previous section, which tend to give the false impression that the specific grinding energy is an invariant material characteristic, as if it were a thermodynamic property, whereas it is clearly dependent on correct design and operation of a mill.

Charles' "Law"; Schuhmann Size Distributions and Rosin-Rammler Size Distributions

Charles' Law (Charles, 1957) is another attempt to fit a pattern to the results of Fig. 1. As can be seen from that figure, an appreciable portion of the size distributions plotted on log–log scales can be fitted by parallel straight lines. The equation of the straight line region is

$$P(x) = a_s x^{\alpha_s}, \qquad 0 \le P(x) \le 1 \qquad (10)$$

which is called the *Schuhmann* size distribution; α_s is called the *distribution coefficient* and it is the slope of the straight line regions of the plots; a_s is a function of grind time $a_s = a_s(t)$. Let the symbol k_C represent the size at which $P(x) = 1$ in Eq. 10, that is, the extrapolation of the straight lines to $P(x) = 1$: note that $1 = a_s k_C^{\alpha_s}$, and hence $P(x) = (x/k_C)^{\alpha_s}$. Charles pointed

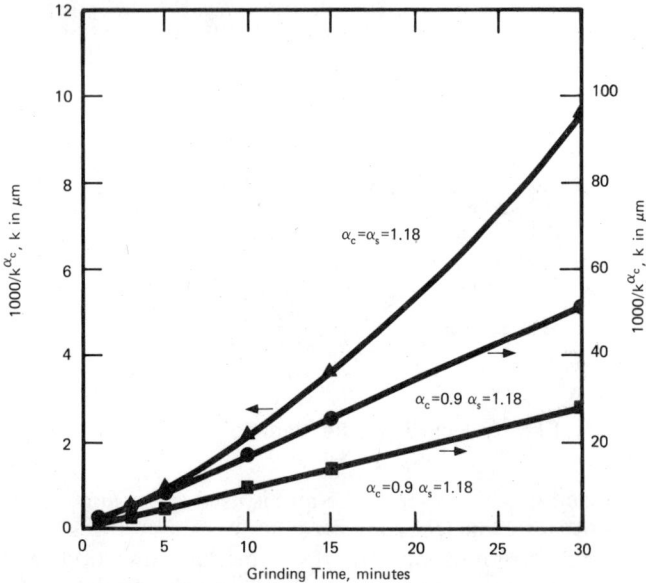

Fig. 8. Test of Charles' Law for ball milling of 850 × 600 μm (20 × 30 mesh) quartz: ■, dry, $\alpha_c = 0.9$, $\alpha_s = 1.18$; ●, wet, $\alpha_c = 0.9$, $\alpha_s = 1.18$; ▲, dry, $\alpha_c = 1.18$, $\alpha_s = 1.18$.

out that in the region of parallel straight lines, $k_C(t)$ may vary with specific grinding energy (proportional to grinding time) according to the empirical relation

$$E_c = m_p(t_2 - t_1)/W = C_C\left(\frac{1}{k_{C_2}^{\alpha_s}} - \frac{1}{k_{C_1}^{\alpha_s}}\right) \qquad (11)$$

where k_{C_1} is $k_C(t_1)$, etc. This equation is one form of *Charles' Law*. (We use a starting time of t_1 because it may not be possible to get a value of k_C for zero time if the feed distribution does not have the form of Eq. 10.) The upper curve in Fig. 8 shows the data of Fig. 1 plotted in this way. In general, it is found that the value of α_s in Eq. 11 is *not* always exactly the same as that in Eq. 10. Thus the variation of k_C with grind time is better described by

$$E_c = m_p(t_2 - t_1)/W = C_C'\left(\frac{1}{k_{C_2}^{\alpha_c}} - \frac{1}{k_{C_1}^{\alpha_c}}\right) \qquad (12)$$

where α_c is approximately equal to α_s, but is not exactly equal to it. The bottom two lines in Fig. 8 illustrate the better fit obtained using Eq. 12. The concept of two different values for α has also been proposed by Agar and Somasundaran (1974).

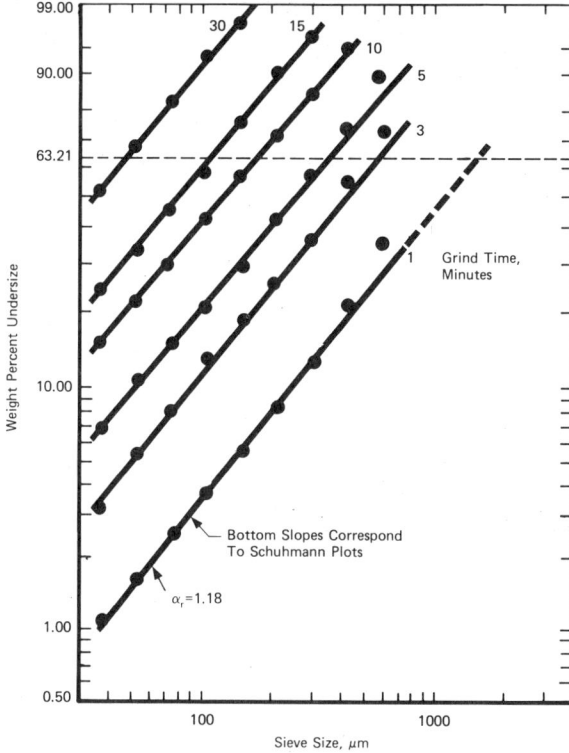

Fig. 9. Rosin-Rammler plot for data of Fig. 1.

This result is an empirical experimental observation and it is not based on any concept of fracture energy; it is not to be concluded that C_C is constant for a given material except under the particular test conditions used. Thus Charles' Law has *two* required components: (1) the slopes of the Schuhmann plots must be constant; and (2) the value of $k_C(t)$ varies with grind time or energy input according to Eq. 11 (or 12). The equation enables the time of grind or the specific grinding energy to be calculated for desired values of k_{C_1} and k_{C_2}, if α_s, α_c and C_C have been determined experimentally.

Similarly, plotting the results of Fig. 1 on Rosin-Rammler paper gives Fig. 9; again an extended region of the results can be fitted by a series of parallel straight lines. The *Rosin-Rammler* equation for the straight line region is

$$1 - P(x, t) = R(x, t) = \exp\left[-(x/x_o)^{\alpha_r}\right] \qquad (13)$$

where $R(x, t)$ is the weight fraction oversize and x_o is the value (extrapolated from the straight line region if necessary) of size at $R = 0.3679$, 63.21% undersize; α_r is close to α_s. Fig. 10 shows that a plot of $1/x_o^{0.90}$ gives

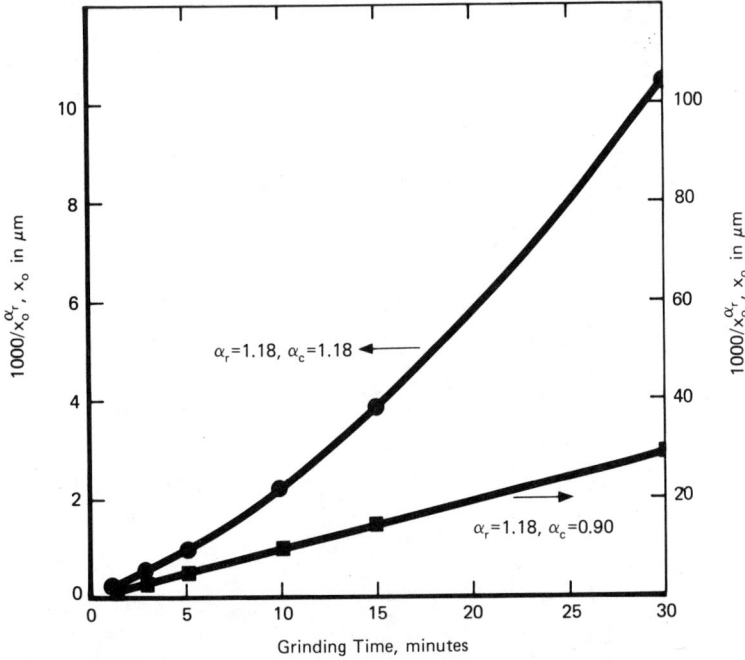

Fig. 10. Test of Charles'-Rosin-Rammler Law in the form of Eq. 14 for dry ball milling of 850 × 600 μm (20 × 30 mesh) quartz.

a reasonable straight line and, as before,

$$E_{Cr} = m_p(t_2 - t_1)/W = C_{Cr}\left(\frac{1}{x_{o_2}^{\alpha_c}} - \frac{1}{x_{o_1}^{\alpha_c}}\right) \tag{14}$$

This is a form of Charles' Law based on Rosin-Rammler size distributions instead of Schuhmann size distributions.

The Rate-of-Production Pattern or the *Zero Order* Rule

Another consistent pattern in the results of Fig. 1 is demonstrated in Fig. 11, where the percentage by weight less than selected sizes is plotted vs. time of grinding. It can be seen that for a size which is small compared to the feed size, the *rate of production* of less-than-size material is approximately *constant* for a protracted period of grinding. This rule must eventually break down as the grinding proceeds with time, since the rate of production of material less than a specified size must finally tend to zero, as demonstrated in the figure. Mathematically,

$$\left.\begin{aligned} \partial P(x,t)/\partial t &= \text{constant}, & P(x,t) &< 0.5 \\ P(x,t) - P(x,0) &= A_z(x)t, & P(x,t) &< 0.5 \end{aligned}\right\} \tag{15}$$

or

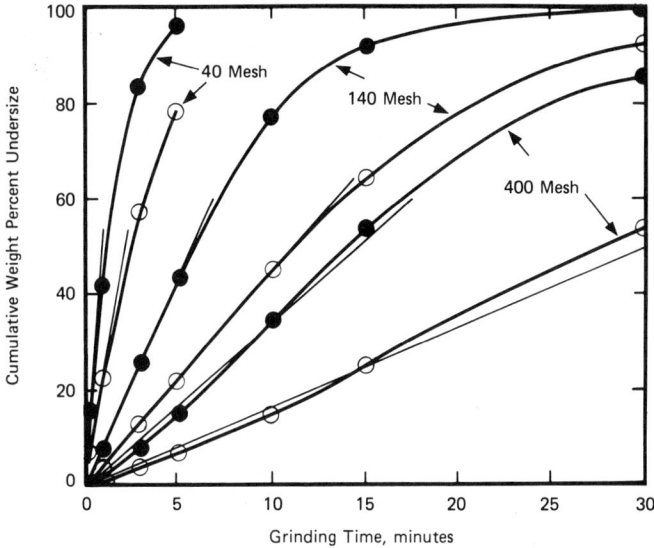

Fig. 11. Test of zero order production ball milling for 850 × 600 μm (20 × 30 mesh) quartz: ○, dry grinding; ●, wet grinding; see Fig. 1.

where $A_Z(x)$ is the *fractional rate of production* of material less than size x, with units of fraction (or %) per minute. The rate of production of less-than-size-x material in the mill containing a mass W is given by $A_Z(x)W$ and a specific energy of grinding can be defined by choosing a convenient value of x, say x^*, then the specific grinding energy is $E = m_p t/W$ and

$$E_Z = m_p t/W = C_Z\left[P(x^*, t) - P(x^*, 0)\right] \tag{16}$$

where $C_Z = m_p/A_Z(x^*)W$. C_Z is a function of the size x^* being considered, and becomes greater as x^* is smaller since $A_Z(x^*)$ decreases as x^* decreases (see Fig. 11).

Thus, this particular pattern in the results states that the rate of production of fine material (material less than a specified small size) is *independent* of the relative *amount* or *size* of material larger than this size, within appropriate limits. This has sometimes been called the *zero order* grinding law (Herbst and Fuerstenau, 1968). It can apply only under a relatively narrow range of feed and product conditions, but within these limitations it offers a very simple way of sizing mills (see Chapter 3).

Summary

In this chapter, a number of empirical relationships which have evolved over a period of years are discussed. These relationships or *laws* are sometimes linked with pseudoscientific reasoning invoked to explain certain patterns that exist in selected sets of grinding data (such as that of Fig. 1). All of the laws serve as rules for estimating the effect of changing the feed and/or product specifications on the time of grinding required or on the specific energy of grinding (energy per unit mass). The laws also predict that it takes longer (and hence, more specific energy) to grind a given mass of coarse feed to a fine product than a less coarse feed to a less fine product.

If the proportionality constant C and slope α that relate to the specific energy term E are determined from test data, the laws can be put into the following commonly used forms:

$$E_R = C_R\left(\frac{1}{x_P} - \frac{1}{x_F}\right) \qquad pseudo\text{-}Rittinger \qquad (S1)$$

$$E_B = C_B\left(\frac{10}{x_P^{1/2}} - \frac{10}{x_F^{1/2}}\right) \qquad Bond\text{-}Batch \qquad (S2)$$

$$E_C = C_C\left(\frac{1}{k_{C_2}^{\alpha_s}} - \frac{1}{k_{C_1}^{\alpha_s}}\right) \qquad Charles\,(Form\ 1) \qquad (S3)$$

$$E_C = C_C'\left(\frac{1}{x_P^{\alpha_c}} - \frac{1}{x_F^{\alpha_c}}\right) \qquad Charles\,(Form\ 2) \qquad (S4)$$

where x_P, x_F values are 80%-passing sizes of product and feed respectively (picked by extrapolation of the straight line region of data such as given by Fig. 1), and k_C is the value of x from the extrapolation to $P(x, t) = 1$. The Rosin-Rammler equation for the straight line region of the curves of Fig. 1 is

$$1 - P(x, t) = R(x, t) = \exp\left[-(x/x_o)^{\alpha_r}\right] \quad Rosin\text{-}Rammler \quad (S5)$$

where $R(x, t)$ is the fraction oversize, x_o is the value of size (extrapolated from the straight line region if necessary) corresponding to $R(x, t) = 0.3679$, and α_r the slope of the straight line. The Charles Law approach using the Rosin-Rammler slope gives

$$E_{Cr} = C_{Cr}\left(\frac{1}{x_{oP}^{\alpha_r}} - \frac{1}{x_{oF}^{\alpha_r}}\right) \quad Charles\,(Form\ 3) \qquad (S6)$$

The Zero Order rule gives

$$E_Z = C_Z\left[\psi(x^*) - F(x^*)\right] \quad Zero\ Order \qquad (S7)$$

where x^* is some control size and $\psi(x)$, $F(x)$ are the weight fractions less than this size of product and feed, respectively. For coarse crushing the

Kick Law may apply

$$E_K = C_K \left[\ln(1/x_P) - \ln(1/x_F) \right] \quad pseudo\text{-}Kick \qquad (S8)$$

The units of C are the same as E (e.g., kWh/t) in all of the laws if the 1 in the $1/x_P$ and $1/x_F$ terms is taken to be 1 μm, the 10 in the $10/x_P^{1/2}$ and $10/x_F^{1/2}$ terms to be $(100 \ \mu m)^{1/2}$, and the 1 in $1/x_P^\alpha$ and $1/x_F^\alpha$ terms is taken as $(1 \ \mu m)^\alpha$.

All of the laws are of the form

$$E = m_p t/W = C_X [\text{function of product size} - \text{function of feed size}]$$

$$(S9)$$

It then follows that the plotting of the actual data using the appropriate functionality of the right-hand side of the above equation vs. t (for constant m_p/W) indicates the validity of each law for the test conditions of the data. The use of these equations for approximate sizing of mills is discussed in Chapter 3.

References

Agar, G. E. and Somasundaran, P., 1974, "Rationalization of Energy-Particle Size Relationships in Comminution," *Proceedings of the Tenth International Mineral Processing Congress*, M. J. Jones, ed., Institution of Mining and Metallurgy, London, pp. 3–21.

Andreasen, A. H. M. and Jenson, I. H., 1955, *Berichte der Deutschen Keramischen Gesellschaft*, Vol. 32, p. 232.

Austin, L. G. and Klimpel, R. R., 1964, "Theory of Grinding Operations," *Industrial and Engineering Chemistry*, Vol. 56, pp. 18–29.

Austin, L. G., 1973, "A Commentary on the Kick, Bond and Rittinger Laws of Grinding," *Powder Technology*, Vol. 7, pp. 315–318.

Bond, F. C., 1952, "The Third Theory of Comminution," *Trans. AIME*, Vol. 193, pp. 484–494.

Bond, F. C., 1960, "Crushing and Grinding Calculations," *British Chemical Engineering*, Vol. 6, pp. 378–391, 543–548.

Charles, R. J., 1957, "Energy-Size Reduction Relationships in Comminution," *Trans. AIME*, Vol. 208, pp. 80–88.

Herbst, J. and Fuerstenau, D. W., 1968, "Zero Order Production of Fine Sizes in Comminution and Its Implications in Simulation," *Trans. SME-AIME*, Vol. 241, pp. 538–548.

Hukki, R. T., 1961, "Proposal for a Solomonic Settlement Between the Theories of Rittinger, Kick and Bond," *Trans. AIME*, Vol. 220, pp. 403–408.

Kick, F., 1883, *Dinglers Polytechnisches Journal*, Vol. 247, pp. 1–5.

Kick, F., 1883a, *Dinglers Polytechnisches Journal*, Vol. 250, p. 141.

Rittinger, R. P. von, 1857, Lehrbuck der Aufbereitungskunde, Ernst u. Korn, Berlin.

Rose, H. E., 1967, *Proceedings*, 2nd European Symposium Zerkleinern, H. Rumpf and W. Pietsch, eds., Dechema Monographien 57, Nr. 993-1026, Verlag Chemie, Weinheim, pp. 27–62.

Walker, W. H., et al., 1937, *Principles of Chemical Engineering*, McGraw-Hill, NY, p. 255.

3

Conventional Grindability Tests and Mill Sizing: The Bond and Other Methods

Introduction

In principle, it is possible to predict the desired size of a production mill for a given capacity from the results of a small-scale continuous test which is a scaled-down version of the full-scale plant, knowing appropriate scale-up laws. In practice, it is difficult to get exact *similitude* of the production mill conditions (ball size mixture, hold-up, classifier action, etc.) on a laboratory scale and the tests are not easy to run. When the test mill is made big enough to give a good similitude with the large scale, it becomes a pilot-scale test system. To avoid the cost of constructing and operating a pilot-scale system, approximate mill sizing methods have been developed, which are discussed in this chapter.

The current methods of sizing ball mills, using data from laboratory grinding tests, are largely a matter of applying empirical equations or factors based on accumulated experience. Different manufacturers use different methods and it is difficult to check the validity of the sizing estimates when estimates from different sources are widely divergent. It is especially difficult to teach mill sizing and circuit design to engineering students because of the apparent lack of a logical engineering foundation for the empirical equations discussed in Chapter 2. However, it will be demonstrated in Chapter 4 that some of these empirical relations are compatible with a logical treatment of batch grinding using size-mass rate balances.

The problem normally posed is this: A feed material of a known weight-size distribution is to be milled to a finer product at a desired rate of Q tph; what size of mill should be used? A satisfactory product can be defined in several ways; to start we will use the simplest criterion that some weight percentage ψ must pass a specified size x^*. As part of the problem, the mill product may be classified into two streams, with the coarse being returned to the feed. Is this desirable, and if so, what amount of circulation should be used?

The Bond Method for Sizing Ball Mills

Outline

This method is discussed first, and in most detail, because it has found wide acceptance in the mineral industries. It has two major engineering advantages: (1) it is very simple, and (2) experience shows that it works for many (but not all) circumstances. Philosophically, it contains five major components:

1) A standardized *grindability* test on the material,
2) An empirical equation which converts the test result to the observed results in a 2.44 m (8 ft) i.d. wet overflow ball mill operated in closed circuit with a circulating load of 350%.
3) An empirical equation of the form of Eq. 7 in Chapter 2 to allow for the overall reduction ratio in closed circuit operation.
4) Scale-up relations to predict the result for larger mills.
5) A series of empirical correction factors based on experience to allow for other milling conditions.

The original papers by Bond were summarized in an important publication (1960), which unfortunately contained several errors. The original papers tend to confuse valuable empirical results with dubious scientific reasoning, but a recent excellent article by Rowland and Kjos (1980) gives a very clear discussion of the practical application of the method and the reader is referred to this article for complete details. A summary is presented here.

STEP 1: The Standard Bond Grindability Test

The material is reduced to prepare a feed of 100% < 3350 μm (6 mesh) and about 80% ≤ 2000 μm. 700 cm³ of this feed (tapped down according to a standard procedure to give a reproducible bulk density) is ground dry in a standard cylindrical test mill of 305 mm i.d. by 305 mm long (12 in. × 12 in.) with rounded corners, run at a fixed speed of 70 rpm (85% of the critical speed). The ball charge consists of a specified number of balls ranging from 15.2 to 44.4 mm (0.6 to 1.75 in.) diam (43 of 1.75, 67 of 1.17, 10 of 1, 71 of 0.75, 94 of 0.61 in.) with the total ball load weighing 20.1 kg. The procedure involves grinding the charge for a short time (100 revolutions, for example), sieving at a desired screen size to remove the undersize and replacing the weight of undersize with an equivalent weight of original feed. This new mixed feed is reground and the process continued, using the net production of undersize per revolution to estimate suitable grinding times (mill revolutions), until a constant 350% circulating load (circulation ratio = 2.5) is reached. When this is achieved the net grams of undersize produced per mill revolution is denoted by Gbp. Screen analysis is performed on the undersize product and the returned material.

STEP 2: Conversion to 2.44 m (8 ft) i.d. Wet Overflow Mill in Closed Circuit (350% Circulating Load)

By comparison of the laboratory test results with results from a wet overflow mill of 2.44 m i.d. operated in closed circuit at 350% circulating

APPARENT OVERALL MILL

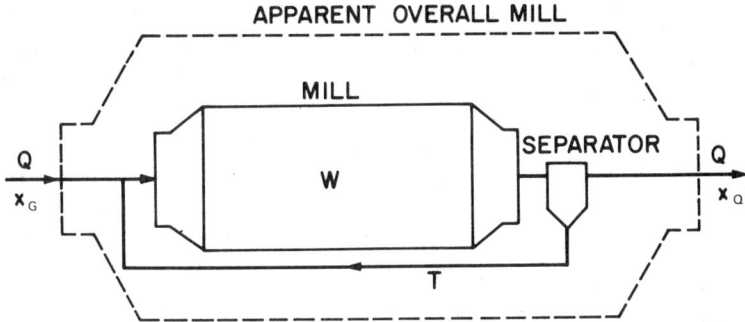

Fig. 1. Bond treatment of production mill and classifier with recycle as an equivalent open-circuit overall mill.

load, Bond concluded that a material could be characterized by the *Work Index Wi* defined by the following empirical equation

$$Wi_{\text{test}} = (1.10)(44.5) \Big/ \left\{ \left(p_1^{0.23} \right)(\text{Gbp})^{0.82} \left[\frac{10}{\sqrt{x_{QT}}} - \frac{10}{\sqrt{x_{GT}}} \right] \right\}, \text{kWh/t} \quad (1)$$

where p_1 is the opening of the classifying screen used in preparing the product and return to the mill in the Bond test, in micrometers; x_{QT} is the 80%-passing size of the product in micrometers and x_{GT} the 80%-size of the original feed, which is near 2000 μm. (Note that the 10 is $\sqrt{100\ \mu m}$, so that $10/\sqrt{x}$ is dimensionless.) The factor 1.10 converts the Bond Work Index in kWh/st to kWh/t.

The Work Index determined in this way can be almost independent of the classifying screen used in the test, but a 75 μm (200 mesh) screen is most often employed (see "Discussion of the Bond Method").

STEP 3: Allowance for Size Reduction Ratio
Bond concluded that over some size ranges the influence of the size of the feed and the size of the product on the grinding energy in closed circuit could be described by the relation

$$E = Wi \left(\frac{10}{\sqrt{x_Q}} - \frac{10}{\sqrt{x_G}} \right), \text{kWh/t} \quad (2)$$

where E is specific grinding energy (based on shaft power, $E = m_p/Q$) and x_Q, x_G are 80%-passing sizes of *circuit* product and *circuit* feed, respectively. Eq. 2 is *not* the same as Eq. 7 in Chapter 2, because it is applied to the overall system, not to the actual feed and product from the mill. Thus the circuit is treated *as if* it were an equivalent open-circuit mill operating in plug flow, as illustrated in Fig. 1. The relation has to be modified for low reduction ratios (x_G/x_Q small), coarse feed sizes (large x_G), and fine grinding (small x_Q), as discussed below.

The physical meaning of Wi is usually described as "the specific grinding energy to go from a large size (x_G = large) to 80%-passing 100 μm, x_Q = 100 μm," that is,

$$E = Wi \left[\frac{10}{10} - \frac{10}{\text{large}} \right] \text{kWh/t}$$

However, this interpretation is misleading because the relation of Eq. 2 does not apply as x_G becomes large, and it only applies under prescribed circumstances. Eq. 2 does apply for x_G = 900 μm, so Wi can be more precisely interpreted as "1.5 times the specific grinding energy to go from a makeup feed of x_G = 900 μm to a circuit product of x_Q = 100 μm in the 2.44 m i.d. wet overflow mill operated at C = 2.5":

$$E = Wi \left[\frac{10}{10} - \frac{10}{30} \right]$$

that is,

$$Wi = (3/2)E$$

STEP 4: Scale-up To Larger Mills

To convert to a larger mill with an i.d. of D, the value of Wi to be used in Eq. 2 is scaled according to

$$\left. \begin{array}{ll} Wi = (Wi_{\text{test}})(2.44/D)^{0.2}, & D < 3.81 \text{ m} \\ Wi = (Wi_{\text{test}})(0.914), & D > 3.81 \text{ m} \end{array} \right\} \quad (3)$$

If no further corrections are necessary (see Step 5) the value of E for a desired x_Q from a given x_G is then calculated from Eq. 2. Then, for a desired mill capacity Q in tons per hour, the required shaft mill power m_p is

$$m_p = QE, \text{ kW} \quad (4)$$

The Bond relation of shaft mill power to tons of grinding media is readily converted to mill power as a function of mill dimension (D = internal diameter, L = length) and loading conditions for wet overflow mills (using tons of media = $\pi D^2 L J \rho_b (1 - \varepsilon)/4$, where bed porosity, ε is taken as 0.4, see Chapter 11):

$$m_p = 7.33 J \phi_c (1 - 0.937J) \left(1 - \frac{0.1}{2^{9-10\phi_c}} \right) (\rho_b L D^{2.3}), \text{ kW} \quad (5)$$

where ρ_b is the true density of the grinding media (t/m^3), ϕ_c is the fraction of critical speed, and J the formal ball loading. Hence, knowing m_p (and the specifications of J, ϕ_c, ρ_b) the values of L and D necessary to give this power can be calculated. (See Chapter 11 for correction of this equation for large ball and mill diameters.)

STEP 5: Corrections for Other Milling Conditions, Large Feed, Fine Grinding, Etc.

Various corrections are applied to the value of Wi before it is used in the calculation of Eq. 2. For example, in order to correct the Wi value to *wet open circuit* grinding, Table 1 is used. Wi is determined from the value given by Eq. 3 by multiplying it by the multiplier given in the table for the appropriate $\psi(p_1)$. p_1 is again the sieve control size used in the Work Index test (Eq. 1) and $\psi(p_1)$ is the percent less than this size which is desired in the product from the open circuit. Overall, then, the desired product size occurs twice in the calculation, first in the multiplier for $\psi(p_1)$ and second in the $10/\sqrt{x_Q}$ term in Eq. 2,

$$Wi_{\text{wet open circuit}} = (\text{multiplier})Wi_{\text{wet closed circuit}} \qquad (6)$$

To obtain values for *dry* grinding the corresponding Wi values for wet grinding are multiplied by 1.30. The m_p from Eq. 5 is multiplied by 1.08.

To allow for *oversized feed*, defined as $x_G > 4000\sqrt{(1.10)(13)/Wi_{\text{test}}}$, Wi is corrected by multiplying by the factor

$$1 + \frac{\left[(Wi_{\text{test}}/1.10) - 7\right]\left[\left(x_G/4000\sqrt{(1.10)(13/Wi_{\text{test}})}\right) - 1\right]}{(x_G/x_Q)} \qquad (7)$$

where Wi_{test} is in kWh/t.

A *fineness-of-grind* correction is used for $x_Q < 75\ \mu$m. The multiplier is

$$\frac{x_Q + 10.3}{1.145x_Q}, \ 15\ \mu\text{m} \leq x_Q \leq 75\ \mu\text{m dry grinding} \qquad (8)$$

Table 1. Conversion of Wet Closed Circuit Bond Work Index to Wet Open Circuit Values

$\psi(p_1)$	Multiplier
50	1.035
60	1.05
70	1.10
80	1.20
90	1.40
92	1.46
95	1.57
98	1.70

For fine wet grinding, this correction can be allowed to go as high as 5, but no higher.

A *low reduction ratio* correction is used for $(x_G/x_Q) < 6$. The multiplier is

$$1 + \frac{0.13}{(x_G/x_Q) - 1.35} \tag{9}$$

Calculation Procedure and Illustrative Results

The sizing calculation can be performed in the stepwise fashion outlined above. However, the influence of mill diameter, rotational speed, and ball loading on mill capacity is conveniently seen by rearranging Eqs. 2, 3, 4, and 5 to

$$Q = 6.13 \frac{(D^{3.5})(L/D)(\rho_b)(J - 0.937J^2)\left(\phi_c - \dfrac{0.1\phi_c}{2^{9-10\phi_c}}\right)}{\zeta Wi_{test}\left(\dfrac{10}{\sqrt{x_Q}} - \dfrac{10}{\sqrt{x_G}}\right)}, \text{ tph};$$

$$D < 3.81 \text{ m} \quad (10a)$$

$$Q = 8.01 \frac{(D^{3.3})(L/D)(\rho_b)(J - 0.937J^2)\left(\phi_c - \dfrac{0.1\phi_c}{2^{9-10\phi_c}}\right)}{\zeta Wi_{test}\left(\dfrac{10}{\sqrt{x_Q}} - \dfrac{10}{\sqrt{x_G}}\right)}, \text{ tph};$$

$$D > 3.81 \text{ m} \quad (10b)$$

where L/D is the length-to-diameter ratio for the mill; $\rho_b \approx 7.9$ t/m³ for steel balls; ζ is the overall correction factor given by the product of the applicable multipliers from Step 5, including 1/1.08 to convert to dry grinding if applicable. The required shaft power for this size of mill, m_p, follows from Eq. 5 and the specific grinding energy from $E = m_p/Q$.

Fig. 2 shows the predicted results for conditions of $J = 0.35$, $L/D = 1.5$, $\phi_c = 0.7$, steel balls, and a constant test Work Index of 10 kWh/t, for various feed sizes. Table 2 gives the correction factors for other values of J and ϕ_c. For example, suppose a closed wet grinding circuit with a 5-m diam ball mill is to produce a circuit product analyzing 80%-passing 150 μm. If the feed to the circuit analyzes 80% less than 2 mm, the production rate is 500 tph (the specific energy is 6.2 kWh/t). If the ball charge is lowered to $J = 0.30$ and the critical speed increased to $\phi_c = 0.75$, then the production rate is 500 (0.91)(1.06) or 480 tph. If the material to be ground is hematite ($Wi = 15$, see Table 3), then the production rate will be 480 (10/15) or 320 tph.

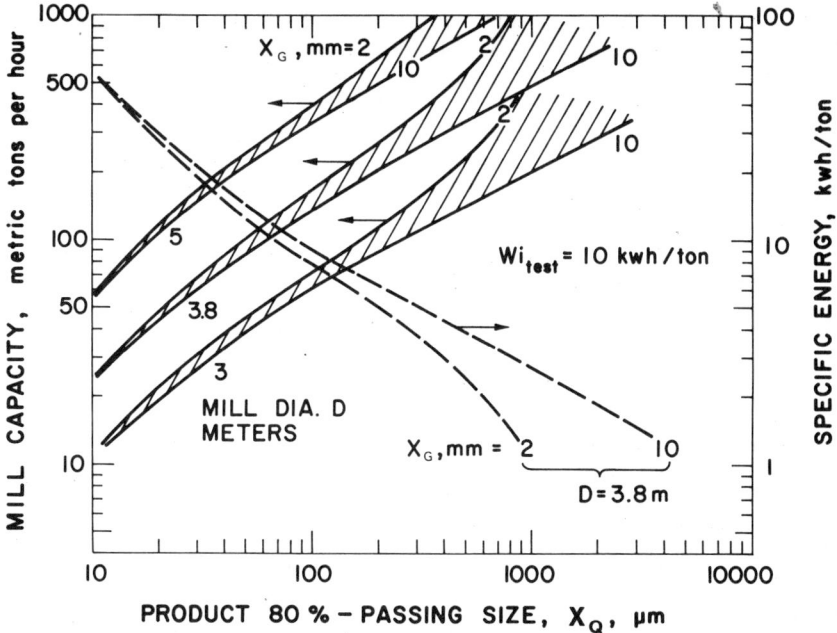

Fig. 2. Mill capacity for closed circuit wet grinding ($C = 2.5$) from Bond Method: $L / D = 1.5$, $J = 0.35$, 70% critical speed, for test Work Index = 10 kWh / t.

Table 3 gives typical values of Wi_{test} for various materials. Fig. 3 shows variation with test screen size for some materials.

Discussion of the Bond Method

There are two logical problems involved in the Bond sizing method. First, the specific grinding energy required to take a feed with a certain x_{80} to a product with a certain x_{80} cannot be the same for a batch test, the standard Bond locked-cycle test, or a steady-state continuous mill with a real mill residence time distribution. The complete size-mass balance treatments of these three cases show that the shape of the product size distribution and the associated specific grinding energy is different for the three cases. It is not possible to correlate the three different values exactly without using the size-mass balance method. However, it must be recognized that these differences are avoided in the Bond sizing method by making an empirical match of the standard test results to actual plant data (on the 2.44 m i.d. mill).

Second, because the Bond method is empirical it is not possible to assign physical meanings to the variation of capacity with x_Q given in Fig. 2. In the section titled, "A Comparison of Circuit Simulations with the Bond Method" in Chapter 7, we will show from the simulation that the capacity including the *fineness of grind* correction factor is actually the expected normal result. On the other hand, the region of higher capacity (lower reduction ratio) probably contains an effect due to overfilling of the mill, leading to direct inefficiency. The Bond method is based on data fitting, not on physical phenomenon.

There are other disadvantages of the Bond sizing method. First, it appears to be based on the mean empirical fit of data from a number of mills and materials run under normal conditions and there will be a range of error for any specific mill and material and set of operating conditions. It does not explicitly include several factors which are obviously important:
1) Recycle ratio and classifier efficiency.
2) Mixture of ball sizes in the mill.
3) Variations of residence time distributions with mill geometry and slurry density.

**Table 2. Multiplying Correction Factors to Q Value of Fig. 2
(Wet Overflow Closed-Circuit, $C = 2.5$) for Other Values of J and ϕ_c**

J	$J - 0.937J^2$	Factor	ϕ_c	$\phi_c - \dfrac{0.1\phi_c}{2^{9-10\phi_c}}$	Factor
0.20	0.16	0.69	0.60	0.59	0.87
0.21	0.17	0.71	1	0.60	0.88
0.22	0.17	0.79	2	0.61	0.90
0.23	0.18	0.77	3	0.62	0.91
0.24	0.19	0.79	4	0.63	0.92
0.25	0.19	0.81	5	0.64	0.94
0.26	0.20	0.84	6	0.65	0.95
0.27	0.20	0.86	7	0.66	0.96
0.28	0.21	0.88	8	0.67	0.97
0.29	0.21	0.90	9	0.67	0.99
0.30	0.22	0.91	0.70	0.68	1.00
0.31	0.22	0.94	1	0.69	1.01
0.32	0.22	0.95	2	0.70	1.02
0.33	0.23	0.97	3	0.71	1.04
0.34	0.23	0.99	4	0.72	1.05
0.35	0.24	1.00	5	0.72	1.06
0.36	0.24	1.01	6	0.73	1.07
0.37	0.24	1.03	7	0.74	1.08
0.38	0.24	1.04	8	0.75	1.09
0.39	0.25	1.05	9	0.75	1.10
0.40	0.25	1.06	0.80	0.76	1.11
0.41	0.25	1.07	1	0.77	1.12
0.42	0.25	1.08	2	0.77	1.13
0.43	0.26	1.09	3	0.78	1.14
0.44	0.26	1.10	4	0.78	1.15
0.45	0.26	1.10	5	0.79	1.16
0.46	0.26	1.11	6	0.79	1.16
0.47	0.26	1.12	7	0.80	1.17
0.48	0.26	1.11	8	0.80	1.18
0.49	0.27	1.13	9	0.81	1.18
0.50	0.27	1.13	0.90	0.81	1.19

4) The influence of lifter design.
5) The influence of slurry density and slurry rheology on breakage rates, and chemical effects on rheology.
6) Variations caused by different under or overfilling of the mill as flow rate is changed, especially for grate or peripheral discharge mills which will not act the same as overflow mills.

Again, it is known that the specific grinding energy E is *not* independent of ball loading J, whereas the use of Eq. 4 explicitly assumes that E is not a function of J. Industrial practice and laboratory tests (see Chapters 5, 11, and 12) show that the specific grinding energy (to go from a specified feed to a specified product) is less for lower ball loadings than the ball loading for maximum mill capacity.

The method uses only the 80%-passing sizes of circuit feed and product to characterize the size distributions, whereas it is clear that mill capacity in general must depend on the *shape* of the feed size distribution and the product size distribution. The prime example of this is the use of the reverse closed circuit shown in Fig. 4, which is advantageous when the makeup feed contains a significant quantity of material already fine enough to meet

Table 3. Representative Bond Work Indices*

Material	Bond Work Index, kWh / t
Bauxite	11
Cement clinker	16
Corundum	33
Dolomite	14
Feldspar	13
Ferrosilicon	12
Flint	32
Fluorspar	11
Granite	12
Gypsum rock	8
Hematite	15
Limestone	15
Magnetite	12
Phosphate rock	12
Pyrite	11
Quartz	16
Silicon carbide	32
Zircon sand	28

*These values are given only as a guide to the magnitude of Bond Work Indices. The Bond Work Index for a given material often varies with the test screen size. In addition, any type of material will usually have a range of *Wi*. For example, different cement clinkers have a range of values depending on the chemical composition and firing conditions. Limestone has grinding properties varying from soft chalky material to dense hard marble.

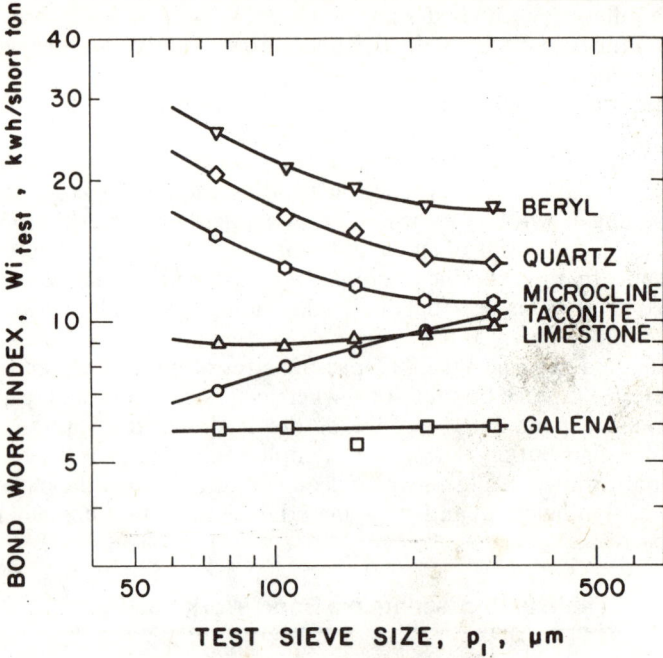

Fig. 3. Variation of test Work Index with test screen size (Smith and Lee, 1968).

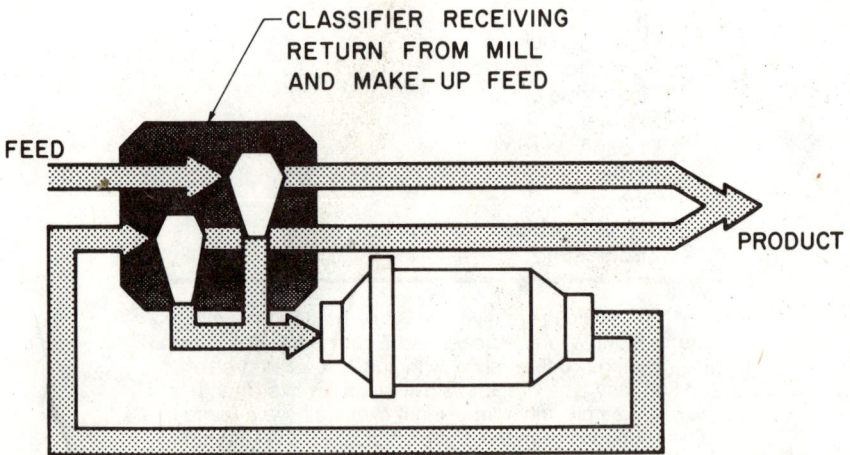

Fig. 4. The reverse closed circuit treated as two identical classifiers.

product specifications. Conceptually, this circuit can be treated *as if* there were two identical classifiers, one classifying the makeup feed and the other classifying the mill product (see Chapter 17). The underflow from the first classifier is the effective makeup feed to the normal closed circuit. In principle, the Bond calculation should be performed on the normal closed circuit part of this circuit which would require a knowledge of the classification action of the classifier on the makeup feed. The reversed circuit can sometimes be more efficient than the normal circuit for given makeup feed and product specifications (x_G and x_Q) because the *shape* of the final product size distribution is different, containing a smaller proportion of fines for the reversed circuit, under suitable circumstances.

However, the simulations given in the section titled, "Comparison of Mill Capacities by Bond and Simulation Methods" in Chapter 7 show that the capacity and size distribution of a reversed circuit is often almost identical to that of a normal circuit, all other factors being equal. This occurs because the real circulation ratio through the mill is lower than for a normal circuit if both use the same classifier, which compensates for the advantageous effect of removing fines from the makeup feed. Thus, the Bond calculation is performed *as if* the circuit were a normal closed circuit, using the 80%-passing sizes of the makeup feed and of the final circuit product. The correct way to design a reverse closed circuit to take advantage of classification of the makeup feed is discussed in that section in Chapter 7.

The application of the method to open circuit milling involves a logical problem. Table 1 reduces the capacity (increases the Work Index) by a factor which depends on the percentage less than the test screen size (p_1), desired in the open-circuit product. However, Fig. 3 shows that the test Work Index does not change significantly with p_1 for some materials. For a specified mill product, the percentage less than p_1 changes as different values are chosen for p_1, and if the test Work Index does not change with p_1 to compensate for the different multipliers from Table 1, a different mill size is predicted for each calculation, which is illogical.

The method cannot be used to characterize the case where an ore contains large fractions of mineral which liberate on breakage, since the circuit is then acting on several components (see the section on "Grinding of Mixtures of Ores" in Chapter 12).

Finally, the empirical oversize correction factor applied to a feed with e.g., $x_F = 10$ mm gives a very large reduction in mill capacity at low reduction ratios, especially for materials with a high Work Index. The mill capacity according to the calculation procedure is almost independent of x_P over a substantial range. This means that a mill operated at a fixed flow rate would give large changes in product fineness with minor fluctuations in feed rate or material grindability, which is clearly not in accord with plant practice or common sense.

Thus, the Bond method does not incorporate a number of important second order effects, and it cannot be used as a guide to the fine-tuning or optimization of a given system, either from the operating point of view or the economic point of view. It is valid as a gross method for mills operating under normal conditions. The function of the more advanced treatment by the size-mass balance method is to enable these disadvantages to be overcome: the effect of the classification action is precisely described by considering the total size distributions around the circuit (see Chapters 4, 6, and 7). In addition, the size-mass balance analysis shows that first-order breakage is normal breakage, but that as fines accumulate in the charge the breakage becomes non-first order, and slows down. In wet grinding the slowing-down is associated with the development of a viscous rheological character of the slurry as particle sizes become small. It is clearly a great advantage to be able to define the conditions at which the breakage rates start to slow down.

The *Operating* Work Index

Rowland (1973) has used the concept of *operating* work index, Wi_{op}, which is defined as the work index necessary in the calculations to make the calculated result agree with the actual operating result from an industrial mill. There are different ways in which this concept can be applied, since the measured capacity from a mill will not in general agree exactly with that predicted and the power drawn by the mill may also not agree exactly. The definition used by Rowland is based on the "kWh/t of capacity," that is, E:

$$Wi_{op} = (Wi)\left(\frac{\text{actual } E}{\text{predicted } E}\right) \qquad (11)$$

Since $E = m_p/Q$, the ratio of the operating work index to the predicted work index can vary from 1 if the measured capacity or the measured mill power is different from that predicted (Eq. 5), or any combination of these two factors. If the case is considered where the power drawn by the mill is correctly predicted, then the ratio of Wi_{op}/Wi is the ratio of predicted capacity to measured capacity.

A mill can be operating efficiently but because of the influences of classifier efficiency, feed size distribution, ball size distribution etc., the Wi_o/Wi ratio can vary from 1. Rowland has given results for ball mills in rod-ball mill circuits which show variation of the ratio from 0.78 to 1.29, with a mean of 0.945; see Table 4. As will be shown in Chapter 7, this range of variation is consistent with the effects of variations in classifier efficiency, primary breakage parameters, ball mix, etc., as predicted by complete mill simulations for efficient mill operating conditions. If the ratio becomes much larger than the upper limit of 1.3, it is an indication that mill conditions are not correct and direct inefficiency is occurring.

Other Conventional Sizing Methods

There are several other simple approximate sizing methods which use information from batch tests in laboratory mills, based on the relations discussed in the Summary of Chapter 2. In order to use these relations to size mills it is necessary to scale-up the descriptive parameters. This can be done by using one or more of the following three empirical and approximate relations. For grinding conditions constant except for mill diameter it is

Table 4. Comparison of Test and Operating Work Indices for Closed-Circuit Ball Mills in Rod-Ball Mill Circuits*

Mill Diameter Inside Liners (m)	(ft)	Feed (x_G, μm)	Product (x_Q, μm)	Operating Work Index, Wi_o	Work Index from Test, Wi	Wi_o / Wi	Sets of Data
3.0	10	1280	165	14.5	14.61	0.99	1
3.5	11 1/2	1150	230	11.48	8.9	1.29	1
3.8	12 1/2	1330	35.3	10.71	11.2	0.96	1
3.8	12 1/2	1123	38.0	9.77	11.2	0.87	1
3.8	12 1/2	1226	36.6	10.24	11.2	0.91	2
3.0	10	1568	121	5.34	5.99	0.89	6
3.0	10	1321	107	5.96	6.26	0.95	6
3.0	10	1444	114	5.65	6.12	0.92	12
3.7	12	1264	181	11.78	13.34	0.88	4
3.7	12	1135	185	13.17	13.18	1.00	4
3.7	12	1200	183	12.45	13.26	0.94	8
						0.945	24

*After Rowland, 1973.

found that shorter grind times are needed (to produce a desired size from a given size of feed) as mill diameter increases:

$$\text{Grind time } \tau \propto 1/D^{n_1} \tag{12}$$

where n_1 is about 0.5. If the specific loading of powder in the mill is W_o mass per unit volume and the mill volume is V, then $W = W_o V$, $\tau = W/Q$ and Eq. 12 can also be expressed in the forms

$$\text{Mill capacity } Q \propto VD^{n_1} \tag{12a}$$

$$Q \propto LD^{2+n_1} \tag{12b}$$

or, defining specific mill capacity as capacity per unit mill volume,

$$\text{Specific mill capacity} \propto D^{n_1} \tag{12c}$$

The second relation is

$$\text{Specific grinding energy} \propto 1/D^{\Delta} \tag{13}$$

where Δ is 0 to 0.2. (It is understood that this *comparative* specific grinding energy is the specific energy required to go from the same feed to the same product in two different mills.) When Δ is zero the same comparative specific grinding energy is used in all mills, while $\Delta > 0$ states that less comparative specific grinding energy is needed for a larger mill diameter.

The third relation (see Chapter 11) is

$$\text{Mill power } m_p \propto (\text{mill volume})(D^{n_2}) \tag{14}$$

or

$$\text{Mill power } m_p = K\rho_b LD^{2+n_2} \tag{14a}$$

or

$$\text{Specific mill power} \propto D^{n_2} \tag{14b}$$

where n_2 is 0.3 to 0.5; ρ_b is the ball density and K is a constant for set mill conditions. We will refer to Eq. 14 as a *mechanical* mill power law, because it is purely a function of the mechanics of raising balls in the rotating cylinder (see Chapter 11).

The relation between these three sets of equations is as follows. The comparative specific grinding energy is equal to mill power multiplied by grinding time divided by hold-up, that is, $1/D^{\Delta} \propto D^{n_2} \times 1/D^{n_1}$. Thus

$$\Delta = n_1 - n_2 \tag{15}$$

For $n_1 = 0.5$ and $n_2 = 0.5$, $\Delta = 0$; for $n_1 = 0.5$ and $n_2 = 0.3$, $\Delta = 0.2$.

In order to apply these relations in a simple way, the laboratory test should ideally be performed with mill conditions identical to those to be used in the production mill, except for mill diameter. This requires that the test mill should be big enough to take the same mixture of balls and feed sizes as the production mill and allow them to tumble freely. This type of test can be called a *breakage similitude* test, as distinct from a *standard* test such as the Bond test.

The use of the above scale-up laws is illustrated by considering first the *zero-order* relation of Chapter 2. The fact that the rate of production of fine material varies approximately linearly with time in a laboratory batch test leads to a particularly simple method of mill sizing if a satisfactory mill product size is defined by some specified weight fraction (or percentage) ψ less than a control size x^*. The test mill is treated as a reactor which takes some time to produce a satisfactory product and hence, a specific rate of production can be calculated. Then the scaling factor gives an estimate of the rate of production of a large mill and a mill size is chosen to give the required output. The test should be an exact simulation of the production mill conditions except for mill diameter.

Let the laboratory test results give a fractional rate of production of less-than-size x^* of A_Z. Expressed as mass per unit time per unit of mill volume,

$$\text{Specific rate of production of less-than-size } x^* = r_1 = A_Z W_o$$

where W_o is the specific loading of powder in the test mill, i.e., kg/m^3 or lb/ft^3 of mill volume. It is now assumed that the specific rate of production, r_2, in a large batch ball mill of diameter D_2 is related to that in a small batch ball mill of diameter D_1 by the simple exponent scale-up factor of Eq. 12c,

$$r_2 = r_1 (D_2/D_1)^{n_1}$$

The next assumption is either that the large mill is also batch, or if it is continuous, that it is almost in *plug flow*, so that all material in the feed will leave the mill at time $t = \tau$ after entering, and thus is treated *as if* batch ground for a time τ, where $\tau = W/Q$ with Q being the desired mass flow of product.

Then, to produce Q in the big mill requires a specific rate of production of less-than-size-x^* material in the large mill given by the mass balance:

Rate of less-than-x^* material out of mill minus rate of less-than-x^* material in feed to mill equals (specific rate of production of less-than-x^* material) (mill volume).

The rate of less-than-x^* material out of the mill is obviously $Q\psi(x^*)$ and the rate in is $QF(x^*)$, where $F(x^*)$ is the fraction of the feed which is less than size x^*. The mass balance statement in symbolic form is thus,

$$Q\psi(x^*) - QF(x^*) = (r_2)(L_2 \pi D_2^2/4)$$

Replacing r_2 by $r_1(D_2/D_1)^{n_1}$ the production mill length L_2 is obtained from

$$\left(\pi L_2 D_2^2/4\right) = \frac{Q[\psi(x^*) - F(x^*)]}{A_Z W_o}(D_1/D_2)^{n_1} \tag{16}$$

Choice of an appropriate L_2/D_2 ratio, based on the standard sizes available from manufacturers, completes the calculations. Normally, a substantial safety factor is included to give overcapacity, by using a value of Q which is 10 to 20% greater than the actual feed rate.

Another method of mill sizing is the *global specific energy* method, which is based on two major assumptions. It is assumed that the specific grinding energy varies with feed size and product size according to a known simple relation, such as

$$E_1 = C_Z[\psi(x^*) - F(x^*)]$$

for example, for both the test and full-scale mill. The value of C_Z is determined from the test data, knowing test mill power m_p and hold-up $W(C_Z = m_p/r_1 W)$. It is then assumed that the comparative specific grinding energy is independent of mill conditions but varies with mill diameter according to Eq. 13,

$$E_2/E_1 = \frac{C_{Z2}}{C_{Z1}} = (D_1/D_2)^{\Delta}, 0 \le \Delta \le 0.2$$

The desired production rate in the large mill is Q, therefore the mill power for the large mill is $m_{p2} = QE_2$, and

$$m_{p2} = QE_1(D_1/D_2)^{\Delta} \tag{17}$$

The size of the mill then follows, for the desired large-scale mill conditions, from one of the mechanical laws of the form

$$m_p = K\rho_b L D^{2+n_2} \tag{14a}$$

where K contains the effect of mill filling J, rotational speed, etc.; n_2 is close to 0.5 and is usually taken as 0.3 to 0.5.

It will be seen from the section, "Analysis of the Batch Grinding Equation" in Chapter 4, that the assumption that the comparative specific grinding energy is constant irrespective of mill conditions is the same as SW/m_p being constant, and it will be seen from Chapter 5 that this is approximately true for a range of powder filling (see Fig. 5 in Chapter 5), ball densities, and rotational speeds. It is *not* true for too low or too high powder filling, for a wide variation of ball diameter, or for a wide variation in ball loading. The assumption of the simple relation of E to feed and product size is *not* true if the mill feed contains particle sizes which are too big for normal breakage.

It is interesting to prove the correspondence between Eqs. 16 and 17. *If the test conditions in the mill duplicate the production mill in all respects except mill diameter, then K and ρ_b in Eq. 14a will be the same between the two mills.* Thus, $K\rho_b L_2 D_2^{2+n_2} = QE_1(D_1/D_2)^\Delta$, using $E_1 = C_Z[\psi(x^*) - F(x^*)]$; $C_Z = m_{p_1}/A_Z(x^*)W_1$, $W_1 = W_o V_1$, $m_{p_1} = V_1 K\rho_b D_1^{n_2} 4/\pi$,

$$\pi L_2 D_2^2/4 = \frac{Q[\psi(x^*) - F(x^*)]}{A_Z W_o}(D_1/D_2)^{n_2+\Delta}$$

Comparing with Eq. 16

$$\Delta = n_1 - n_2 \tag{15}$$

as previously assumed. It is thus demonstrated that the two methods of sizing are identical when an exact mill replication is used between the test and production mill ($K\rho_b$ the same). However, the use of the global energy approach is more general because it does not *necessarily* assume complete replication of mill conditions between the test and production mill.

In the above example, the relation used between E_1, feed size, and product size was that of the *Zero Order Law*. However, the same general line of reasoning can be applied using *any* of the batch grinding *laws* described in Chapter 2. As we have seen they are all of the form (see Table 5),

$$E = C_X f(\psi, F)$$

where ψ stands for a point specifying the desired product size distribution; F a point specifying the feed distribution; $f(\psi, F)$ is a different function for each law; and C_X is a different constant for each law. The general scale-up factor is that a larger diameter ball mill requires the same or a smaller

Table 5. Expression of the Various Laws of Batch Grinding in the Form *Specific Energy $E = C_X f(\psi, F)$*

Name	$f(\psi, F)$	ψ	F	C_X (kWh / t)
Zero-Order	$\psi(x^*) - F(x^*)$	$\psi(x^*)$	$F(x^*)$	C_Z
Charles	$\dfrac{1}{k_P^{\alpha_c}} - \dfrac{1}{k_F^{\alpha_c}}$	k_P	k_F	C_C
Bond Batch	$\dfrac{1}{x_P^{1/2}} - \dfrac{1}{x_F^{1/2}}$	x_P	x_F	C_B
Pseudo-Kick	$\log(x_P / x_F)$	x_P	x_F	C_K
Pseudo-Rittinger	$\dfrac{1}{x_P} - \dfrac{1}{x_F}$	x_P	x_F	C_R

specific grinding energy to go from a given feed to a given product, according to Eq. 13

$$E_2 = E_1(D_1/D_2)^\Delta \tag{18}$$

Then for a production mill in *plug flow* the batch grinding relation of Eq. S9 in Chapter 2 can still apply and, since $QE_2 = m_{p_2}$

$$m_{p_2} = QC_X f(\psi, F)(D_1/D_2)^\Delta \tag{19}$$

and the mill is then sized using a mechanical mill equation for m_p (see Chapter 11), such as Eq. 5.

The errors and uncertainties of the above sizing methods are as follows. It is implicit that the large-scale mill is giving size distributions with exact *similitude* between the small and large scale, which may not be true. Using values of C_X determined in a batch test for calculations on a continuous mill assumes that the continuous mill operates at plug flow, which is not true. The time of grind to go from the same feed to the same product in the full-scale vs. the laboratory mill is a constant factor, $(D_2/D_1)^{n_1}$, for exact similitude, but not otherwise. In practice, the mill product material not taken as final product is usually recycled to the feed. If this returned material broke in the same manner as the original feed, the new composite feed would produce the same set of size distributions as in Fig. 1 in Chapter 2, and the assumption of similar size distributions would be correct. However, this is obviously only a crude approximation to the truth.

The calculation obviously does not incorporate classification or make any allowance for classifier efficiency. It is clearly not possible, except by the more detailed analysis discussed later in Chapter 7, to judge the precise effect of recirculation and efficient/inefficient classification. As will be seen later, the use of classification and recirculation to give a closed mill circuit instead of simple batch or open circuit grinding enables the product size distribution to be varied (within limits) even when it passes through a preselected $\psi(x^*)$ point. The output from the circuit is different for the different distributions; a high degree of recycle with efficient classification avoids the consumption of energy in overgrinding fines, but this fact does not appear in the linear rate methods for sizing mills just illustrated.

All errors and uncertainties inherent in these approaches apply equally well whatever function or law is used. The choice of the function depends on the data and on personal preferences.

Mill Sizing by the Size-Mass Balance Method:
A Comment

Using the breakage relations described in Chapter 5 and the circuit relations described in Chapter 6, the size-mass rate balance method gives an exact simulation of mill behavior for prescribed milling conditions. It includes, of course, the effect of *oversize* feed material (which is to the right of the maximum in S values shown in Fig. 8 in Chapter 5); it can be extended to very fine grinding; it includes the real recycle ratio, the residence time distribution, etc., etc. For a given D, ball load J, ball size mix, and classifier behavior, the simulation will compute a value of Q/W for a desired product specification. Mill volume for a desired capacity Q then follows from $W = W_o V$,

$$V = \pi L D^2/4 = (Q/W_o)/(Q/W)_{\text{simulation}} \qquad (20)$$

The value of m_p is then obtained from the mechanical equation, e.g., Eq. 5 and the design is completed.

The simulation automatically gives the variation in output for different makeup feed, changed classifier parameters, etc., for a given size of mill. Alternatively, the effect of these variables on mill size for a given Q also follows automatically. This technique sizes the reactor (the mill) based on the rates of reaction (the breakage rates) occurring in the reactor. The energy input necessary to run the reactor (the mill) then follows from the size and mechanics necessary to turn the device. This seems to us a far more logical manner to perform the design than to estimate the energy first (using some empirical correlation with size distributions and test energies) followed by making the mill big enough to consume this amount of energy.

The Bond energy approach would be logical if *specific grinding energy* were a thermodynamic quantity (like heat content or entropy), which would be constant for different milling conditions, but this is not true. The grinding energy depends on whether the mill is operated with the correct filling conditions, speed, ball sizes, etc.; it depends on the degree and efficiency of recycle and classification and on the degree of overfilling at high flow rates. Bond's attempts (1960) to justify his laws by fracture mechanics are certainly incorrect and full of logical flaws. It is very dangerous to consider specific grinding energy as a fixed thermodynamic quantity since this draws attention away from the development of better mill designs and better operating conditions.

Summary

There are two broad categories of laboratory grindability tests: *standard* tests and *similitude* tests. Standard tests are useful when test results have been empirically correlated with full-scale plant performance, e.g., the Bond Work Index test.

The standard Bond test involves: (1) performing the standard test in a locked-cycle manner, so that the specific grinding energy is more comparable to a closed circuit mill; and (2) automatically scaling the standard test result to a 2.44 m (8 ft) i.d. wet overflow mill operating in closed circuit at a circulating load of 350% by empirical match of the test results to specific energy measured on such a mill. The specific energy is expressed in a standardized form as the Bond Work Index, Wi_{test}, in kWh/t.

The Bond scale-up law to any other mill diameter uses

$$Wi = Wi_{\text{test}}(2.44/D)^{0.2}, \quad D \leq 3.81 \text{ m} \Big\} \\ Wi = (0.914)Wi_{\text{test}}, \quad D \geq 3.81 \text{ m} \Big\} \qquad \text{(S1)}$$

The specific grinding energy is then calculated using

$$E = Wi\left(\frac{10}{\sqrt{x_Q}} - \frac{10}{\sqrt{x_G}}\right) \qquad \text{(S2)}$$

where x_G, x_Q are 80%-passing sizes of the makeup feed and desired circuit product. For a desired mill circuit capacity of Q tph

$$m_p = QE \qquad \text{(S3)}$$

The Bond mechanical equation for shaft mill power is

$$m_p = 7.33(L/D)J\phi_c(1 - 0.937J)\left(1 - \frac{0.1}{2^{9-10\phi_c}}\right)(\rho_b D^{3.3}), \text{ kW} \qquad \text{(S4)}$$

for D in m, ρ_b in t/m³. Therefore, for specified mill conditions (J, ϕ_c, ρ_b, L/D), the mill diameter follows. Various empirical corrections are used for other mill conditions such as open circuit, dry, coarse or fine grinding, or oversized feed.

The method is approximate because it does not explicitly incorporate a number of factors which influence mill capacity, such as shape of feed size distribution, recycle ratio and classifier efficiency, the effect of ball size, variations in residence time distribution, lifter design, slurry density and rheology, and the hold-up level in the mill. It also assumes that E is not a function of ball loading J, which is not in accord with laboratory and industrial evidence. The *operating* work index, Wi_o, is defined as the work index calculated from actual mill performance. The ratio Wi_o/Wi varies from plant to plant because of these variable factors. The Bond method cannot be used as a guide to optimization of mill conditions, especially for conditions which depart from normal conditions.

Other methods of approximate mill sizing based on laboratory batch grinding data use the relations of Chapter 2 plus scale-up laws. The tests are performed with the same breakage conditions (ball load, ball mixture, hold-up, fraction of critical speed, etc.) in the mill as expected in the production mill. These similitude tests can be used to estimate the effect of changing mill conditions, but they rely on the accuracy of the assumption of similitude and the scale-up laws used. The empirical scale-up laws are:

$$\text{Grind time} \propto 1/D^{n_1} \qquad (S5a)$$

or

$$\text{Mill capacity} \propto LD^{2+n_1}, \ n_1 \approx 0.5 \qquad (S5b)$$

$$\text{Specific grinding energy } E \propto 1/D^{\Delta}, \ \Delta \approx 0 \text{ to } 0.2 \qquad (S6)$$

and

$$\text{Net mill power} \propto LD^{2+n_2}, \ n_2 = 0.3 \text{ to } 0.5 \qquad (S7)$$

The relation between these laws is

$$\Delta = n_1 - n_2 \qquad (S8)$$

The global energy method assumes that the specific grinding energy varies with mill feed and product size by a known simple relation

$$E = C_X f(\psi, F) \qquad (S9)$$

where ψ is a point specifying the desired product [e.g., 60% < 208 μm (65 mesh)] and F stands for a point describing the feed. The relations $f(\psi, F)$ are those of Chapter 2, and C is determined from the laboratory similitude test. The general scale-up factor is that the production mill requires the same or smaller specific energy according to

$$E_2 = E_1 (D_1/D_2)^{\Delta}, \ \Delta = 0 \text{ to } 0.2 \qquad (S10)$$

Eqs. S3 and S4 complete the calculation.

Again, these methods give approximate answers only, because they do not allow for the real residence time distribution in a continuous mill or for the effect of recycle and classifier efficiency.

The size-mass balance method described in this book (especially see Chapters 6, 7, and 12) gives simulation of mill behavior under all conditions and shows the effect of all operating and design parameters on the capacity reported as $(Q/W)_{\text{simulation}}$. The simulation can be used for mill sizing since

$$V = (Q/W_o)/(Q/W)_{\text{simulation}} \qquad (S11)$$

where V is mill volume and W_o is specific hold-up (mass of solid per unit mill volume) and mill power follows from an appropriate mechanical equation. The simulation does not assume that specific grinding energy is independent of the filling levels, mixture of ball sizes, classifier efficiency, or other factors.

References

Austin, L. G. and Brame, K., 1983, "A Comparison of the Bond Method for Sizing Wet Tumbling Ball Mills with a Size-Mass Balance Simulation Model," *Powder Technology*, Vol. 34, pp. 261–274.

Bond, F. C., 1960, "Crushing and Grinding Calculations," *British Chemical Engineering*, Vol. 6, pp. 378–391, 543–548.

Rowland, C. A., Jr., 1973, "Comparison of Work Indices Calculated from Operation Data with Those from Laboratory Test Data," *Proceedings of the Tenth IMPC*, M. J. Jones, ed., Institute of Mining & Metallurgy, London, pp. 47–61.

Rowland, C. A., Jr. and Kjos, D. M., 1980, "Rod and Ball Mills," *Mineral Processing Plant Design*, A. L. Mular and R. B. Bhappu, eds., 2nd ed., AIME, New York, pp. 239–278.

Smith, R. W. and Lee, K. H., 1968, "A Comparison of Data from Bond Type Simulated Closed-Circuit and Batch Type Grindability Tests," *Trans. SME-AIME*, Vol. 241, pp. 91–99.

4

The Batch Grinding Equation: The Size-Mass Rate Balance

Introduction

It was seen in Chapter 2, Fig. 1, that results of batch grinding tests in laboratory tumbling ball mills have a basic pattern and there are several empirical and limited relations which can be applied to the results with more or less accuracy. In this chapter we will describe a more detailed analysis of this basic pattern using the concepts of specific rates of breakage and primary progeny fragment distributions in a complete size-mass balance. This detailed description is far more useful in analyzing test data, and the concepts involved will eventually enable us to perform precise simulations of milling circuits, and to describe very positively the influence of the variables in the grinding process. Fig. 1 in Chapter 2 shows an excellent agreement between computed and test data and this chapter will show the basis for the computed curves.

Historically, R. L. Brown (1941) appears to have been the first to attempt to construct a differential size-mass balance to describe the grinding process, but his formulation was clumsy and cumbersome to apply. His work was extended by Broadbent and Callcott (1956), who used matrix algebra and the concepts of stages of breakage to describe the process, and by Epstein (1947) who used a continuous probability formulation. Independently, Sedlatschek and Bass (1953) gave basically the formulation which we will apply in this book. Filippov (1961), Gaudin and Meloy (1962), and Gardner and Austin (1962), also independently of one another, extended and verified the concepts. The papers by Gardner and Austin were the first to demonstrate convincingly the applicability of the concepts to experimental data.

It has become the vogue to refer to the balances described here as population balances. However, this term is somewhat misleading since the balances are performed always on *mass*, not number and the concepts are much more akin to chemical reactor design for multiple first-order reactions than to conventional population balances. The basic literature quoted above certainly predates the first papers discussing population balances.

The First-Order Grinding Hypothesis

Consider a simple batch test mill to be a well-mixed container holding an amount of mass W of powdered material, with the material receiving a variety of breakage actions when the mill is running. It is convenient to represent the size distribution of the powder in the mill as shown in Fig. 4 in Chapter 1, where the size intervals are a geometric screen series ($\sqrt[4]{2}$ or $\sqrt{2}$). If the starting feed is all within the top size interval, numbered interval 1, then $w_1(0) = 1$. This feed is ground for a set time t_1, a sample removed, and the fraction still within the original size interval determined by sieving and weighing. The sample is returned to the mill, the mill run for additional time to give a total grind time of t_2, reanalyzed, and so on. It would appear reasonable that the rate of disappearance of size 1 might fit a *first-order* law:

Rate of disappearance of size 1 due to breakage $\propto w_1(t)W$

or

$$-\frac{d[w_1(t)W]}{dt} \propto w_1(t)W$$

Since the total mass W is constant, this becomes:

$$dw_i(t)/dt = -S_1w_1(t) \tag{1}$$

where S_1 is the proportionality constant and is called the *specific rate of breakage*, with units of time^{-1}. Then, if S_1 does not vary with time,

$$w_1(t) = w_1(0)\exp(-S_1t) \tag{2}$$

that is,

$$\log[w_1(t)] = \log[w_1(0)] - S_1t/2.3$$

Fig. 1 shows a typical experimental result. It should be recognized that there is no fundamental reason why the first-order hypothesis should apply in any given milling situation and several cases of deviation from the rule are discussed later, but very frequently it is an excellent approximation to the truth. Experimental verification of the hypothesis by results such as those in Fig. 1 proves that the build-up of fines does not affect the specific rate of breakage of the top size material. However, it does *not* prove that finer material will also break in a first-order manner, in the presence of a varying quantity of coarser material. This basic check was performed by Gardner and Austin (1962) using a radiotracing technique. If some size is traced, then the disappearance with time of the traced material from that size fraction can be distinguished from the appearance in that size of untraced products of fracture of larger sizes. Then

$$w_j^*(t)/w_j^*(0) = \exp(-S_jt) \tag{3}$$

where $w_j^*(t)$ is the fraction of traced material of size j.

Fig. 2 shows the result they obtained, which demonstrates first-order breakage in an environment in which the amounts of both smaller and larger sizes are changing with time. If this is true, then it also appears reasonable to assume that (for given milling conditions and fixed W) the batch grinding test can be repeated with a smaller size as the feed top size and a value of S for this size determined as in Fig. 1 to give the same result as a tracer test. This technique is known as the *one-size-fraction* technique. Fig. 3 shows a typical set of results for grinding in a laboratory ball mill. Note that we will adopt the convention of plotting the S value of a size interval against the *upper* size of the interval and we will always denote the size by the *upper* size.

The analogy of the specific rates of breakage to specific rates of reaction is obvious, and $w_i(t)$ has a meaning comparable to that of the partial molar concentration in a batch chemical reactor.

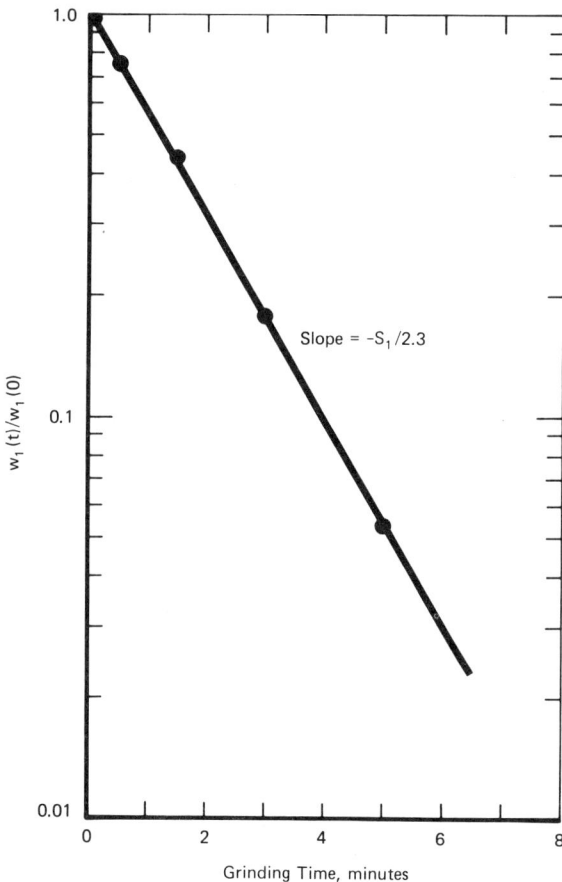

Fig. 1. Example of first-order plot: 1.18 mm × 850 μm (16 × 20 mesh) anthracite in a 0.6-m (2-ft) diam ball mill.

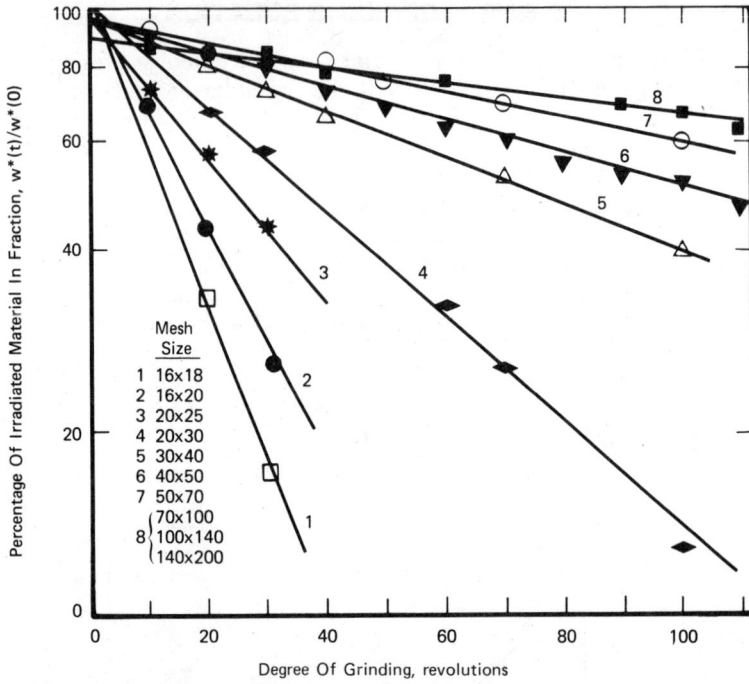

Fig. 2. First-order plot for traced coal ground in a standard Hardgrove machine (20 rpm).

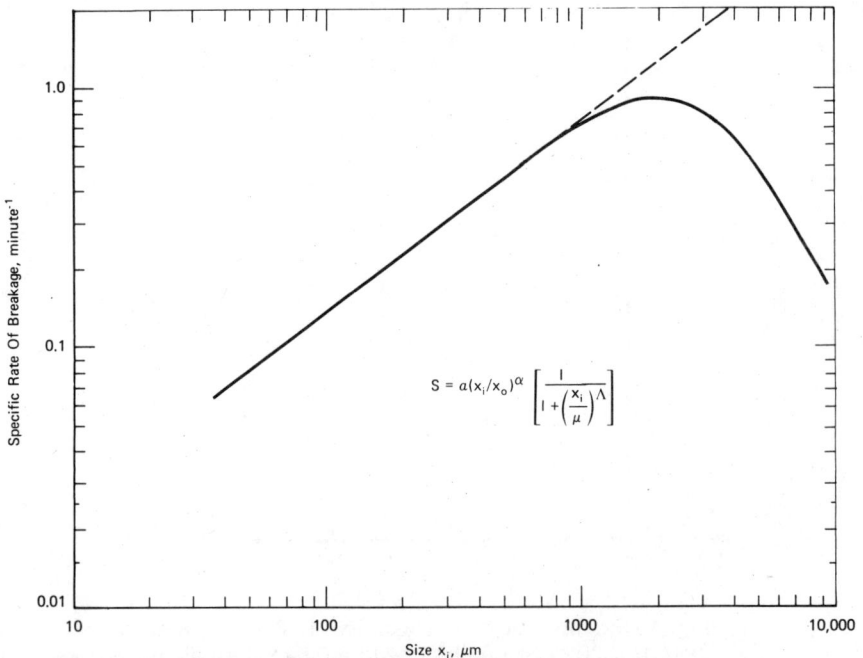

$$S = a(x_i/x_o)^\alpha \left[\frac{1}{1 + \left(\frac{x_i}{\mu}\right)^\Lambda} \right]$$

Fig. 3. Example of variation of specific rates of breakage with particle size: quartz ground in a 200 mm (8-in.) diam ball mill with 25.4 mm (1-in.) steel balls; S values are for $\sqrt{2}$ size intervals (see Fig. 1 in Chapter 2).

64

The Primary Breakage Distribution Function, or Progeny
Fragment Distributions

Clearly, grinding of even a single size produces a whole range of product sizes. To describe the grinding process it is necessary to describe this distribution of sizes. In the sense used here, breakage is defined as occurring only when the particles are broken out of their original size range. Thus, in a $\sqrt{2}$ size interval, say 1.18 mm × 850 μm (16 × 20 US mesh), the material must be broken less than 850 μm (20 mesh) to be considered as broken, and hence the products of breakage are defined as appearing in sizes less than 850 μm (20 mesh).

Primary breakage is defined as follows: Material breaks and the fragments produced are mixed back in with the general mass of powder in the mill. If this distribution of fragments can be measured before any of the fragments are *reselected* for further breakage, then the result is the *primary breakage distribution* (see Fig. 4). The term primary does not necessarily mean that the fragments are produced by a single fracture propagation, but only that they are produced by breakage actions occurring before they are remixed back into the bulk. It should also be noted that the values measured in milling situations are presumably the mean of a whole variety of breakage actions on many particles and cannot be expected to compare directly with results from compressive tests on single particles.

There are two symbolisms convenient for characterizing the primary progeny fragment distributions. First, if material of size 1 is broken the

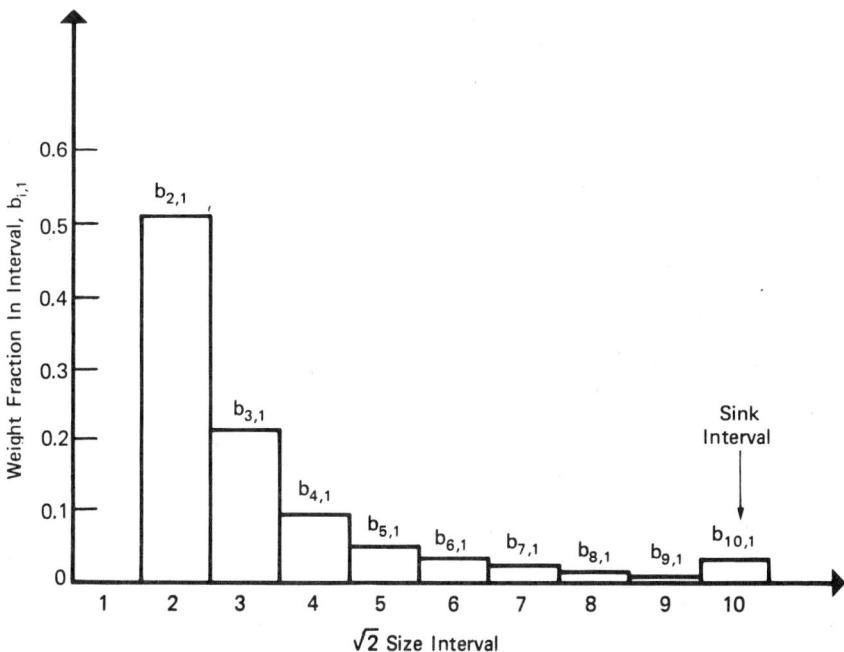

Fig. 4. Typical bar graph of primary progeny fragment distribution.

"weight fraction of the products which then occur in the size interval i is called $b_{i,1}$." The set of numbers $b_{i,1}$, where i ranges from 2 to n, then, describes the distribution of fragments produced from size 1. In general, the matrix of numbers $b_{i,j}$ is needed to describe the breakage of all sizes of interest; that is, the set $b_{i,1}$ with $n \geq i \geq 2$; plus the set $b_{i,2}$ with $n \geq i \geq 3$, etc. The second symbolism in common use is to accumulate the b values from the bottom interval and let $B_{i,1}$ be the "cumulative weight fraction of material broken from size 1 which appears *less than* the upper size of size interval i." Table 1 gives a typical example. Again, in general, a complete matrix of $B_{i,j}$ values is required for complete characterization of all breakage actions.

As will be discussed in Chapter 9, values of B can be determined from one-size-fraction tests at short grind times, where approximate corrections to allow for reselection for breakage of the primary fragments are reasonably valid. It is inherent in this technique that the values of B do not change with the grinding time in the mill. Put another way, if 16×20 mesh feed material is breaking toward the end of the grinding time then it is assumed that the $B_{i,16 \times 20}$ values are the same as for this size breaking at the start of the grind, even though the size environment has changed. This, again, was proved by the radiotracing experiments carried out by Gardner and Austin (1962).

It might seem an impossibly complicated task to measure the matrix of B values for all materials under all milling conditions. However, it is often found (Shoji, Lohrasb, and Austin, 1979) that the B values are insensitive to the precise mill conditions, at least in the normal operating range of milling conditions. In addition, B values for all the materials we have tested show a similar general form (see Chapter 5). Further, it is often found that the B values are dimensionally normalizable; that is, the fraction which appears at sizes less than, say, half of the starting size is independent of the starting size. For this reason, it is common practice to plot B values against dimensionless (normalized) size, as shown in Fig. 5. If the B values are normalized, the matrix of B values is reduced to a vector, as illustrated in Fig. 6, so that $b_{i,j}$ can be replaced with b_{i-j}.

Table 1. Typical Set of Primary Progeny Fragment Sizes

Size US mesh	Interval Number i	$b_{i,1}$	$B_{i,1}$
18 × 25	1	0.0	1.0
25 × 35	2	0.52	1.0
35 × 45	3	0.21	0.48
45 × 60	4	0.10	0.27
60 × 80	5	0.05	0.17
80 × 120	6	0.031	0.12
120 × 170	7	0.021	0.086
170 × 230	8	0.015	0.064
230 × 325	9	0.0115	0.049
< 325	10	0.038	0.038

It is now apparent why geometric size intervals are preferred over linear size intervals and, in fact, are essential to apply the simplification of normalizability. However, it requires an infinite number of such intervals to go down to zero size and it is convenient to use geometric size intervals down to the $n - 1$th interval and let the nth interval contain all finer material. The nth interval is thus abnormal as it is a *sink*, and this fact must always be remembered: for example, whatever the relation between S and particle size, S_n must always be zero because material cannot be broken out of the final nth interval; and $b_{n,j} = B_{n,j}$, by definition (see Table 1). Note that for notational simplicity we will often use $b_{i,j} = b_{ij}$, and $B_{i,j} = B_{ij}$.

As far as the rates of production of fine material are concerned, the b values appear as:

Rate of production of size i from breakage of larger size j

$$= (\text{fraction to } i \text{ from } j)(\text{rate of breakage of size } j)$$

$$= b_{ij}S_j w_j(t)W$$

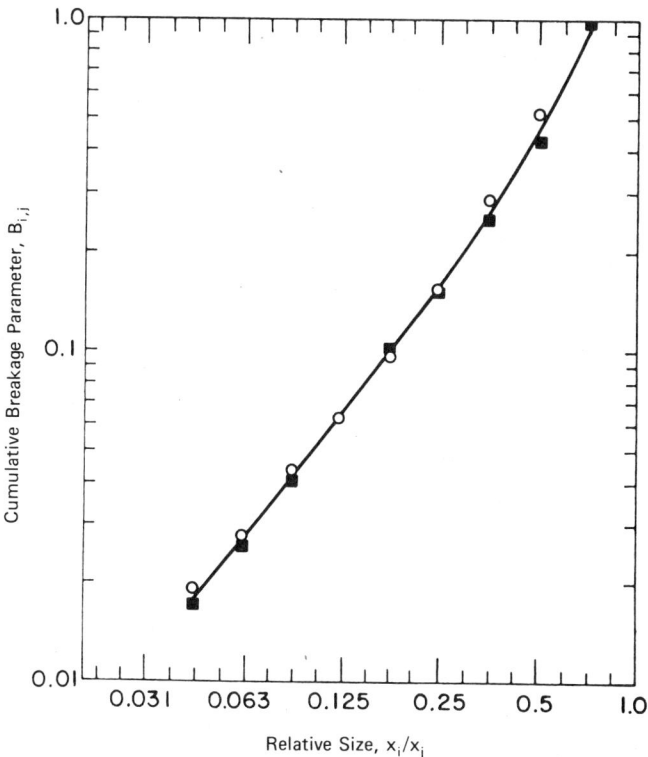

Fig. 5. Cumulative breakage distribution parameters for ball milling of 850 × 600 μm (20 × 30 mesh) quartz (see Fig. 1 in Chapter 2): ■, dry; ○, wet (45% solid by volume).

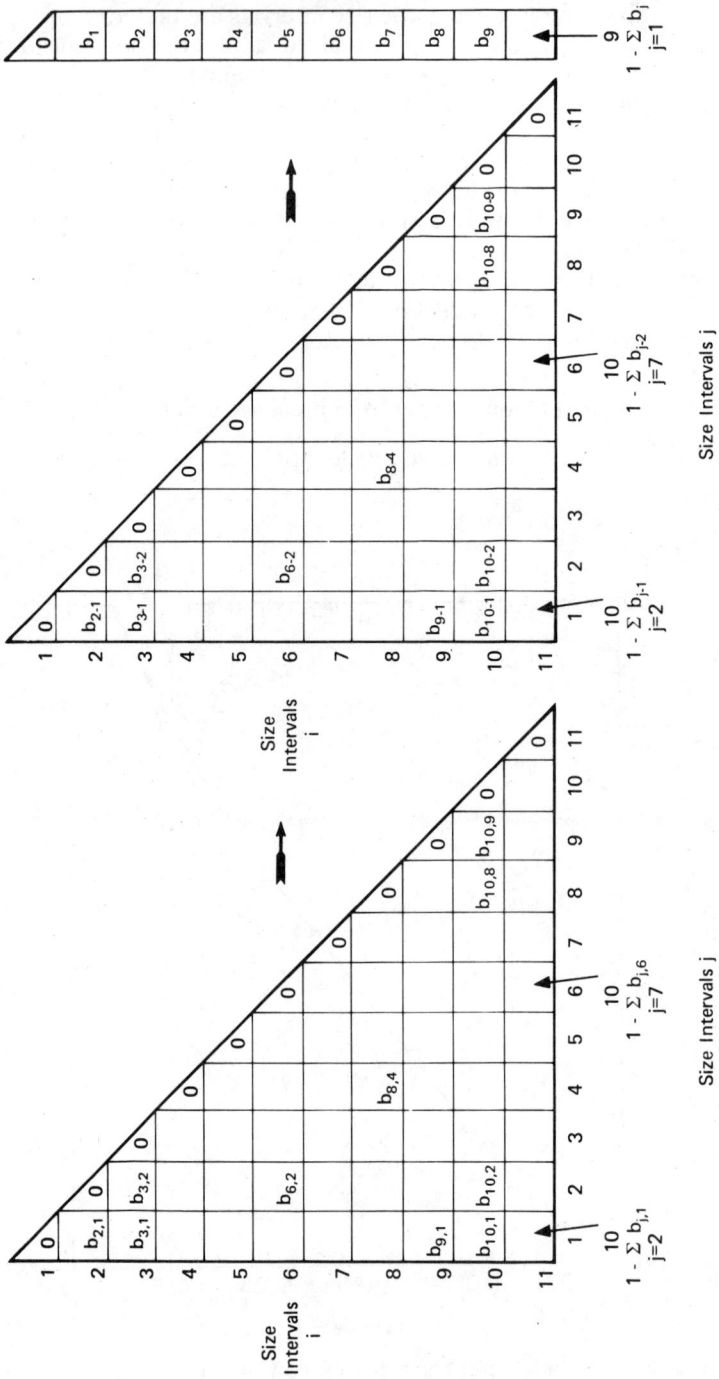

Fig. 6. Illustration of breakage distribution matrix transformed into normalized form.

Similarly,

Rate of production of less-than-size i from breakage of larger size j

$$= B_{ij}S_jw_j(t)W.$$

The Size-Mass Balance

We are now in a position to perform a complete size-mass balance or population balance on the batch grinding system. The concept used in the formulation of grinding equations is basically that of a rate-mass balance on each particle size interval present; it can be considered as a population balance but it is usually mass which is experimentally measured rather than numbers of particles, and mass is conserved whereas number is not, so it is far more convenient to work in terms of mass. In addition, the concept of *first-order breakage* leads to simpler solutions and is physically real in many cases, although it is not necessary in the balances.

The statements involved are:

1) The rate of disappearance of size j material by breakage to smaller sizes $= S_jw_j(t)W$.
2) The rate of appearance of size i material produced by fracture of size j material $= b_{i,j}S_jw_j(t)W$.
3) The rate of disappearance of size i material by breakage to smaller sizes $= S_iw_i(t)W$.
4) The net rate of production of size i material equals the sum rate of appearance from breakage of all larger sizes minus the rate of its disappearance by breakage.

Symbolically, this final balance is:

$$\frac{d[w_i(t)W]}{dt} = -S_iw_i(t)W + \sum_{\substack{j=1 \\ i>1}}^{i-1} b_{ij}S_jw_j(t)W$$

or

$$dw_i(t)/dt = -S_iw_i(t) + \sum_{\substack{j=1 \\ i>1}}^{i-1} b_{ij}S_jw_j(t), \quad n \ge i \ge j \ge 1 \qquad (4)$$

This is the fundamental size-mass rate balance for fully mixed batch grinding, and this set of n differential equations describes the grinding process; it gives, of course, Eq. 1 when $i = 1$. In general, the equation set can be solved by numerical computation (see Chapter 17) but if the values of S are independent of grinding time, a closed analytical solution exists (see Chapter 6); the solution must, of course, start from a known feed size distribution $w_i(0)$. Solution for various grinding times generates the value of $w_i(t)$, from which values of $P(x_i, t)$ are readily calculated by accumulation. Fig. 7 illustrates the mass balance.

Fig. 1 in Chapter 2 shows the computed solution for various grinding times, using the normalized B values of Fig. 5 and the S values of Fig. 3, for

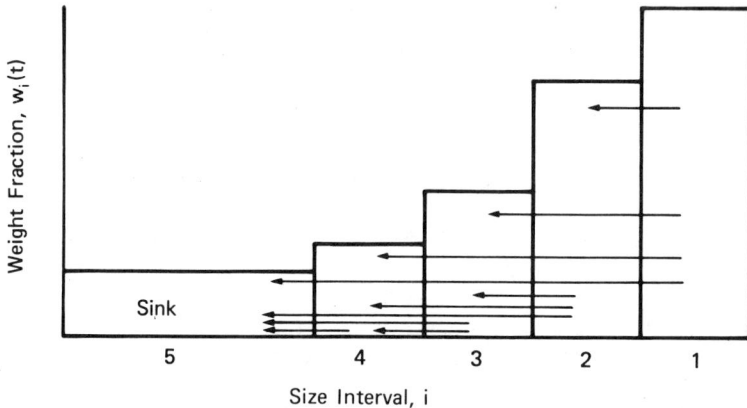

Fig. 7. Illustration of mass balance action in fully mixed batch grinding: size interval 2 is receiving material from size 1; size interval 3 is receiving material from sizes 1 and 2, etc.; and the finest sink interval is receiving material from all larger sizes.

the feed shown. Also shown are experimental values determined by sieving, and an extension to subsieve sizing using the Sedigraph (see Chapter 9). *It is clear that the agreement between computed and experimental results is excellent.* This is strong confirmation that the assumptions made in measuring and applying the values of S and B to the solution are correct for this set of data: Grinding *is* first-order and the B values *are* constant with time. It is concluded that the set of S and B values (extrapolated to fine sizes if desired) can be used in the solution of the batch grinding equation to give an exact and complete simulation of the size distributions.

There are several other assumptions implicit in this formulation. It is assumed that no regrowth of particles, smaller or larger, occurs by cold welding. It is also implicit that the fracture properties of a given size j in the products of breakage are the same as size j material in the raw feed; this is not always true since the source history of the feed may affect the breakage properties. It is also implicit, of course, that the material being broken is *homogeneous* from the breakage point of view; that is, it does not consist of a mix of stronger and weaker components nor does it develop such a mix as it breaks. It is surprising how homogeneous, from this point of view, most materials are: rocks with obvious visual inhomogeneities often give excellent first-order breakage. Later chapters will deal with some cases where these assumptions are not strictly valid.

The technique of setting up the size-mass rate balances can be illustrated again, from a slightly different point of view. The basic balance can be expressed as:

Rate of production of material less than size x_i equals the sum of the rates of production of material less than size x_i by breakage from all larger sizes.

Symbolically,

$$\frac{dP(x_i, t)}{dt} = \sum_{\substack{j=1 \\ i>1}}^{i-1} B_{i,j} S_j w_j(t), \quad n \geq i \geq j \geq 1 \left.\begin{array}{c} \\ \\ \\ \end{array}\right\} \tag{5}$$
$$= 0, \qquad\qquad i = 1$$

This is readily derived algebraically from Eq. 4 using $P(x_i, t) = \sum_{k=n}^{i} w_k(t)$, and requires the same inputs for solution. It is a more convenient form, however, for obtaining a simple analytical solution to the batch grinding equation that will be used later in a number of places (see Chapter 8).

In general, the computed solutions of Eqs. 4 or 5 can be represented as

$$P(x_i, \theta) = P(x_i, \theta, \underline{S}, \underline{B}, \underline{w}(0)) \tag{6}$$

where θ is the grind time, and $\underline{S}, \underline{B}, \underline{w}(0)$ the appropriate sets of input values.

In reading the literature on grinding equations, considerable confusion can arise because of the variety of symbolisms and mathematical forms which have been used. Austin has reviewed (1971–72) this literature with the specific purpose of pointing out the mathematical identities implicit in the various forms. Table 2 is taken from this paper. The forms above the center line are size-continuous forms, which assume that the first-order hypothesis applies to a differential size interval. There is no experimental evidence that this is true and several good reasons why it is unlikely (see Chapter 8). Therefore, in this book we will use only the finite size interval forms, based on $\sqrt[4]{2}$ or $\sqrt{2}$ size interval. The four forms below the line are equivalent to density function forms, and corresponding cumulative forms are readily derived. In two of the equations the specific rate of breakage is expressed as the fractional breakage, π_j, occurring in a single *stage of breakage*, that is, one pass through the machine for a once-through device such as a roll crusher. If breakage in a reservoir mill were considered as a series of stages of breakage, the connection between π_j for a single stage and the usual specific rate of breakage is clearly $S_j = \pi_j/\Delta t$, Δt being the time in the single breakage stage.

Analysis of the Batch Grinding Equation

Some very important conclusions concerning milling can now be made. First, because solution of the batch grinding equation set gives virtually identical results to the experimental data of Fig. 1 in Chapter 2, all the empirical relations deduced by applying pseudo-Rittinger, Charles', and Bond's laws to the data must also be consequences of the first-order hypothesis of grinding combined with the shape of S_i and $B_{i,j}$ values as a function of particle size. Similarly, the applicability of the constant rate-of-production or zero order law to the data also follows from the same considerations. There is no need, therefore, to look for fundamental reasons why these relations apply: they are the fortuitous result of a reasonable (and

Table 2. Comparison of Equations of Batch Grinding*

Conditions	Equation
Time-continuous, size-continuous	**Cumulative forms** $$\partial P(x,t)/\partial t = \int_{y=x}^{x_{max}} S(y)B(x,y)d_y P(y,t)$$ or $$P(x,\tau) = P(x,0) + \int_0^{\tau}\int_{y=x}^{x_{max}} S(y)B(x,y)d_y P(y,t)$$ or $$\partial^2 P(x,t)/\partial t\,\partial x$$ $$= \int_{y=x}^{x_{max}} S(y)\frac{\partial B(x,y)}{\partial x}d_y P(y,t) - S(x)\partial P(x,t)/\partial x$$ **Density function form** $M(x,t) = \partial P(x,t)/\partial x$ $b(x,y) = \partial B(x,y)/\partial x$ $$\partial M(x,t)/\partial t = -S(x)M(x)$$ $$+\int_{y=x}^{x_{max}} S(y)b(x,y)M(y,t)dy$$
Size-continuous, discrete time interval or stage-of-breakage	**Cumulative form** $$P_p(x) = P_{p-1}(x) + \int_{y=x}^{x_{max}}\pi(y,p)B(x,y)dP_{p-1}(y)$$ p is number of stages of breakage. $\pi(y,p)$ is specific breakage of size y in the pth stage. **Density function form** $$M_p(x) - M_{p-1}(x)$$ $$= -\pi_{p-1}M_{p-1}(x) + \int_{y=x}^{x_{max}}\pi(y,p)b(x,y)M_{p-1}(y)dy$$
Time-continuous, discrete size interval (usually geometric intervals)	**Summation form** $$\partial w_i(t)/\partial t = \left[\sum_{\substack{j=i-1\\i>1}}^{1} S_j b_{i,j}w_j(t)\right] - S_i w_i(t)$$ **Matrix forms** $$dw(t)/dt = -[I-B]Sw(t)$$ where $w(t)$ is a column vector, and the rest column-row matrices.
Discrete time interval, discrete size interval	**Summation form** $$w_{i,p} - w_{i,p-1}$$ $$= \left[\sum_{\substack{j=i-1\\i>1}}^{1} S_j b_{i,j}w_{j,p-1}\right] - S_i w_{i,p-1}$$ **Matrix forms** $$p_p = D^p f = Dp_{p-1}$$ where p, f are column vectors and $D = BS + I - S$

*Source: Austin, 1971 - 72.

experimentally proven) hypothesis and the form of the S and B values. Instead, the problem of fundamental understanding becomes: "Why do the values for S and B vary with size and mill conditions in the observed manner?" (see Chapter 5).

Second, the family of size distributions obtained in Fig. 1 in Chapter 2 depended on the feed size chosen. However, the basic size-mass rate balance can be solved with the S and B values for *any feed size distribution* and thus gives a general simulation of batch grinding. This is illustrated in Fig. 8, where an unnatural feed size distribution is used. Clearly, the old empirical relations discussed in Chapter 2 cannot be applied to such data, except at long grind times when the influence of the unnatural feed has been damped out.

Third, supposing two milling situations are compared in which B values are unchanged but the S values change by a constant factor,

$$S_i' = \kappa S_i, \quad n \geq i \geq 1 \tag{7}$$

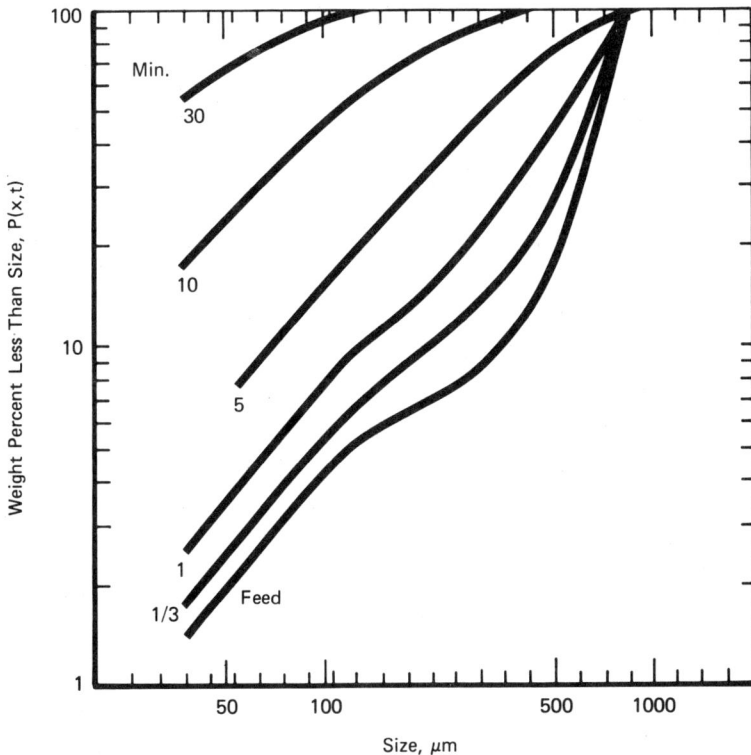

Fig. 8. Computation of size distributions of batch grinding with an unnatural feed-size distribution.

The two batch grinding equations are

$$dw_i/dt = -S_iw_i + \sum_{\substack{j=1 \\ i>1}}^{i-1} b_{ij}S_jw_j \tag{A}$$

$$dw_i/dt = -S_i'w_i + \sum_{\substack{j=1 \\ i>1}}^{i-1} b_{ij}S_j'w_j \tag{B}$$

Substituting for S_i' in Eq. B gives

$$\frac{dw_i}{d(\kappa t)} = -S_iw_i + \sum_{\substack{j=1 \\ i>1}}^{i-1} b_{ij}S_jw_j, \quad n \geq i \geq j \geq 1 \tag{C}$$

If Eqs. A and C are solved with the same feed size distribution for a total grind time θ for A and θ' for C, it is clear that the solutions will produce *identical* results when $\kappa\theta'$ in case C equals θ in case A, because the equations are *identities* if κt is replaced by t in Eq. C. Put another way, if case B has everything identical to case A except that all specific rates of breakage are doubled (in general, increased by κ), it is clear that the solution to case A for grind time θ of 5 min, say, is identical to the solution of case B for $\theta' = 2.5$ min ($\theta' = \theta/\kappa$, in general). In the terminology of Eq. 6, $P(x_i, \theta, \underline{S}, \underline{B}, w(0)) = P(x_i, \theta/\kappa, \underline{S}', \underline{B}, w(0))$ and the grinding time τ to go from a given feed to a desired product is

$$\tau \propto 1/\kappa \propto 1/S \tag{8}$$

This is an extremely useful conclusion, because if it is found that two different grinding tests give the same family of curves but shifted only by a *time-scale* factor, it can be assumed that B values are the same, and the variation of S with size is the same, and only a scale factor of S is different. This is illustrated in Fig. 9, where it is clear that the size distributions vary identically with time, but shifted by a time factor of 1.85, that is, the larger diameter mill produces the same size distribution in $1/1.85$ of the time. Considering tumbling ball milling, the mill conditions include: (1) fraction of mill filled by balls, J; (2) fraction of mill filled by powder, f_c; (3) mill diameter, D; (4) mill speed, rpm; (5) ball size, d; (6) ball density, ρ_b; etc. A change in each of these may possibly change *only* the time-scale factor in S.

Fourth, if the concept of a constant specific energy to accomplish a certain grind (given feed to desired product) is valid, then it follows that S values must be proportional to the mill power per unit powder mass in the mill. That is, doubling the energy input rate per unit mass in the mill must lead to a doubling of S values, a halving of the time to produce a given grind and hence, the same energy per unit mass. This has been confirmed under certain conditions in laboratory tumbling ball mills by Malghan and Fuerstenau (1976), as will be discussed in Chapter 5. The implication is that grinding of a given material in a ball mill is an identical process for different mill conditions in all respects except the time-scale factor of S values.

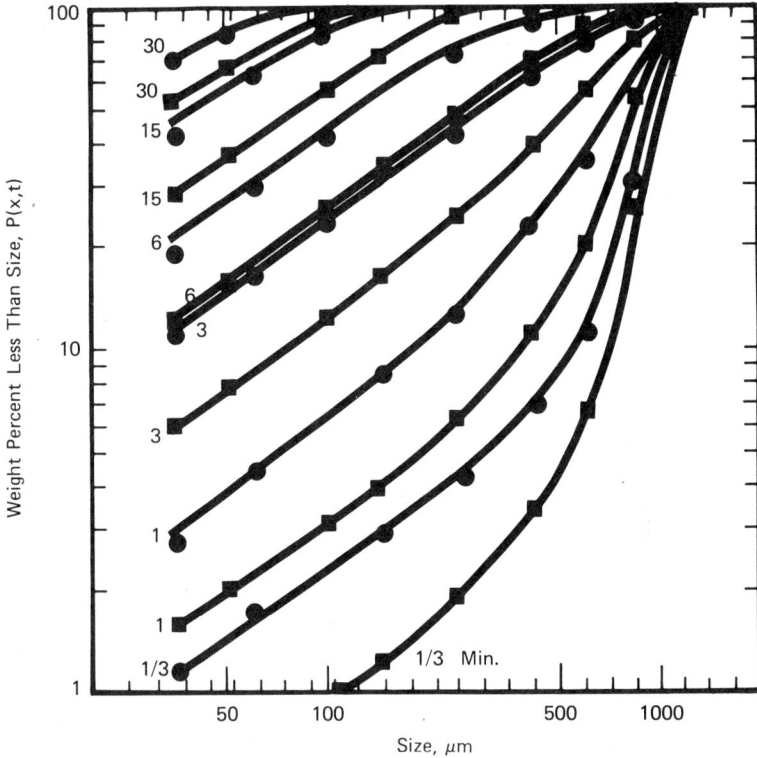

Fig. 9. Comparison of size distribution of 1.18 mm × 850 μm (16 × 20 mesh) petroleum coke ground in: ●, 200 mm (8-in.) diam ball mill; ■, 0.6-m (2-ft) diam ball mill ($U = 1$, $J = 0.3$, $\phi_c = 0.7$, 25.4 mm (1-in.) diam balls).

These statements can be formalized by considering the two important factors in grinding: the mill capacity, which is the output rate Q of desired products, and the specific grinding energy E, which is the energy per ton to grind to the desired product size. If the grinding time taken to give a desired product size distribution from a given feed is τ, the mill capacity is clearly $Q = W/\tau$. Defining *specific mill capacity* as the capacity per unit mass of powder in the mill, that is, Q/W,

$$\text{specific capacity } Q/W = 1/\tau \qquad (9)$$

Since τ is proportional to $1/S$,

$$\text{specific capacity } Q/W \propto S \qquad (10)$$

Thus the specific mill capacity is proportional to S values (for constant B values). The specific grinding energy is given by $E = m_p/Q = \tau m_p/W$ where m_p is the mean power draw by the mill and then,

$$E \propto m_p/SW \qquad (11)$$

Thus the specific energy of grinding is constant as mill conditions are changed *if m_p/SW is constant* for all the conditions.

We have tried throughout this book to avoid the use of the term *grinding efficiency* because the degree of conversion of energy to useful size reduction is an ill-defined concept. In addition, the only efficiency which is of real interest is the meeting of product specifications at the minimum of cost. However, a grinding operation can be called "more efficient" when the specific grinding energy is reduced in comparison to some other condition. Thus efficient breakage actions occur when m_p/SW is a minimum: wrong mill conditions which increase m_p/SW cause *direct* inefficiency.

Summary

The establishment and utilization of appropriate size-mass balance equations for different grinding equipment configurations is the central theme of this book. A basic concept of this approach is the assumption of *first-order breakage* or rate of disappearance of each size fraction within the mass being ground. Thus for a simple fully mixed batch mill:

Rate of disappearance of size interval j due to breakage $\propto w_j(t)W$ (S1)

or
$$dw_j^*(t)/dt = -S_j w_j^*(t) \qquad \text{(S2)}$$

Thus
$$w_j^*(t)/w_j^*(0) = \exp(-S_j t) \qquad \text{(S3)}$$

where S is called the *specific rate of breakage*, with units of time^{-1}, and $w_j^*(t)$ is the weight fraction of original material of size j at time $t = 0$ still remaining at time t. The size intervals are chosen in $1/\sqrt[4]{2}$ or $1/\sqrt{2}$ progression and numbered starting with 1 for the largest size interval, down to n for the smallest.

Experimentally this first-order hypothesis has been shown to be an excellent approximation for many materials in several types of mill. Thus a plot of $\log[w_j^*(t)/w_j^*(0)]$ vs. t for each size interval j gives a straight line with slope of $-S_j/2.3$ (e.g., Fig. 2). When only the disappearance of the largest size fraction is followed, called the *one-size-fraction* method, the use of Eq. 3 is greatly simplified. This is because no special effort is required to identify (or trace) the original material of size 1 at time $t = 0$.

A complementary concept is that of describing the *progeny fragment distribution* once a given size j material is broken. The set of numbers $b_{i,j}$ is called the *primary breakage vector*, where $b_{i,j}$ is defined as the fraction of broken j material which appears in size interval i, where i is a smaller size than the size of j. It is implicit that these breakage fragments are measured before any of the particles are reselected for further breakage. This requires special tracing of each starting size j other than the top size. Also the distributions are assumed to be constant over the time of grinding. Fig. 4 illustrates the description of the set of $b_{i,1}$ values resulting from breakage of size interval 1 material.

Thus, to describe the complete pattern of breakage products that result from the simultaneous breakage of n sizes ($j = 1, 2, \ldots, n$) requires a matrix of values. This matrix is denoted by $b_{i,j}$ where $n \geq i \geq j$, $b_{j,j} = 0$, and n is the number of the *sink interval* containing all fragments less than some specified smallest measured size. Obviously, $\sum_{i=j}^{n} b_{i,j} = 1$ by definition. Also a cumulative breakage distribution $B_{i,j}$ can be constructed so that $B_{i,j} = \sum_{k=n}^{i} b_{k,j}$ where $B_{j,j} = 1.0$, $B_{j+1,j} = 1.0$ and $B_{n,j} = b_{n,j}$ (see Fig. 5). It has been experimentally shown that the $b_{i,j}$ matrix is relatively insensitive to milling conditions for any given material. Even further, it has been determined in many cases that the $b_{i,j}$ matrix is dimensionally normalizable so that the $b_{i,j}$ matrix reduces to a vector b_{i-j} where $b_0 = 0$. This means that breakage of all sizes from 1 to $n-1$ gives rise to the same progeny fragment size distribution based on relative size. This concept is illustrated in Fig. 6.

The approach used in formulating grinding equations is that of performing a rate-mass balance on each particle size interval present. For example, the set of n differential equations describing fully mixed batch grinding can be formulated as:

The net rate of production of size i material equals the sum rate of appearance of size i from breakage of all larger sizes minus the rate of its disappearance by breakage. (S4)

Symbolically, this is

$$dw_i(t)/dt = -S_i w_i(t) + \sum_{\substack{j=1 \\ i>1}}^{i-1} b_{i,j} S_j w_j(t), \quad n \ge i \ge j \ge 1 \quad \text{(S5)}$$

Fig. 7 illustrates the mass balance concept. The development of the above equation assumes that there is no particle regrowth and that the material being broken is homogeneous in its breakage character.

The solution of Eq. S5, using independent experimentally determined sets of S_j and $b_{i,j}$ values, can be shown to agree well with experimental data (e.g., see Fig. 1 in Chapter 2). This general agreement leads to a number of important conclusions. The first is the obvious inference that all the empirical relations described in Chapter 2 are the fortuitous result of the (experimentally proven) first-order hypothesis and the form of the S and B values. Thus to explain grinding, the question becomes: "Why do the values for S and B vary in the manner observed?"

A second conclusion is that the use of the mass balance concept allows for the accurate description of the effect of varying feed size distribution. The mass balance concept allows for the accurate estimation of energy required to go from any feed to any product fineness.

Third, if it is found that two different batch grinding tests give the same family of curves but shifted only by a time-scale factor, it can be assumed that the B values are the same, the variation of S with size is the same, and only a scale factor, κ, of S is different

$$S_i' = \kappa S_i \quad \text{(S6)}$$

This condition makes correlations of capacity and energy with varying mill conditions much simpler.

A fourth conclusion relates to the assumption of a constant specific energy to accomplish a certain grind (from a given feed to a desired product). If the specific energy is constant, it follows that S values must be proportional to the mill power per unit powder mass in the mill, as follows. In a batch grind it is clear that the specific mill capacity Q/W is proportional to S. Since $Q = W/\tau$, $Q/W = 1/\tau$ and hence,

$$1/\tau \propto S \tag{S7}$$

where τ is the time taken to give a desired product size distribution from a given feed assuming $b_{i,j}$ values are constant. Specific grinding energy is given by

$$E = \tau m_p/W = m_p/Q \tag{S8}$$

where m_p is the average power draw by the mill. Then

$$E \propto m_p/SW \tag{S9}$$

Thus, the specific energy of grinding is constant *only* if m_p/SW is constant for all grinding conditions. Later chapters will discuss the validity of this assumption and show where it breaks down. Wrong mill conditions which increase m_p/SW represent direct mill inefficiency.

References

Austin, L. G., 1971–72, "A Review Introduction to the Description of Grinding as a Rate Process," *Powder Technology*, Vol. 5, pp. 1–17.

Broadbent, S. R. and Callcott, T. G., 1956, "A Matrix Analysis of Processes Involving Particle Assemblies," *Philosophical Transactions of the Royal Society of London*, Vol. A249, pp. 99–123.

Broadbent, S. R. and Callcott, T. G., 1956a, "Coal Breakage Processes," *Journal of the Institute of Fuel*, Vol. 29, London, pp. 524–528.

Broadbent, S. R. and Callcott, T. G., 1956b, *Journal of the Institute of Fuel*, Vol. 29, London, pp. 528–539.

Broadbent, S. R. and Callcott, T. G., 1957, *Journal of the Institute of Fuel*, Vol. 30, London, pp. 13–17.

Brown, R. L., 1941, "Broken Coal-III Generalized Law of Size Reduction," *Journal of the Institute of Fuel*, Vol. 14, London, pp. 129–134.

Epstein, B., 1947, *Journal of the Franklin Institute*, Vol. 244, pp. 471–477.

Epstein, B., 1948, *Industrial and Engineering Chemistry*, Vol. 40, pp. 2289–2291.

Epstein, B. and Lowry, H. H., 1948, "Some Aspects of the Breakage of Coke," *Blast Furnace, Coke Oven and Raw Materials Conference*, reprint, AIME, Apr.

Filippov, A. F., 1961, "Distribution of the Sizes Which Undergo Splitting," *Theory of Probability and Its Applications*, Vol. 6, English translation, USSR, pp. 275–280.

Gardner, R. P. and Austin, L. G., 1962, "A Chemical Engineering Treatment of Batch Grinding," *Proceedings*, 1st European Symposium Zerkleinern, H. Rumpf and D. Behrens, eds., Verlag Chemie, Weinheim, pp. 217–247.

Gaudin, A. M. and Meloy, T. P., 1962, "Model and a Comminution Distribution Equation for Repeated Fracture," *Trans. SME-AIME*, Vol. 223, pp. 43–50.

Malghan, S. G. and Fuerstenau, D. W., 1976, "Scale-Up of Ball Mills Using Population Balance Models and Specific Power Input," *Proceedings*, 4th European Symposium Zerkleinern, H. Rumpf and K. Schonert, eds., Dechema Monographien 79, Nr. 1576-1588, Verlag Chemie, Weinhein, pp. 613–630.

Sedlatschek, K. and Bass, L., 1953, "Contribution to the Theory of Milling Processes," *Powder Metallurgy Bulletin*, Vol. 6, pp. 148–153.

Shoji, K., Lohrasb, S., and Austin, L. G., 1979, "The Variation of Breakage Parameters with Ball and Powder Loading in Dry Ball Milling," *Powder Technology*, Vol. 25, pp. 109–114.

5

What Laboratory Tests Tell Us About Breakage in Ball Mills

Introduction

Our current knowledge of ball milling is based on a combination of past experience with pilot-scale or large-scale mills and results from small-scale laboratory mills. As discussed in Chapter 1, the description of a continuous large-scale mill has to include the residence time distribution in the mill. Batch tests in a laboratory mill, however, can focus solely on the factors which affect breakage without the complicating effect of mass transfer. In addition, it is possible to examine quantitatively the influence of each factor which affects breakage because the experiments are much easier and quicker to perform on the laboratory scale and can be more precisely controlled. On the other hand, there is no *a priori* guarantee that results from a small-scale mill will be identical or similar to those in a larger mill. Correspondence between the laboratory scale and the pilot or large scale must be proved experimentally.

There is no doubt that there is some degree of correlation between laboratory results and larger scale results; otherwise the use of laboratory grindability tests (see Chapter 3) to estimate the required size of large-scale mills would not give correct answers. However, the earlier sizing methods have relied heavily on empirical matching of laboratory results under standard conditions with full-scale results or on pilot-scale tests in fairly large continuous mills. There are two reasons for the difficulty of applying small-scale results directly to the larger scale. First, the transfer of small-scale results to the larger scale can only be done via the detailed size-mass balance method, in order to take into account *all* of the differences between the laboratory test and the full-scale operation. The theory and information necessary to do this has only become available in recent years. Second, it seems likely that assumptions which are good for some materials, some feed sizes, and some mills will not apply for all materials, sizes, and mills. In other words, there may be exceptions to general rules which will cloud the picture.

The following sections present information from laboratory tests which undoubtedly corresponds *qualitatively* with large-scale results. It is believed

79

that in many cases the equations developed from laboratory results can be *quantitatively* extended to large-scale mills, using the methods of Chapter 12.

The Mode of Operation of a Tumbling Ball Mill

It is instructive to restate the mode of operation of a tumbling ball mill. Concentrating on breakage behavior, the mill operation is as follows. The rotation carries balls and powder round the mill as illustrated in Fig. 1. As the balls tumble down, they strike powder nipped against other balls. In addition, the general movement of balls in the bed will rub particles between balls. Crabtree et al. (1964) distinguish several different types of fracture which can occur. First, a massive impact will give complete disintegration of a particle (*fracture*); Steier and Schonert (1971) have photographed the fragmentation which occurs and the violence with which particles are thrown out from the fracture region. Second, a glancing blow can chip off a corner (*chipping*); this mechanism rounds irregular rock into roughly spherical pebbles in autogenous milling. Third, rubbing will give wear of surfaces (*abrasion*); again, in autogenous milling the roughly rounded pebbles formed by chipping will wear to give smooth pebbles similar to beach pebbles. Chipping and abrasion will lead to production of fine material and their combined effects are sometimes referred to as *attrition*.

In any ball mill operating under normal conditions, all of these mechanisms of size reduction will be at work. The measured values of specific rates of breakage are the net effect of the sum of the mechanisms. The measured values of the primary progeny fragment distribution are the total of the fragments produced by each mechanism. Because of the complete range of the types of impact present in this type of mill loaded with heavy steel balls, it is even probable that each of the mechanisms overlap, forming a con-

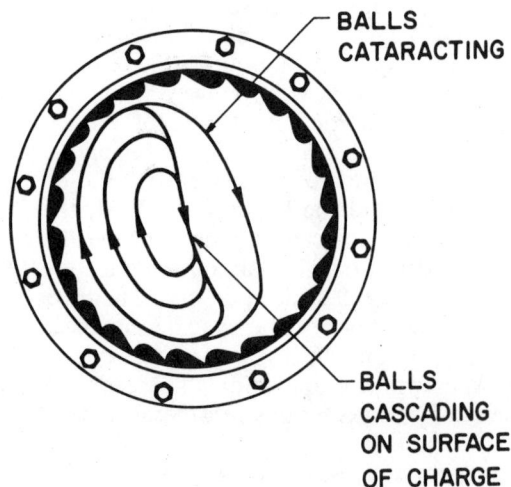

Fig. 1. Illustration of ball movement in a ball mill at normal operating rpm.

tinuum of breakage actions. If this continuum shifts from one mechanism towards another as conditions in the mill are changed, it is to be expected that the B values will change, because the primary fragment distributions produced by the three mechanisms are different.

At low rotational speeds the balls have a relatively gentle tumbling action and in fact, there is a tendency for the mass of balls to be raised by the rotating case and to slip back as a locked mass. As speed is increased the tumbling action increases and the bed appears as an inclined surface from which balls are emerging, rolling down, and reentering the surface. The bed is expanded, allowing particles or slurry to penetrate between the balls. The series of collisions with other balls as a ball tumbles down is the major method of transferring stress to particles. The bed is in a *cascading* state. At higher rotational speeds, more of the balls are ejected from the surface at the top of the mill and a *cataract* of balls is formed. The fraction of critical speed at which these processes occur depends on the filling conditions and the type of lifters (liners) (Marstiller, 1979). Bar lifters can carry some fractions of the balls up to cataract, while wave or beveled lifters require higher rotational speeds to give the same degree of cataracting. Wear of lifters can change mill performance with time.

The power required to turn the mill passes through a maximum as rotational speed increases, corresponding to a maximum in the rate of lifting and the mean height of lift (see Chapter 11). For the case of breakage of normal particle sizes, which are not too large to be completely fractured, it appears that the large number of ball-powder-ball impacts caused by cascading is optimum for breakage. Thus the maximum rates of breakage are obtained approximately at the speed of maximum power draw, normally close to 75% of critical speed, depending on the ball load and type of lifters.

Variation of Breakage with Particle Size

Fig. 2 shows a typical result for the variation of specific rates of breakage with particle size x_i (upper size of ith interval) for a ball charge consisting of a single size. For the smaller particles,

$$S_i = ax_i^\alpha, x_i \ll d \tag{1}$$

The theory of fracture implies that smaller particles are relatively stronger because larger Griffith flaws exist in larger particles and they are broken out as size is reduced. In addition, however, it is certain that it is more difficult to nip a given mass of smaller particles in a mill in comparison to larger particles so that a geometric effect is present. The fact that the specific rates of breakage are a simple power function of size has not been adequately explained on a theoretical basis, but it has been amply demonstrated by many experiments. The value of α is a positive number, normally in the range 0.5 to 1.5, which is characteristic of the material (providing the test conditions are in the normal operating range) but the value of a will vary with mill conditions. The units of a in Eq. 1 will vary for different values of α, so for comparison of one material with another under standard test

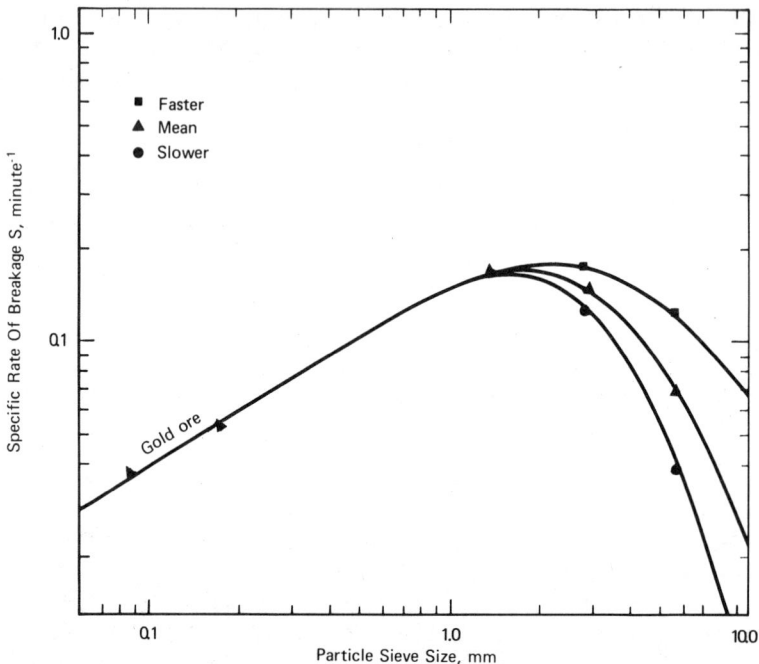

Fig. 2. Specific rates of breakage of a South African gold ore as a function of particle size [$\sqrt[4]{2}$ intervals; 200-mm diam ball mill, 25.4-mm balls].

conditions it is convenient to use Eq. 1 in the form

$$S_i = a(x_i/1000 \ \mu\text{m})^{\alpha} \qquad (1a)$$

where a now has dimensions of time^{-1}. Physically, it is $S_{18 \times 25}$ where $S_{18 \times 25}$ is the *normal* specific rate of breakage (extrapolated from the straight line region of Fig. 2 if necessary) of the 1.0 mm × 710 μm (18 × 25 mesh) fraction, and x_i is in micrometers. If x_i is expressed in mm, $S_i = a(x_i/1 \text{ mm})^{\alpha}$.

For larger sizes, it is found that the disappearance of material from a given top size interval is often not first order, but appears to consist of a faster initial rate and a slower following rate (see Chapter 16). Some of the particles are too big and strong to be properly nipped and fractured by the balls and have a slow rate of breakage. In addition, the accumulation of finer material appears to cushion the breakage of these larger sizes. We refer to the first-order breakage of smaller sizes as *normal* breakage and to the non-first-order breakage of larger sizes as the *abnormal* breakage region. In this region a mean effective specific rate can be defined by the time required to break 95% of the material, and this value is also shown in Fig. 2. It is seen that the mean specific rates of breakage of larger sizes start to decrease, so that S_i passes through a maximum at a certain size, x_m. The presence of a maximum is quite logical because large lumps obviously will be too strong to be broken in the mill. The presence of abnormal breakage in a given mill situation represents *direct inefficiency*: The particles are too big for the energy of the tumbling balls to be used efficiently in causing fracture.

To allow for the slower mean rate of breakage of the larger sizes, correction factors Q_i are introduced to Eq. 1

$$S_i = a(x_i/1000)^\alpha Q_i \tag{2}$$

$Q_i = 1$ for smaller sizes and becomes small for large sizes. The experimentally determined values of Q_i can be fitted by the empirical function

$$Q_i = \frac{1}{1 + (x_i/\mu)^\Lambda}, \qquad \Lambda \geq 0 \tag{3}$$

where μ is the particle size at which the correction factor is $1/2$ and Λ a positive number which is an index of how rapidly the rates of breakage fall as size increases (the higher the value of Λ, the more rapidly the values decrease). Again, Λ is found to be primarily a characteristic of the material but μ will vary with mill conditions. Table 1 gives the characteristic values of α and Λ for a number of different materials measured under standard conditions.

It seems probable that lifters that give more cataracting will increase the rates of breakage of large sizes because of the larger impact forces produced by cataracting: thus, μ would be expected to be larger for this type of lifter, at a given rotational speed. Similarly, higher rotational speeds would have the same effect for the same reason. However, no quantitative relations for such effects are available to date.

The values of the size x_m at which S is a maximum varies somewhat from one material to another, being larger for weaker solids which break more readily. x_m is related to μ since both are dependent on where the curve of S vs. x starts to bend. Inserting Eq. 3 into Eq. 2, differentiating, and setting $dS/dx = 0$ at $x = x_m$,

$$\mu = \left(\frac{\Lambda - \alpha}{\alpha} \right)^{1/\Lambda} x_m \tag{4}$$

where Λ is always greater than α.

The primary *progeny* fragment distributions for the normal breakage region have the form shown in Fig. 3, where the values are plotted in the cumulative form, $B_{i,j}$, vs. size as a fraction of the breaking size, x_i/x_j. Three important features are to be noted. First, these B values do not seem to be sensitive to milling conditions such as powder load, ball load, mill diameter, etc. There has been no satisfactory explanation of this fact, but it has been verified experimentally in many tests. The result suggests that the mean breakage action caused by a ball-ball collision is the same for different mill diameters, which in turn implies that *cascading* is the main action. A ball

Table 1. Breakage Properties of Some Materials (D = 190 mm, mill volume = 5250 cm³, ball diameter = 26 mm)

Breakage Characteristic	Parameters	North Carolina Quartz	Ores	
			Copper Ore	Pennsylvania Limestone
Rate of breakage	α^*	0.80	0.95	0.90
	a^*, min⁻¹, dry	0.40	0.70	0.95
	a^*, min⁻¹, wet	1.00	1.20	N.D.
	μ^*, mm	1.90	1.40	1.40
	$\Lambda^\dagger$	3.7	2.7	2.0
B values, small sizes	γ^*	1.10 dry, 1.30 wet	0.70	0.65
	$\beta^\dagger$	5.8	4.3	3.2
	Φ_{18-25}	0.65 dry, 0.58 wet	0.40	0.28
	δ	0.00	0.00	0.34
Experimental conditions:	weight of powder, g	328.0	343.0	338.0
	specific gravity*	2.7	2.7	2.7
	fractional ball filling, J	0.20	0.20	0.20
	operational speed, ϕ_c	0.72	0.75	0.75

*Values rounded to nearest 0.05.
†Values rounded to nearest 0.1.
Note: See also Table 1 in Chapter 7, Tables 1 and 10 in Chapter 12, and Table 2 in Chapter 15.

Table 1. (continued)

Breakage Characteristics	Parameters	Coals				
		Shamokin Anthracite	Western Kentucky No. 9	Belle Ayre So. Wyoming Coal	Ohio No. 9	Lower Freeport
Rate of breakage	α^*	0.75	0.80	0.80	0.80	1.05
	a^*, min^{-1}, dry	0.95	1.60	1.80	1.40	2.50
	a^*, min^{-1}, wet	0.95	2.0	2.2	2.0	4.7
	μ^*, mm	1.80	3.60	3.70	2.40	3.40
	$\Lambda^\ddagger$	3	3	3	3	3
B values, small sizes	γ^*	1.00	0.90	0.90	0.95	0.80
	$\beta^\dagger$	3.1	2.8	2.8	3.5	2.3
	$\Phi^*_{18 \times 25}$	0.40	0.40	0.40	0.50	0.50
	δ	0.0	0.0	0.0	0.0	0.0
Experimental conditions:	weight of powder, g	120.0	120.0	120.0	120.0	120.0
	specific gravity*	1.45	1.40	1.30	1.60	1.60
	fractional ball filling, J	0.20	0.20	0.20	0.20	0.20
	operational speed, ϕ_c	0.72	0.72	0.72	0.72	0.72
	Hardgrove Grindability Index	35	55	58	65	88

*Values rounded to nearest 0.05.
†Values rounded to nearest 0.1.
‡A value of $\Lambda = 3$ is a sufficient approximation for the coals.

Table 1. (continued)

Breakage Characteristic	Parameters	Cements				
		Clinker L	Clinker M	Clinker P	Clinker S	Slag Clinker P
Rate of breakage	α^*	0.90	0.90	1.10	0.95	1.60
	a^*, min^{-1}, dry	0.80	0.85	1.20	0.80	1.68
	a^*, min^{-1}, wet	N.D.	N.D.	N.D.	N.D.	N.D.
	μ^*, mm	1.75	1.70	1.75	2.05	1.50
	Λ^*	2.50	4.05	3.35	3.60	4.20
B values, small sizes	γ^*	0.75	0.90	0.85	0.80	1.25
	$\beta^\dagger$	4.0	4.0	4.0	3.3	4.3
	$\Phi_{16 \times 20}$	0.34	0.51	0.34	0.28	0.58
	δ	0.23	0.20	0.25	0.22	0.0
Experimental conditions:	weight of powder, g	300.0	300.0	300.0	300.0	300.0
	specific gravity†	3.2	3.2	3.2	3.2	3.2
	fractional ball filling, J	0.20	0.20	0.20	0.20	0.20
	operational speed, ϕ_c	0.80	0.80	0.80	0.72	0.80

*Values rounded to nearest 0.05.
†Values rounded to nearest 0.1.

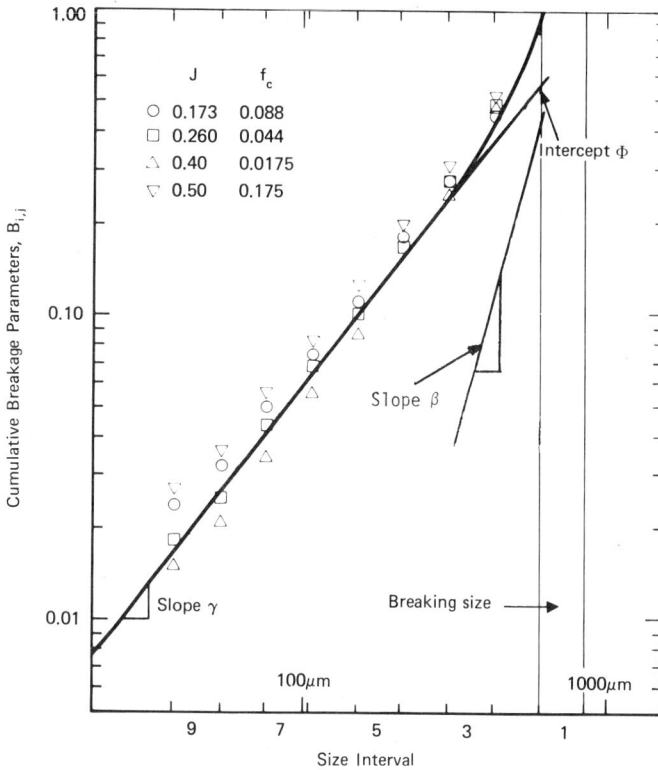

Fig. 3. Cumulative progeny fragment distributions from the breakage of 850 × 600 μm (20 × 30 mesh) quartz under various mill load conditions (D = 195 mm, d = 25.4 mm, ϕ_c = 0.7).

cascading in a large diameter mill bounces down from top to bottom with a series of small impacts of the same magnitude as in a smaller mill, whereas a cataracting ball would have a much greater impact force in a large diameter mill. The rubbing force between balls as they rise in the bed would be expected to give attrition. Second, for some materials the curves of $B_{i,j}$ fall on top of one another for all values of j. This is referred to as the case of *normalized B*, and it means that all of the particles break into a fragment distribution with dimensional similarity, that is, the weight fraction of product less than, say, 1/2 of the breaking size is constant.

Third, the values of $B_{i,j}$ can be closely fitted by an empirical function made up of the sum of two straight lines on log-log paper, that is,

$$B_{i,j} = \Phi_j \left(\frac{x_{i-1}}{x_j} \right)^\gamma + \left(1 - \Phi_j \right) \left(\frac{x_{i-1}}{x_j} \right)^\beta, 0 \le \Phi_j \le 1 \qquad (5)$$

where Φ_j, γ, and β are defined in Fig. 3 and are characteristic of the material. The function of Eq. 5 can be called the *primary breakage distribution function*. If the B_{ij} values are not normalized the degree of nonnormalization can often be characterized by the additional parameter δ defined by $\Phi_{j+1} = \Phi_j R^{-\delta}$, $\delta \geq 0$, where $R = 1/\sqrt{2}$ (see "Nonnormalized B Values" in Chapter 9). The values of γ lie between 0.5 and 1.5 and β is typically in the range 2.5 to 5 (see Table 1). Broadbent and Callcott (1956) used an equation which gave the same B values for all materials, but we do not find this to be applicable; especially, the product size distribution from a mill is sensitive to the value of γ.

The primary progeny fragment distributions for larger sizes, to the right of the maximum in S values, will sometimes have different B values. Generally, the value of γ is less for the abnormal breakage region. This is probably because the mean breakage action in this region contains bigger components of chipping and abrasion, leading to relatively more of the finest material and hence, a low value of γ. The shape of the distributions tend to be somewhat different, with a plateau region in the mid-size range. They can then be fitted by the more general function (Miles, Shah, and Austin)

$$B_{i,j} = \Phi_j \left\{ \frac{1 - \exp\left[-\left(\dfrac{x_{i-1}/x_j}{k} \right)^\gamma \right]}{1 - \exp\left[-(1/k)^\gamma \right]} \right\} + (1 - \Phi_j)\left(\frac{x_{i-1}}{x_j} \right)^\beta \qquad (5a)$$

where k is an additional parameter which gives the finer sizes the form of the Rosin-Rammler distribution. This reduces asymptomatically to the simpler form for k large for realistic values of γ (also see "Ball Size Effects on B Values" in Chapter 16).

Appendix A1 gives illustrations of the effects of different characteristic breakage parameters on the size distributions produced by batch grinding. Grinding of a single size fraction as feed gives Schuhmann slopes close to γ, but the slopes tend to change toward the α value. A distributed feed size distribution gives Schuhmann slopes which change toward γ. Feed sizes which are too big for normal breakage give a characteristic plateau region in the size distribution, since the rates of breakage of smaller and larger sizes are less than the sizes in this region.

Rotational Speed

Typical variation of the net power required to turn a mill is shown in Fig. 4 as a function of rotational speed. The normal specific rates of breakage vary with speed in the same way. However, the maximum in power occurs at different fractions of critical speed from one mill to another, depending on the mill diameter, the type of lifters, the ratio of ball to mill diameter, and the ball and powder filling conditions. The maximum is usually found in the range of 70 to 85% of critical speed, with 70 to 75% being the usual range for mills of large diameter with full ball loading ($J \approx 0.4$).

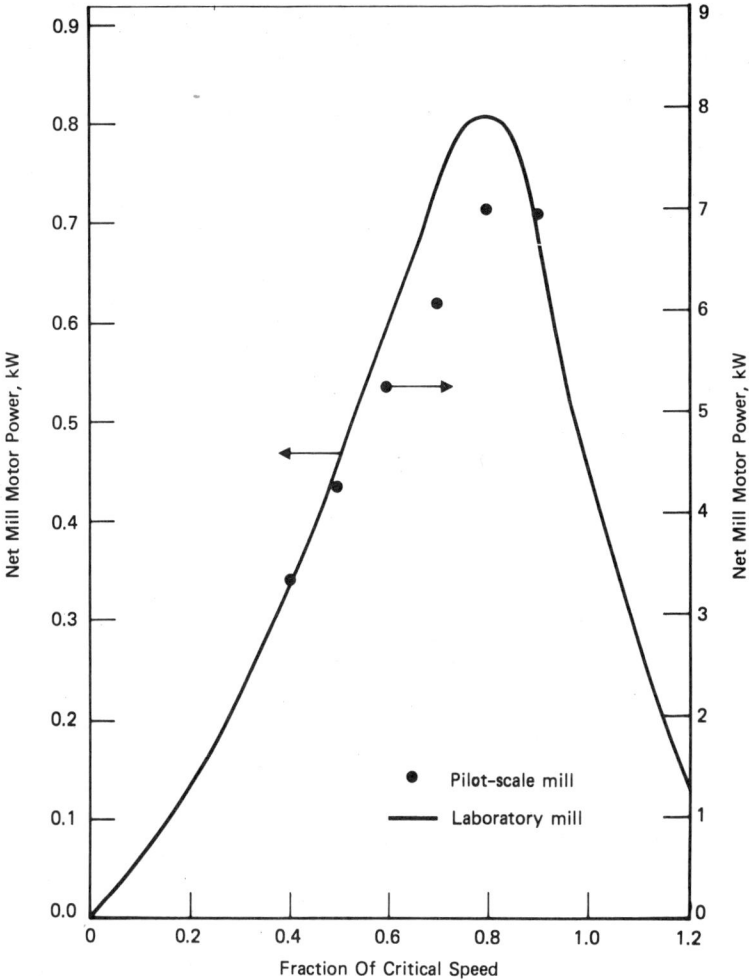

Fig. 4. A typical variation of net mill power with rotational speed for a laboratory mill fitted with lifters ($D = 0.6$ m, $L = 0.3$ m, $J = 0.35$, $d = 25.4$ mm) and a pilot scale mill ($D = 0.82$ m, $L = 1.5$ m, $J = 0.35$, mixture of ball sizes).

Within the range of speed near the maximum power draw there are relatively small changes in the normal specific breakage rates with rotational speed. The test results described in the previous and following sections were obtained with mill speeds within this range. There is no significant variation in B values with rotational speed within this range. An empirical fit to data gives the approximate expression for mill power as

$$m_p \propto (\phi_c - 0.1)\left(\frac{1}{1 + \exp[15.7(\phi_c - 0.94)]}\right), \quad 0.4 < \phi_c < 0.9 \quad (6)$$

This equation can be used to correct S values from one rotational speed to another,

$$S_i \propto (\phi_c - 0.1)\left(\frac{1}{1 + \exp[15.7(\phi_c - 0.94)]}\right), \quad 0.4 < \phi_c < 0.9 \quad (6a)$$

The possible effect of rotational speed on the specific rates of *abnormal* breakage was mentioned in the previous section.

Ball and Powder Loading

If mill tests are run at a constant hold-up mass W or fractional volume filling f_c, the values of the specific rates of breakage are a direct index of the ability of the mill to break material. However, if tests are compared in which f_c is a variable then it is necessary to include the amount of material being acted upon. Thus, it is more informative in this case to compare the *absolute rates of breakage*, defined by S_iW or by S_if_c. S_if_c has the physical meaning of the volume of powder broken per unit time per unit mill volume if all of the powder is of size i. For example, a mill of volume V loaded at $f_c = 0.2$ with a specific rate of breakage of top size of $S_1 = 1$ min^{-1} has a total breakage rate of the top size of $S_1 f_c V \rho_p w_1 = 0.2V\rho_p w_1$ mass per minute, where ρ_p is the formal bulk density of the powder and w_1 is the fraction of size 1. At another filling condition of $f_c = 0.1$, the value of S_1 might be for example 1.75 min^{-1}, giving a total breakage rate of the top size of $0.175V\rho_p w_1$ mass per minute. Clearly the first case is a higher breakage rate at the same w_1 even though the *specific* rate of breakage is smaller than in the second case.

It is found that for a given ball loading J it is undesirable both to underfill or overfill the mill with powder. At low powder filling much of the energy of the tumbling balls is taken up in the steel-to-steel contact giving low values of Sf_c. The mill power is the same whether the powder filling is low or normal, so a low value of Sf_c gives a low energy efficiency of grinding, since low values of m_p/SW give *direct inefficiency*. In addition, the values of α are smaller than normal, which has the effect of grinding fine material faster than normal, giving overgrinding of fines. The relative powder-to-ball loading is defined by U (where $U = f_c/0.4J$, see Eq. 7 in Chapter 1) and normal values of α and normal values of the primary progeny fragment distributions are obtained for U greater than about 0.2 to 0.3.

On the other hand, at high powder fillings, the powder cushions the breakage action and Sf_c is again lower than normal. This also gives rise to non-first-order breakage, with the rate slowing down as fines accumulate in the bed. Fig. 5 shows the variation of absolute rate of breakage as a function of J and f_c, for the dry and wet grinding of quartz in laboratory mills fitted with small lifter bars, in the region of normal breakage.

The general shape of the curve of breakage rate vs. powder filling at a set ball load is explained as follows. A low filling of powder obviously gives a small rate of breakage. As the amount of powder is increased, the collision

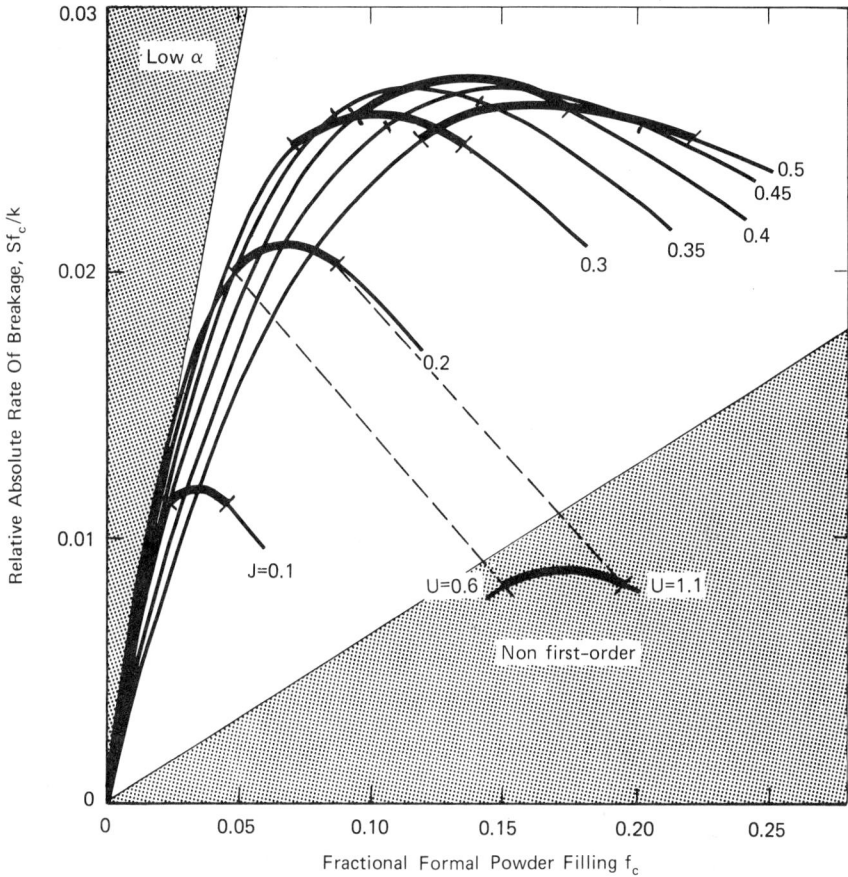

Fig. 5a. Variation of relative absolute rate of breakage with powder and ball filling for dry grinding in a laboratory mill.

spaces between the balls are filled and higher rates of breakage are obtained. When all the effective spaces in which collisions between tumbling balls are occurring are filled with powder, the rates of breakage reach a maximum. Further addition of powder increases the hold-up but does not give increased breakage because the collision zones are already saturated and further powder just enters a *reservoir* of powder. A plateau of almost constant breakage rates is obtained. Eventually, overfilling leads to deadening of the collisions by powder cushioning, the ball-powder bed expands to give poor ball-ball-powder nipping collisions, and the breakage rates decrease.

The results of a number of different workers in small mills at fixed ball loads were summarized by Shoji, Lohrasb, and Austin (1979) in the empirical equation

$$S(f_c) \propto a \propto (2.80e^{-4.1U} + e^{-0.8U}), 0.3 \leq U < 2$$

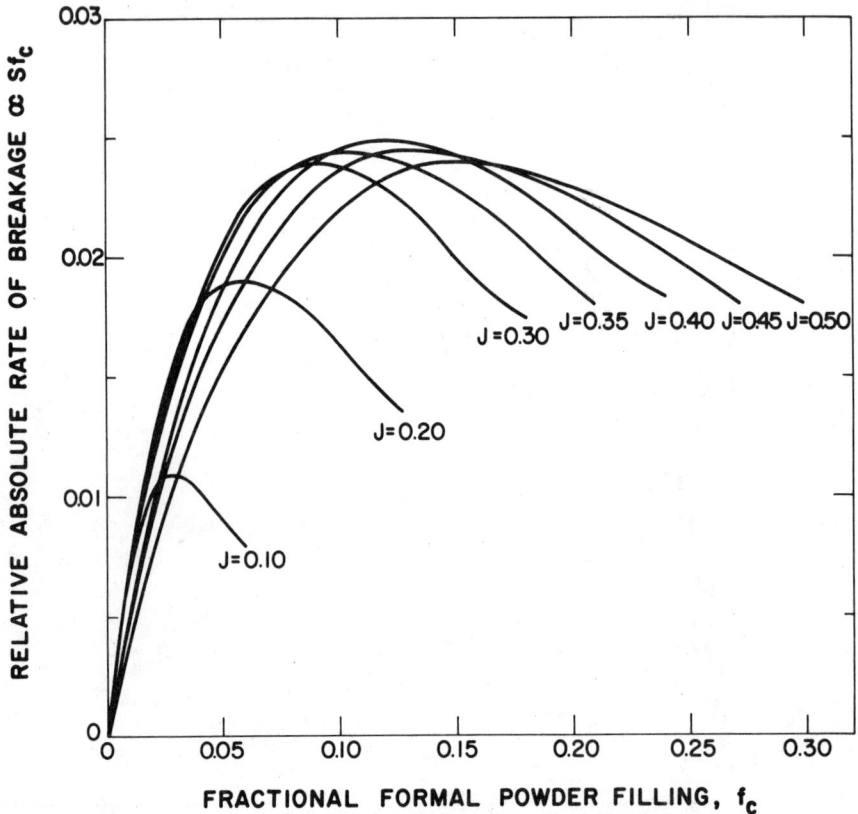

Fig. 5b. Variation of relative absolute rate of breakage for wet grinding in a laboratory mill.

where a is the preexponential factor in Eq. 1. Later work (Shoji et al., 1982) showed that a somewhat simpler expression could be used over the normal filling region, and the variation with ball load was also incorporated to give

$$S(f_c, J) \propto a \propto \frac{1}{1 + 6.6J^{2.3}} \exp[-cU], 0.5 \leq U \leq 1.5, 0.2 \leq J \leq 0.6$$

(7)

where c is 1.20 for dry grinding and 1.32 for wet grinding at normal slurry densities (the effect of high slurry density on the optimum filling level is discussed in the section, "Effect of Environment in the Mill"). The plot of Eq. 7 is shown in Fig. 5. Differentiating Sf_c with respect to U and setting equal to zero shows that the maximum absolute rates of breakage occur at a value of $U = 1/c$. It is concluded that $U = 0.6$ to 1.1 is the optimum filling condition for the maximum breakage rates at all ball loads. However, mills

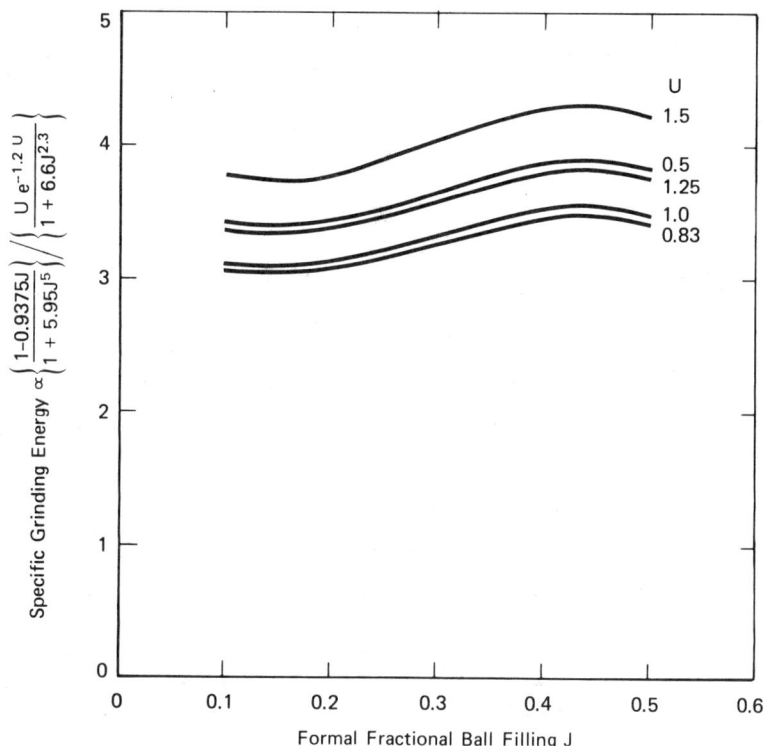

Fig. 6. Relative specific grinding energy as a function of ball filling: dry grinding in a laboratory mill.

will normally be run at the higher end of this range to avoid the excess ball wear caused by lower powder filling. The maximum mill capacity is obtained at ball fillings of 40 to 45% (for small mills).

The net mill power as a function of ball load was fitted by the empirical function

$$m_p \propto \frac{1 - 0.937J}{1 + 5.95J^5}, 0.2 \leq J \leq 0.6 \tag{8}$$

Combining Eqs. 7 and 8 gives the result shown in Fig. 6. Although capacity of a laboratory mill is a maximum at 40 to 45% ball load, the relative specific grinding energy m_p/SW is a minimum at about 15 to 20% ball load. In practice, ball loads less than 25% are not normally used because low ball loads can give excessive liner wear. In addition, mill capacity is clearly lower for lower ball loads (see "Optimization of Mill Power and Ball Loading for Tumbling Ball Mills" in Chapter 11).

Ball Diameter, Hardness, and Density

Considering a representative unit volume of mill, the rate of ball-on-ball contacts per unit time will increase as ball diameter decreases because the

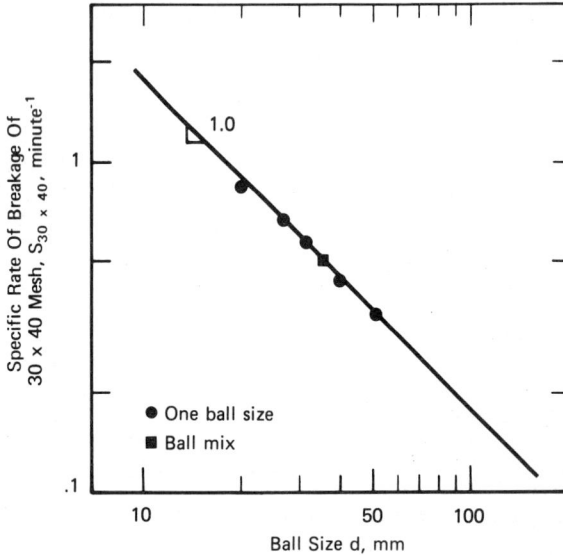

Fig. 7. Variation of specific rate of breakage with ball diameter ($J = 0.2$, $U = 0.5$, $\phi_c = 0.7$, $D = 0.6$ m).

number of balls in the mill increases as $1/d^3$. Thus the rates of breakage of smaller sizes are higher for smaller ball diameters. Fig. 7 shows the effect of ball diameter in a 0.6 m (2-ft) diam mill, which gives the relation

$$a \propto 1/d \qquad (9)$$

where a is the preexponential factor of Eq. 1, which depends on mill conditions, and d is ball diameter.

However, detailed analysis of the data shows that the B values also change in a systematic manner (see "Ball Size Effects on B Values" in Chapter 16). For example, the best current estimates of the variation of B parameters for quartz are given in Table 2. It appears that the greater impact force of a collision involving a larger ball gives a somewhat bigger proportion of fines, that is, γ is lower, Φ is higher. Thus, the lower specific rate of breakage due to larger balls is partially compensated by the production of a bigger proportion of fine fragments. It will be shown in Chapter 16 that overall mean $\overline{B}_{i,j}$ values can be used as a reasonable approximation for a mixture of balls. For a Bond ball mix (see "Ball Wear and Ball Size Distributions" in Chapter 16) with a top size of 51 mm, the overall $\overline{\gamma}$ value is 0.91 times the γ for 25.4 mm balls, and $\overline{\Phi}$ is 1.06 times the Φ for 25.4 mm balls.

For flexibility, Eq. 9 can be put as

$$a \propto 1/d^{N_o} \qquad (9a)$$

**Table 2. _B_ Parameters as a Function of Ball Size,
Dry Grinding of Quartz ($\beta = 5.8$)**

Ball Size d, mm	In.	γ	γ Factor Compared to 25 mm (1 in.) Ball	Φ	Φ Factor Compared to 25 mm (1 in.) Ball
19	3 / 4	1.10	1.02	0.51	0.81
22	7 / 8	1.09	1.01	0.58	0.92
25	1	1.08	1.00	0.63	1.00
32	1 1 / 4	1.05	0.97	0.68	1.08
38	1 1 / 2	1.00	0.93	0.69	1.10
44	1 3 / 4	0.95	0.88	0.70	1.11
51	2	0.88	0.81	0.70	1.11
64	2 1 / 2	0.78	0.72	0.70	1.11

The exponent N_o for milling in general is in doubt. For example, small (200 mm i.d.) laboratory mills give $N_o \approx 0$, that is, no effect of ball size.

It is obvious that it is not desirable to feed a mill with particle sizes which are too large, because the S_i values for these sizes will be to the right of the maximum shown in Fig. 2, giving low rates of breakage. It has long been known that bigger diameter balls will break large particles more efficiently. In terms of the specific rates of breakage, this concept can be quantified (Austin et al.) by the empirical equation

$$x_m \propto d^2 \tag{10}$$

where x_m is the size at which the maximum value of S occurs (for a given set of conditions) and d is ball diameter. Eq. 10 states that the position of the maximum in S moves to larger particle sizes as ball diameter is increased or, put another way, as ball diameter is increased the mill can efficiently break a feed containing larger sizes of particles.

The values of a and x_m are dependent on the material and milling conditions. However, the value of α does not appear to vary with ball diameter in the normal breakage region. a shows a wide variation from soft (weak) to hard (strong) materials, but the range of x_m is relatively narrow; x_m depends not only on the strength of the large particles but also on the geometry of nipping of particles between balls, which is a factor which does not change much with material type. Combining Eqs. 2, 3, 4, 9, and 10 gives the typical results shown in Fig. 8.

The overall effect of a mixture of ball sizes in the normal breakage region is taken to be the linear weighted sum

$$\bar{S}_i = \sum S_{i,k} m_k \tag{11}$$

where m_k is the weight fraction of balls of size interval denoted by k and

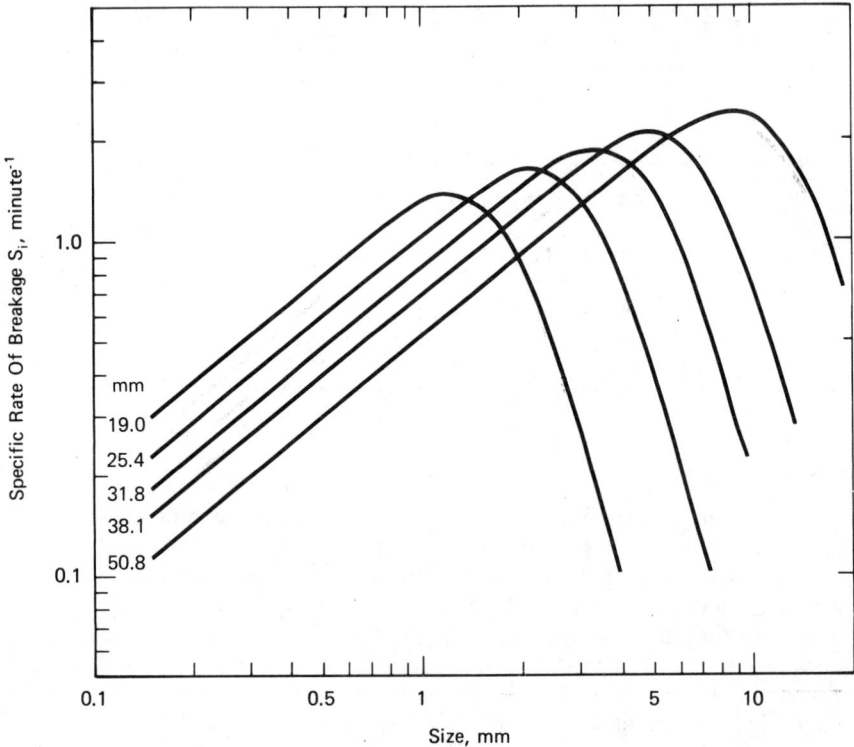

Fig. 8. Predicted variation of S values with ball diameter for dry grinding of quartz ($J = 0.2$, $f_c = 0.04$, $\phi_c = 0.7$, $D = 0.6$ m).

$S_{i,k}$ is the specific rate of breakage of size i by balls of this size. If the feed sizes are small enough that Eq. 1 applies for all the ball sizes in the mill

$$\bar{S}_i \propto x_i^\alpha \sum_k \left(\frac{m_k}{d_k^{N_o}} \right) \tag{11a}$$

and the mill behaves as if it had a single mean ball size $\bar{d}$ defined by

$$\bar{S}_i \propto \frac{1}{\bar{d}^{N_o}} x_i^\alpha = x_i^\alpha \sum_k \left(\frac{m_k}{d_k^{N_o}} \right)$$

that is, for $N_o = 1$,

$$1/\bar{d} = \sum_k \frac{m_k}{d_k} \tag{11b}$$

Note that this is a specific surface area mean ball diameter, that is, the ball diameter that will give the same sum ball area per unit mass as the mixture of balls. Fig. 7 shows the S value determined for a mixture of 50% of 27 mm balls and 50% of 50 mm balls. The result falls on the line for a ball size of 36 mm. Eq. 11 gives $1/\bar{d} = (0.5/27) + (0.5/50)$ or $\bar{d} = 35$ mm. This demonstrates the simple additivity in the normal breakage region. However, it is not possible to define a mean ball size for conditions where Eq. 2 applies, because no single ball size can duplicate the breakage action of a mixture of balls on large particle sizes.

One method of allowing for the uncertainties of the effect of d on breakage parameters and the additivity of mixtures of balls, is to determine S and B values in a mill containing the mixture of balls to be used in practice. This cannot be done in mills which are too small and we recommend batch tests in a mill of at least 0.6 m i.d. if balls up to 60 mm diam are to be used.

The mill power does not vary much with ball diameter (down to 19 mm diam), so that the wrong choice of ball diameter, leading to low S_i values, gives *direct inefficiency*, lowering $S_i W / m_p$ and increasing the comparable specific grinding energy. However, due to ball wear in most ball milling applications it is necessary periodically to add larger balls to establish an appropriate steady-state mixture of ball sizes in the mill (see "Ball Wear and Ball Size Distributions" in Chapter 16). In addition, a charge of too small balls leads to slumping of the charge and low power draw plus excessive wear rates of the balls. The optimum choice of ball sizes clearly depends on the feed size distribution, the desired product size, and the balance between energy cost and steel cost. This optimization is considered in more detail in Chapter 16.

Rose and Sullivan (1958) showed that ball hardness does not affect mill capacity providing the ball is above a reasonable hardness. Von Seebach (1969) has shown by using hollow steel balls that the effect of ball density on specific breakage rates is linear in dry grinding,

$$S_i \propto a \propto \rho_b \tag{12}$$

where ρ_b is the true ball density. Note that mill power is also directly proportional to ρ_b, so that a mill with lower density grinding media will have both a lower capacity (t/h) and a lower power draw, giving the same comparable specific grinding energy as one with high density balls. However, our own tests in a mill of 0.6 m diam give the ratio of S_i values for alloy steel balls ($\rho_b = 7.8 \times 10^3$ kg/m^3) to ceramic balls ($\rho_b = 3.7 \times 10^3$ kg/m^3) as 1.75, which is less than the ball density ratio of 2.1, although mill power was in the ratio of ball density. This suggests that it is better to determine S_i values for the ball materials of interest rather than estimate it using Eq. 12.

Mill Diameter

There have been very few direct measurements of S and B values in large diameter mills, and back-calculation of values from continuous large-scale mill data is subject to large errors (see Chapter 10). Therefore, it is necessary to extrapolate results from smaller mills and also to infer results from the variation of industrial mill capacity with mill diameter. Austin (1973); Shoji, Austin, and Luckie; and Malghan and Fuerstenau (1976) have demonstrated that $B_{i,j}$ values are often the same for a given material in mills of 150 mm to 600 mm (2 ft) diam, and that the exponent α is the same. Under identical filling conditions the specific rates of breakage increase as D^{N_1} where N_1 is close to 0.5 (see Fig. 9) thus,

$$S_i \propto a \propto D^{N_1} \tag{13}$$

As in "Variation of Breakage with Particle Size," this result suggests (Austin, 1973) that cascading gives the normal breakage action in a mill, using the following reasoning (see also "The Theory of Mill Power for Tumbling Ball Mills" in Chapter 11). An S_i value is a fraction broken per unit time and therefore represents the breakage per unit mill volume. The mean number of balls raised and tumbled *per mill revolution* per unit mill volume is constant irrespective of mill diameter, but the mean number of *impacts* which a ball makes as it cascades down the mill charge is proportional to D. However, at a given fraction of critical speed, the number of mill revolutions *per unit time* is proportional to $1/\sqrt{D}$. Combining these factors assuming that each impact contributes to breakage gives $S_i \propto D/\sqrt{D}$, that is, $S_i \propto D^{0.5}$.

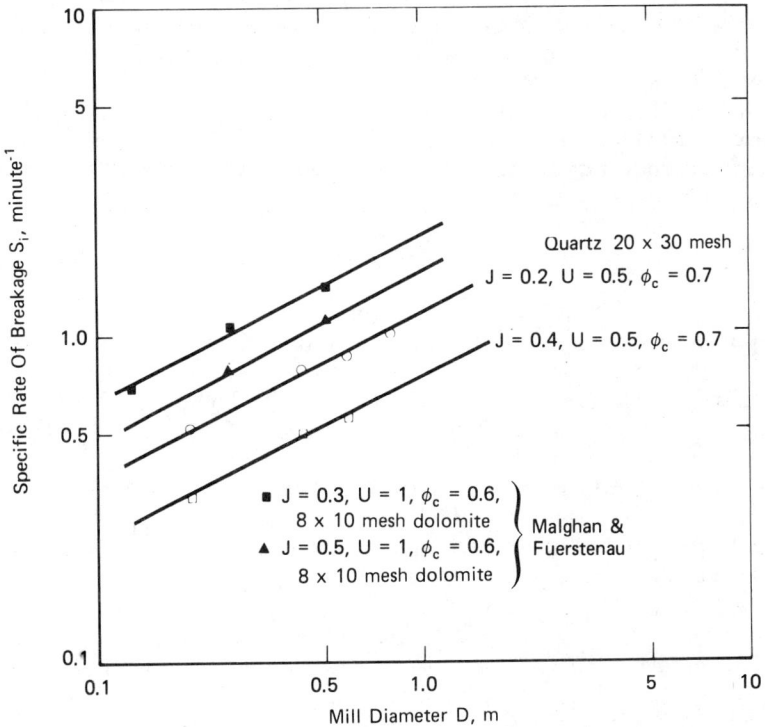

Fig. 9. Variation of S_i values with mill diameter (normal breakage, $d = 25.4$ mm).

Assuming that α and $B_{i,j}$ values stay the same for even larger mills, the mill capacity as a function of mill size (to go from the same feed to the same product under the same loading conditions) would be

$$Q \propto \frac{\pi}{4} L D^{2+N_1} \tag{14}$$

or

$$Q_2/Q_1 = \left(\frac{L_2}{L_1}\right)\left(\frac{D_2}{D_1}\right)^{2+N_1} \tag{14a}$$

where L is mill length. This assumes also that mill capacity per unit length is constant, that is, that there is negligible effect of the end walls of the cylinder. The well-known empirical rule for mill capacity is indeed

$$Q \propto \frac{\pi}{4} L D^{2.5}$$

Bond (1960) states that mill capacity for mills greater than 3.8 m in diameter is proportional to $\pi/4L(D/3.8 \text{ m})^{2.3}$, which suggests that N_1 decreases somewhat for large mill diameters. Eq. 14a becomes

$$Q_2/Q_1 = \left(\frac{L_2}{L_1}\right)\left(\frac{D_2}{D_1}\right)^2\left(\frac{3.8}{D_1}\right)^{0.5}\left(\frac{D_2}{3.8}\right)^{0.3}, D \geq 3.8 \text{ m} \tag{14b}$$

Chapter 11 shows that mill power also varies as $LD^{2.5}$ for small mills, so that the comparable specific grinding energy is constant with mill diameter for tests in small mills.

In addition, it is to be expected that a larger mill diameter will shift the maximum in S to somewhat larger particle sizes, for a given ball diameter. We have used the empirical expression

$$x_m \propto \mu \propto D^{N_2} \tag{15}$$

where N_2 is about 0.2. Thus the maximum size of ball for a given maximum particle size in the feed can be reduced for larger mill diameters. The wear and damage of liners by impact of large balls is aggravated by large mill diameter, so it is also desirable for this reason to reduce the size and amount of the largest balls for large values of D.

Effect of Environment in the Mill

It is well-known that ball milling in water gives higher mill capacities than dry grinding, providing the ratio of solid to water (slurry density) is not so high that the mill charge becomes thick and viscous. Bond (1960) states that the capacity of wet grinding in the industrial scale is 1.3 times that for dry grinding under otherwise comparable conditions. Austin et al.

(1981) showed that the $B_{i,j}$ and α values were approximately the same between wet and dry grinding in a small laboratory mill (at least for the materials investigated). The values were also the same for different pulp densities, providing the pulp density was kept in the fluid region. However, the ratio of the S values, that is, the factor a between dry and wet grinding varied from 1.1 to 2.0 for different materials. The ratio was 2.0 for quartz. When the $B_{i,j}$ and α values do not change, the mill capacity will be directly proportional to these ratios. More detailed investigations indicate that the B values for wet grinding have somewhat larger γ values, indicating proportionally less production of fines.

In laboratory batch grinding tests it is the breakage action which is being studied, not mass transfer along a mill, and the results therefore show that breakage occurs more rapidly in the presence of water. In addition, the wet and dry comparison is made in the normal, first-order breakage region where cushioning (see below) is not evident. Thus the water does not

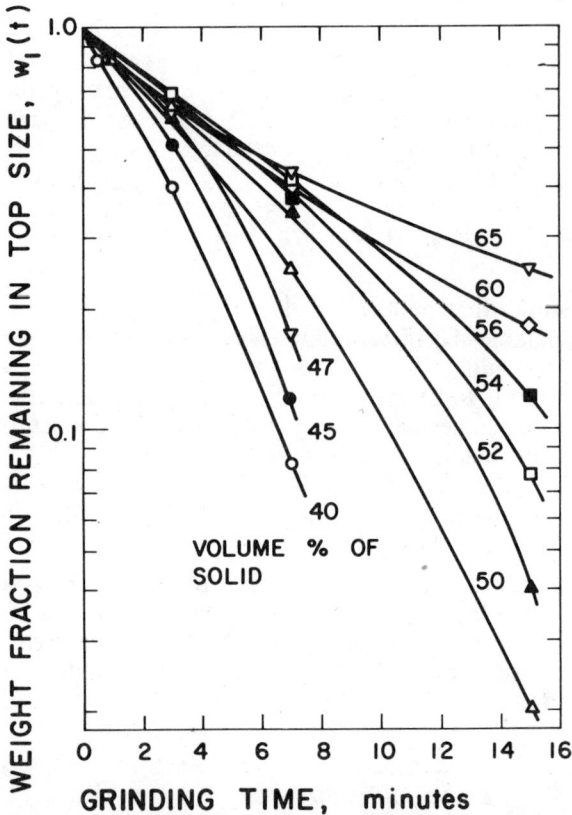

Fig. 10. Variation of batch breakage rate of the top size [850 × 600 μm (20 × 30 mesh) quartz], with time for various slurry densities ($D = 200$ mm, $d = 25$ mm, $J = 0.3$, $U = 1.0$).

act primarily by preventing cushioning, coating of the balls, or reagglomeration of fines. On the other hand, as Chapter 15 will show, chemical grinding additives do not change the breakage rates until the pulp density is high, in which case they operate by affecting the fluidity of the mass. Thus, the influence of the water seems to be primarily to allow better transfer of the mechanical action of the tumbling balls to the stressing of the particles, leading to higher rates of breakage but yet giving the same type of breakage, and consequently, approximately the same primary progeny fragment distribution.

A detailed study of wet grinding in a laboratory mill using analysis via S and B values has been performed by Tangsathitkulchai and Austin. Their results for quartz ground in distilled water can be summarized as follows. The breakage of a batch of single feed size of 850 μm $\times$ 600 μm (20 $\times$ 30 mesh) quartz gave a systematic non-first-order breakage, as shown in Fig. 10. The acceleration or deceleration of breakage rate of the top size was produced by the accumulation of fine material, since it could be obtained also by starting with a feed of 50% of 850 μm $\times$ 600 μm (20 $\times$ 30 mesh) quartz plus 50% fine material. Whether the first-order plot for the top size shows acceleration, deceleration, or approximate first-order depends on the ball and powder load and mill diameter, in addition to slurry density.

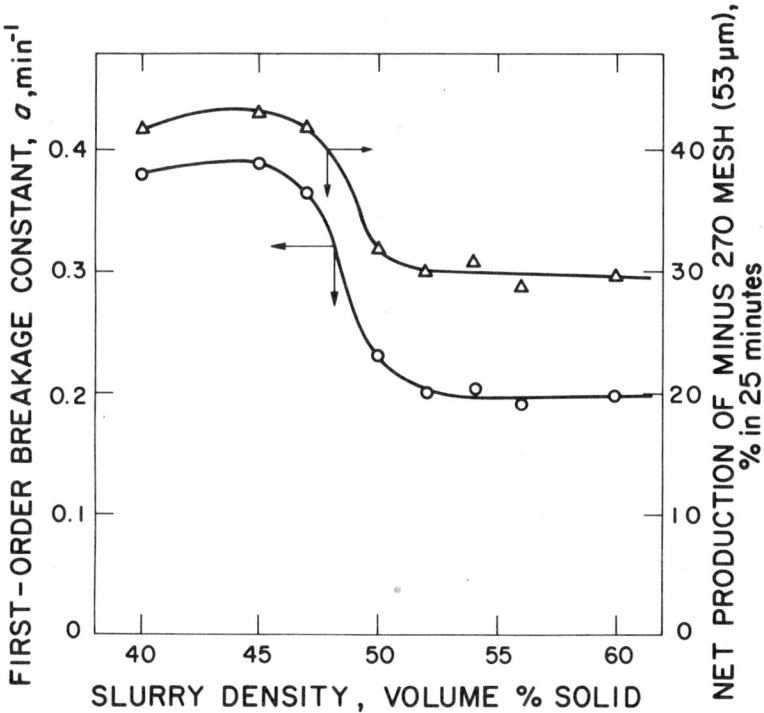

Fig. 11. Effect of slurry density on breakage rates of quartz (see Fig. 10).

However, the breakage of smaller sizes was essentially first-order (with constant $\alpha = 0.8$), once a natural broken size distribution accumulated in the mill. Fig. 11 shows the variation of the a value for this normal breakage plus the net rate of production of minus 53 μm (270 mesh) material. It is concluded that there is a small maximum at the 45 volume % slurry density, which gives the maximum breakage rates. The increase in breakage rate at the maximum was found to be a function of the γ value which defines the relative production of fine material on primary breakage: the effect is greater for γ values in the region of 0.6 to 0.8, and is less for larger γ values (see Chapter 15). A rapid decrease in breakage rate occurs over a short range of higher slurry density. Fig. 12 shows the B values for these conditions. Within the range of experimental reproducibility, the values are constant (and normalized) for normal slurry density but change to a different set of values (also normalized) for the high slurry density, with proportionally higher fines production. The net rate of production of fines varies with slurry density in the same way as a, but the match is not exact because of the change in B values.

Fig. 13a shows the size distributions produced at normal and high slurry densities, starting with 850 μm × 600 μm (20 × 30 mesh) quartz. It shows

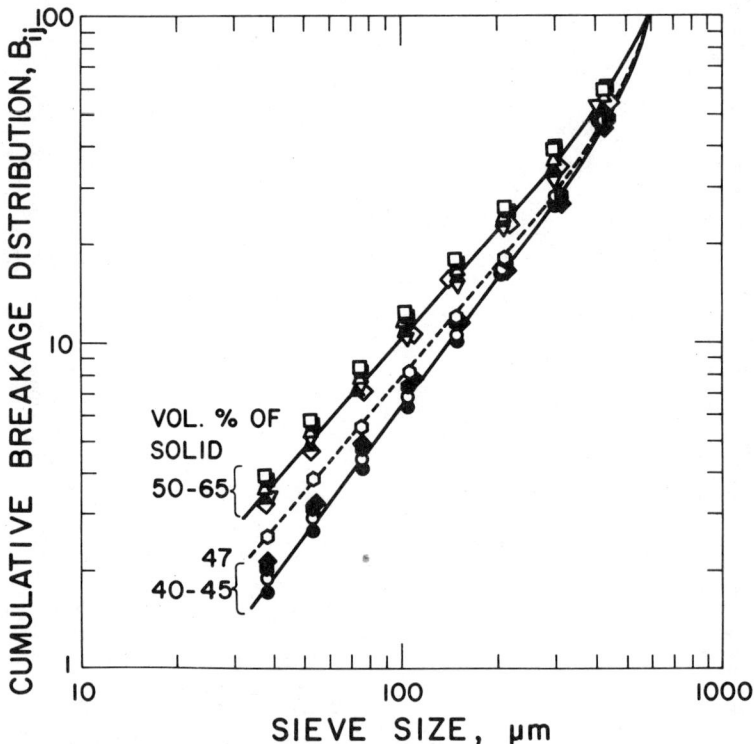

Fig. 12. Variation of *B* values with slurry density (see Fig. 10).

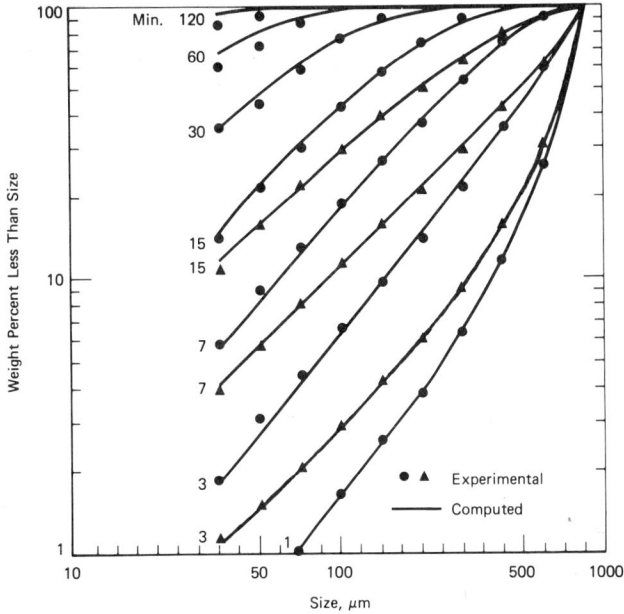

Fig. 13a. Comparison of size distributions from batch grinding of 850 × 600 μm (20 × 30 mesh) quartz at normal and high slurry density (D = 200 mm; d = 26 mm; J = 0.3; U = 1.0; ϕ_c = 0.7); ●, 40 volume %; ▲ 54 volume %.

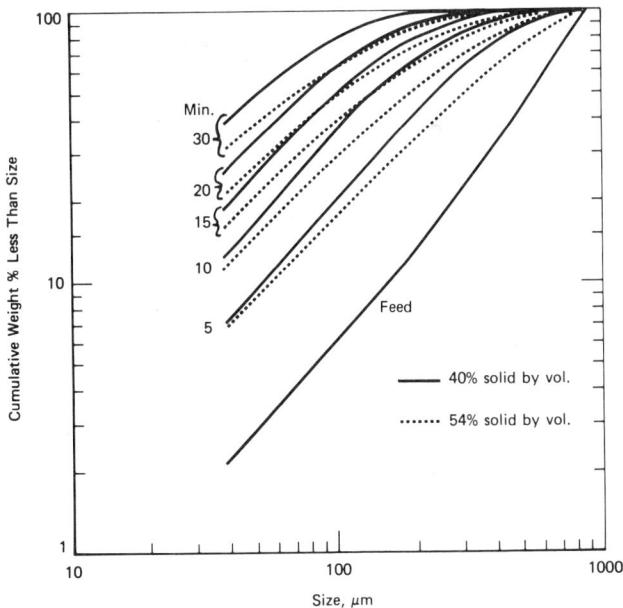

Fig. 13b. Comparison of size distributions for 40 volume percent and 54 volume percent solids for natural feed (see Fig. 13a).

that breakage slows down at long grinding times, because the experimentally determined size distribution is not as fine as that predicted by the first-order simulation. Fig. 13b shows the result of starting with a natural size distribution: the change in slope of the plots due to the change in B values is quite clear.

Essentially similar conclusions have been previously reported by Klimpel (1982, 1982a, and 1983) and Katzer, Klimpel, and Sewell (1981), based on the rate of production of fine sizes rather than breakage rates. Fig. 11 shows that the rate of production of fine sizes [e.g., less than 53 μm (270 mesh)] varies in the same way as breakage rates. On this basis, the results by Katzer, Klimpel, and Sewell (1981) with respect to mill filling and slurry density are given in Fig. 14, where the values have been fitted by a form of Eq. 7,

$$\text{net rate of production} = kU\exp(-cU) \tag{7a}$$

It appears that the optimum filling level at a given slurry density shifts to higher values of U as slurry density increases. Fig. 15 shows the effective production of fines at: (a) optimum filling level, and (b) constant mass

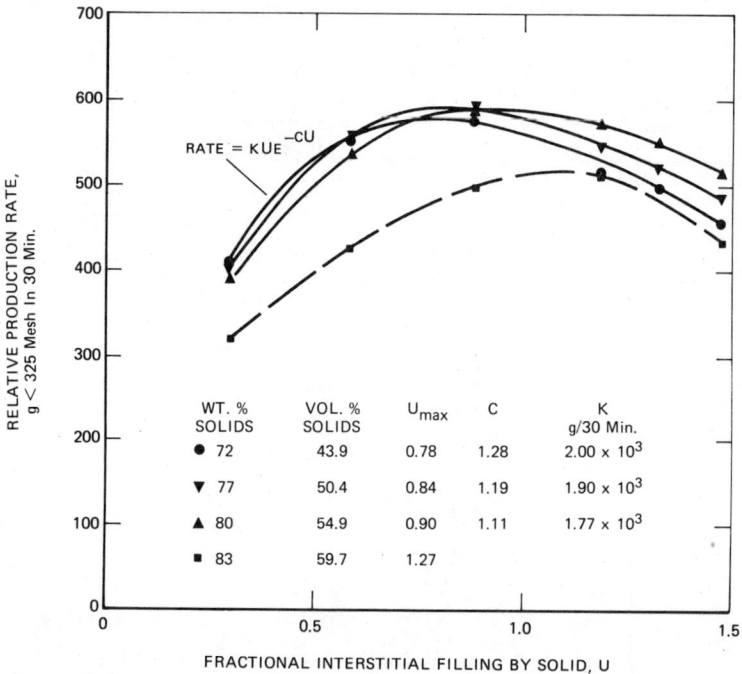

The graph contains the following legend table:

WT. % SOLIDS	VOL. % SOLIDS	U_{max}	C	K g/30 Min.
● 72	43.9	0.78	1.28	2.00×10^3
▼ 77	50.4	0.84	1.19	1.90×10^3
▲ 80	54.9	0.90	1.11	1.77×10^3
■ 83	59.7	1.27		

RATE = KUE^{-CU}

RELATIVE PRODUCTION RATE, g < 325 Mesh In 30 Min.

FRACTIONAL INTERSTITIAL FILLING BY SOLID, U

Fig. 14. Variation of breakage rates with mill filling and slurry density.

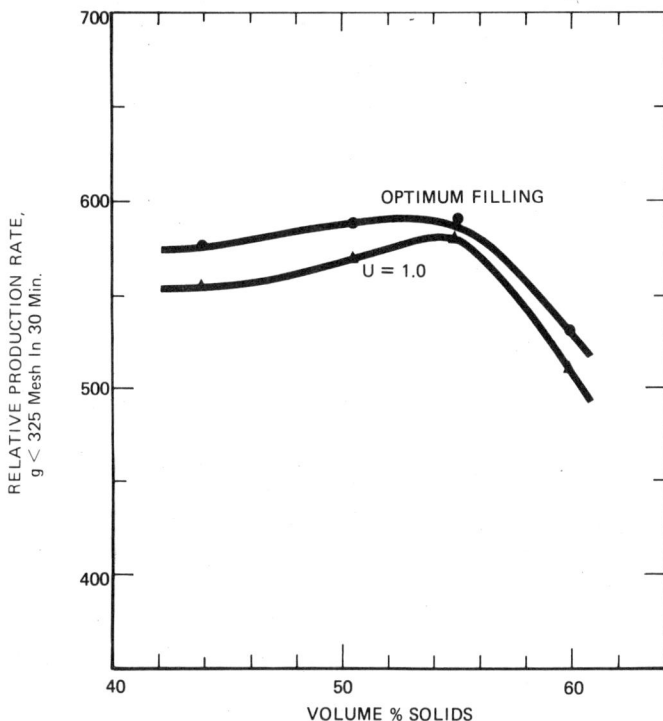

Fig. 15. Variation of breakage rates with slurry density at: (a) optimum filling levels; (b) a hold-up of $U = 1$.

hold-up of $U = 1$. The slight maximum also occurs at constant slurry volume, of slurry sufficient to fill the interstitial ball voids. It is concluded that the maximum production rate obtained at each optimum filling condition increases slightly to an optimum slurry density, then the rate drops very significantly at high slurry density.

Unlike dry grinding, Katzer, Klimpel, and Sewell (1981) found that the net power to the batch test mill increased and decreased with filling level and slurry density very similarly to the variation in breakage rates, except at extreme conditions of underfilling, overfilling, or high slurry density. This means that the comparative specific grinding energy was almost constant and the optimums in mill performance were optimums of capacity, not specific energy.

Klimpel (1982, 1982a, and 1983) and Katzer, Klimpel, and Sewell (1981) have also performed measurements of slurry rheology at various stages in the grinding process. They describe three regimes of slurry viscosity: Regime A, where slurry viscosity is low and does not increase much as percent solids is increased; Regime B, where viscosity starts to increase rapidly with small increases in percent solids; and Regime C, where very large increases in viscosity occur for small increases in percent solids. In addition, they characterize the rheological character in the three regions as: Regime A, dilatant; Regime B, near-Newtonian pseudoplastic with low yield stress; and Regime C, pseudoplastic with high yield stress (see Fig. 16a). The match of these regimes to slurry density is roughly that A describes the low slurry density behavior, B describes the slurry density close to the maximum breakage rates, and C describes the slurry rheology at high slurry density, leading to slowing-down effects as grinding proceeds to fine sizes.

The best slurry density for maximum breakage rates depends on the additional factors which affect the optimum filling levels and the slurry rheology. These are: (1) the fineness of the size distribution; (2) the shape of the size distribution, that is, the ability of particles to pack within one another in the slurry; and (3) the degree of dispersion or flocculation of the particles and pH and chemical effects on fluidity (see Chapter 15). Fig. 16b is a summary of work by Klimpel (1984) on the rheological properties of ore slurries in water as a function of the shape of the size distribution and the volume percent of solids. Size distributions were prepared by blending size fractions to give Schuhmann size distributions characterized by the slope, α_s, and the maximum size k (see Chapter 2, discussion of Charles' Law). The effect of slurry density and Schuhmann slope in causing transition from Regime A to B to C is clearly delineated in the figure. At a given Schuhmann slope, the finer size distributions move from dilatant to pseudo-plastic to pseudoplastic-with-yield at lower slurry densities. Above a top size of 600 μm, the effect of characteristic top size was not pronounced.

The onset of non-first-order breakage in the *slowing-down region* can occur at any slurry density providing the size distribution becomes fine enough. For example, Fig. 17 shows the variation of slurry rheology as the size distribution becomes finer during batch grinding (Klimpel, 1982), for circumstances where breakage of all sizes is initially first-order and slowing-down appears as grinding proceeds. The development of yield stress is clearly indicated.

As a first approximation, the slowing-down of specific breakage rates applies equally to *all sizes* in the mill charge. This leads to the important fact that the family of size distributions produced by batch grinding remains unchanged in the presence of the slowing-down effect if the B values do not change, but the grinding time required to reach a given size distribution is longer. Thus (Austin and Bagga, 1981),

$$S_i'(t) = KS_i(0) \tag{16}$$

where $S_i(0)$ is the set of normal S values, and $S_i'(t)$ is the *mean* value of S_i from time zero to time t. K is a reduction factor ($0 \leq K \leq 1$) which becomes smaller as the percentage of fines increases. Letting θ be the equivalent first-order grinding time (the *false* time) to reach the distribution,

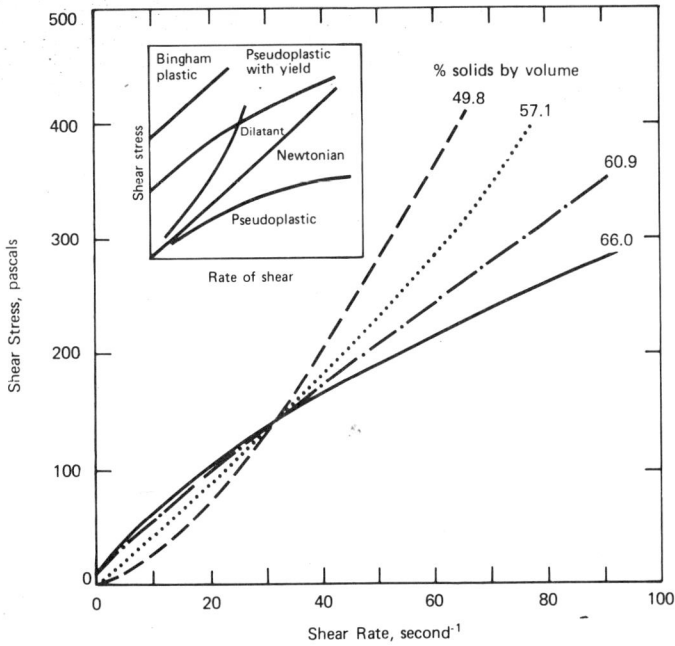

Fig. 16a. Variation of slurry rheology for 850×600 μm (20×30 mesh) coal in water as a function of slurry density.

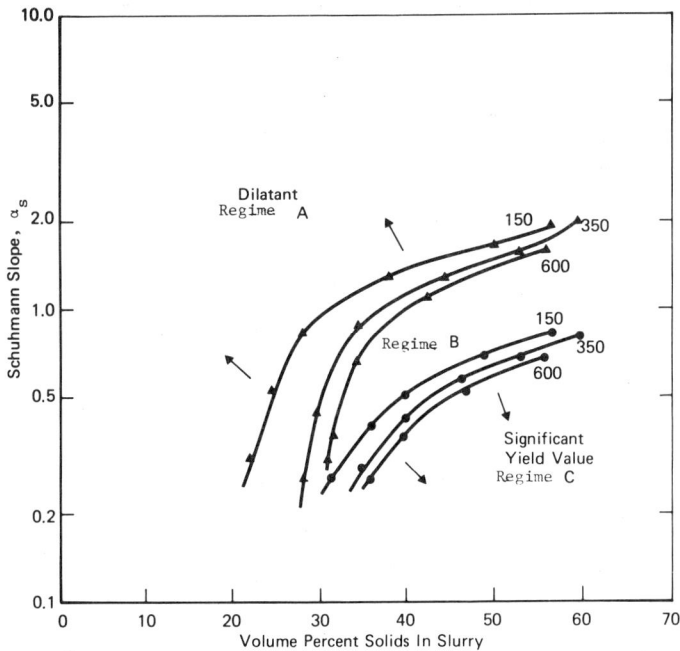

Fig. 16b. The influence of particle size distribution on the slurry rheology of an iron ore where the size distributions are prepared to follow a Schuhmann plot with maximum sizes of 150, 350, and 600 μm. ▲ indicates point of transformation from dilatant to pseudoplastic; ● from pseudoplastic to pseudoplastic with significant yield value.

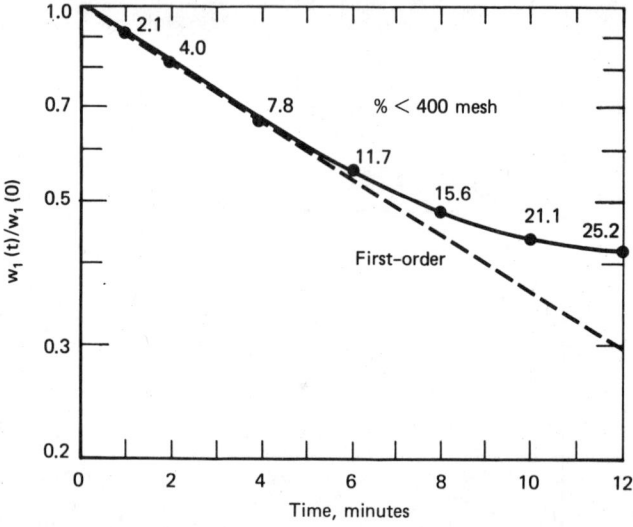

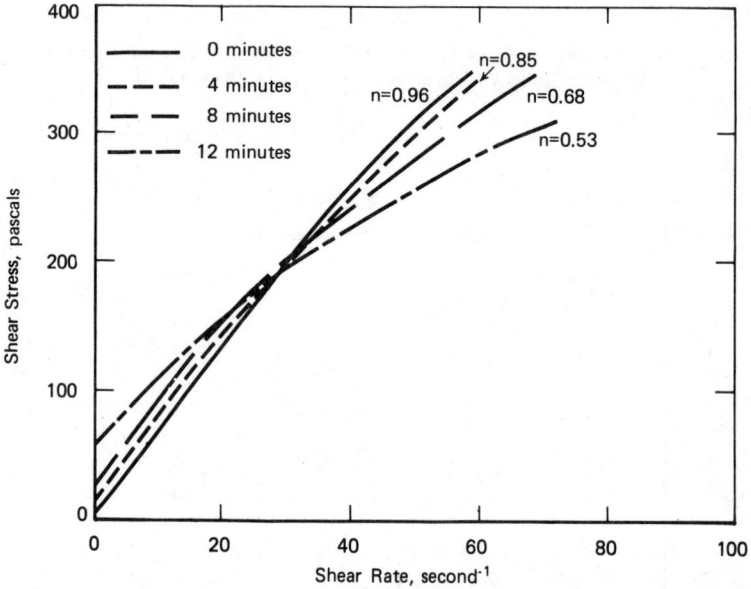

Fig. 17. Variation of slurry rheology with fineness of grinding for a starting feed of 300 × 250 μm (50 × 60 mesh) coal in water [shear stress $\propto$ (shear rate)^n].

$w_1(t)/w_1(0) = \exp[-S_1(0)\theta] = \exp[-KS_1(0)t]$, therefore,

$$K = \theta/t \tag{17}$$

The instantaneous value of S_i at time t can be represented by

$$S_i(t) = \kappa S_i(0) \tag{18}$$

where κ is also a reduction factor ($0 \leq \kappa \leq 1$) which is a function of the fineness of grind (hence a function of t): since $dw_1(t) = -S_1(0)w_1(t)\,d\theta$ and $dw_1(t) = -S_i(t)w_1(t)\,dt$,

$$\kappa = d\theta/dt \tag{19}$$

Knowing the variation of θ with t enables κ to be determined by graphical differentiation. The relation between K and κ is clearly $Kt = \int_0^t \kappa\,dt$. Fig. 18 shows the result for wet grinding of quartz. Breakage proceeds at normal rates ($\kappa = 1$) until the size distribution reaches an 80%-passing size of about 150 μm, then the rate falls to a new lower value at 80%-passing size of about 30 μm, at 40 volume percent solids. The decrease occurs at coarser grinds for higher slurry density and the values of κ decrease to small values.

The exact physical reasons for the various breakage regions have not been clearly elucidated to date. The slowing-down region is explainable on the basis that the development of increasing viscosity as grinding proceeds causes the slurry to be able to absorb impact with reduced fracture, thus leading to slowing-down. A high yield stress means that the slurry requires substantial energy to produce small strains. The maximum in breakage rates is conventionally explained by postulating that a sufficiently thick slurry coats the balls, leading to efficient ball-particle-ball collisions. However, the increase in net mill power also suggests that the slurry rheology aids the lifting of balls, which will also give matching higher breakage rates.

It is also found (Austin and Bagga, 1981) that dry grinding to very fine sizes can cause an action which slows the whole grinding process. Typical results are shown in Fig. 18. This does not appear to be due primarily to ball coating, since no coating was observed with quartz. Possibly a bed of cohesive fine particles develops almost liquid-like properties, so that particles flow away from the ball-ball collision region and insufficient stress is transmitted to individual particles for fracture to occur. It is quite clear from the figure that different materials show this effect to different extents, possibly due to major differences in cohesive forces between different materials.

It is well-known (Benjamin, 1976; Austin et al., 1981) that dry ball milling of material for extended periods of time (many hours) can lead to pelletizing and cold-welding of fines into larger particles. However, the slowing-down effect occurs at shorter times and does not show the incorporation of traced fine material into larger sizes (Austin et al., 1981).

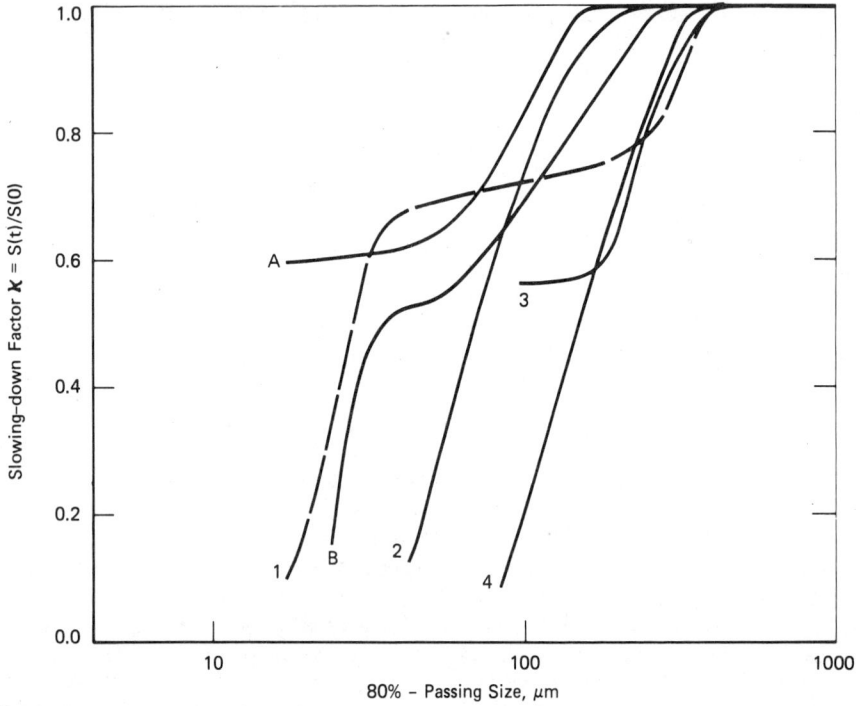

Fig. 18. The slowing-down factor κ as a function of fineness of grinding 850 × 600 μm (20 × 30 mesh feed) and slurry density (D = 200 mm, d = 26 mm, ϕ_c = 0.7).
Wet grinding: (J = 0.3, U = 1.0).
  A = 40 volume % quartz in water
  B = 54 volume % quartz in water
Dry grinding: (J = 0.2, U = 0.5)
  1 = Quartz
  2 = Western Kentucky #9 coal, 52 Hardgrove Grindability Index
  3 = Cement clinker
  4 = Lower Kittanning coal (113 H.G.I.).

Mass Transport and Hold-up

Fig. 5 and Eq. 7 show that the rates of breakage decrease as the mill is overfilled. Recent measurements of residence time distributions in full-scale mills (Weller, 1980) (also see Chapter 14) have enabled hold-up in the mills to be estimated. Table 3 shows the results expressed as equivalent fractional filling of the mills by powder for a formal porosity of 0.4. Since the ball load is approximately 35 to 40% for these mills, a filling of about 0.16 corresponds to U = 1, and it is clear that mills are often operated in the overfilled condition ($U > 1$).

In order to allow for this in simulation models it is necessary to have a mass transfer law to relate hold-up and flow rate. Unfortunately, this law will depend on the rheological properties of the slurry, which will be a

Table 3. Hold-up of Wet Continuous Ball Mills

Mill Diameter m	L / D	Solid Flow through Mill Ft / h	Slurry Density % wt Solids	Specific Gravity Solids ρ_s	τ min	Hold-up W, t	Fractional Mill Filling by Powder, f_c
0.3	1	0.08	67	2.65	1.72	0.0023	0.21
0.3	1	0.14	70	3.8	(2.27)	0.0053	0.49
1.83	2	114	76	3.9	3.5	6.6	0.29
2.03	1.5	65	63	3.8	1.45	1.6	0.07
		122	71	3.2	1.72	3.5	0.16
		44	76	3.2	2.44	1.8	0.08
2.21	5.5	133	70	2.7	8.6	19.0	0.25
2.30	0.93	155	75	3.5	2.00	5.2	0.28
232	0.75	47	65	3.0	1.67	1.3	0.10
		100	52	4.2	1.03	1.7	0.10
		232	68	3.0	0.90	3.5	0.27
2.34	0.78	60	72	2.9	1.85	1.85	0.13
2.34	1.2	185	79	3.9	2.72	8.4	0.28
2.70	1.37	82	70	4.2	4.72	6.5	0.12
2.93	0.8	295	78	3.7	3.13	15.4	0.45
		375	75	3.7	1.70	10.6	0.30
3.20	1.34	388	81	3.6	3.30	21.0	0.29
		163	71	3.4	6.58	18.0	0.25

function of the slurry density and the fineness of grind. In addition, it will depend on the type of mill discharge, the size mixture of balls in the mill, etc. Marchand, Hodouin, and Everell (1980) concluded from analysis of industrial scale data that the relation was of the form

$$W/W_1 = (F/F_1)^{0.5} \tag{20}$$

where the exponent of 0.5 was determined empirically. Mori (1973) also concluded from flow studies on a dry ball mill of 300 mm by 300 mm that the relation was

$$W = kF^{0.5} \tag{20a}$$

Hogg, Shoji, and Austin (1974, 1975) concluded that a model for mass transfer of dry powder in rotating cylinders might be extended to mass flow in ball mills. The model states that the flow rate through a mill under given fractional filling conditions is related to mill diameter D and rotational speed ω by

$$F \propto \omega D^4 \tag{21}$$

At a given fraction of critical speed, $\omega \propto 1/\sqrt{D}$, so $F \propto D^{3.5}$. This gives flow rate per unit of mill volume as proportional to $D^{0.5}$, to give a particular fractional filling. Since rates of breakage per unit of mill volume also scale by $D^{0.5}$ (for suitably small particle size), the combination leads to the conclusion that a small mill grinding a given feed to a given product will have the same fractional filling (hold-up) as a larger mill, with the value of $\tau \propto 1/D^{0.5}$, all other conditions being identical.

In the lack of a rigorous theory of mass transfer in mills, we will later use these equations in the form

$$U = (F/F_1)^{0.5}, F > F_1 \tag{22}$$

where F_1 is the flow rate (through a given mill under given conditions) which experimentally gives $U = 1$. This is conveniently put as

$$U = [(F/W_1)/(F_1/W_1)]^{0.5}$$

or

$$U = [(F/W_1)\tau_1]^{0.5} \tag{22a}$$

where W_1 is the hold-up for $U = 1$ and $\tau_1 = W_1/F_1$ (see "Allowance for Variation in Filling Level" in Chapter 6).

Scale-up of Laboratory Batch Grinding Results

The empirical equations which predict how first-order S_i values change with ball and mill diameter, ball and powder loading, and rotational speed are Eqs. 2, 3, 4, 8, 10, 13, and 15; they can be joined into

$$S_i(d) = a_T(x_i/x_o)^\alpha \left(\frac{1}{1 + (x_i/C_1\mu_T)^\Lambda} \right) C_2 C_3 C_4 C_5 \tag{23}$$

where

$$C_1 = (D/D_T)^{N_2}(d/d_T)^2$$

$$C_2 = (d_T/d)^{N_o}$$

$$C_3 = (D/D_T)^{N_1}, D \le 3.81 \text{ m}$$

$$= \left(\frac{3.81}{D_T} \right)^{N_1} \left(\frac{D}{3.81} \right)^{N_1 - \Delta}, D \ge 3.81 \text{ m}$$

$$C_4 = \left(\frac{1 + 6.6J_T^{2.3}}{1 + 6.6J^{2.3}} \right) \exp[-c(U - U_T)]$$

$$C_5 = \left(\frac{\phi_c - 0.1}{\phi_{cT} - 0.1} \right) \left(\frac{1 + \exp[15.7(\phi_{cT} - 0.94)]}{1 + \exp[15.7(\phi_c - 0.94)]} \right)$$

where the subscript T refers to the laboratory test mill conditions and results. The values of a_T and μ_T for the ball diameter and ball and powder loading of interest are material characteristics, as are α, Λ, and the characteristic parameters of the $B_{i,j}$ values, namely, γ, β, Φ. Insufficient work has been done to determine whether N_o, N_1, and N_2 are always the same for different materials. Combining this calculation with Eq. 5 and Eq. 11 enables the computation of $\bar{S}_i$ and $B_{i,j}$ values, which are the values required for the solution of the batch grinding equation, Eq. 4 in Chapter 4.

Alternatively, batch tests can be used to determine $\bar{S}_i$ values in a reasonably large laboratory mill (e.g., 0.6 m i.d.) with the mixture and loading of balls anticipated in the full-scale mill, and the values scaled for mill diameter using C_3 and if necessary, C_4 and C_5. In this case, the values of μ_T and Λ are for the $\bar{S}_i$ values produced by the mixture of balls, and $d = d_T$ in C_1 and C_2. This technique has the advantage of avoiding assumptions concerning the additivity of ball size effects.

Summary

Experience suggests that there is a correlation between small-scale grinding results and large-scale results. However, accurate prediction of operating full-scale plant results from laboratory ball mill tests requires detailed analysis using the size-mass balance approach. The breakage occurring in a tumbling media mill includes components from disintegrative fracture, chipping, and abrasion. The tumbling action in a ball mill at normal rpm (e.g., 70–80% critical speed) consists primarily of a stream of balls rolling down the inclined bed surface (cascading) plus free flight of some balls from top to bottom (cataracting). Maximum rates of breakage of normal particles occur at the cascading condition, due to many repeated ball-ball impacts, which is at or near to the condition of maximum mill power.

The specific rates of breakage of material in each $\sqrt{2}$ size interval pass through a maximum as particle size increases, according to a modified power law (see Fig. 2)

$$S_i = a(x_i/x_o)^\alpha Q_i \tag{S1}$$

where x_o is 1 mm and α is characteristic of the material. Q_i is a correction factor which is 1 for smaller sizes (*normal* breakage) and less than 1 (*abnormal* breakage) for particles too large to be nipped and fractured properly by the ball size in the mill. In the abnormal breakage region, each size behaves as if it has some fraction of weak (faster-breaking) material and a remaining fraction of stronger material. Using a mean value for S_i in this region, values of Q_i are empirically described by

$$Q_i = \frac{1}{1 + (x_i/\mu)^\Lambda}, \quad \Lambda \geq 0 \tag{S2}$$

where μ, Λ are characteristic parameters for the material and mill conditions tested. The maximum rate of breakage occurs at a size x_m given by

$$x_m = \mu\left(\frac{\alpha}{\Lambda - \alpha}\right)^{1/\Lambda}, \Lambda > \alpha \qquad (S3)$$

It seems likely that μ and x_m are increased by lifter designs which increase the degree of cataracting in the mill, due to the higher impact forces of cataracting; higher rotational speed would have the same effect. Thus cataracting gives higher breakage rates of large sizes in the abnormal breakage region.

It is convenient to represent the primary progeny fragment distributions in the cumulative form (the *primary breakage distribution function*), which can be fitted by the empirical function

$$B_{i,j} = \Phi_j\left(\frac{x_{i-1}}{x_j}\right)^{\gamma} + (1 - \Phi_j)\left(\frac{x_{i-1}}{x_j}\right)^{\beta} \qquad (S4)$$

where Φ_j, γ, and β are characteristic of the material for normal breakage, that is, the same for different ball fillings, mill diameters, etc. The value of Φ_j, γ, and β can be different for abnormal breakage and a more complex equation may be required. If Φ_j is constant for all breaking sizes, the B values are dimensionally *normalized*.

Fig. 5 shows the variation of relative values of Sf_c with f_c as a function of ball and powder loading for normal breakage. Maximum breakage occurs at ball fillings of $J = 0.40$ to 0.45. A flat maximum occurs at filling levels of $U = 0.6$ to 1.1. Underfilling or overfilling the mill is undesirable. The specific rates of breakage vary with solid hold-up according to

$$S_i \propto \exp[-cU], 0.5 \le U \le 1.5 \qquad (S5)$$

The variation of rates of breakage of quartz with ball diameter d is given by

$$S_i \propto 1/d^{N_o}, N_o \approx 1.0 \qquad (S6)$$

for the normal breakage region. The value of x_m is assumed to vary with ball diameter according to

$$x_m \propto d^2 \qquad (S7)$$

Thus smaller balls are more efficient for breaking smaller particles but larger balls are more efficient for breaking larger particles. However, the proportion of fine material in the primary progeny distribution for normal breakage can be higher for larger balls, which partly counteracts the smaller rates of breakage of larger balls on small particles.

The rates of breakage of a mixture of ball sizes is the weighted sum of the breakage of each ball size. The mean B values are also a weighted sum of the B values for each ball size. The problem of the correct choice of ball mix to go most efficiently from a given feed to a desired product is discussed later (Chapter 16).

Rates of breakage are independent of ball hardness for reasonably hard balls and directly proportional to ball density for hollow steel balls; however, it is recommended that each ball material be tested if accurate results are desired.

Under otherwise identical conditions, the rates of breakage are proportional to mill diameter D raised to a power,

$$S_i \propto a \propto D^{N_1} \tag{S8}$$

where N_1 is 0.5 for small laboratory mills, 0.5 for larger mills and, according to Bond, 0.3 for mills greater than 3.8 m in diameter. Without allowance for different residence time distributions, different classification efficiencies, etc., this gives a simple scaling law of mill capacity

$$Q \propto \frac{\pi}{4} L D^{2 + N_1} \tag{S9}$$

The values of x_m increase with mill diameter according to

$$x_m \propto D^{N_2} \tag{S10}$$

where N_2 is about 0.2.

Combination of these equations enable the calculation of specific rates of breakage for any batch grinding conditions, and hence, the simulation of first-order batch grinding size distributions.

Wet grinding gives almost the same breakage characteristics as dry grinding, but the specific rates of breakage are uniformly higher by a factor varying from 1.1 to 2. Slurry density does not much affect the breakage, for a given solid mass in the mill, until the pulp density is made high enough to change the rheological character of the slurry (from dilatant to pseudoplastic with yield stress). Too high a pulp density gives first-order breakage but with lower rates of breakage. Between these extremes there is a small range of slurry density where rates of breakage are slightly higher with increased percentage of solids. The rates of breakage vary with solid hold-up according to Eq. S5, but the value of U for maximum breakage shifts to higher values at higher slurry density (c decreases). Thus maximum breakage rates occur at an optimum combination of hold-up and slurry density. The rheology at this condition is pseudoplastic but without high yield stress.

The development of a more viscous slurry as fines are produced during grinding eventually causes a slowing-down of breakage rates. This occurs at coarser size distributions for higher slurry density and is associated with the development of rheology with yield stress.

A slowing-down effect is also seen for fine dry grinding. The slowing-down effect occurs at different degrees of fineness for different materials. The effect is tentatively explained by cohesive forces between particles leading to almost liquid-like properties; it is then more difficult to impact the particles.

An empirical law which relates fractional interstitial ball filling by powder, U, to flow rate F is

$$U = \left[(F/W_1)/(F_1/W_1) \right]^{0.5}, F > F_1 \tag{S11}$$

where W_1 is the mill hold-up at a flow rate F_1 through the mill which gives $U = 1$; then $U = W/W_1$, of course. This is conveniently put as

$$U = \left[(F/W_1)\tau_1 \right]^{0.5} \tag{S12}$$

where $\tau_1 = W_1/F_1$ must be determined experimentally for a given mill and conditions. It is expected that $\tau_1 \propto 1/D^{0.5}$, all other factors being constant. Grinding of a single size of feed gives Schuhmann slopes close to the slope γ of the B values, but the slopes tend to change in the direction of the α value. The shape of the size distributions is dependent most significantly on B values, but the degree of grinding is dominated by S values. On the other hand, a distributed starting feed gives Schuhmann slopes lying between the original feed slope and γ. Feed sizes which are too big for normal breakage give a characteristic plateau region in the size distribution, deficient in sizes near the maximum of S values, since the rates of breakage of larger and smaller sizes are less than in this region. On the other hand, breakage of particles of sizes to the left of the maximum in S values gives the same shape of size distributions regardless of the starting size.

References

Austin, L. G., 1973, "Understanding Ball Mill Sizing," *Industrial and Engineering Chemistry, Process Design and Development, Vol.* 12, *pp.* 121–129.

Austin, L. G. and Bagga, P., 1981, "An Analysis of Fine Dry Grinding in Ball Mills," *Powder Technology*, Vol. 28, pp. 83–90.

Austin, L. G., Celik, M., and Bagga, P., 1981, "Breakage Properties of Some Materials in Laboratory Ball Mills," *Powder Technology*, Vol. 28, pp. 235–241.

Austin, L. G., et al., 1981, "An Analysis of Ball-and-Race Milling. Part I. The Hardgrove Mill," *Powder Technology*, Vol. 29, pp. 263–275.

Austin, L. G., et al., "The Effect of Ball Size on Breakage Parameters in Tumbling Ball Mills," *Powder Technology*, in press.

Benjamin, J. S., 1976, "Mechanical Alloying," *Scientific American*, Vol. 234, No. 5, pp. 40–48.

Bond, F. C., 1960, "Crushing and Grinding Calculations," *British Chemical Engineering*, Vol. 6, pp. 378–391, 543–548.

Broadbent, S. R. and Callcott, T. G., 1956, "A Matrix Analysis of Processes Involving Particle Assemblies," *Philosophical Transactions of the Royal Society of London*, Vol. A249, pp. 99–123.

Crabtree, D. D., et al., 1964, "Mechanisms of Size Reduction in Comminution Systems," *Trans. SME-AIME*, Vol. 229, pp. 201–210.

Hogg, R., Shoji, K., and Austin, L. G., 1974, "Axial Transport of Dry Powders in Horizontal Rotating Cylinders," *Powder Technology*, Vol. 9, pp. 99–106.

Hogg, R., Shoji, K., and Austin, L. G., 1975, "Flow of Particles through Small Continuous Dry Ball Mills," *Trans. SME-AIME*, Vol. 258, pp. 194–198.

Katzer, M., Klimpel, R., and Sewell, J., 1981, "Example of the Laboratory Characterization of Grinding Aids in the Wet Grinding of Ores," *Mining Engineering*, Vol. 33, No. 10, pp. 1471–1476.

Klimpel, R., 1982, "Laboratory Studies of the Grinding and Rheology of Coal-Water Slurries," *Powder Technology*, Vol. 32, pp. 267–277.

Klimpel, R., 1982a, "Slurry Rheology Influence on the Performance of Mineral/Coal Grinding Circuits—Part I," *Mining Engineering*, Vol. 34, No. 12, pp. 1665–1668.

Klimpel, R., 1983, "Slurry Rheology Influence on the Performance of Mineral/Coal Grinding Circuits—Part 2," *Mining Engineering*, Vol. 35, No. 1, pp. 21–26.

Klimpel, R., 1984, "The Influence of Material Breakage Properties and Associated Slurry Rheology on Breakage Rates in the Wet Grinding of Coal/Ores in Tumbling Media Mills," *Particulate Science and Technology*.

Malghan, S. G. and Fuerstenau, D. W., 1976, "Scale-Up of Ball Mills Using Population Balance Models and Specific Power Input," *Proceedings*, 4th European Symposium Zerkleinern, H. Rumpf and K. Schonert, eds., Dechema Monographien 79, Nr. 1576-1588, Verlag Chemie, Weinheim, pp. 613–630.

Marchand, J. C., Hodouin, D., and Everell, M. D., 1980, "Residence Time Distribution and Mass Transport Characteristics of Large Industrial Grinding Mills," *Proceedings*, 3rd IFAC Symposium, J. O'Shea and M. Polis, eds., Pergammon Press, pp. 295–302.

Marstiller, S., 1979, "Cost and Up-Time Considerations for Selection of Mill Liners and Grinding Balls," Continuing Education Course, "Ball Milling," The Pennsylvania State University, Sept.

Miles, T. I., Shah, I., and Austin, L. G., "A Primary Breakage Function for Abnormal and Autogenous Breakage," *Powder Technology*, in press.

Mori, Y., 1973, "The Residence Time Distribution and the Size Distribution of Products Flowing through a Continuous Type Ball Mill," *Proceedings*, 1st International Conference in Particle Technology, R. Davies, ed., Chicago Press, pp. 217–223.

Rose, H. E. and Sullivan, R. M. E., 1958, *Rod, Ball and Tube Mills*, Chemical Publishing Co., NY.

Shoji, K., Lohrasb, S., and Austin, L. G., 1979, "The Variation of Breakage Parameters with Ball and Powder Loading in Dry Ball Milling," *Powder Technology*, Vol. 25, pp. 109–114.

Shoji, K., et al., 1982, "Further Studies of Ball and Powder Filling Effects in Ball Milling," *Powder Technology*, Vol. 31, pp. 121–126.

Shoji, K., Austin, L. G., and Luckie, P. T., "Effect of Mill Diameter on Breakage Parameters in Tumbling Ball Mills," *Powder Technology*, in press.

Steier, K. and Schonert, K., 1971, *Proceedings*, 3rd European Symposium Zerkleinern, H. Rumpf and K. Schonert, eds., Dechema Monographien 69, Nr. 1292-1326, Verlag Chemie, Weinheim, pp. 167–192.

Tangsathitkulchai, C. and Austin, L. G., "The Effect of Slurry Density in Wet Grinding in a Laboratory Mill," *Powder Technology*, in press.

von Seebach, H. M., 1969, "Effect of Vapors of Organic Liquids in the Comminution of Cement Clinker in Tube Mills," Research Institute Cement Industry, Dusseldorf, West Germany.

Weller, K. R., 1980, "Hold-Up and Residence Time Characteristics of Full Scale Grinding Circuits," *Proceedings*, 3rd IFAC Symposium, J. O'Shea and M. Polis, eds., Pergammon Press, pp. 303–309.

6

Grinding Circuit Simulation: Formulation and Computation

Introduction

In this chapter the concepts of Chapter 4 are expanded to encompass the *simulation* of complete simple grinding circuits. A simulation of a physical process is a mathematical model which behaves on computation in a manner identical to that of the real process. Generally, a simulation is only an approximation to the real behavior, especially for a process as complicated as milling, and the mathematical models can be more or less complex depending on how closely one wishes to simulate the real situation. For example, the fundamental elements of even simple steady-state milling circuit simulation *in detail*, are as follows.

1. The values of S and B as a function of the complete size range involved, from feed to product.

2. The variation of S and B as a function of the mill conditions: e.g., for ball milling this involves the factors of mill rotational speed, ball filling, powder filling, mill diameter, mill length, mixture of balls of different diameter, ball density, pulp density in wet milling, and influence of liner shape.

3. The mass transfer laws of flow of material through the mill; in particular, for a given mass flow, what is the hold-up and what is the residence time distribution? In general, the mass transfer laws plus any influence of classifying liners within the mill (which give preferential separation of the different sizes of ball along the mill) determine the variation of the factors which affect S and B along the mill.

4. The splitting laws of the size classifier, that is, how it splits its feed stream into product and recycle streams, *for every size interval involved*, as a function of the rate of mass throughput, the feed size distribution, and classifier operating variables (such as pulp density and underflow spigot size for hydrocyclones and fan speeds for air separators).

Other factors can be included, such as classification at the mill discharge, nonhomogeneous components, and the accumulation of a hard component. To incorporate every conceivable factor into a milling simulation is a long-term research project, and in practice, therefore, certain elements in the total are held constant or empirically measured to reduce the simulation to manageable proportions. In this chapter concepts of Chapter 4 are extended to continuous milling, and in Chapter 7, the most important elements of the total are used to demonstrate how circuits behave. The fine detail of more complete analyses is postponed to later chapters. The results of this somewhat simplified treatment are still sufficiently valid to be of great practical importance.

Continuous Milling and Residence Time Distribution

The general problem of continuous milling is reduced to a simpler case of considerable practical importance by assuming that the levels of balls and powder in the mill are virtually constant along the mill. This is likely to be a reasonable approximation for an overflow ball mill, illustrated in Fig. 1. It can be assumed that the set of S and B values are known for the mill conditions, for the material being ground. To construct a simulation model for the steady-state continuous operation of such a mill, it is necessary to join the breakage equation (Eq. 23 in Chapter 5) with a description of the distribution of residence time in the mill.

Since there are no good theoretical descriptions of mass transfer in mills, it is not possible at present to deduce the residence time distribution (RTD) of an overflow mill from *a priori* reasoning. However, the residence time distributions of ball mills have been measured using various tracer tests (see Chapter 14) and at this stage it can be assumed that information on RTDs is available. It is assumed that all of the material entering the mill flows and mixes in the mill independently of particle characteristics such as size, shape, and density; experimental evidence appears to support this simplifying assumption (Chapter 14). Before treating the general case, it is useful to examine the two extremes of RTD behavior: *plug flow* and *fully mixed*, as follows.

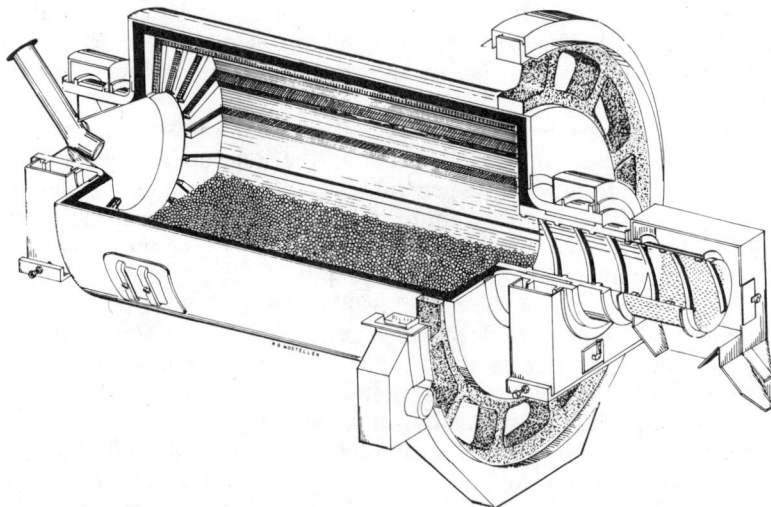

Fig. 1. Illustration of wet overflow ball mill (courtesy of Kennedy Van Saun Corp.).

For a retention mill in which conditions along the mill are virtually constant, plug flow for a residence time of τ gives exactly the same answer as batch grinding for time τ, that is,

$$\frac{dw_i(t)}{dt} = -S_i w_i(t) + \sum_{\substack{j=1 \\ i>1}}^{i-1} b_{ij} S_j w_j(t), \quad n \geq i \geq j \geq 1 \tag{1}$$

is the descriptive set of equations (see Chapter 4) to be integrated from $t = 0$ to $t = \tau$. This is perhaps more readily visualized by considering the value of w_i at a position l along the mill of total length L, $w_i(l)$; see Fig. 2. A size-mass balance on an element at l to $l + dl$ gives:

Rate of size i material leaving the element equals rate of size i material coming into the element plus rate of production of size i from breakage of larger sizes within the element minus rate of disappearance by breakage of size i material within the element. (2)

Symbolically this is

$$Fw_i(l + dl) = Fw_i(l) + \sum_{j=1}^{i-1} b_{ij} S_j w_j(l)(Wdl/L) - S_i w_i(l)(Wdl/L)$$

because for a constant level along the mill the amount of powder in the section is Wdl/L. Since $w_i(l + dl) = w_i(l) + (dw_i(l)/dl) \, dl$ and $\tau = W/F$,

$$\frac{dw_i(l)}{dl} = -S_i w_i(l)(\tau/L) + \sum_{j=1}^{i-1} b_{ij} S_j w_j(l)(\tau/L)$$

The time taken to reach position l is given by $t/l = \tau/L$, therefore $dt/dl =$

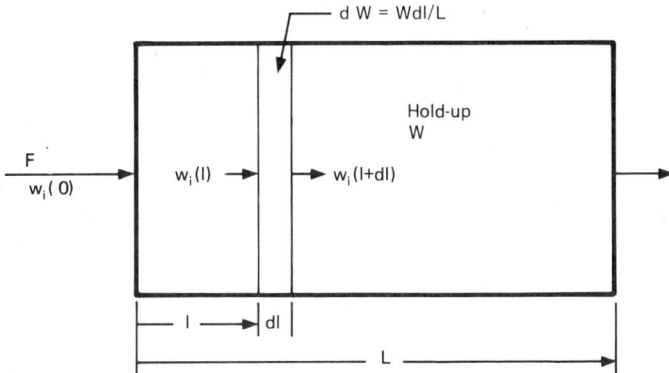

Fig. 2. Nomenclature for one-dimensional continuous mill.

τ/L and the above expression becomes, as before,

$$\frac{dw_i(t)}{dt} = -S_i w_i(t) + \sum_{\substack{j=1 \\ i>1}}^{i-1} b_{ij} S_j w_j(t), \quad n \geq i \geq j \geq 1 \quad (batch/plug) \quad (1)$$

To avoid introducing an extra equation, Eq. 1 is used and referred to as the *batch/plug* equation. For plug flow, the initial feed set is $w_i(0) = f_i$, and the final product set is $w_i(\tau) = p_i$, assuming no size classification at the mill exit. For the time being, the solution (see next section) of this set of equations is given in the symbolic form

$$p_i = w_i(\tau) = \sum_{j=1}^{i} d_{ij}(\tau) f_j, \quad n \geq i \geq j \geq 1 \quad (3)$$

where d_{ij} has the physical meaning of the fraction of feed size j *transferred* to size i in the product via the repeated steps of the breakage process over time τ: the set of d_{ij} values is the *mill transfer function*.

The other RTD extreme is actually simpler to treat (Austin and Gardner, 1962) and the equation of steady-state, fully mixed, continuous grinding is derived from the verbal statement of Eq. 2 applied over the mill as a single fully mixed element,

$$p_i = f_i + \tau \left(\sum_{\substack{j=1 \\ i>1}}^{i-1} b_{ij} S_j w_j \right) - S_i w_i \tau \quad n \geq i \geq j \geq 1, \quad (fully\ mixed) \quad (4)$$

Note that this is not a differential equation requiring integration, but only a set of simple algebraic equations which can be readily recursively calculated starting with $i = 1$, etc. If it is assumed as before that no classification occurs at the mill exit, so that w_i in the fully mixed mill equals the product size distribution p_i,

$$p_i = \left(f_i + \tau \sum_{\substack{j=1 \\ i>1}}^{i-1} b_{ij} S_j p_j \right) / (1 + \tau S_i), \quad n \geq i \geq j \geq 1 \quad (4)$$

A more general treatment for the partially back-mixed case, also using conventional RTD techniques, was first given by Reid (1965). If the feed entering the mill were tagged in some way over a very short period of time, it would be found that some of this tagged material would appear in the product after a short residence time, more after a longer time and so on, until the final traces were removed at long periods of time. Let the fraction of the feed which has appeared in the removed product up to time t after admission be $\Phi(t)$ and let $\phi(t) = d\Phi(t)/dt$. Then the fraction of the feed which has a residence time of t to $t + dt$ is $\phi(t)\,dt$. If grinding is perfectly first-order, with constant S and B for a given size at any position along the mill (*lumped parameter model*), this fraction of the feed behaves as if it had been ground in a batch mill for time t, giving a product mass-size distribution of $w_i(t)$. However, this is only a fraction $\phi(t)\,dt$ of the total product, so the mass-size distribution of the overall steady-state product is

$$p_i = \int_{t=0}^{\infty} w_i(t)\phi(t)\,dt, \quad n \geq i \geq j \geq 1 \qquad (5)$$

or

$$P(x) = \int_{t=0}^{\infty} P(x,t)\,d\Phi$$

or, from Eq. 3,

$$p_i = \int_{t=0}^{\infty} \left[\sum_{j=1}^{i} d_{ij}(t)f_j \right] \phi(t)\,dt, \quad n \geq i \geq j \geq 1 \qquad (6)$$

The case where S and B are not constant along the mill (*distributed parameter model*) requires more complicated equations, since integration along the mill must then be performed (see, in Chapter 17, "Distributed Parameter Models"). The formulation of more complex grinding equations is postponed to later chapters since the equations presented so far enable important conclusions to be drawn.

Solution of Grinding Equations; Classification and Mill Circuits

Although there is a complete chapter on the solution of grinding equations and circuit equations, it is convenient at this point to demonstrate the basic methods of solution. The batch/plug flow equation was solved by Reid (1965) as follows. For $i = 1$,

$$dw_1(t)/dt = -S_1 w_1(t)$$

gives upon integration

$$w_1(t) = w_1(0)\exp(-S_1 t)$$

For $i = 2$,

$$dw_2(t)/dt = -S_2 w_2(t) + b_{21} S_1 w_1(t)$$

or

$$\frac{dw_2(t)}{dt} + S_2 w_2(t) = b_{21} S_1 w_1(0)\exp(-S_1 t)$$

Multiplying through by the integrating factor $\exp(S_2 t)$,

$$\exp(S_2 t)\frac{dw_2(t)}{dt} + S_2 w_2(t)\exp(S_2 t) = b_{21} S_1 w_1(0)\exp\left[-(S_1 - S_2)t\right]$$

or

$$d\left[w_2(t)\exp(S_2 t)\right]/dt = b_{21}S_1 w_1(0)\exp\left[-(S_1 - S_2)t\right]$$

and, hence, for $S_2 \neq S_1$, by separation and integration,

$$w_2(t) = \frac{b_{21}S_1 w_1(0)}{S_2 - S_1}\exp(-S_1 t) - \frac{b_{21}S_1 w_1(0)\exp(-S_2 t)}{S_2 - S_1} + w_2(0)\exp(-S_2 t)$$

Proceeding similarly, for $i = 3$, $i = 4$, etc. collecting terms and deducing the general term,

$$w_i(t) = \sum_{j=1}^{i} a_{ij} e^{-S_j t}, \quad n \geq i \geq 1 \tag{7}$$

$$a_{ij} = \begin{cases} 0, & i < j \\[2mm] w_i(0) - \displaystyle\sum_{\substack{k=1 \\ i>1}}^{i-1} a_{ik}, & i = j \\[4mm] \dfrac{1}{S_i - S_j} \displaystyle\sum_{k=j}^{i-1} S_k b_{ik} a_{kj}, & i > j \end{cases}$$

This is referred to as the *Reid Solution*; in this form the values of a_{ij} do not depend on grind time but they do depend on feed size distribution. Table 1 gives the first few terms of the solutions. The number of terms expands rapidly as i becomes larger. (For the case of $S_i = S_j$, see "Simple Batch Grinding" in Chapter 17).

Table 1. The First Three Terms in the Reid Solution of the Batch Grinding Equation

$$w_1(t) = w_1(0)e^{-S_1 t}$$

$$w_2(t) = \frac{S_1 b_{21}}{(S_2 - S_1)}w_1(0)e^{-S_1 t} + w_2(0)e^{-S_2 t} - \frac{S_1 b_{21}}{(S_2 - S_1)}w_1(0)e^{-S_2 t}$$

$$w_3(t) = \frac{S_1 b_{31}}{(S_3 - S_1)}w_1(0)e^{-S_1 t} + \frac{S_1 b_{21}}{(S_2 - S_1)}\frac{S_2 b_{32}}{(S_3 - S_1)}w_1(0)e^{-S_1 t}$$

$$+ \frac{S_2 b_{32}}{(S_3 - S_2)}w_2(0)e^{-S_2 t} - \frac{S_1 b_{21}}{(S_2 - S_1)}\frac{S_2 b_{32}}{(S_3 - S_2)}w_1(0)e^{-S_2 t}$$

$$+ w_3(0)e^{-S_3 t} - \frac{S_1 b_{31}}{(S_3 - S_1)}w_1(0)e^{-S_3 t} - \frac{S_1 b_{21}}{(S_2 - S_1)}\frac{S_2 b_{32}}{(S_3 - S_1)}w_1(0)e^{-S_3 t}$$

$$- \frac{S_2 b_{32}}{(S_3 - S_2)}w_2(0)e^{-S_3 t} + \frac{S_1 b_{21}}{(S_2 - S_1)}\frac{S_2 b_{32}}{(S_3 - S_2)}w_1(0)e^{-S_3 t}$$

Regrouping the terms in a different manner Luckie and Austin (1972) showed that the above equation can be put in the form:

$$w_i(t) = \sum_{j=1}^{i} d_{ij} w_j(0), \quad n \geq i \geq 1 \tag{8}$$

$$d_{ij} = \begin{cases} 0, & i < j \\ e^{-S_i t}, & i = j \\ \sum_{k=j}^{i-1} c_{ik} c_{jk} (e^{-S_k t} - e^{-S_i t}), & i > j \end{cases}$$

$$c_{ij} = \begin{cases} -\sum_{k=i}^{j-1} c_{ik} c_{jk}, & i < j \\ 1, & i = j \\ \dfrac{1}{S_i - S_j} \sum_{k=j}^{i-1} S_k b_{ik} c_{kj}, & i > j \end{cases}$$

This form is more convenient than the Reid Solution for some applications, because the set of d_{ij} values represent the *transfer function* for feed to product. The d_{ij} values depend on grinding time but not feed size distribution.

In general, it is convenient to program the solution (see Chapter 17):

$$p_i = \sum_{j=1}^{i} d_{ij} f_j, \quad n \geq i \geq 1 \tag{9}$$

$$d_{ij} = \begin{cases} e_j, & i = j \\ \sum_{k=j}^{i-1} c_{ik} c_{jk} (e_k - e_i), & i > j \end{cases}$$

$$c_{ij} = \begin{cases} -\sum_{k=i}^{j-1} c_{ik} c_{jk}, & i < j \\ 1, & i = j \\ \dfrac{1}{S_i - S_j} \sum_{k=j}^{i-1} S_k b_{ik} c_{kj}, & i > j \end{cases}$$

$$e_j = \int_0^{\infty} e^{-S_j t} \phi(t)\, dt$$

where f_j is the vector of values of mill feed size distribution and p_i the vector of values of mill product size distribution. The grinding time τ is introduced via the RTD term, $\phi(t)$, in the expression for e_j. For batch/plug flow,

$$e_j = \exp(-S_j\tau) \qquad (10a)$$

For fully mixed grinding,

$$e_j = 1/(1 + S_j\tau) \qquad (10b)$$

It is common in chemical reactor engineering to treat an RTD *as if* it arose from a series of fully mixed reactors. For m equal reactors,

$$e_j = 1/(1 + S_j\tau/m)^m \qquad (10c)$$

It is found that the experimentally measured RTDs of wet overflow ball mills can be treated to a reasonable approximation as the equivalent of one large fully mixed reactor (τ_1) followed by two equal smaller fully mixed reactors (τ_2); see Chapter 14. This model gives

$$e_j = 1/(1 + S_j\tau_1)(1 + S_j\tau_2)^2 \qquad (10d)$$

where $\tau = \tau_1 + 2\tau_2$.

The basic working equations are 8, 9, and 10. Clearly, these are not difficult equations to compute, but the numbers of terms involved for a typical case where the number of size intervals is 20 or more normally makes it necessary to use a digital computer for the solution. Other RTD models are treated in Chapters 14 and 17. Note that $S_n = 0$ since there can be no breakage out of the *sink* interval; also $b_{n,j} = B_{n,j}$.

The subject of classifiers will receive a chapter to itself, but at this stage it is necessary to introduce the concepts of *first-order classification*, *selectivity values*, and *circulation ratio*. Classification is the splitting of a stream of particles into two streams, with separation by size (and/or density) e.g., one stream with mainly large particles, one stream with fine particles. For example, a screen is a classifier (a strict size classifier under ideal conditions) which sends oversize to one stream and undersize to another. The circulation ratio, C, is defined by the relative mass flow rates of the return (coarser) stream to the product (finer) stream, $C = T/Q$. Note that it is also common to see the *circulation ratio* as defined here multiplied by 100 and reported as a percentage. Also, the term *circulating load* is often used, defined as $F/Q = 1 + C$, again often expressed as a percentage instead of a fraction; for example $C = 1.5$ will be expressed as 150% or a circulating load of 250%. In performing the mass balances it is convenient to consider the proportional mass flow rates as illustrated in Fig. 3, where p_i, q_i, t_i represent the respective weight fractions in size interval i.

The classifier action is characterized by a set of numbers, one for each size interval, which describes how each size is split into the two product

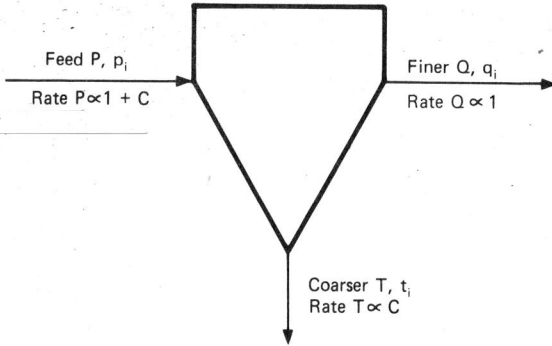

Fig. 3. Mass flow rates around classifier.

streams. The *selectivity* number for size i is defined by

s_i = *fraction* of size i in feed to the classifier which is sent to the coarser stream (tailings, recycle, underflow, sands, etc.)

$$= \frac{t_i T}{p_i P}$$

$$= \frac{C t_i}{(1 + C) p_i} \tag{11}$$

For a given set of test conditions on a classifier the values of t_i, q_i, p_i, and C are determined (see Chapter 13) and, s_i values are determined. The fraction sent to the finer stream is obviously $1 - s_i$. Describing the classifier action in this way *assumes* that the process of size selection in the classifier is first-order since the amount of size i selected to send to one stream is proportional to the amount of size i presented to the classifier. A plot of selectivity numbers vs. size is called a *Tromp* curve or *partition* or *selectivity* curve (see Figs. 6 and 8 in Chapter 13, for examples).

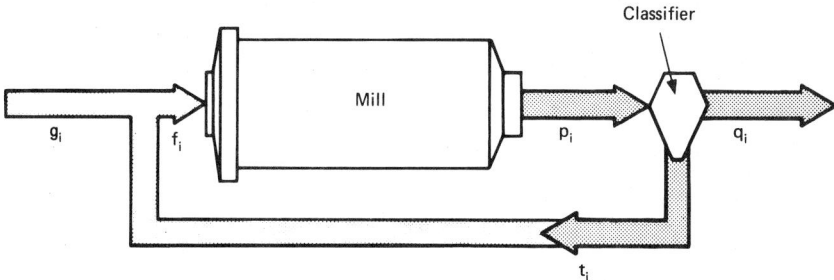

Fig. 4. Nomenclature of size distributions in closed circuit milling.

It is now possible to formulate the solution of the simple mill circuit shown in Fig. 4. The first-order mill model is (Eqs. 8, 9, 10, etc.),

$$p_i = \sum_{j=1}^{i} d_{ij} f_j$$

The mill feed is related to the circuit feed g_i (makeup feed) by

$$f_i(1 + C) = s_i p_i(1 + C) + g_i \tag{12}$$

For a given makeup feed size distribution, the value of the circulation ratio and the mill feed size distribution are not known and appear as results of the simulation for a specified τ. These unknowns are handled as follows: multiplying both sides of the mill model by $1 + C$ and letting $p_i(1 + C) = p_i^*$,

$$p_i^* = \sum_{j=1}^{i} d_{ij} f_j(1 + C)$$

The unknown mill feed is eliminated by substituting for $f_i(1 + C)$ from Eq. 12

$$p_i^* = \sum_{j=1}^{i} d_{ij} s_j p_j^* + \sum_{j=1}^{i} d_{ij} g_j$$

$$= d_{ii} s_i p_i^* + \sum_{j=1}^{i-1} d_{ij} \left(s_j p_j^* + g_j \right) + d_{ii} g_i$$

which gives

$$p_i^* = \frac{d_{ii} g_i + \sum_{\substack{j=1 \\ i>1}}^{i-1} d_{ij} \left(s_j p_j^* + g_j \right)}{1 - d_{ii} s_i}, \quad n \geq i \geq j \geq 1 \tag{13}$$

This set of equations is the basic equation set of the simple mill circuit, which enables the product size distribution to be calculated from a known makeup feed for a set value of τ (set values of d_{ij}). It is solved *sequentially* starting with $i = 1$, since then $p_1^* = d_{11} g_1 / (1 - d_{11} s_1)$. This value of p_1^* is used in the next expression $p_2^* = [d_{22} g_2 + d_{21}(s_1 p_1^* + g_1)]/(1 - d_{22} s_2)$ and so on. Then, of course, C follows from

$$1 + C = \sum_{j=1}^{n} p_j^* \tag{14}$$

(since $\Sigma_{j=1}^{n} p_j = 1$), and the values of p_i follow from

$$p_i = p_i^*/(1 + C)$$

A mass balance on the classifier gives the circuit product size distribution

$$q_i = (1 - s_i)p_i(1 + C) \qquad (15)$$

Other size distributions are readily calculated from Eq. 12 and others. The mill output rate is derived from the mean residence time used in the calculations of d_{ij}: to get d_{ij} values, a value of τ has been set and since $\tau = W/F = W(1 + C)Q$,

$$Q = \frac{W}{(1 + C)\tau} \qquad (16)$$

It is important to realize that if S_j values are changed to new values by $S_j' = kS_j$ (that is, every value is changed by a constant factor k) then the solution of the equations with S_j' is *identical* to that with S_j for a new τ given by $\tau_{\text{new}} = \tau_{\text{old}}/k$. Thus Eq. 16 can be put as

$$Q = \frac{kW}{(1 + C)\tau_{\text{old}}} \qquad (16a)$$

and it is not necessary to recompute the circuit, only to increase Q by the appropriate factor k.

There are two immediate applications for this concept. First, the hold-up W in a mill changes with flow rate and, hence, S_j values change. If, however, the hold-up changes in the region where $S_jW \approx$ constant, then constant values of S_j and W can be used to give Q, for then $Q = k(W/k)/(1 + C)\tau_o = W/(1 + C)\tau_o$. Similarly, in real mill systems the hold-up W may not be known, but it can be treated as equivalent to a reasonable formal value of W, e.g., W_1 corresponding to $U = 1.0$. Then the residence time used in the simulation is the corresponding *formal* residence time τ_f defined by $\tau_f = W_1/F$, and

$$Q = \frac{W_1}{(1 + C)\tau_f} \qquad (16b)$$

Second, it is found that the RTDs of wet overflow mills correspond fairly closely to a fully mixed condition. Thus the mean size distribution in the mill is not too different from the size distribution leaving the mill. In this case, the mill environment for calculating the slowing-down factor κ (see "Effect of Environment in the Mill" in Chapter 5) can be taken as the mill product. Eq. 16 can be used as

$$Q = \frac{\kappa W}{(1 + C)\tau_o} \qquad (16c)$$

where τ_o is the first-order residence time necessary to give the desired product size. The real τ mean residence time value is, of course, τ_o/κ.

Allowance for Variation in Filling Level

If a mill is operated in closed-circuit with a high circulation ratio to give a relatively low size reduction ratio, then the mass flow through the mill will be high. This will tend to overfill the mill (see Table 2 in Chapter 5). The results from laboratory investigations (see "Ball and Powder Loading" in Chapter 5) indicate that overfilling will reduce the breakage rates in the mill. In order to allow for this effect the empirical mass transfer relation of Eq. 20 in Chapter 5 can be used:

$$W = W_1 (F/F_1)^{0.5}$$

The computation of the effect of the new hold-up level is conveniently done as follows. This equation can be put as

$$U = [(F/W_1)\tau_1]^{0.5}, \quad F \geq F_1$$

where τ_1 is defined as the residence time for $U = 1$, $\tau_1 = W_1/F_1$. The actual τ allowing for overfilling is related to a false τ_f for $U = 1$ by $\tau_{real} = (a_1/a_{real})\tau_f$. The real value of W is related to W_1 by $W = UW_1$. Thus, since $Q = W/(1 + C)\tau_{real}$,

$$Q = K_o W_1/(1 + C)\tau_f = K_o Q_f \qquad (17)$$

where Q_f is the false mill capacity assuming $U = 1$, and $K_o = U a_{real}/a_1$. From Eq. 7 in Chapter 5, $a \propto \exp[-cU]$ where $c = 1.32$ for wet grinding, therefore $a_{real}/a_1 = \exp[-1.32(U - 1)]$, and

$$K_o = U \exp[-1.32(U - 1)] \qquad (18)$$

Combining Eqs. 17 and 18 with Eq. 22a in Chapter 5, $U = [(1 + C)Q\tau_1/W_1]^{0.5} = [K_o\tau_1/\tau_f]^{0.5} = \{U \exp[-1.32(U - 1)](\tau_1/\tau_f)\}^{0.5}$, or

$$\tau_1/\tau_f = U \exp[1.32(U - 1)], \quad \tau_1/\tau_f \geq 1 \qquad (19)$$

Fig. 5 shows how K_o can be graphically determined for a known τ_1 for any τ_f given by the simulation: K_o is taken as 1 for lower flow rates (longer residence times), that is, $U = 1$ for $F \leq F_1$.

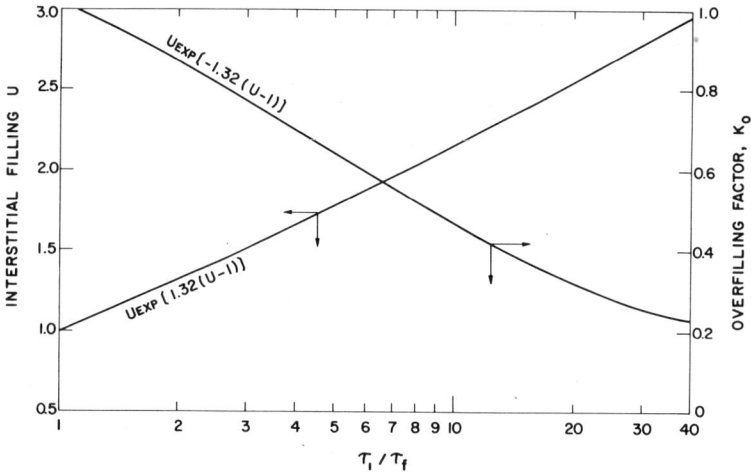

Fig. 5. Graphical method for obtaining overfilling factor from simulations based on $U = 1$.

The Ultimate Limiting Case

In simulations in closed circuit it will sometimes occur that a maximum capacity (in the absence of overfilling) exists, and the programs will not compute if given a τ value which is too low. This occurs in closed circuit with a classifier for which $s_i = 1.0$ above a certain cutoff size. This condition can be calculated directly following the approach of Austin and Perez (1977) and Klimpel and Austin (1971). Consider the mill and its classifier to be a *black box*, as in Fig. 6. The black box produces the least proportion of fines when the classifier is 100% efficient at separating at the desired top size, that is, *cutting* at the desired size, the top size of interval i_s, for example. One hundred percent efficient means that anything larger than i_s cannot leave the cycle and anything less than or equal to i_s immediately leaves the black box, like a perfect screen discharge. If also the internal or

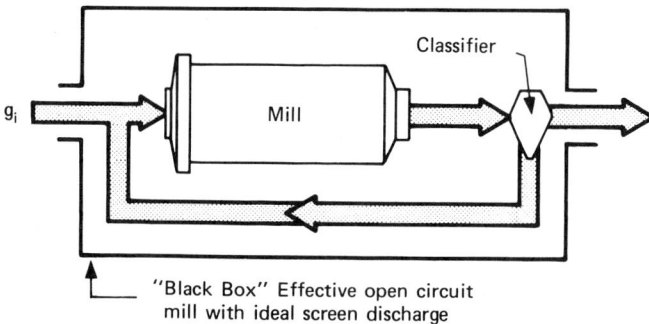

Fig. 6. Illustration of closed-circuit grinding as an equivalent screen discharge *black box* in open circuit.

external circulating load is very high, the crushed material is presented to the classifier at a high rate, and no build-up of small material occurs in the system, so no extra breakage of desired product size occurs. In other words, the limit occurs when there is no build-up of less-than-cut-size material in the mill.

Under these conditions, material of size 1 entering the box must break to i_s or smaller before it can leave. The first breakage of unit quantity of size 1 material produces b_{21} of size 2, b_{31} of size 3, etc., and $B_{i,1}$ less than the top size of i. In turn, the b_{21} material breaks to $b_{32}b_{21}, b_{42}b_{21}, \ldots, B_{i,2}b_{21}$; and b_{31} material breaks to $b_{43}b_{31}, b_{53}b_{31}, \ldots, B_{i,3}b_{31}$; and so on until no material greater than i_s is left. Thus, the unit quantity of size 1 material distributes itself to

$$\bar{B}_{i,1} = B_{i,1}a_{11} + B_{i,2}a_{21} + b_{i,3}a_{31} + \cdots$$

$$+ B_{i,i_s-1}a_{i_s-1,1}; \ n \geq i \geq i_s \geq 1$$

where $a_{11} = 1$, $a_{21} = b_{21}a_{11}$, $a_{31} = b_{31}a_{11} + b_{32}a_{21}$, etc.

Similarly, in general,

$$\left. \begin{array}{l} \bar{B}_{i,j} = B_{i,j}a_{jj} + B_{i,j+1}a_{j+1,j} + \cdots \\ \\ \qquad \cdots + B_{i,i_s-1}a_{i_s-1,j}; \ n \geq i \geq i_s \geq j \geq 1 \\ \\ \text{where} \\ \\ a_{ij} = \begin{cases} \displaystyle\sum_{k=j}^{i-1} b_{i,k}a_{k,j}, & i > j \\ 1, & i = 1 \\ 0, & i < j \end{cases} \end{array} \right\} \qquad (20)$$

The cumulative size distribution is given by

$$Q(x_i)_{\text{steepest}} = \sum_{k=i}^{n} g_k + \sum_{j=1}^{i_s-1} \bar{B}_{i,j}g_j, \quad n \geq i \geq i_s \geq 1 \qquad (21)$$

where the first term on the RHS is the cumulative amount of small makeup feed sizes up to i, and the second term the sum of breakage products from all larger sizes.

It is concluded that providing the classifier has a sharp cut at some upper size i_s (that is, it sends all larger material to recycle) the steepest possible size distribution is dependent only on the values of B (that is, the basic breakage pattern of the material) and the feed size distribution g_i. Improvement of classifier efficiency, variations in the specific rates of breakage, or RTD, are irrelevant as far as the limiting steepest size distribution is concerned. It can only be varied by changing the method of fracture to give different slopes to the B curves. The limiting size distributions tend to reach a slope identical to the Schuhmann slope of the B values.

Applying the same concepts to the rates of breakage, it is apparent that all material larger than the size of i_s can leave the mill classifier system only by breakage to smaller sizes. In addition, the high circulating load necessary to get fines out as soon as they are formed means that the system behaves in this limit as a fully mixed reactor. From Eq. 4, for all sizes larger than the cut size

$$0 = g_i + \left(\sum_{\substack{j=1 \\ i>1}}^{i-1} b_{ij}\gamma_j \right) - \gamma_i, \quad i_s > i \geq j \geq 1$$

where $\gamma_i = S_i w_i \tau$. Hence,

$$\gamma_i = g_i + \sum_{\substack{j=1 \\ i>1}}^{i-1} b_{ij}\gamma_j, \quad i_s > i \geq j \geq 1 \tag{22}$$

which enables the values of γ_i to be sequentially calculated starting at $i = 1$. Then, because the values of w_i in the mill only exist for sizes larger than x_{i_s}, $\sum_1^{i_s-1} w_i = 1$ and

$$\tau = \sum_1^{i_s-1} \frac{\gamma_i}{S_i} \tag{23}$$

from which

$$(Q/W)_{\text{max}} = 1 \Big/ \left(\sum_1^{i_s-1} \frac{\gamma_i}{S_i} \right) \tag{23a}$$

Eqs. 22 and 23 enable this maximum rate to be readily calculated.

The above physical concept can be expressed verbally as follows: A mill with a classifier that gives zero throughput of material above a certain size (x_{i_s}) has an ultimate maximum rate of throughput for a given feed determined *only* by the b and S values of the larger $(> x_{i_s})$ feed sizes. This concept will be used in Cases 3 and 8 discussed in the next chapter.

Summary

The simulation of a continuous milling circuit at steady state can be performed with the simplifying assumptions that the effective values of S, B, and hold-up are constant along the mill, plus the assumption that an experimental mill RTD has been determined and is the same for all particle sizes in the mill. The batch grinding equation integrated for time τ is the same as that for plug flow in a continuous mill. In general,

$$p_i = \int_{t=0}^{\infty} w_i(t)\phi(t)\,dt \qquad (S1)$$

where p_i is the weight fraction in size i leaving the mill, $w_i(t)$ is the solution of the batch grinding equation for time t for feed $w_i(0) = f_i$ and $\phi(t)\,dt$ is the fraction of feed which leaves after a residence of time t to $t + dt$.

In general, it is convenient to program the solution (see Chapter 17):

$$p_i = \sum_{j=1}^{i} d_{ij}f_j, \quad n \geq i \geq 1 \qquad (S2)$$

$$d_{ij} = \begin{cases} e_j, & i = j \\ \sum_{k=j}^{i-1} c_{ik}c_{jk}(e_k - e_i), & i > j \end{cases}$$

$$c_{ij} = \begin{cases} -\sum_{k=i}^{j-1} c_{ik}c_{jk}, & i < j \\ 1, & i = j \\ \dfrac{1}{S_i - S_j}\sum_{k=j}^{i-1} S_k b_{ik}c_{kj}, & i > j \end{cases}$$

$$e_j = \int_0^{\infty} e^{-S_j t}\phi(t)\,dt$$

where f_j is the vector of values of mill feed size distribution and p_i the vector of values of mill product size distribution. The grinding time τ is introduced via the RTD term, $\phi(t)$, in the expression for e_j. For batch/plug flow,

$$e_j = \exp(-S_j\tau) \qquad (S3a)$$

For fully mixed grinding,

$$e_j = 1/(1 + S_j\tau) \qquad (S3b)$$

It is common in chemical reactor engineering to treat an RTD *as if* it arose from a series of fully mixed reactors. For m equal reactors,

$$e_j = 1/(1 + S_j\tau/m)^m \qquad (S3c)$$

For one large fully mixed reactor followed by two smaller equal fully mixed reactors

$$e_j = 1/(1 + S_j\tau_1)(1 + S_j\tau_2)^2 \tag{S3d}$$

where $\tau = \tau_1 + 2\tau_2$.

Solutions are usually performed on a digital computer because of the number of computations involved for 20 or more size intervals.

The first-order action of a classifier is described by the set of selectivity numbers s_i, defined as the fraction of size i in the feed which is sent to the coarse (recycle) stream. A plot of s_i vs. x_i is called a Tromp or partition curve. The circulation ratio is defined as T/Q and, hence, mass balance around the classifier on any size i gives

$$t_i C = p_i s_i (1 + C) \tag{S4}$$

The feed to the mill in closed circuit is

$$f_i(1 + C) = s_i p_i(1 + C) + g_i \tag{S5}$$

where g_i is the makeup feed.

In a closed-circuit simulation for a given residence time τ and makeup feed g_i the values of f_i and C are not known. Using the substitution $p_i^* = (1 + C)p_i$ in Eq. S2, plus Eq. S5 gives

$$p_i^* = \frac{d_{ii}g_i + \displaystyle\sum_{\substack{j=1 \\ i>1}}^{i-1} d_{ij}\left(s_j p_j^* + g_j\right)}{1 - d_{ii}s_i}, \quad n \geq i \geq j \geq 1 \tag{S6}$$

which enables p_i^* to be recursively calculated for known d_{ij}. Then $1 + C = \sum_{j=1}^{n} p_j^*$, $p_i = p_i^*/(1 + C)$, $q_i = (1 - s_i)p_i(1 + C)$, etc. The mill capacity Q is obtained from $\tau = W/F$ or

$$Q = \frac{W}{(1 + C)\tau} \tag{S7}$$

Allowance for the effect of overfilling the mill at high flow rates can be made by performing simulations assuming $U = 1.0$ to give a false mean residence time τ_f and a false capacity Q_f. If the actual mill filling is related to mass flow by

$$U = \left[(F/W_1)\tau_1\right]^{0.5}, \quad F > F_1 \tag{S8}$$

where W_1 is hold-up at $U = 1$ and flow rate F_1, then the corrected capacity is

$$Q = K_o Q_f \tag{S9}$$

where .

$$K_o = U \exp[-1.32(U - 1)] \tag{S10}$$

$$\tau_1/\tau_f = U \exp[1.32(U - 1)], \quad \tau_1/\tau_f \geq 1 \tag{S11}$$

A graphical solution for K_o knowing τ_1 and τ_f is given.

There is a maximum achievable mill capacity for a closed-circuit with a classifier which recycles all material above a certain size. This is because material can only be fed to the mill as fast as the larger sizes break down to less than the cut size, and maximum achievable rate occurs at high circulation ratio with the mill filled with these large sizes. This maximum rate corresponds to the steepest cumulative size distribution which can be obtained. Equations are given for this ultimate limiting rate and the corresponding size distribution.

References

Austin, L. G. and Gardner, R. P., 1962, "Prediction of Size-Weight Distribution from Selection and Breakage Data," *Proceedings*, 1st European Symposium Zerkleinern, H. Rumpf and D. Behrens, eds., Verlag Chemie, Weinheim, pp. 232–248.

Austin, L. G. and Perez, J. W., 1977, "A Note on Limiting Size Distributions from Closed Circuit Mills," *Powder Technology*, Vol. 16, pp. 291–293.

Klimpel, R. R. and Austin, L. G., 1971, "Mathematical Modeling and Optimization of an Industrial Rotary-Cutter Milling Facility," *Proceedings*, 3rd Symposium Zerkleinern, H. Rumpf and K. Schonert, eds., Dechema Monographien 69, Nr. 1292–1326, Verlag Chemie, Weinheim, pp. 449–473.

Luckie, P. T. and Austin, L. G., 1972, "A Review Introduction to the Solution of the Grinding Equations by Digital Computation," *Minerals Science and Engineering*, Vol. 4, pp. 24–51.

Reid, K. J., 1965, "A Solution to the Batch Grinding Equation," *Chemical Engineering Science*, Vol. 20, pp. 953–963.

7

Circuit Simulations: Results and Conclusions

A Comparison of Circuit Simulations
with the Bond Method

It will be remembered from Chapter 3 that the Bond mill sizing method gives a gross description of mill capacity for normal closed circuit wet overflow ball mills operated with a circulation ratio of $C = 2.5$ (see Fig. 2 in Chapter 3). It is not possible to perform an exact simulation corresponding to the Bond method, because important variables such as the feed size distribution, the mixture of ball sizes in the mill, the slurry density, the RTD, and the classifier partition curve are not specified in the Bond method. However, reasonable estimates were made of these unspecified factors and model simulations performed for comparison with the Bond results. The material chosen for study was a quartzitic base copper ore, since it had a Bond Work Index of 15 kWh/t, which is a representative average value (it was assumed that Wi_{test} did not vary with screen size; see Fig. 3 in Chapter 3). The characteristic breakage parameters are given in Table 1.

These values were scaled to a mill diameter of 3.05 m i.d. and $L/D = 1.6$, for a ball load of $J = 0.35$, a rotational speed of 70% of critical, and a formal interstitial filling of $U = 1$, giving $W = 7.57$ t. It will be noted that scaling with the equations of Chapter 5 gives essentially the same variation of capacity with mill diameter, L/D, and J as the Bond method (in this region of D, L/D, and J) so any reasonable values can be used for comparison purposes. The ball mixture was taken as the equilibrium mixture for a make-up ball size of 50.8 mm diam using the Bond wear law ($\Delta = 0$ in Eq. 14 in Chapter 16); see Table 2.

The residence time distribution used was the one-large/two-small model with relative sizes of 0.73, 0.135, and 0.135 (see Chapters 12 and 14). The feed size distribution assumed was 80%-passing 1 mm, with a Rosin-Rammler slope of 0.50. Finally, the classifier selectivity values were calculated using the logistic function (Eq. 9 in Chapter 13) with a bypass value of 0.3 and a Sharpness Index of 0.5 (see Chapter 13). Simulations were performed to give a range of 80%-passing sizes in the circuit product at $C = 2.5$, searching for the values of Q and d_{50} to achieve the desired specification, assuming $U = 1$ for each condition.

Table 1. Breakage Parameters for Copper Ore

Weight solid	= 1.36 kg
True specific gravity	= 2.65
Slurry density	= 72% wt solid
	= 49% vol. solid
Ball charge:	d = 25.4 mm
	J = 0.325
Mill:	D = 203 mm
	V = 5790 cm^3
	rpm = 60
Breakage parameters:	α = 0.91
	γ = 0.61
	Φ = 0.63
	β = 2.9
	δ = 0
	$S_{12\times16}$ = 0.57 min^{-1}
	a = 0.35 min^{-1}

Fig. 1 shows the results compared to the Bond calculation. As expected, the computations do not match exactly because the *standard* conditions chosen for the simulations required guesses of a number of parameters. An appropriate small change in any one of these parameters would give an exact match. Multiplying the model values by 0.80 gave a reasonable match for x_Q values smaller than about 26 μm, that is, mill capacity less than about 20 t/h. (It was later found that the Work Index of the ore was probably lower than 15 kWh/t: a value of 13 kWh/t would require a correction factor of 0.92.) In this region the Bond fineness-of-grind factors are very significant, so it is seen that these factors are a *natural consequence* of the simulation model. However, at high capacities (low residence times) the simulation assuming $U = 1$ gives capacity values up to three times as large as the Bond method. This is undoubtedly due to the neglect of the effect of overfilling of the mill at such high flow rates; it is clearly impossible to pass flow rates of 1000 t/h ($Q = 286$ t/h) through a 3-m diam mill without leading to overfilling.

This effect was incorporated into the simulation by assuming a hold-up vs. solid mass flow rate of the form of Eq. 20 in Chapter 5:

$$W = W_1(F/F_1)^{0.5}, F > F_1$$

The value of U at any solid flow rate F through the mill is then W/W_1, and Eq. 7 in Chapter 5 was used to calculate the change in S_i values and, hence, change in capacity, as discussed in "Allowance for Variation in Filling Level" in Chapter 6. For the case considered here, $U = 1$ at $\tau_1 \approx 4.8$ min, that is, $F_1 \approx 70$ t/h, $Q \approx 20$ t/h. The mill overfills at higher flow rates and the simulated Q_f values based on $U = 1$ must be multiplied by K_o to obtain the actual W values. Fig. 1 shows that allowing for overfilling now gives a reasonable agreement between model and Bond results. Using the same

Table 2. Values of Breakage Parameters of the Test Ore Scaled to a 3.05-m i.d. Mill ($J = 0.35$, $U = 1.0$, $\phi_c = 0.7$)

Ball Size Range, mm—mm	Weight Fraction of Balls	Mean Ball Size, mm	Particle Size Intervals, I.S.O. mesh	Overall $\bar{S}$, min^{-1}
50.70 – 44.45	0.411	47.625	3 / 8 in. × 0.265 in.	3.0
44.45 – 38.70	0.252	41.275	0.265 in. × No. 4	3.1
38.10 – 31.75	0.186	34.925	4 × 6	2.7
31.75 – 25.40	0.091	28.575	6 × 8	2.3
25.40 – 19.05	0.043	22.225	8 × 12	1.8
19.05 – 12.70	0.016	15.875	12 × 16	1.4
			16 × 20	1.05
			20 × 30	0.79
			30 × 40	0.58
			40 × 50	0.43
			50 × 70	0.31
			70 × 100	0.23
			100 × 140	0.17
			140 × 200	0.12
			200 × 270	0.090
			270 × 400	0.066
			38 × 26 μm	0.048
			26 × 18.5 μm	0.035
			18.5 × 13 μm	0.026

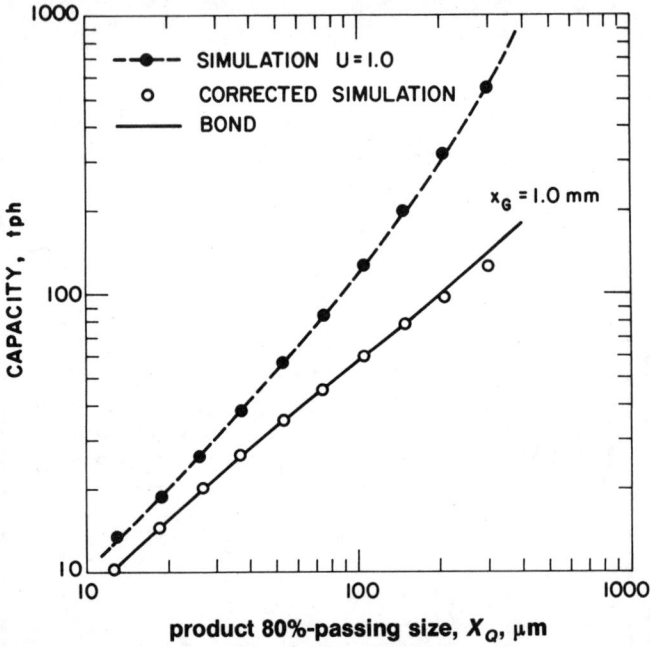

Fig. 1. Comparison of model simulation and Bond prediction for mill diameter of 3 m.

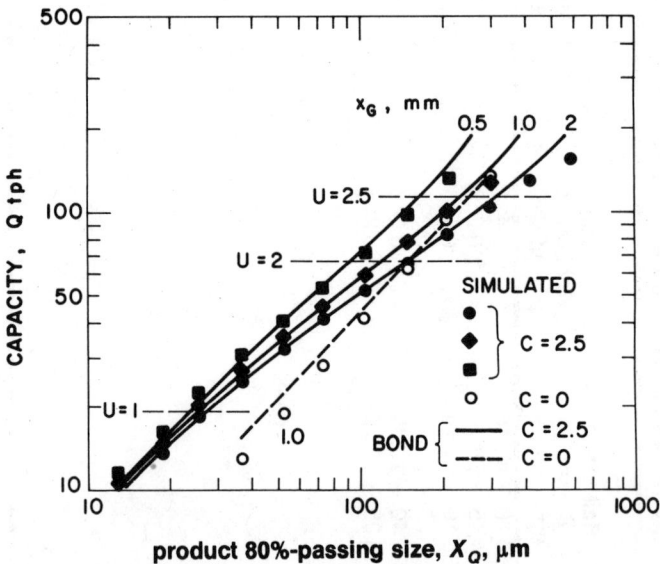

Fig. 2. Comparison of model simulations with Bond predictions allowing for overfilling at high flow rates (see Fig. 1).

correction factors gives reasonable results for Rosin-Rammler feeds with $x_G = 0.5$ and 2.0 mm, as also shown in Fig. 2. Note that overfilling represents a direct inefficiency

The model simulation and the Bond calculation also gave reasonable agreement for fine grinding of coarse feeds, but the simulation predicts higher capacities at very low reduction ratios of coarse feeds. This is not a region of much practical importance in conventional ball milling.

Extending the comparison to open circuit milling demonstrates a logical problem with the Bond method. The Bond correction factors given in Table 1 in Chapter 3 are based on the percentage of open circuit product less than the size used to determine Wi_{test}, but the value of Wi_{test} often varies only slightly with the screen used in its determination. Thus the Bond method can predict different mill capacities for the same material under the same conditions in the same mill, depending on the size used in determining Wi_{test}.

As the Bond correction factor from closed circuit ($C = 2.5$) to open circuit is 1.2 for 80% less than the screen used to determine Wi_{test}, the screen size was chosen as 150 μm (100 mesh) since this gave a factor of approximately 1.2 between the simulated closed and open circuit results [the screen size used to determine the Work Index was actually 75 μm (200 mesh)]. Fig. 2 shows the Bond and open circuit capacity prediction using the percentage minus 150 μm of the simulated open circuit size distributions to determine the Bond factor. The agreement between simulated results and Bond results is fairly close. The ratio between closed circuit capacity at $C = 2.5$ and open circuit capacity is approximately 2.0 for fine grinding where the flow rates are low enough to avoid overfilling. This is in agreement with the result quoted by Taggart (1945). Table 3 gives the mean factors for conversion of the closed circuit data to open circuit conditions, based on the simulated capacities. (The open circuit capacities are obtained from the closed circuit values by dividing by the Work Index multiplier; see Eq. 10 in Chapter 3.) Similar conclusions were obtained by Austin and

Table 3. Conversion of Wet Closed Circuit Capacities to Wet Open Circuit Values from Simulation Results

% < 150 μm (100 mesh)	Work Index Multiplier
50	0.9
60	0.9
70	1.00
75	1.10
80	1.20
85	1.35
90	1.60
92.5	1.65
95.0	1.85
97.5	2.05
> 99	2.20

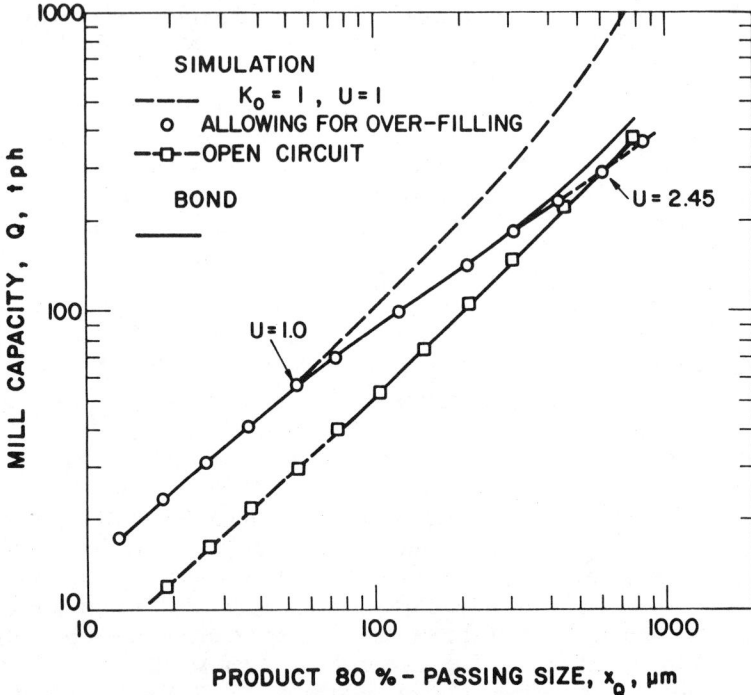

Fig. 3. Comparison of mill capacities by Bond and simulation methods ($D = 3.8$ m; $L/D = 1.5$; $J = 0.35$; 70% critical speed): feed 80%-passing 2 mm, quartz $Wi_{test} = 19$ kWh / t.

Brame (1983) who performed the simulations for quartz of test Work Index = 19 kWh/t. Their results are shown in Fig. 3.

It must be clearly understood that this approximate agreement between the simulation model and the Bond method does not mean that the so-called Bond "third law of comminution" (Eq. 7 in Chapter 2)

$$\text{specific grinding energy} = Wi\left(\frac{10}{\sqrt{x_Q}} - \frac{10}{\sqrt{x_G}}\right)$$

is a general law based on theoretical reasoning. The agreement is only obtained when the various Bond correcting factors are incorporated to modify this equation, and it is only obtained when the feed size distribution, classification parameters, etc. are appropriate values. The same Bond Work Index can be obtained for materials with different combinations of breakage parameters α and γ, but the effect of these parameters on the variation of mill capacity with x_G and x_Q is not compensatory and different results are obtained. Materials which break with a higher proportion of fines (low γ

value) or feeds with a high proportion of fines will probably give more viscous rheological properties to the mill contents and lead to higher overfilling at a specified flow rate. The Bond equation applies as an empirical approximation to a complete model over a limited range of breakage properties and x_G, x_Q values, at standard operating conditions. The effect of changing operating conditions on capacity is explored in the next sections.

The Behavior of Different Designs of Mill Circuits

We have now come to the principal thrust of this chapter: What does simulation tell us about mill circuit behavior?

Since the slowing-down and overfilling effects are a cause of *direct inefficiency*, varying with material and slurry density, the following discussion will treat both: (1) no slowing-down or overfilling effects so that the conclusions illustrate changes in *indirect efficiency* only; and (2) overfilling effects assuming the mass transfer relation given in "Allowance for Variation in Filling Level" in Chapter 6, applicable to wet overflow mills. A mill which does not overfill to the same extent due to a different discharge arrangement would give a result between these two extremes. The mill conditions used in the previous section are used for illustrative purposes, usually with a representative feed of 80%-passing 1 mm with a Rosin-Rammler slope of 0.75, with overfilling occurring at flow rates greater than 90 tph.

First (*Case 1*), consider a ball mill run at open circuit, with known values of S and B_{ij}, a constant hold-up W, and a constant feed. Fig. 4 shows the computed size distributions for plug flow, fully mixed, and for an RTD corresponding to three reactors in series, of relative size 0.7 : 0.15 : 0.15, for a value of τ of 5 min. The output rate Q is the same for each, since $Q/W = \tau$ for open circuit. It is concluded that plug flow gives a finer size distribution, with less coarse material and is, therefore, to be preferred. Physically, this arises because a more fully mixed system gives a greater chance for large material to escape before it is ground and the size distribution acted upon in the mill is that of the product size. On the other hand, in plug flow the mill is acting on coarser size distributions along most of its length and is more efficient because the breakage of larger sizes is faster. The size distributions converge at very fine sizes.

Second (*Case 2*), consider the same mill with an RTD corresponding to the three reactors in series run under open circuit conditions at different mean residence times. The result is shown in Fig. 5. As expected, longer residence time (smaller output rate Q) gives finer grinding.

Third (*Case 3*), consider the same mill run under closed circuit conditions with a classifier having a fixed set of s_i values ($a = 0.3$, S.I. $= 0.5$, $d_{50} = 150$ μm, see Chapter 13). The result is shown in Fig. 6. Although the classifier parameters are set, the system still has one degree of freedom because changing the makeup feed rate changes the residence time in the mill. For very low feed rates, the grind is so fine that the classifier hardly operates on it, so the result is identical to that of Fig. 5. As feed rate is

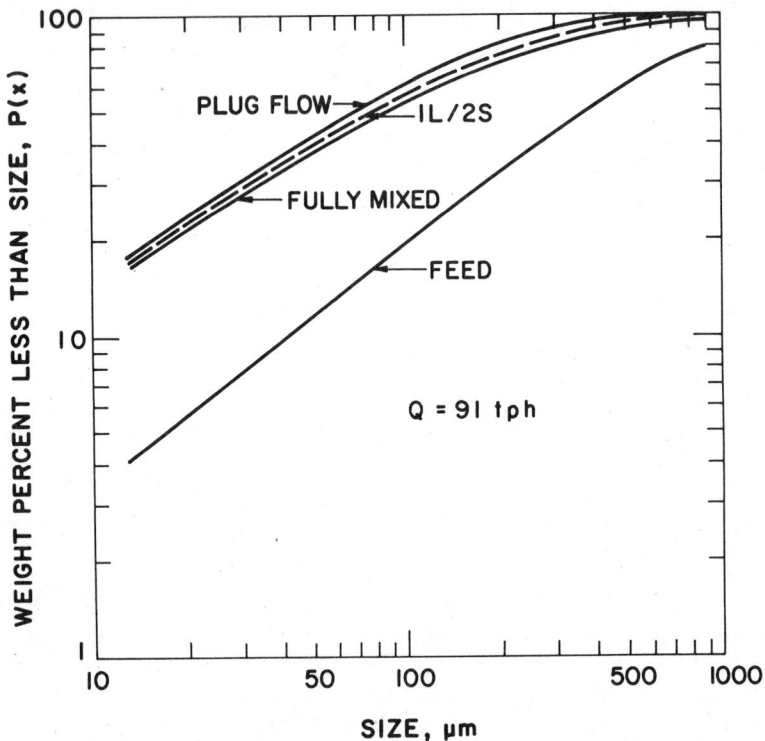

Fig. 4. *Case 1*. Comparison of size distributions from open circuit grinding for $\tau = 5$ min, for different residence time distributions.

increased, the grind is coarser, the classifier sends more to recycle, the circulation ratio goes up, and the circuit product becomes coarser. A very important point emerges from the simulation. The typical set of s_i values has values of 1 for larger sizes, which means that the classifier will not allow large material to leave the circuit. An upper cutoff size exists which behaves like a grate or screen preventing exit of larger material. Thus, as the makeup feed rate is increased a *maximum capacity* is reached, controlled by the rates at which feed sizes larger than the cutoff size can just break to less than the cutoff size. Any attempt to feed the circuit at higher rates will lead to blockage of the mill by the accumulation of unbroken large material (see "The Ultimate Limiting Case" in Chapter 6). The circulation ratio at this stage becomes large. The capacity at which this occurs is, of course, lower in the presence of the overfilling effect.

Further, it can now be seen that it is not possible to compare the result of using a closed circuit vs. using an open circuit without defining the conditions more carefully, because the size distributions are changed. Thus, the important simulation concept is reached that to compare two systems, they should be run to produce a *desired product*. To start, we can use a

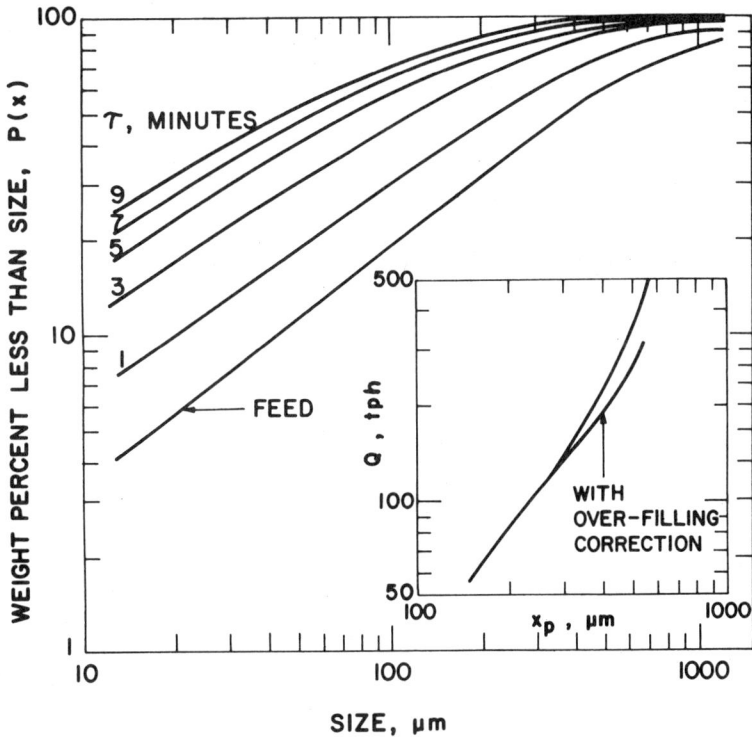

Fig. 5. *Case 2.* Size distributions for open circuit grinding at various residence times; three reactors (0.7 : 0.15 : 0.15) RTD.

single *control point*; that is, the circuits are set to produce a size distribution which passes through a specified point, the *one-point match*. Thus *Case 4* is a repetition of the open circuit results of Fig. 4, but with the output rate varied to obtain $80\% < 75\ \mu$m, giving the result shown in Fig. 7. It is obvious that there is a unique value of feed rate which gives the one-point match. In other words, the one degree of mill operating freedom that was available, namely setting the makeup feed rate, is removed by specifying the one-point match. It is clear that plug flow gives a much higher output rate than a fully mixed mill, and the fully mixed mill has a much higher percentage of fines. Accurate mill simulation can be obtained only if a reasonably accurate model for RTD is used, for this case. The overfilling factor is not significant for this fineness of grind at open circuit.

When this is repeated on the mill with the set classifier parameters used in *Case 3* for a given set of s_i values, there is again only one value of feed rate (and, hence, C) which will meet the one-point criterion. However, if we design the classifier differently, or if it is an adjustable classifier, another degree of freedom is introduced since the set of s_i values is then controllable. As will be shown later (Chapter 13), it is convenient to represent the change in classifier setting by a change in d_{50} value. Allowing d_{50} to vary allows the

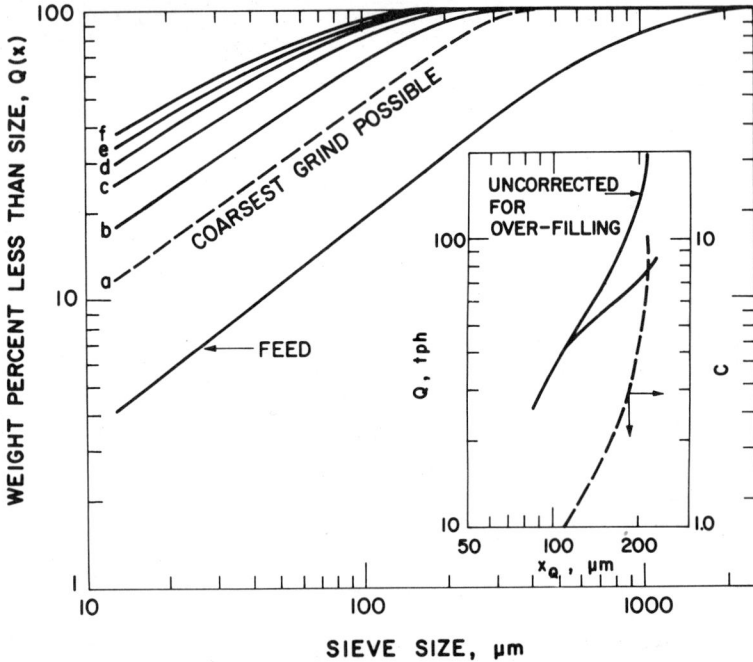

Fig. 6. *Case 3.* Size distributions for closed circuit grinding with set classifier selectivity numbers, as a function of feed rate Q; three reactor (0.7 : 0.15 : 0.15) RTD.

different mill RTDs to be compared at a given circulation ratio, $C = 2.5$, *Case 5.* Comparing the three types of RTD (see Fig 8), it will be seen that the influence of closing the circuit is to *greatly reduce* the differences between the size distributions and the output rates for the three different RTDs as compared to open circuit operation. This means that for closed circuit simulations it is not necessary to know the RTD with great accuracy. More important, comparison with *Case 4* shows that closing the circuit produces steeper size distributions and gives higher Q. The physical reason for this is that circulation removes fines and recycles coarse material, so that the residence time τ necessary to produce the one-point match is shorter for the closed circuit. The mill is acting on a mixture of sizes in the mill which is coarser on the average for closed circuit than for open and breakage energy is not used for overgrinding of fines.

Next, *Case 6* emerges. Setting the classifier at a smaller d_{50} means that it will operate with a smaller cutoff size, so that the maximum size in the product stream is reduced and more fine material is recycled. Since an extra degree of freedom has been introduced, it is possible to obtain the one-point match for a range of makeup feed rates (hence, circulation ratios), each value of Q and C corresponding to a new value of d_{50}. Fig. 9 shows the result, which is one of the most important conclusions from mill circuit

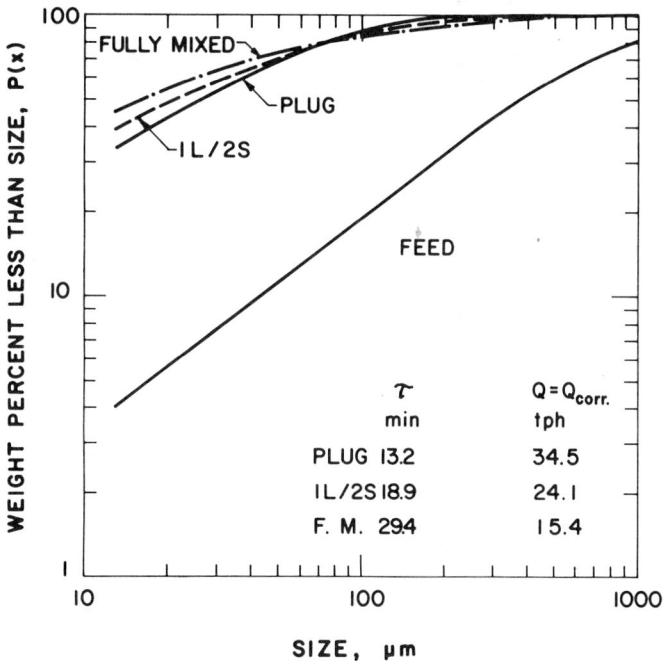

Fig. 7. *Case 4.* Comparison of size distributions for open circuit grinding to pass through the control point 80% less than 75 μm (200 mesh); see Fig. 4.

simulations. A high value of d_{50} allows the circuit to run at open circuit and the appropriate value of Q gives a *flat* size distribution through the control point. A low value of d_{50} gives very high C, and the appropriate value of Q gives a *steep* size distribution through the control point; thus, a *permitted band* of size distributions exists, between $C = 0$ and $C = \infty$. Any attempt to obtain a size distribution passing through the first control point and also through a second point lying outside of the permitted band cannot succeed. Obviously, a high value of C gives higher output rates, Q.

If the overfilling factor is included, another very important point emerges. Attempting to improve *indirect* efficiency by using high recycle will lead to increased overfilling and a reduction in *direct* efficiency: thus an optimum circulation ratio exists. (Note also that the existence of classifier bypass *a* means that there is a minimum value of C when the circuit is closed; see Chapter 13.) The increase in capacity given by closing the circuit at the optimum circulation ratio varies from a factor of about 2.0 for a fine grind to only about 10% for a coarse grind.

Let us now consider *Case 7*, where *two* control points are specified. If these points are reasonably spaced, any size distribution passing through the two points is almost identical to any other size distribution through the two points. Obviously, the second control point must lie within the permitted

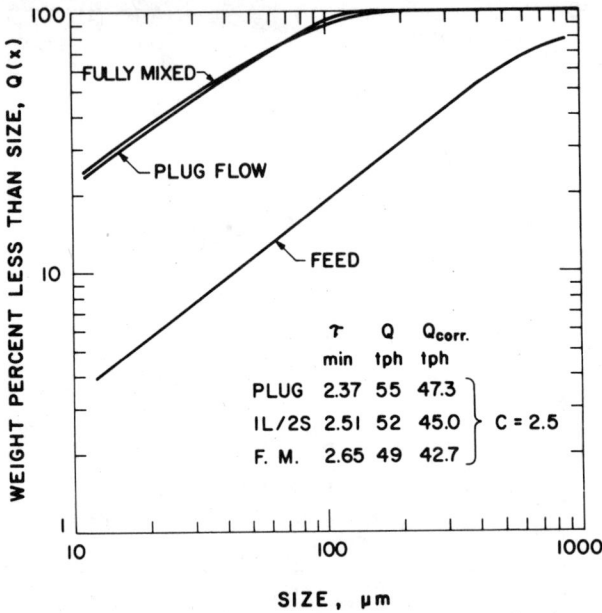

Fig. 8. *Case 5.* Size distributions through control point 80% < 75 μm at $C = 2.5$.

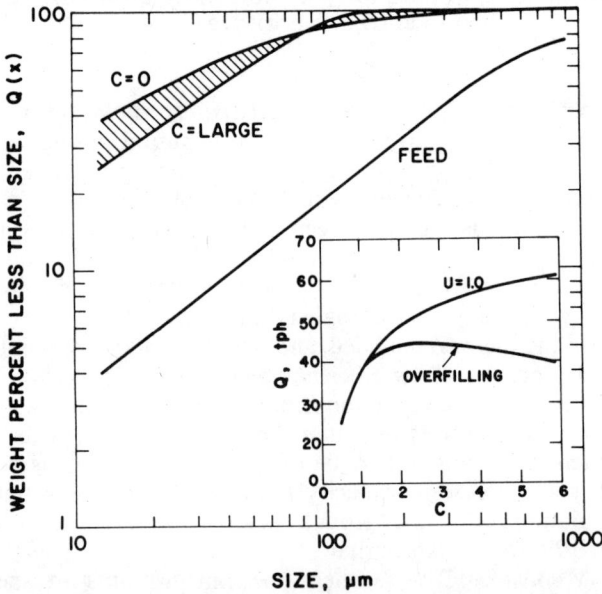

Fig. 9. *Case 6.* Permitted band of size distributions through control point 80% < 75 μm with variable d_{50} of classifier; three reactor RTD (see Fig. 6).

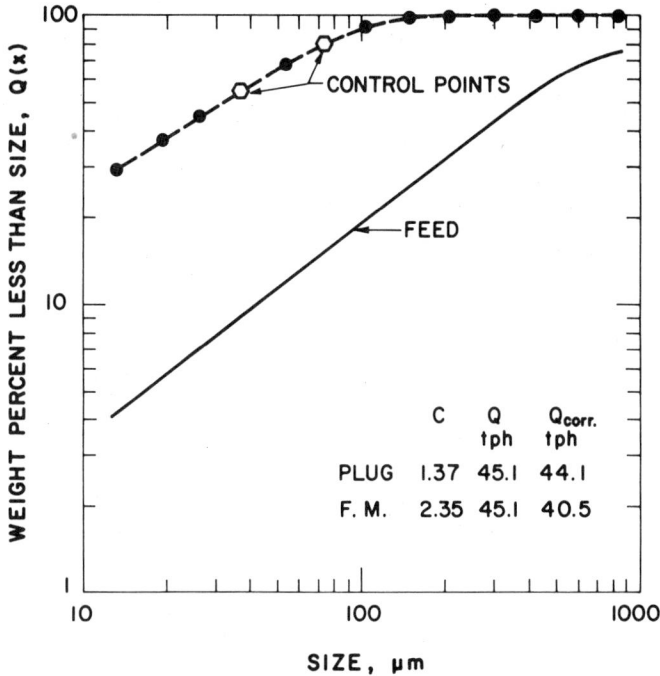

Fig. 10. *Case 7.* Size distributions through two control points
80% < 75 μm, 55% < 38 μm.

band. Specifying a feasible second point removes a degree of freedom, so that there is only one value of d_{50} (with corresponding C and Q values) which enables the two-point match to be obtained. Fig. 10 shows the result, which is again an extremely important one. It says that *if* a circuit can be run to satisfy the desired two points, *the value of Q is virtually identical for any RTD.* For a mill nearer to the fully mixed condition, a higher C is necessary to reach this state, but the Q is virtually the same as for ideal plug flow. (Note that it can be shown theoretically, as in Chapter 19, Eqs. 37 and 45, that the expressions for specific energy are not identities between the different RTDs, but in practice the computations give almost identical specific energies.) However, in the presence of the overfilling effect, the higher flow for the fully mixed case causes a bigger correction factor and the capacity is reduced; again, the effect is greater for coarser grinds.

It has been assumed in these simulations that the comparisons are made between the same geometry of mill with fixed S, B, etc. Thus, the comparisons are all made at constant power input to the mill. The specific grinding energy for the circuit product is given by $E = m_p/Q$, where m_p is constant. Consequently, in the absence of overfilling, the results of *Case 5*, the one-point match, state that E is lower for plug flow than fully mixed. The result of *Case 6*, a one-point match with varied d_{50}, states that E is lower for

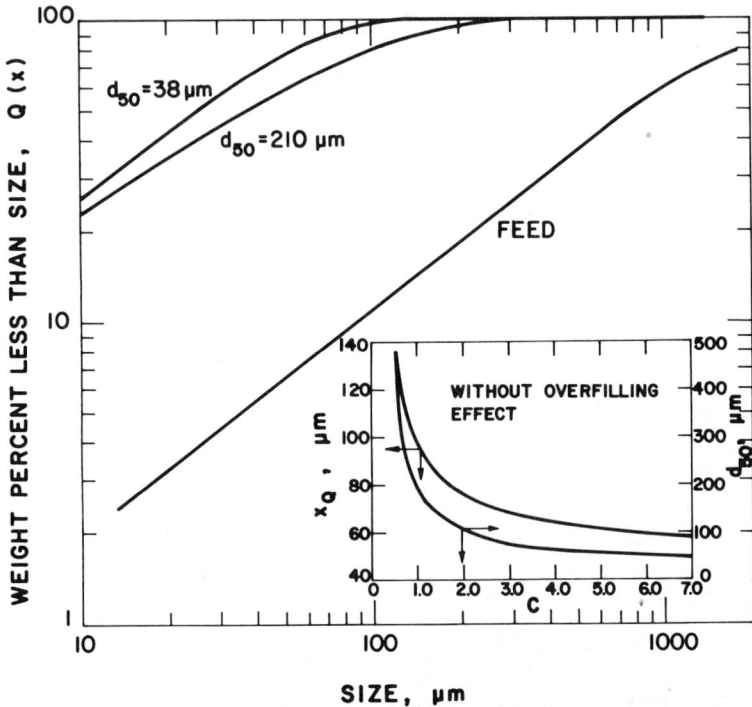

Fig. 11. *Case 8.* Variation of product size distribution at constant makeup feed rate by variation of d_{50} ($Q = 45.5$ tph); three reactor RTD (no overfilling effects included).

a higher circulation ratio. The result of *Case 7*, the two-point fit, states that output and specific energy are virtually identical whatever the RTD or circulation ratio. Thus, to produce the same size distribution takes virtually the same energy in a given mill, regardless of the RTD used, if only indirect efficiency is considered. In the presence of overfilling, a plug flow mill is more efficient than a fully mixed mill, especially for coarser grinds.

If the circuit is run with constant makeup feed Q, the size distribution can be varied by varying d_{50} (*Case 8*). The results are shown in Fig. 11. As expected, decreasing d_{50} at constant Q gives higher circulating loads, steeper size distributions, and somewhat smaller values of the 80%-passing size. Again, as discussed in *Case 3*, a limit exists where the upper cutoff size of the classifier acts like a discharge grate and the value of Q is the maximum which can be passed through the mill classifier circuit. For a constant Q, this limit is reached when d_{50} reaches a limiting small size, 38 μm, as indicated in the figure. Including the overfilling effect, the lower value of d_{50} and consequent higher value of C gives more overfilling, which reduces the fineness of grinding. Thus the variation in size distribution produced by classifier variation at a fixed Q is quite limited (except for very fine grinding at low flow rates).

The Effect of Design Classifier Efficiency

It will be shown in Chapter 13 that the Tromp curves for most industrial wet classifiers can be represented by a three parameter equation:

$$\left. \begin{aligned} s_i &= a + (1 - a)c_i \\ c_i &= f(d_{50}, \lambda) \end{aligned} \right\} \tag{1}$$

where the three descriptive parameters are a, d_{50}, and λ. For example, a suitable function for c_i could be

$$c_i = \cfrac{1}{1 + \left(\cfrac{x_i}{d_{50}}\right)^{-\lambda}} \tag{2}$$

The value of a is an apparent bypass fraction, that is, this fraction of all feed sizes is sent to the coarser stream, even the finest sizes. The smaller is a, the more efficient is the classifier, since less fines are recirculated to the mill feed. Defining a *Sharpness Index* by the ratio of the sizes at which $c_i = 0.25$ and 0.75 gives

$$\text{S.I.} = (1/9)^{1/\lambda}, 0 \le \text{S.I.} \le 1 \tag{3}$$

or

$$\lambda = 0.954/\log[\text{S.I.}]$$

for the relation of Eq. 2. This is a somewhat more readily visualized parameter than λ, since 1 represents a vertical line and zero a horizontal line. The value of d_{50} is the factor which sets where the Tromp curve lies in the particle size scale. Fig. 12 illustrates the effect of varying a and S.I. The steeper the slope of the Tromp curve, the more sharply the classifier cuts and the dotted line shows the ideally *efficient* selectivity curve, S.I. = 1. A horizontal line is an ideally *inefficient* classifier, since it represents just a stream splitter with no preferential size classification.

Let us now ask the question: What effect does classifier efficiency have on circuit performance and specific grinding energy? Figs. 13 and 14 show the results of varying the classifier efficiency parameters on the same mill as considered previously. As expected, for a one-point match at a given circulation ratio, improvement in classifier efficiency steepens the size distributions, increases output rate Q, and reduces the specific energy E, both with and without the overfilling effect. The value of bypass a has a particularly strong influence on capacity. As before, the presence of the overfilling effect decreases the predicted capacity at higher circulation ratios. In addition, when the classifier is less efficient (either via low S.I. or high a) then the optimum circulation ratio is higher. This is because the decrease in output caused by the lower classifier efficiency means that a higher circula-

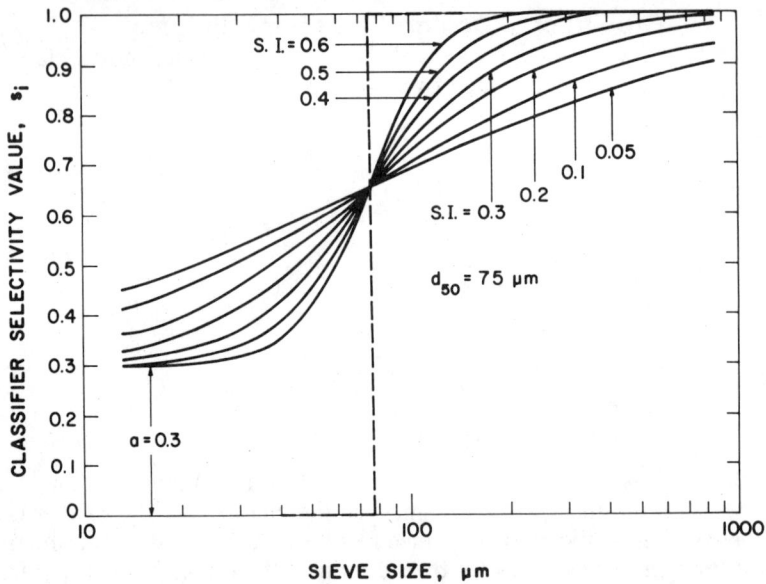

Fig. 12a. The variation of classifier selectivity values as the classifier parameter S.I. is varied: d_{50} and bypass held constant at 75 μm, $a = 0.3$.

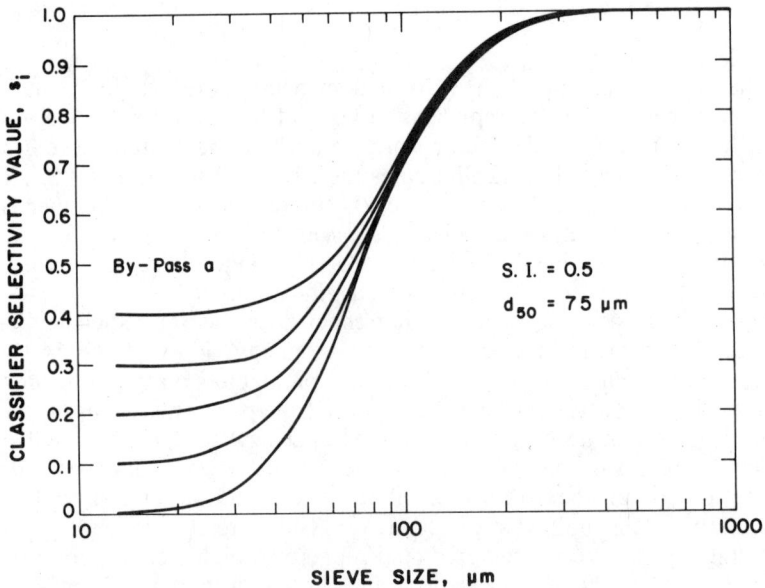

Fig. 12b. The variation of classifier selectivity values as the apparent bypass fraction a is varied: d_{50} and S.I. held constant ($d_{50} = 75$ μm, S.I. = 0.5).

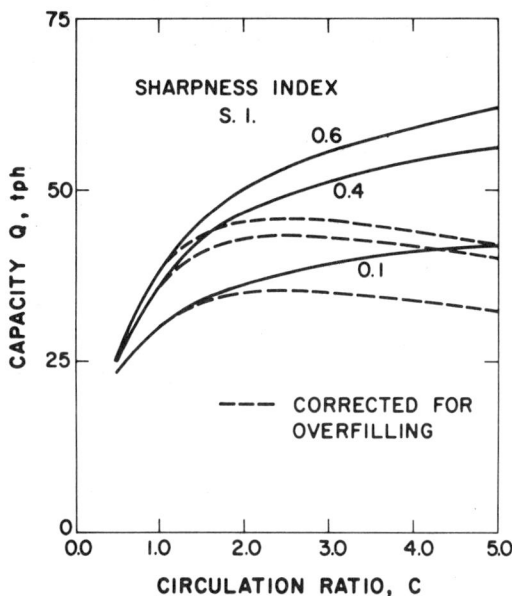

Fig. 13. Effect of Sharpness Index of classifier on circuit capacity (see previous figures for mill conditions); $a = 0.3$, product 80%-passing 75 μm.

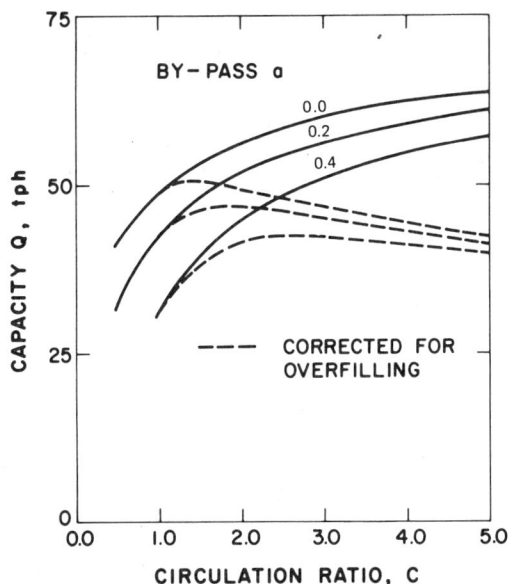

Fig. 14. Effect of bypass of classifier on circuit capacity (see previous figures for mill conditions); S.I. = 0.5, product 80%-passing 75 μm.

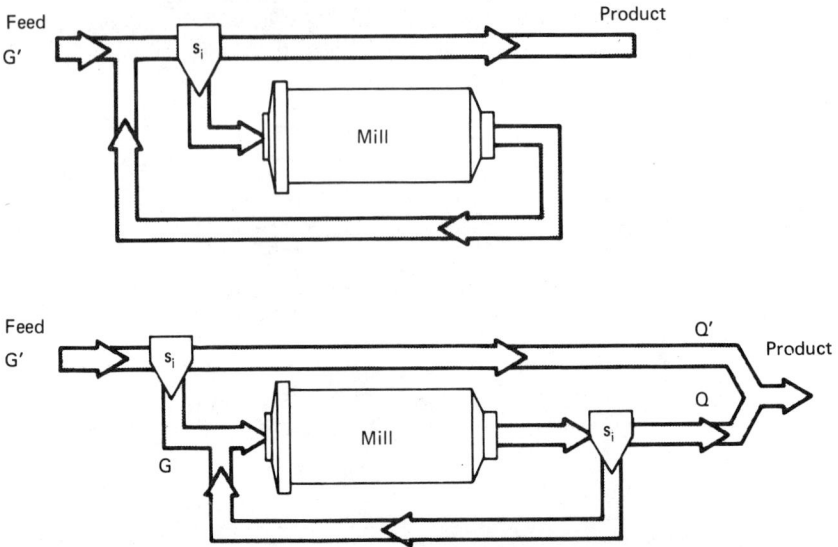

Fig. 15. The upper sketch is the reversed closed circuit. This can be treated as shown in the lower sketch where the pre- and post-classifiers have the same selectivity values. The circuit is advantageous over normal closed circuit only if G' contains a significant amount of material already fine enough so that the circuit avoids overgrinding of the fine feed material.

tion ratio is required to give the flow rates through the mill which produce overfilling.

Finally, it should be realized that it is easy to extend the simulations to the well-known type of circuit shown in Fig. 15. The incoming makeup feed G' is subjected to the classification action and the coarse product is the actual makeup feed G to the mill. Then the mill is treated exactly as described above, and the overall product is the sum of the two hypothetical streams Q and Q'. Clearly, this type of circuit is only advantageous when the incoming makeup feed contains a substantial fraction of material which is already fine enough and overgrinding of this material is to be avoided.

Simulations of this type of circuit in comparison with the same feed and product specifications and the same mill used in normal closed circuit show the following features. In a normal closed circuit operating at optimum throughput and circulation ratio, the flow rate of solid through the mill is given by $P = (1 + C)Q$, where Q is the circuit capacity. In the reversed closed circuit, only a fraction of Q is obtained from the mill, so the mill flow rate and circulation ratio are different and not necessarily optimum. If the same classifier settings are used, the lower rate of flow through the mill leads to a finer mill product and less recycle. This should be compensated by reducing d_{50} to increase the flow through the mill to bring it back to optimum. This is illustrated in Table 4, where the flow rate through the mill is held constant in the comparison. The optimum output is 38.5 t/h

Table 4. Comparison of Reverse Closed Circuit with Normal Closed Circuit: Feed 80%-passing 1 mm, Product 80%-passing 75 μm (S.I. = 0.5, a = 0.3, d_{50} varied)

Circuit	τ, min	d_{50}, μm	Q, tph	P, tph	1 + C, (P/Q)	Circuit Product %-minus: 26μm	75 μm
Normal	6.0	145	28.2	56	1.98	48.6	80
Reverse	6.0	125	30.4	56	1.83	47.0	80
Normal	3.7	111	35	90	2.58	45.3	80
Reverse	3.7	104	37.3	90	2.42	44.1	80
Normal	2.5	92.5	37.1 (39.8*)	124 (134*)	3.36	43.4	80
Reverse	2.5	87	38.5 (41.1*)	124 (134*)	3.25	43.1	80

*Without overfilling factor.

obtained in reverse closed circuit at a circuit circulating load of 3.25, for this specification of fineness of product. The coarser the product specification the bigger the advantage of the reverse circuit for the same feed. Increased output is obtained by reduction in the amount of very fine material (indirect efficiency) if the overfilling factor is made the same by operating at the same mill flow rate.

Classifier Design: The Hydrocyclone

In the two previous sections it has been assumed that the designer can specify a classifier which will give the appropriate d_{50} required by the simulation. In this section the design of the classifier to give a desired d_{50} will be illustrated using the Arterburn (1982) (Krebs) model for a hydrocyclone, described in "Quantitative Models for Hydrocyclones" in Chapter 13, that is, Eqs. 20, 22, 23, and 24 in Chapter 13.

There are five operational parameters that need to be established: the volume fraction of solids in the cyclone feed (c_f), overflow (c_o) and underflow (c_u); the circulation ratio (C); and the water split (a'). However, these are not independent of one another; the equations relating them are

$$c_u = \frac{1}{1 + a'\left[\frac{1+C}{C}\right]\left[\frac{1-c_f}{c_f}\right]}$$

and

$$c_o = \frac{1}{1 + (1-a')(1+C)\left[\frac{1-c_f}{c_f}\right]}$$

Therefore, once three of them are defined, the other two parameters must follow. However, there are additional constraints; e.g., a roping condition must be avoided. Mular's criterion (1980) for avoiding the roping condition

(see Chapter 13) is

$$c_u \leq 0.5385c_o + 0.4931$$

In addition, the volume percent solid in the feed realistically should not exceed 40%. Because of the interrelationships and the constraints, it is convenient to determine the operating parameters from a graphical presentation, such as Fig. 16.

For example, suppose it is desired to produce an overflow between 25-30 weight percent solid from a primary normal closed circuit for a solid of specific gravity $\rho_s = 2.65$. From Fig. 16, we get $c_u \leq 0.56$ for $c_o = 0.125$. For a desired circulation ratio of $C = 3$, the water split would be about 0.25 and

VOLUME FRACTION SOLIDS IN FEED, c_f

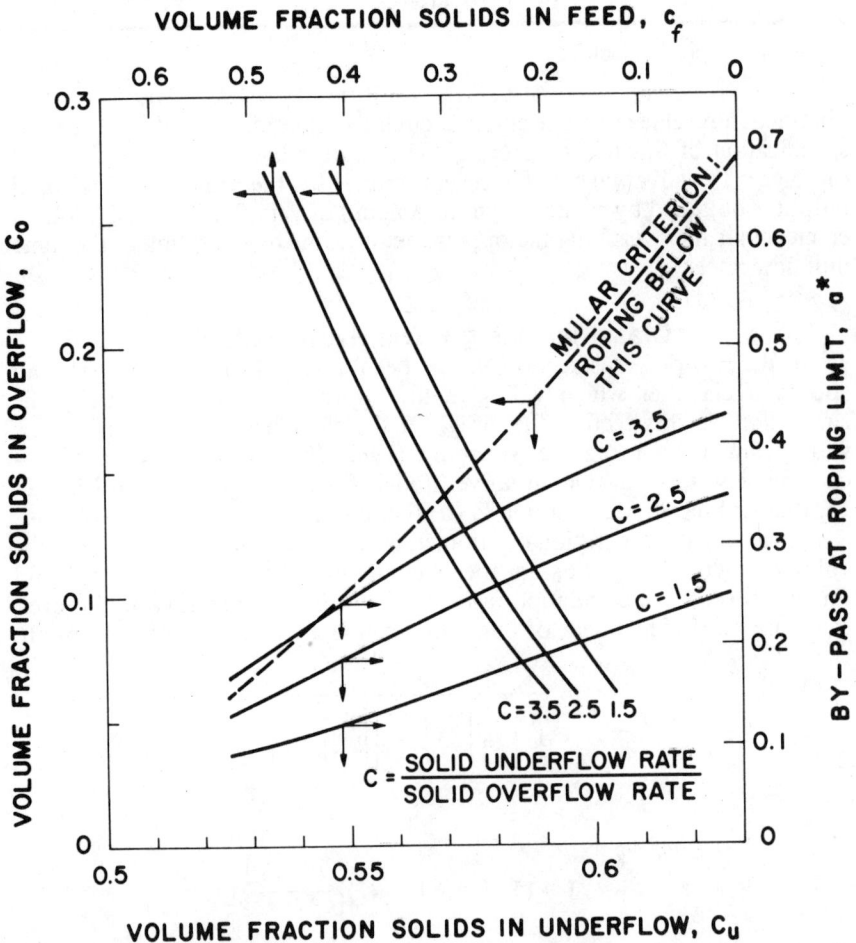

Fig. 16. Mass balances around a hydrocyclone at Mular roping limit; weight % = $100c\rho_s / [c\rho_s + (1 - c)\rho_w]$.

the feed solids content would be around $c_f = 0.3$. Now, if we assume that the apparent bypass of the cyclone is approximately equal to the water split, then the mill circuit simulator can be used to determine the required d_{50} value in a normal closed circuit configuration to achieve a one-point match while operating at $C = 3$ and $a = 0.25$. Suppose the d_{50} value determined with the mill circuit simulator is 117 μm. This value can be entered into Eq. 22a from Chapter 13 in order to calculate the cyclone diameter:

$$(\text{cyclone diam, cm}) = 0.0144(d_{50}, \mu\text{m})^{1.5}(\text{pressure drop, kPa})^{0.42}$$

$$\times (\rho_s - \rho_l)^{0.75}(1 - 1.9c_f)^{2.145}$$

Using a mid-range operating pressure drop of 69 kPa gives a cyclone diameter of 26 cm. Since a 25-cm cyclone is commercially available, this is selected.

The pulp split to underflow for the cyclone can be calculated as Eq. 20a in Chapter 13:

$$a'' = a'(1 - c_f) + c_f \left[\frac{C}{1 + C} \right]$$

which gives 0.4. From Eq. 24 in Chapter 13, the ratio of the apex diameter to the cyclone diameter is calculated as

$$\left(\frac{\text{apex diam}}{\text{cyclone diam}} \right)^2 = \frac{(0.0025)a''\sqrt{\Delta P}}{0.00215\Delta P + 0.02265}, \ \Delta P \text{ in kPa}$$

giving a ratio of 0.22, which lies between the ratio of 0.1 and 0.35 required by Arterburn (1982) for his standard cyclone geometry.

Finally the number of cyclones in parallel that would be required for a circuit output of Q is calculated as Eq. 23 in Chapter 13:

$$\text{number of cyclones} = \frac{(111.2)(1 + C)Q}{\rho_s c_f \sqrt{\Delta P} (\text{cyclone diam, cm})^2}$$

for Q in t/h. For this example

$$\text{number of cyclones} = 0.10777Q$$

which is 11 for $Q = 100$ t/h.

If a larger diameter cyclone is desired in order to reduce the number of required cyclones, then some changes in the five operating parameters are required. For example, suppose 51-cm diam cyclones are desired instead of the 25-cm diam cyclones, but the operating circulation ratio of $C = 3$ is still desired. Since doubling the operating pressure drop across the cyclone only increases the cyclone diameter by 1/3, the cyclone feed volume percent solids is the variable to change. Reducing it from 0.3 to 0.215 allows use of a 51-cm diam cyclone. However, this means that the c_o value will be reduced. For example, resetting c_u to 0.53 to take some of the increased water input gives a lower a' value of 0.18 and a c_o value of 0.077. The reduced bypass will increase the d_{50} given by the mill simulation and the computation is repeated until a reasonable set of values is obtained with respect to number and size of cyclones and cyclone feed, underflow, and overflow concentrations.

The Behavior of Operating Mill Circuits in Closed Circuit

In the previous sections it has been assumed throughout that the designer can set the conditions of design to obtain the circuits being compared. On the other hand, for an existing mill with an existing classifier the situation is quite different. The mill operator can vary the circuit behavior only within the strict limitations of the classifier response to changing conditions: varying the flow rate to an existing classifier may change the d_{50}, S.I., and a in a manner not capable of control. To allow for the variation of the classifier performance with operating conditions requires a model of the classifier which is capable of predicting the appropriate variation in the a, S.I., and d_{50} values for all operating conditions, including any necessary constraints to prevent over or underloading of the classifier. The mill classifier simulation outlined in the previous section can be used to demonstrate the effect of changing flow rate in an operating circuit with a fixed classifier.

There are two cases where this knowledge might be valuable. First, having performed the design simulation given in the previous section, a simulation with varying operating conditions can predict the range of stable operating conditions. Second, a mill classifier model can be constructed for an existing circuit (see Chapter 12) using data from a plant trial, and the simulation used to predict circuit behavior under different conditions. This case requires that the actual classifier behavior be fitted to an appropriate model. In the example below, we will use the same hydrocyclone model employed for the design simulation of the previous section.

The simulation proceeds by varying the τ value in the mill according to a search procedure until a specified C value is attained. During the search, the operating ΔP is determined from $\Delta P = \Delta P_o (\tau_o/\tau)^2$, where ΔP_o and τ_o are the design values, because the solid feed rate to each cyclone is inversely proportional to τ. The resulting value of ΔP is used in Eq. 22a from Chapter 13 to calculate the new d_{50}, assuming the slurry density to the cyclone to be held constant.

The value of the fractional pulp split to underflow then follows from Eq. 24 in Chapter 13, for the fixed cyclone geometry. An estimate of the water split to underflow follows from Eq. 20a in Chapter 13 using the specified C value. The apparent bypass a, is taken as a', and the new s_i values calculated and used in each mill circuit computation during the search iteration.

Table 5 and Fig. 17 give a typical result (in the absence of overfilling). If the slurry density to the cyclone is not changed and if the number of operating cyclones is not changed, there is only a limited range of variation

Table 5. Illustration of Operating Mill Simulation (ρ_s = 2.7) for Fixed Set of Hydrocyclones

Feed, vol. %	C	ΔP, kPa	U-Flow, wt%	O-Flow, wt%	d_{50}, μm	a, %	75 μm (200 mesh), % <	38 μm (400 mesh), % <	Q, t/h
24	2.5	69	76.6	23.0	126.8	18.6	69.7	45.1	71.9
24	3.1	92	74.6	21.0	116.4	22.0	71.5	46.1	70.85
24	3.4	10.3	73.5	20.3	112.2	23.7	72.5	46.5	70.2
22.5	1.5	34.5	75.6	27.0	145.7	15.2	69.9	46.3	65.4
22.5	2.0	51	75.3	24.0	129.2	17.1	71.4	46.8	66.5
22.5	2.5	69	74.0	21.8	118.0	19.7	73.1	47.5	66.35
22.5	3.4	104	71.0	19.1	104.3	24.8	76.1	49.0	64.8
20	1.5	34	71.0	24.4	130.3	16.5	74.8	50.1	57.6
20	1.9	47.6	70.9	22.1	118.0	18.2	76.3	50.5	58.5
20	2.5	69	69.4	19.7	105.4	21.3	78.6	51.5	58.5
20	3.4	104	66.3	17.3	93.1	26.5	81.8	53.2	57.2

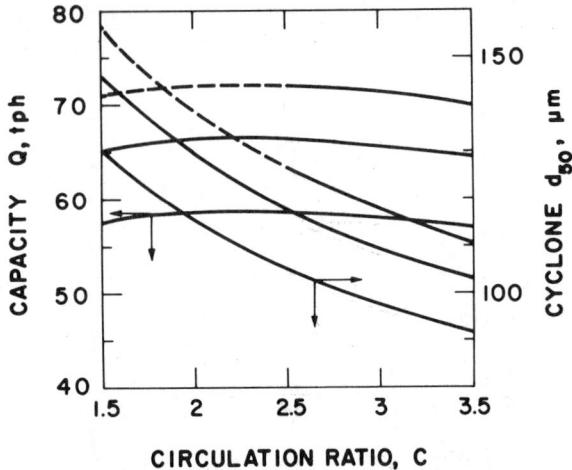

Fig. 17. Circuit capacity and variation of d_{50} for an operating mill simulation (see Table 5).

of flow rate through the circuit before: (1) the cyclone ΔP falls lower or higher than the limiting operating conditions, or (2) the cyclone underflow reaches too high a slurry density and starts to rope. This is checked by comparing the conditions to the Mular roping criterion.

If the design condition (24% by volume feed slurry for this example) approaches the roping condition (a proper design strategy), then the C value can only be increased until the upper ΔP limit is reached. This gives a decrease in the d_{50} and a values and increase in the fineness of the circuit product, plus a corresponding small decrease in the relative production rate (less than 3% overall). The water split to underflow increases and the cyclone moves away from the roping condition.

If the design condition used a lower volume percent in the feed, e.g., 22.5%, then the C value can be also decreased until the lower ΔP limit is reached. The d_{50} values decrease over the range from the lower ΔP limit to the upper ΔP limit, as does the fineness of the circuit product, and a increases. However, as shown in Fig. 17, the relative production rate increases with increasing C from the lower ΔP limit until it reaches a maximum (an increase of a little over 1%) and then decreases until the higher ΔP limit is reached (a decrease slightly less than 3%).

It is concluded that an *existing* circuit with a fixed slurry density of feed to the cyclone is relatively insensitive to variation in circulation ratio, since the increased flow through the cyclone counteracts the normal advantage of higher circulating load by increasing the bypass of fines back for overgrinding. In practice, of course, the number of cyclones, the dilution of the slurry feed to the cyclone, and the apex diameter can all be altered to extend this range. For this type of simulation it is valuable to have the simulation operated in an interactive mode with visual display of the size distributions, so that the effect of simultaneously varying the controllable parameters can be investigated rapidly.

Summary

A comparison with the Bond sizing method assuming reasonable values for unknown factors, and using the breakage characteristics of an ore of 13–15 kWh/t Work Index, showed that the simulation model agreed with the Bond method for low flow rates (fine grinds). However, in order to match the Bond result at higher flow rates it was necessary to invoke overfilling of the mill using a mass transfer law. The empirical law used was

$$U = (F/F_1)^{0.5}, \quad F \geq F_1 \atop U = 1.0, \qquad\quad F \leq F_1 \Bigg\} \tag{S1}$$

where F_1 is the flow rate through the mill which gives $U = 1$. In order to scale the same way as the Bond calculation it is necessary for F_1 to scale with mill diameter by

and

$$F_1 = (D^{3.5})(L/D), \quad D \leq 3.81\text{m} \atop F_1 = (D^{3.3})(L/D), \quad D \geq 3.81\text{m} \Bigg\} \tag{S2}$$

where F_1 is in units of metric tons per hour.

Simulations were performed to investigate variation of *indirect* inefficiency with recycle, classifier efficiency, etc. (assuming no slowing-down or large particle size effects) and these results were compared to those including the overfilling effect (*direct* inefficiency). The conclusions were:

Case 1. For open circuit operation at a given capacity, an RTD closer to plug flow gives less coarse sizes (a steeper size distribution) by comparison to more fully mixed RTDs.

Case 2. A given mill at open circuit gives a finer grind for longer residence time and lower capacity.

Case 3. For set classifier parameters, a mill in closed circuit gives coarser size distributions as capacity, and, hence, circulation ratio is increased. A *maximum allowable capacity* exists when the feed rate of sizes larger than the upper cutoff size of the classifier equals the rate at which these sizes break to less than this size in the mill.

Case 4. At open circuit, if the capacity is adjusted to give a size distribution with a *one-point match* (e.g., 80% minus 75 μm), a plug flow RTD gives the steepest size distribution and about double the capacity of a fully mixed RTD.

Case 5. In closed circuit for a *one-point match* with classifier parameters changed to give the same circulation ratio, the difference in capacity due to RTD is greatly reduced, being about 10% higher for plug flow than for fully mixed. The capacity compared to open circuit (*Case 4*) is about 70% higher (for a realistic RTD) because the size distribution is steeper and overgrinding of fines is avoided (indirect inefficiency is reduced). However, in the presence of the overfilling effect at low reduction ratios, the ratio of closed circuit capacity to open circuit capacity becomes smaller, approaching 1 at high capacity. This is because the overfilling in the mill depends on the actual flow rate through the mill, which is higher for closed circuit operation. The decrease in indirect inefficiency is counteracted by the increase in direct inefficiency.

Case 6. Classifier design permits different d_{50} values to be employed, giving different classifier selectivity effects. It is then possible to obtain a *one-point match* at varying circulation ratios. It is found that there is a *permitted band* of size distributions. Capacity is highest, and the size distribution steepest, for high circulating loads obtained by reducing the d_{50} value. It is not possible to obtain a steeper size distribution than that given by the natural breakage characteristics of the material. Capacity is lowest for high values of d_{50}, which gives open circuit operation.

In the presence of overfilling, the permitted band of size distributions is the same, but the variation in capacity with circulation ratio is much less. It is possible for capacity to reach a maximum at an optimum circulation ratio, since the overfilling caused by higher circulating load overweighs the decrease in overgrinding of fines.

Case 7. A second matching point can be specified within the permitted band, corresponding to a particular d_{50}, which defines a complete size distribution (*two-point match*). In this case mill capacities and specific grinding energies are virtually independent of RTD for fine grinding with no overfilling. In the presence of overfilling (low reduction ratio) the higher circulation ratio of the fully mixed RTD required to give the two-point match leads to a lower capacity.

Case 8. Closed circuit operation with a set circuit capacity (constant Q) but varying d_{50} gives the steepest and finest size distribution at higher circulating loads. In the presence of overfilling an optimum circulation ratio exists which gives the finest size distribution.

The influence of classifier efficiency was investigated by varying the apparent bypass fraction *a* and Sharpness Index (S.I.) of the selectivity curve

$$s_i = a + (1 - a)c_i \tag{S3}$$

$$c_i = \frac{1}{1 + (x_i/d_{50})^{-(0.954/\log \text{S.I.})}} \tag{S4}$$

Classification is more efficient for low values of *a* and high values of S.I. For a one-point match, decreasing *a* or increasing S.I. gives steeper size distributions and increased capacity by reducing the circulation of fines and overgrinding. If a two-point match comparison is made, the same capacity and specific grinding energy is obtained for efficient and inefficient classification in the absence of overfilling, but the higher circulating load produced by inefficient classification causes overfilling, reduced capacity, and higher specific grinding at low reduction ratios.

If the makeup feed contains a substantial fraction of material already fine enough, it is advantageous to use a reversed closed circuit.

The above comparisons were based on the ability to design and install classifiers with different d_{50} values. The procedure for combining a classifier model with a mill model to give the classifier design parameters was illustrated using the Arterburn (Krebs) model for hydrocyclones and the Mular criterion for the roping limit. Most efficient operation occurs near the roping limit where bypass is at a minimum.

In an operating hydrocyclone circuit, varying the feed rate to the circuit will give automatic changes to d_{50} and bypass. When the flow rate through the hydrocyclone in increased, the d_{50} and the bypass decrease (if the number of operating cyclones, the slurry density of the feed to the cyclones, the apex diameter, etc. are kept constant). The decrease in d_{50} acts counter to the decrease of grinding time, so that the product does not coarsen as much as expected. The allowable variation is restricted by the cyclone pressure drop reaching too low or too high values for correct operation, or by the cyclone reaching the roping limit. The simulations show that only small changes in mill capacity are feasible and that size distributions do not vary much even though substantial changes in circulating load are produced.

References

Arterburn, R. A., 1982, "The Sizing and Selection of Hydrocyclones," *Design and Installation of Comminution Circuits*, A. L. Mular and G. V. Jergensen, II, eds., AIME, New York, pp. 592–607.

Austin, L. G. and Brame, K. A., 1983, "A Comparison of the Bond Method for Sizing Wet Tumbling Ball Mills with a Size-Mass Balance Simulation Model," *Powder Technology*, Vol. 34, pp. 261–274.

Mular, A. L. and Jull, N. A., 1980, "The Selection of Cyclone Classifiers, Pumps and Pump Boxes for Grinding Circuits," *Mineral Processing Plant Design*, A. L. Mular and R. B. Bhappu, eds., 2nd ed., AIME, New York, pp. 376–403.

Taggart, A. F., 1945, *Handbook of Mineral Dressing: Ores and Industrial Minerals*, Wiley, New York.

8

Links Between the Old and the New: The Simple Solutions to the Grinding Equations

Introduction

The results presented previously have been obtained from solutions of the grinding equations which require a computer to handle the amount of calculation involved. However, it is necessary to examine the size-mass balance equation to see whether it can be solved directly to give Bond's law, Rittinger's law, etc. This chapter, then, will present a brief summary of mathematical investigations of the batch/plug equation and the fully mixed grinding equation. It will be shown that there is a very simple solution to each case when S and B have certain forms. Unfortunately, real values of S and B do not fit the required forms, so the simple solution can only be used as a very approximate model for simulation. However, because they can be applied to plant data without the effort of determining real S and B values in the laboratory, these simple solutions will be used in a number of places in this book and in fact a whole chapter, Chapter 19, is devoted to development of simulations using the simple model. At this stage the following features will be explored:

1. Balances on finite size intervals vs. balances on differential size intervals.

2. Solution of the batch grinding equation, including the simple solution.

3. The relation between this solution and the old *laws* of grinding.

4. Corresponding solutions for steady-state continuous grinding.

Finite Size Intervals vs. Continuous Functions?

In many of the original formulations of the grinding equations it was assumed that the first-order breakage law applies to a *differential* range of particle size, dy, so that, for example,

rate of disappearance of size y to $y + dy$ due to breakage

$$= S(y)\, dP(y, t)W$$

$$= S(y)\frac{\partial P(y, t)}{\partial y}\, dy W$$

[It was also then assumed that the progeny fragment distribution is continuous, so that breakage of size y to $y + dy$ gives a cumulative B function of $B(x, y)$.] It must be immediately noted, however, that there is no experimental proof that this type of first-order relation holds. The experimental evidence for first-order breakage is all from tests on $\sqrt[4]{2}$ or $\sqrt{2}$ size intervals. If it were possible to prepare narrow size fractions, would they break according to a first-order disappearance law? Actually, the strengths of a number of particles all of size y are very *unlikely* to be the same, as crushing experiments always demonstrate (see Fig. 1). Also, when a collection of particles in a narrow size interval is broken in a tumbling ball mill, it seems likely that there is a wide distribution of applied forces, ranging from no more than the weight of part of the bed acting on a particle to an impactive force due to a ball dropping from almost the full height of the mill. It is the overall effect of the distribution of particle strength and the distribution of applied force which produces breakage in a mill.

We are thus faced with a logical problem. It is clear that if too big a size interval is chosen for a first-order disappearance test, then the test will give non-first-order breakage. On the other hand, a very narrow size range should give a distribution of strengths, again leading to non-first-order breakage.

To investigate this dilemma, Austin, Shoji, and Everell (1973) and Gardner and Austin (1975) have explored the effects of applying a distribution of forces to particles with a distribution of strengths. To summarize their discussions in nonmathematical terms, it was concluded that a distribution of basic strengths of the particles is expected to lead to non-first-order effects, with weaker material breaking first. On the other hand, the distribution of applied force counteracts this, because whether a given particle is subjected to a low or high force in the mill is random with time. Similarly, the distribution of strengths noted in crushing tests on irregular particles might be largely due to *orientation* effects; that is, a given particle would break at different strengths depending on how it was oriented in the crushing action. In a mill acting on many particles the orientation effect would be random, again tending to give first-order breakage rates. Another relevant factor is that the crushing strength appears to be proportional to particle volume, so that the wide distribution shown in Fig. 1 can be partly ascribed to the change of volume of $4:1$ between the smallest and largest particles tested.

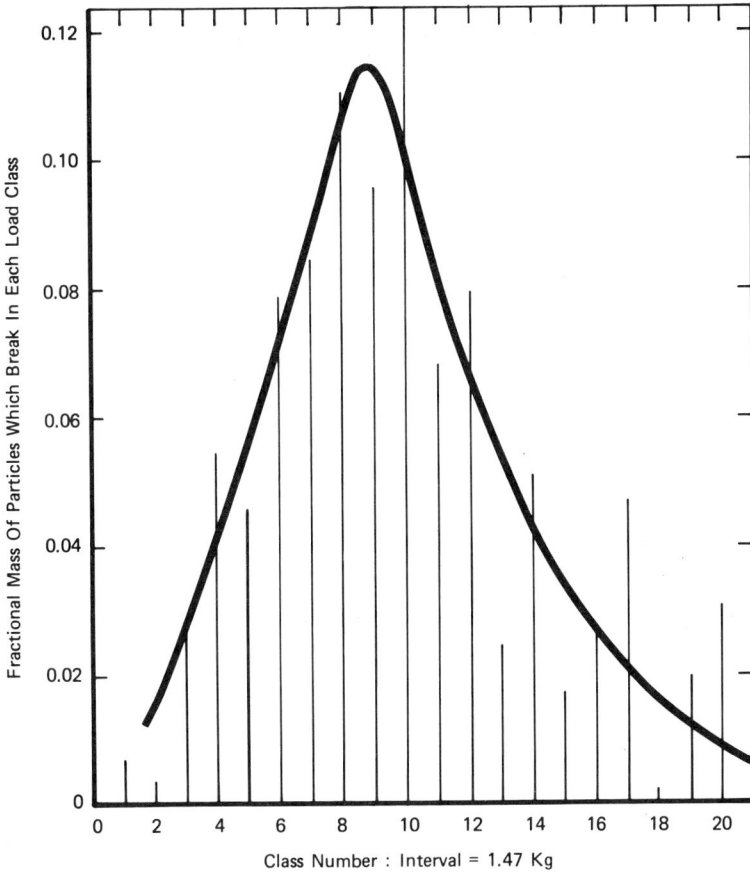

Fig. 1. Experimental distribution of failure loads in slow compression testing of particles of limestone, 6.3 × 4.75 mm, (1/4 in. by 4 US mesh).

To investigate the effect of interval size by simulation, Austin and Luckie (1970–71) divided a normal sieve interval into 10 subintervals and assumed that the values of S and B for each subinterval varied with particle size in the same way as S and B vary between $\sqrt{2}$ sieve intervals; that is $S = ax^{\alpha}$, $B(x, y) = (x/y)^{\beta}$, $x/y > 0.5$ (see "Variation of Breakage with Particle Size" in Chapter 5). Assuming first-order within each subinterval and solving the grinding equation over the 10 subintervals gave the result shown in Fig. 2. The overall rate of breakage for the whole sieve interval appears to *increase* somewhat with time instead of decrease because breakage of the smaller (stronger) particles left at long times throws a bigger fraction of material out of the size interval. *This would clearly compensate a decrease in rate due to weaker material being broken sooner.*

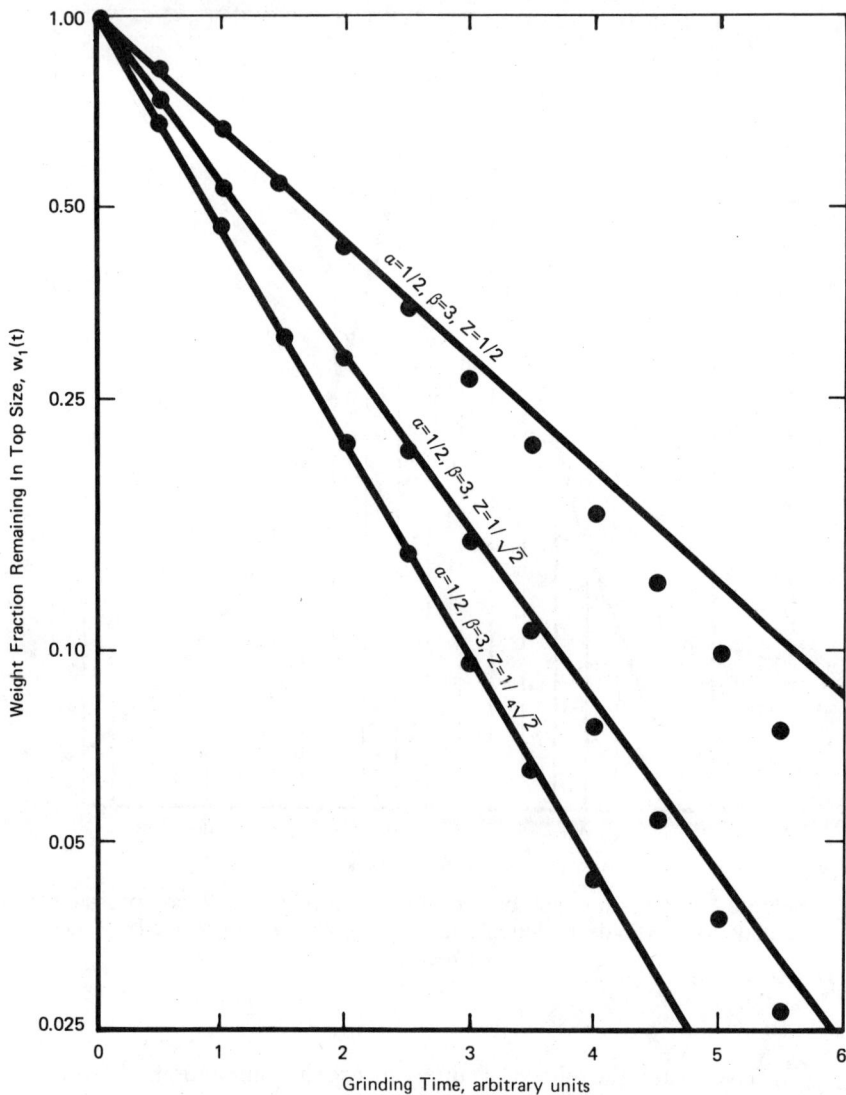

Fig. 2. Simulated deviations from first-order breakage as a function of interval size Z: $\alpha = 1/2$, $\beta = 3$.

The net result of all of these effects appears to be a set of compensations such that breakage of $\sqrt[4]{2}$ or $\sqrt{2}$ size intervals behaves in a first-order manner. This is an empirical fact and it does not follow that the first-order law would apply to a very narrow size range of material. Neither does it prove that the B values from a breaking sieve interval are the same toward the end of grinding as at the beginning: again, however, a set of compensations appear to give this result, at least as a reasonable approximation. On the other hand, there is ample evidence that a $2:1$ interval is normally too big to use for a size-mass balance with constant S and B values.

In practice, the S and B values are usually defined and measured on $\sqrt{2}$ intervals. Although the S values plotted vs. sieve size as in Fig. 2 in Chapter 5 are connected by a smooth curve as if they were a continuous function of size, only the set of n values of S_i are used in the computations. Thus we always use interval values, even though these values are represented by the continuous function $S = ax^\alpha$, because the function is used only in the form $S_i = ax_i^\alpha$ where x_i is the upper size of each of the size intervals.

Solution of the Batch / Plug Flow Grinding Equation:
The *Simple Solution*

If it is *assumed* that S and B values can be defined for differential size intervals, Filippov (1961), Austin, Luckie, and Klimpel (1972) and King (1972–73) have explored solutions to the resulting batch grinding equation (see "The Size-Mass Balance" in Chapter 4)

$$\frac{\partial P(x, t)}{\partial t} = \int_{y=x}^{x_{\max}} B(x, y) S(y) d_y P(t) \tag{1}$$

The solution contains hypergeometric functions which have to be evaluated and it is cumbersome to use. As Gupta and Kapur (1976) have shown, the definition of S and B values as continuous functions $S(y)$ and $B(x, y)$ which do not vary with grind time does *not* lead to values of S_j and B_{ij} which are constant with time, thus violating experimental fact. For this reason, we will not report the solution here but, instead, give the very simple form which results when the Gaudin-Meloy (1962) Herbst-Fuerstenau (1968) *compensation condition* applies.

The starting point is the equation of batch grinding (see Eq. 4 in Chapter 4) in the form

$$\frac{dP(x_i, t)}{dt} = \sum_{j=1}^{i-1} B_{ij} S_j w_j(t), \quad n \geq i > 1 \tag{2}$$

To this we add the *compensation condition*; this was discussed by Gaudin and Meloy, and Herbst and Fuerstenau were the first to use it to develop a method for obtaining S and B values. The condition is that

$$S_j B_{i,j} = \text{function of } i \text{ only}, \quad i \geq j \tag{3}$$

Physically, this means that the specific rate of appearance of material less than size x_i from breakage of size interval j does *not* depend on the size of j. It is a compensation condition because S_j is larger for large particle sizes, but the fraction of breakage to less than size i is smaller, and the two effects might *exactly compensate* for all sizes of j.

For example, this exact compensation would occur if the form of S values fitted the relation

$$S_j = ax_j^\alpha \tag{4}$$

and the form of B was normalized and was fitted by

$$B_{i,j} = (x_{i-1}/x_j)^\alpha, \quad i > j \tag{5}$$

for then

$$S_j B_{ij} = ax_{i-1}^\alpha, \quad \text{not dependent on } j \tag{6}$$

If the values of S and B are such to satisfy the compensation condition, Eq. 3 can be put as $S_j B_{ij} = S_{i-1} B_{i,i-1}$ by letting $j = i - 1$, since any value of j less than i can be chosen. But since $B_{i,i-1} = 1$ by definition, then

$$S_j B_{i,j} = S_{i-1} \tag{7}$$

This can now be inserted into Eq. 2 to give

$$\frac{dP(x_i, t)}{dt} = \sum_{j=1}^{i-1} S_{i-1} w_j(t) = S_{i-1} \sum_{j=1}^{i-1} w_j(t)$$

and since $\sum_{j=1}^{i-1} w_j = 1 - P(x_i)$,

$$\frac{dP(x_i, t)}{dt} = S_{i-1}[1 - P(x_i, t)] \tag{8}$$

Separating and integrating from $t = 0$ to $t = \tau$, with an initial cumulative feed size distribution of $P(x_i, 0)$, gives

$$\left. \begin{aligned} P(x_i, \tau) &= 1 - [1 - P(x_i, 0)]\exp(-S_{i-1}\tau), \quad n \geq i > 1 \\ &= 1, \qquad\qquad\qquad\qquad\qquad\qquad\qquad i = 1 \end{aligned} \right\} \tag{9}$$

This is the only simple closed analytical solution to the equation of batch grinding and we will refer to it as the *simple solution*. It also immediately applies to continuous *plug flow* grinding at constant hold-up level, e.g., an overflow discharge ball mill, as discussed in Chapter 6, for a residence time of τ. Again, the physical meaning of Eq. 8 is quite clear: It states that the rate of production of less-than-size x_i material is dependent *only* on the amount of material, $[1 - P(x_i, t)]W$, greater than this size, *not* on the size distribution making up this amount.

Several workers (Razumov et al., 1969; Kapur, 1970) have used a grinding equation of the form

$$dR(x, t)/dt = k(x)[1 - P(x, t)] = k(x)R(x, t)$$

This is identical to Eq. 8 except that it is implicit that it can be applied at any size x, whereas Eq. 8 should strictly be applied only at the values of x in the size interval (sieve) sequence, since it is S_i which is measured by a direct first-order test on material of size i. If Eq. 9 were exactly applicable to batch grinding, we could drop the index i, set $S_{i-1} = S(x)$, and consider the solution as continuous in $P(x)$ and $S(x)$

$$1 - P(x, \tau) = [1 - P(x, 0)]\exp[-S(x)\tau], \quad 0 \le x \le x_{max} \quad (9a)$$

To use Eq. 9a, it is not necessary that the first-order hypothesis apply to a differential size element. As long as it applies to, say, a $\sqrt{2}$ interval, then the special solution of Eq. 9 is valid for $S_j B_{ij} = S_{i-1}$. Since the definition of an interval is quite arbitrary, then Eq. 9 can be applied to *any* value of x and the equation then acts as a definition of an *apparent* $S(x)$ which has the values of S_{i-1} for $\sqrt{2}$ intervals when $x = x_i$.

To test whether a set of batch grinding data fits Eq. 9, the results can be plotted as $\log[R(x_i, t)/R(x_i, 0)]$ vs. t and should give a straight line for the various values of i: the values of S_{i-1} can be determined from the slopes. Unfortunately, grinding data does *not* fit this expression, except as a rough approximation, and the criterion $S_j B_{ij} = S_{i-1}$ is not satisfied. However, it is often approximately satisfied for milling of small feed sizes. Another informative plot is the Weibull plot, which is a Rosin-Rammler-percent scale vs. a log-time scale. Eq. 9 in this form is

$$\log(2.3\log[R(x, 0)/R(x, t)]) = \log t + \log S(x)$$

and a plot of $R(x, t)/R(x, 0)$ vs. t on these scales would give a slope of -1. Fig. 3 shows the result for the data of Fig. 1 in Chapter 2; since the starting feed is in the top two intervals, the amount greater than size x at zero time is 1 for all other sizes, that is, $R(x, 0) = 1$.

Two important conclusions can be drawn from this figure. First, for small sizes and low grinding times the slope does tend to one. Second, at longer grinding times, the slopes are *always greater* than unity. This means that the forms of S and B are such that $S_j B_{ij}$ *increases* as the breaking size j approaches the size being considered, thus giving a rate of increase of less-than-size-x material higher than that predicted by Eq. 8. This is shown in Table 1. Note that the slope should be exactly 1 for the top size interval because Eq. 9 becomes $R(x_2, t) = R(x_2, 0)\exp(-S_1 t)$, and since $R(x_2, t) = w_1(t)$ then $w_1(t) = w_1(0)\exp(-S_1 t)$. Thus, whether the compensation condition applies or not, the top size interval is independent of B and must give a slope of 1 if breakage is first-order.

Links Between the Old and the New

In Chapter 2, the results of a typical batch grinding test (Fig. 1) were analyzed to see whether they fitted Charles' law, the pseudo-Rittinger relation, the Bond batch grinding law, or others. In Chapter 4, it was shown that this set of data could be accurately computed from the solution of the batch grinding equation, using a set of S values plus a set of normalized B

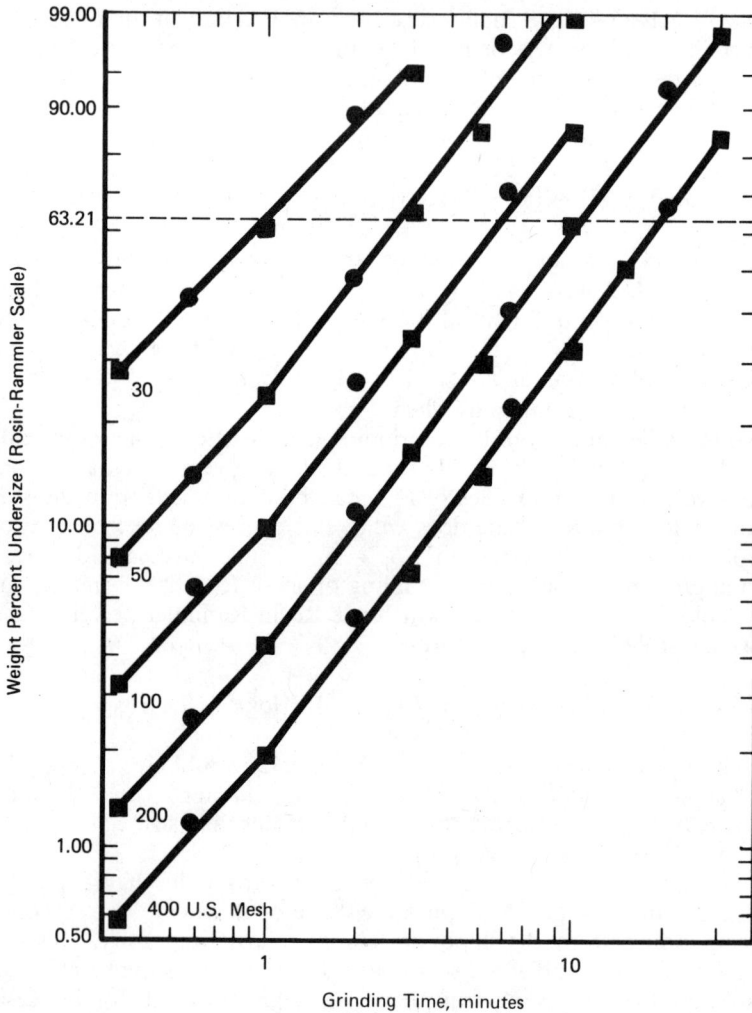

Fig. 3. Weibull plots for wet and dry ball milling of 850 × 600 μm (20 × 30 US mesh) quartz (see Fig. 1 in Chapter 2): ●, wet (45% solid by volume); ■, dry 1.8 time-scale factor applied (wet grind 1.8 times as fast as dry).

Table 1. Values of $S_jB_{7,j}$ for the Data of Fig. 1 in Chapter 2, $j<i$

j	= 1	2	3	4	5	6
S_j, min^{-1}	0.524	0.397	0.301	0.228	0.173	0.131
$B_{7,j}$	0.063	0.102	0.153	0.250	0.432	1.0
$S_jB_{7,j}$, min^{-1}	0.0330	0.0404	0.0461	0.0570	0.0747	0.1310

values. Immediately, then, we have a connection between the old laws and the size-mass balance approach: They both fit the same set of data so they are compatible and not exclusive of one another. In this section, however, the correlation between the two approaches is examined in a somewhat more extensive manner, using the *simple solution* to the batch grinding equation, which is approximately (very approximately) valid for ball milling data.

First, we can show that a Rosin-Rammler size distribution results from this special solution when $S(x) = ax^\alpha$; that is, $S(x)$ is a power function of particle size. Inserting Eq. 4 into Eq. 9,

$$R(x, t) = R(x,0)\exp\left[-(x/x_0)^\alpha\right] \tag{10}$$

where

$$x_o = (1/at)^{1/\alpha} \tag{10a}$$

x_o is the 36.79% above-size point; that is, x_o is the size on the distribution at which 36.79% of the weight is greater than x_o and 63.21% of the weight is less than x_o. For values of x less than the finest size in the feed, $R(x,0) = 1$ and Eq. 10 becomes the well-known Rosin-Rammler size distribution. If the feed size is itself an *R-R* distribution of characteristics α and x_{oF}, then $R(x,0) = \exp[-(x/x_{oF})^\alpha]$ where x_{oF} is the 36.79% above-size point of the feed. Eq. 10 then becomes

$$R(x, t) = \exp\left[-(x/x_o)^\alpha\right] \tag{11}$$

where

$$x_o = \left[\frac{1}{at + (1/x_{oF})^\alpha}\right]^{1/\alpha} \tag{11a}$$

which is again a Rosin-Rammler distribution.

Second, these equations state that x_o varies with grinding time according to

$$a(t_2 - t_1) = \frac{1}{x_{o_2}^\alpha} - \frac{1}{x_{o_1}^\alpha}$$

which is one of the grinding laws examined in Chapter 2. Third, if Eq. 10 applies exactly, then it is readily shown that $2.3\log(1/0.2) = (x_P/x_o)^\alpha$ or $x_P^\alpha = 1.61x_o^\alpha$, and

$$a(t_2 - t_1) = 1.61\left(\frac{1}{x_P^\alpha} - \frac{1}{x_F^\alpha}\right) \tag{12}$$

$$E_B = (1.61m_p/aW)\left(\frac{1}{x_P^\alpha} - \frac{1}{x_F^\alpha}\right) \tag{12a}$$

where x_P is the 80%-passing size at time t_2 and x_F is the 80%-passing size at time t (if t_1 is zero, x_F is the 80%-passing size of the feed). This is, therefore, a general form of Bond's batch grinding equation (Holmes, 1957) retaining α as general instead of taking it as one half for all cases. [Note that the Bond (1965) method of plotting data is only a different form of Rosin-Rammler plot (Austin and Luckie, 1972).] Eq. 12 also gives the pseudo-Rittinger Law for $\alpha = 1$, and it is found experimentally that α is close to 1 for many materials. Other materials, however, give values close to $1/2$ (see Chapter 5).

Fourth, Eq. 10 reduces to

$$1 - R(x, t) = P(x, t) \simeq (x/x_o)^\alpha$$

for small sizes where $x \ll x_o$ and there is no feed material of size x, $R(x, 0) = 1$. From Eq. 10a,

$$P(x, t) \simeq atx^\alpha, \quad x \text{ small} \tag{13}$$

Thus, the Schuhmann distribution results, with $a_s = at$ and since $P(x, t_1) = ax^\alpha t_1$, $P(x, t_2) = ax^\alpha t_2$, and $P(k, t) = 1$,

$$a(t_2 - t_1) = \frac{1}{k_2^\alpha} - \frac{1}{k_1^\alpha}$$

Therefore, it is seen that Charles' law is also a consequence of the *simple solution*, and that the empirical α_c in Charles' law (see "Charles' Law: Schuhmann Size Distributions and Rosin-Rammler Size Distributions" in Chapter 2) is the same as the α in $S(x) = ax^\alpha$.

Finally, Eq. 13 shows that

$$P(x^*, t) = (ax^{*\alpha})t, \quad x^* \text{ small} \tag{14}$$

which means that the production of material less than size x^* is linear with time, or

$$\frac{dP(x^*, t)}{dt} = \text{constant}$$

This is the rate of production or zero order law.

It is very important to realize that although Bond's batch grinding equation can be derived from the *simple solution* as above, with size distributions which are parallel straight lines on Rosin-Rammler plots, it may apply *better* to the 80%-passing points without the rest of the size distributions fitting this pattern. In this case, the values of x_P, x_F are then actual values, not values from the extrapolation of the R-R lines to 80%. α is then just the best number to make Eq. 12 work and is not necessarily the exponent α in $S(x) = ax^\alpha$. Thus the pseudo-Rittinger law can apply with

$\alpha = 1$, and the Bond law with $\alpha = 1/2$, *for the same set of data* because extrapolated x_{80} values are used in the pseudo-Rittinger expression and actual values in the Bond.

Again, because the *simple solution* does *not* fit real data, the derivation of Eq. 13 has to be qualified. Real values of S_j and B_{ij} for ball milling are such that the product $S_j B_{ij}$ *increases* as the breaking size j approaches the size being considered. This has the compensatory effect of making Eq. 13 apply to much larger values of x than the *simple solution* would predict, so that the Schuhmann distribution is experimentally applicable over a wider size range than expected, often up to 70% of the size distribution. In addition, the value of α is not in practice always the same in the expressions $P(x) = a_s x^{\alpha_s}$, $a(t_2 - t_1) = 1/k_2^{\alpha_c} - 1/k_1^{\alpha_c}$ and $S_j = ax_j^\alpha$, although the three values are sometimes fairly close to one another. The same concept applies to the zero order law: the real values of S and B give a product SB which makes this law apply to much larger values of time and higher degrees of breakage than the *simple solution* would predict.

Other *Simple Solutions* for Steady-State, Continuous Grinding

For completeness, three special cases of continuous grinding are also considered: the case of a fully mixed mill, the case of a mill with the RTD described by the semi-infinite solution of the diffusion model, and the three-reactors-in-series case (see in Chapter 14). Consider the case of continuous fully mixed grinding with no size classification of the material as it leaves the mill. Consider a cumulative feed size distribution of $F(x_i)$ and a size distribution in the mill of w_j. A size-mass balance on less-than-size-x material gives

$$P(x_i) = F(x_i) + (W/F) \sum_{j=1}^{i-1} B_{ij} S_j w_j$$

Inserting the compensation condition, using $\tau = W/F$, and noting that $\sum_{j=1}^{i-1} w_j = 1 - P(x_i)$ for no classification at the mill exit,

$$P(x_i) = F(x_i) + \tau S_{i-1}[1 - P(x_i)]$$

or

$$P(x_i) = \frac{F(x_i) + \tau S_{i-1}}{1 + \tau S_{i-1}} \tag{15}$$

In continuous form, Eq. 15 is

$$P(x) = \frac{F(x) + \tau S(x)}{1 + \tau S(x)} \tag{15a}$$

This is the fully mixed equivalent to the batch/plug flow case of Eq. 9.

Chapter 14 shows that the residence time distribution of a mill may sometimes be approximately represented by the semi-infinite solution of the convective-mixing equation, e.g., Eq. 12 in Chapter 14,

$$\phi(t^*) = \frac{1}{2\sqrt{\pi D^* t^{*3}}} \exp\left[-(1-t^*)^2/4D^*t^*\right]$$

where t^* is t/τ and $D^* = D/Lu$. At steady state (with no classification at the mill exit) the product size distribution is given by

$$P(x) = \int_{t=0}^{\infty} P(x,t)\phi(t)\, dt$$

where $P(x, t)$ is the solution of the batch grinding equation. For the simple approximate solution of Eq. 9,

$$P(x) = \int_{t=0}^{\infty} \left\{1 - [1 - P(x,0)]e^{-S(x)t}\right\}\phi(t)\, dt$$

$$= \int_{t^*=0}^{\infty} \left\{1 - [1 - P(x,0)]e^{-S(x)\tau t^*}\right\}\phi(t^*)\, dt^*$$

or

$$1 - P(x) = [1 - F(x)]\int_{t^*=0}^{\infty} e^{-S(x)\tau t^*}\phi(t^*)\, dt^*$$

$$= [1 - F(x)]\exp\left[-\left(\frac{\sqrt{(1 + 4D^*\tau S(x))} - 1}{2D^*}\right)\right] \quad (16)$$

Of course, this reduces to Eq. 9 when D^* corresponds to plug flow, that is, $D^* \to 0$, since $[1 + 4D^*\tau S(x)]^{1/2} = 1 + 2D^*\tau S(x)$ for small values of $4D^*\tau S(x)$.

For the case of a mill which has an RTD equivalent to that from three fully mixed reactors in series, the treatment gives

$$P(x)$$

$$= \frac{F(x) + \left\{\tau_1 S(x) + [1 + \tau_1 S(x)]\tau_2 S(x) + [1 + \tau_1 S(x)][1 + \tau_2 S(x)]\tau_3 S(x)\right\}}{[1 + \tau_1 S(x)][1 + \tau_2 S(x)][1 + \tau_3 S(x)]}$$

$$(17)$$

where $\tau_1 + \tau_2 + \tau_3 = \tau$. Experimental measurements of RTDs in ball mills (see Chapter 14) indicate that they can be approximated by a model of one large reactor in series with two equal smaller reactors. In this case, $\tau_2 = \tau_3 =$

$(\tau - \tau_1)/2$ and

$$P(x) = \frac{F(x) + \tau S(x)\left\{ k + [1 + k\tau S(x)]\left(\dfrac{1-k}{2}\right)\left[2 + \left(\dfrac{1-k}{2}\right)\tau S(x)\right]\right\}}{[1 + k\tau S(x)]\left[1 + \left(\dfrac{1-k}{2}\right)\tau S(x)\right]^2}$$

(17a)

where $k = \tau_1/\tau$.

Experimentally it is known that $S(x)$ becomes small for small values of x. Inserting this limit into Eqs. 15, 16, or 17 gives

$$P(x) \approx \tau S(x)$$

(18)

since $F(x)$ is also $\ll 1$ for small x. This is the same result as Eq. 13. Thus the product size distributions for either plug or fully mixed flow in the mill (and any RTD in between) tend to be the same values at small sizes, for the same mean residence time, if the compensation condition applies exactly.

Summary

It is known that crushing tests on many particles of a given size always lead to a distribution of strengths. Some of the variation in strength arises because a particle will have different strengths depending on the orientation of the applied force. In a mill, there will be a wide range of different forces applied to particles in different orientations. If all particles in a small size interval had the same strength, calculations show that batch grinding of a $\sqrt{2}$ size interval would show an apparent increase in breakage rate at longer times, due to the bigger fraction of breakage products leaving the interval from breakage of the smaller sizes. This acts to compensate for weaker particles breaking more rapidly. In practice, the grinding of a $\sqrt{2}$ size interval experimentally gives first-order breakage, but this does not mean that the concept of first-order breakage can be applied to a differential size interval. A size interval of $2:1$ is usually too large to obtain first-order breakage.

There is one condition that gives a simple analytical solution for the batch grinding equation. It is the *compensation* condition, that,

$$S_j B_{i,j} = \text{function of } i \text{ only}, \quad i \geq j$$

(S1)

For example, this would apply for

$$S_j = ax_j^\alpha$$

and

$$B_{i,j} = \left(x_{i-1}/x_j\right)^\alpha$$

for then

$$S_j B_{i,j} = ax_{i-1}^\alpha, \quad \text{not dependent on } j.$$

Eq. S1 leads to the *simple solution* of the batch grinding equation,

$$P(x_i, \tau) = 1 - [1 - P(x_i, 0)]\exp(-S_{i-1}\tau), \quad n \geq i \geq 1 \qquad \text{(S2a)}$$

This can be put as a continuous function in x_i,

$$P(x, \tau) = 1 - [1 - P(x, 0)]\exp(-S(x)\tau), \quad 0 \leq x \leq x_{\max} \qquad \text{(S2b)}$$

In practice, this equation does not apply accurately to ball milling because the real values of $S_j B_{i,j}$ increase as $j \to i$.

Using the power function for S values, $S(x) = ax^\alpha$, in Eq. S2b gives

$$R(x, t) = R(x, 0)\exp\left[-(x/x_o)^\alpha\right] \qquad \text{(S3a)}$$

where

$$x_o = (1/at)^{1/\alpha} \qquad \text{(S3b)}$$

and $R(x, t)$ is the fraction (or percentage) of the product size distribution remaining on size x. For a coarse feed $R(x, 0) = 1$ for smaller values of product particle size x, so Eq. S3a becomes the Rosin-Rammler size distribution. Similarly, Eq. S3b is Charles' law (see Chapter 2) in the Rosin-Rammler form. Eq. S3a gives the Schuhmann size distribution $P(x) = atx^\alpha$, for x small, and, hence, Charles' law. Using the Rosin-Rammler 80%-passing size to eliminate x_o from Eq. S3 gives for small x

$$E_B = (1.61 m_p/aW)\left(\frac{1}{x_P^\alpha} - \frac{1}{x_F^\alpha}\right) \qquad \text{(S4)}$$

which is a form of Bond's law.

Actually, Charles' law can apply better to real data over a wider size range than expected from the simple solution because of the real form of $S_i B_{i,j}$. Similarly, the Bond batch equation with $\alpha = 1/2$ applies better to actual 80%-passing sizes than those predicted from Eq. S3.

For steady-state continuous fully mixed grinding the compensation condition gives

$$P(x) = \frac{F(x) + \tau S(x)}{1 + \tau S(x)} \qquad \text{(S5)}$$

where $F(x)$ is the cumulative feed size distribution. For the RTD resulting from the semi-infinite solution of the convective-mixing equation (see

Chapter 14) the compensation condition gives

$$P(x, \tau) = 1 - [1 - F(x)] \exp\left[-\left(\frac{\sqrt{1 + 4D^*\tau S(x)} - 1}{2D^*}\right)\right] \quad (S6)$$

where D^* is the dimensionless diffusion coefficient in the mill ($D^* = D\tau/L^2$, D being the effective diffusion coefficient). This equation can be used to derive approximate models for full-scale plant data where it is not possible to measure S and B values.

Experimental RTD measurements on mills indicate that they can be approximated by the equivalent system of one large fully mixed reactor in series with two equal smaller fully mixed reactors. This gives

$$P(x, \tau) = \frac{F(x) + \tau S(x)\left\{k + [1 + k\tau S(x)]\left(\frac{1-k}{2}\right)\left[2 + \left(\frac{1-k}{2}\right)\tau S(x)\right]\right\}}{[1 + k\tau S(x)]\left[1 + \left(\frac{1-k}{2}\right)\tau S(x)\right]^2}$$

$$(S7)$$

where $k = \tau_1/\tau$.

References

Austin, L. G. and Luckie, P. T., 1970–71, "Influence of Interval Size on the First-Order Hypothesis of Grinding," *Powder Technology*, Vol. 4, pp. 109–110.

Austin, L. G. and Luckie, P. T., 1972, "Grinding Equations and Bond Work Index," *Trans. SME-AIME*, Vol. 252, pp. 259–266.

Austin, L. G., Luckie, P. T., and Klimpel, R. R., 1972, "Solutions to the Batch Grinding Equation Leading to Rosin-Rammler Distributions," *Trans. SME-AIME*, Vol. 252, pp. 87–94.

Austin, L. G., Shoji, K., and Everell, M. D., 1973, "An Explanation of Abnormal Breakage of Large Particle Sizes in Laboratory Mills," *Powder Technology*, Vol. 7, Jan., pp. 3–8.

Bond, F. C., 1960, "Crushing and Grinding Calculations," *British Chemical Engineering*, Vol. 6, pp. 378–391, 543–548.

Filippov, A. F., 1961, "Distribution of the Sizes Which Undergo Splitting: Theory of Probability and Its Applications," Vol. 6, English translation, USSR, pp. 275–280.

Gardner, R. P. and Austin, L. G., 1975, "The Application of the First-Order Grinding Law to Particles Having a Distribution of Strengths," *Powder Technology*, Vol. 12, pp. 65–69.

Gaudin, A. M. and Meloy, T. P., 1962, "Model and a Comminution Distribution Equation for Repeated Fraction," *Trans. SME-AIME*, Vol. 223, pp. 43–50.

Gupta, V. K. and Kapur, P. C., 1976, "A Critical Appraisal of the Discrete Size Models of Grinding Kinetics," *Proceedings*, Fourth European Symposium Zerkleinern, H. Rumpf and K. Schonert, eds., Dechema Monographien 79, Nr. 1576-1588, Verlag Chemie, Weinheim, pp. 447–465.

Herbst, J. A. and Fuerstenau, D. W., 1968, "Zero Order Production of Fine Sizes in Comminution and Its Implications in Simulation," *Trans. SME-AIME*, Vol. 241, pp. 538–548.

Holmes, J. A., 1957, "A Contribution to the Study of Comminution—A Modified Form of Kick's Law," *Transactions of the Institution of Chemical Engineers*, Vol. 35, London, pp. 125–141.

Kapur, P. C., 1970, "Part A: Reduction of the Grinding Equation. Part B: An Approximate Solution to the Grinding Equation," *Trans. SME-AIME*, Vol. 247, pp. 299–303; 309–313.

King, R. P., 1972–73, "An Analytical Solution to the Batch Comminution Equation," *Journal of the South African Institute of Mining and Metallurgy*, Vol. 73, pp. 127–131.

Razumov, K., et al., 1969, "Fundamental Concepts of Ball Mill Grinding," *Minerals Processing*, Vol. 10, pp. 8–11.

9

Methods for Direct Experimental Determination of the Breakage Functions *S* and *B*

Introduction

Most of the results presented in this book have been obtained from laboratory or pilot scale mills varying from 200 mm to 1 m in diameter and this chapter will concentrate on methods used to analyze such laboratory data. The art of obtaining accurate values of S and B from large-scale continuous tests is not well developed, mainly because the values have to be inferred from the size distribution at the exit of the mill with the residence time distribution of the mill as a complicating factor. Kelsall, Reid, and Restarick (1967–68; 1968–69) and Kelsall and Reid (1965) described a technique applicable to continuous milling, which involved admitting a pulse of tracer material to the mill feed, e.g., quartz added to a steady feed of calcite. Samples of the exit stream were taken at suitable time intervals after the admission of the tracer. The quartz and calcite in the exit stream were separated by dissolving the calcite in acid and the size distribution of the quartz determined as a function of time spent in the mill. However, this technique requires a great deal of labor and has the disadvantage that reliable size analyses are only obtained after some initial time. This is due to the minimum delay time for significant quantities of material to leave the mill, which makes it difficult to get unequivocal values of B_{ij} (see "Analysis of Data: B Values"). In addition, the hold-up of the mill varied with feed rate so that tests were not performed at a constant mill load. The technique did have the major advantage that the breakage values were determined in a steady-state environment consisting of a complete mixture of sizes.

Gardner and Sukanjnajtee (1973) and Gardner, Rogers, and Verghese (1980) have outlined procedures for using radiotracing to follow breakage and residence time distribution at the same time in large-scale continuous milling, but at the time of writing the technique has been used primarily for RTD (see Chapter 14). Frühwein (1976) suggested a procedure based on sampling the powder lying along a tube mill after suddenly stopping the mill. Again, however, it has not yet been tested in practice. Snow and Meloy

181

(1972) have also attempted measurements on large mills using tracing techniques, but apparently with limited success.

In this chapter emphasis is placed on the batch laboratory test, from which most of the reliable values of S and B have been obtained to date. The discussion will be based primarily on the use of sieving as the size analysis technique. The errors and problems which are encountered will be stressed, since it is usually easy to overcome such problems once their existence is recognized, especially in small-scale laboratory test work.

The application of back-calculation techniques to determine S and B parameters from batch and continuous data is deferred to the next chapter; the methods presented here are direct experimental determination.

The Batch Grinding Test: The One-Size-Fraction Method

This consists of grinding a starting material which is predominantly of one size interval, e.g., a $\sqrt{2}$ sieve interval. This size is sieved out of the starting supply of material; if necessary, larger sizes are broken in a jaw or roll crusher to provide more of the desired size. The larger the test mill, the more tedious it is to prepare sufficient mass of one-size fractions, and it is helpful to have sieving devices which use larger screens than the standard 200-mm (8-in.) sieves [e.g., a 457.2-mm (18-in.) square Tylab system or a 0.46 × 1.22 m (1 1/2 × 4 ft) Derrick vibrating screen].

A sample is taken from this material and subjected to a *blank* sieving test, for the same sieve loading and sieving time to be used after the first grinding time, using the two standard 200-mm (8-in.) sieves describing the size interval. It may be found that a small percentage of material is retained on the top size screen, and this can be considered as *of size* without significant error. It is also nearly always found that some percentage of material passes the lower screen. Most of this is retained on the next smaller geometric size screen, so it is mainly in the second size interval. This is *near-size* material that has poor screening statistics because it has to hit the screen holes in a suitable orientation in order to pass through. The material which passes through to the second interval could erroneously be classed as *broken* even though grinding has not yet been applied, so this error will be referred to as the *incomplete-sieving error* ε; it should not be greater than 5% if accurate B values are to be obtained (see "Analysis of Data: B Values"). If the test is solely for an S value, it is only necessary that the amount of the size under investigation form a substantial part of the test sample; Fig. 1 shows a typical result.

When a suitable amount of material of the desired test size has been prepared and a blank size analysis performed, the test mill is filled with the desired filling of balls and material (plus liquid if wet grinding is being studied) spread uniformly in the mill. It is then ground in the mill for varying times, which are chosen to enable B to be calculated from short-time data and S from longer time data. The mill contents are sampled, size analysis performed, and the sample can be returned to the mill contents for further grinding or a fresh feed used for the longer grind time, whichever is more convenient. For a large mill, the sample may be too small to require its return to the mill, in which case the complete grinding time sequence can be carried out at once, with stops for sampling. For small mills, sampling can be performed by completely tipping the mill contents through a coarse screen to retain the balls, and splitting the charge to a suitable sample size. For larger mills, direct ladle sampling from the stopped mill has been quite satisfactory for dry milling.

Most mills are good mixers and excessive handling in the sampling procedure (by repeated riffling, cone and quartering, etc.) should be avoided as more likely to demix the sample than mix it. For wet milling, however, stopping the mill produces immediate partial separation of solid and liquid, with the finer material being retained to a greater extent in the liquid. It is necessary to tip the entire mill contents, filter, and spread the solid out to dry; the dried material is then mixed and split to obtain a representative sample.

For the determination of S values, only the fraction left in the top size need be determined, so sieving at this screen size is all that is necessary. To determine B values, however, very accurate complete size analysis at short grind times is necessary. In addition, the determined values of S and B are normally used to forward-compute expected size distributions for compari-

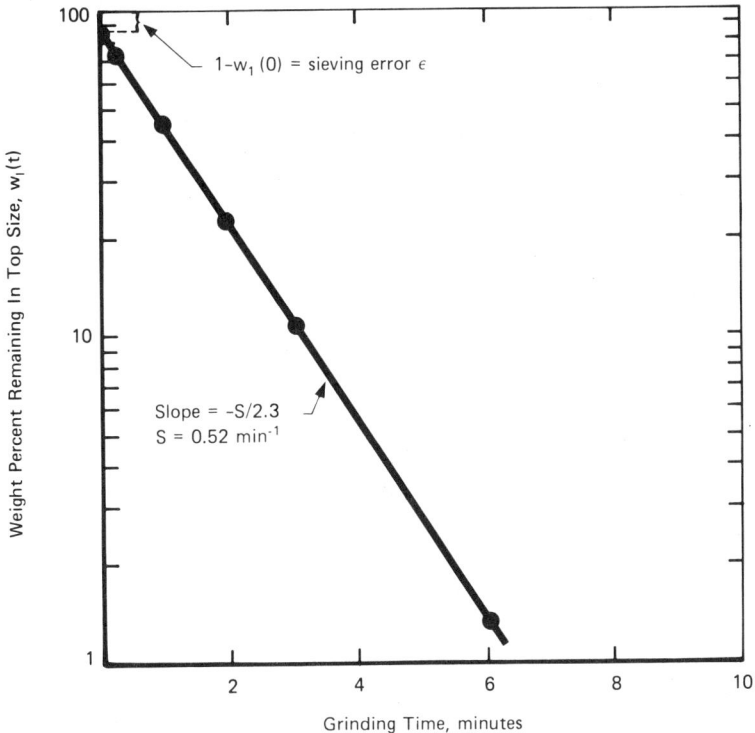

Fig. 1. First-order plot for dry grinding of 1.18 mm $\times$ 850 μm (16 $\times$ 20 mesh) petroleum coke in 200 mm i.d. ball mill.

son with the experimental data in order to check the consistency of the data and complete size distributions are necessary for the comparison. A complete study on a given material under a given set of conditions normally involves measuring S values for three or four starting sizes, getting the size distribution at short grind times to enable B values to be calculated for each of these sizes, and determining the complete family of size distributions resulting from one or two of the starting sizes.

It is a considerable advantage to be able to measure variations of the power into the test mill during the testing, since unusual changes in power will indicate that something is wrong with the mill conditions. It is not usually possible to do this by measuring the wattage of the motor driving the conventional roller table, because the major energy loss is in the drive system and the motor wattage is thus not a sensitive indicator of the power to the mill. The section "Measurements of Mill Power" in Chapter 11 discusses techniques for power estimation.

Although size analysis by sieving seems an uncomplicated and unambiguous process requiring little skill, it has a number of pitfalls. Ideally, batch sieving for long times should remove all of the near-size material. However, in practice it is found that however long a given sample is sieved, a further 10 min of sieving will allow more material to pass through. There are a number of reasons for this (Allen, 1975), but the most disturbing is that near-size particles are continually abraded to a size which will pass the screen. Thus a suitable sieving schedule must be established which is a *compromise* between times sufficient to allow small sizes to fall through yet not long enough to give too much abrasion. The schedule should be established by rate-of-sieving studies, especially on the smaller sizes.

Table 1 gives a typical result, in which five identical samples were sieved for various times. It was concluded that the sieving time should be 30 to 40 min. The sieving schedule established was: 20 min sieving, removing the sample, brushing the underside of the inverted screens to remove material stuck in the screens, replacing the sample, and sieving for another 10 min. The screens should be brushed clean during the final sample collection. Too little or too much screening are equally undesirable and can give false slopes of the lines in Fig. 1 in Chapter 2.

It should also be recognized that sieving fine material through fine screens gives rise to blockage of the screens in inverse ratio to the size involved. Thus 100 g of sample on a 200-mm (8-in.) diam 1.18 mm (16

Table 1. Test of Dry Sieving Schedule for Cement Clinker (10.3 g on nest of three 200-mm (8-in.) screens on Rotap shaker)

Screen size		Wt % Less than Size in Sieving Time, min				
mesh	μm	20	30	40	50	60
200	75	91.46	91.75	92.04	92.14	92.23
270	53	82.91	83.39	83.79	83.98	84.27
400	38	68.83	69.71	70.39	70.68	70.97
− 400 wt % increase			0.88	0.68	0.29	0.29

mesh) screen is quite valid, but at 45 μm (325 mesh) the sample weight reaching and passing the screen should be no more than about 10 g, otherwise the screen may repeatedly block and prevent fines from passing through. On the other hand, the larger the top size of the material the larger the sample weight should be to obtain a representative number of larger particles.

The starting sample for size analysis should always be weighed and the loss in the sum weight after sieving into the screen size fractions should be no more than about 0.4%. This is treated as loss of fine material and added to the measured weight of the sink interval. This correction is only valid for carefully performed sieving and weighing in which the major weight loss is by unavoidable loss of fines. If the weight loss is too high, showing loss of large particles due to careless handling, adding the weight loss to the sink interval makes this point on the size distribution too high, giving the results shown in Fig. 2, point c.

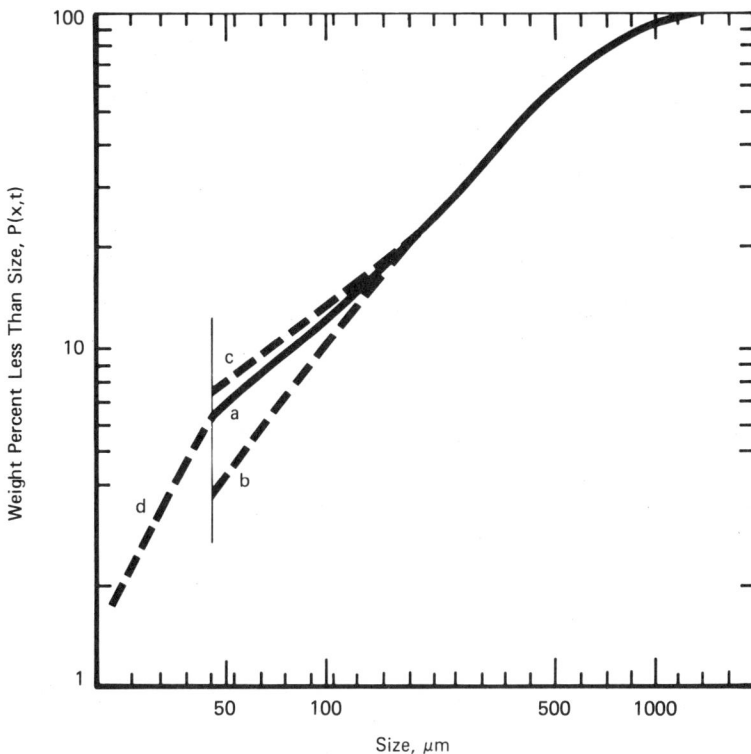

Fig. 2. Illustration of the effect of errors in the measurement of the finest size (sink interval) on the Schuhmann plot of size distribution: (a) correct; (b) incomplete sieving; (c) excess weight losses; (d) added subsieve size distribution where fines have been lost.

For softer materials which abrade more rapidly, or sticky materials, it is better to wet-sieve at 38 μm (400 mesh), followed by drying of the retained material and dry sieving according to a suitable schedule. If vacuum filters are available, the -38 μm (-400 mesh) material can be recovered. Wet screening is more efficient (Aplan, 1975) if the sample is stirred with liquid in a large beaker, with a suitable dispersing agent, if necessary, and allowed to settle; the supernatant liquid contains most of the fines, which can then be run through the screen with a minimum of additional washing. The settled solid is rewashed if necessary, and finally dried and dry screened. This avoids the nuisance of the large amounts of water (or other liquid) necessary to spray-wash the whole sample on the screens and leads to a good cleaning of fines from the larger fractions.

Analysis of Data: S Values and Non-First-Order Effects

Let $w_1(t)$ be the weight fraction of the size interval being examined. First-order grinding gives

$$\log w_1(t) - \log w_1(0) = -S_1 t/2.3 \qquad (1)$$

Thus a plot of $w_1(t)$ on a log scale vs. t on a linear scale should give a straight line (see Fig. 3). The point at zero time is obtained from the blank sieving test, and thus automatically allows for incomplete-sieving error. It is not correct to fit a line passing through zero at $t = 0$ unless the blank shows no incomplete-sieving error. It is clear from Fig. 3 that the test grinding times vary with conditions; they should ideally be chosen to give w_1 values of about 0.8, 0.5, 0.1, and 0.05. The value of S is determined from the slope of the plot. Plotting S_i vs. x_i on log-log scales enables the value of α to be determined.

Fig. 4 defines four regions which can be encountered in the first-order plot. Curve ab shows that rapid breakage occurs at first, followed by slower first-order breakage. This might be due to more efficient tumbling action at the beginning of grinding, and this can be tested by power measurements on the mill, since the initial power will then also be high. More often, however, it is due to some of the starting material being weaker than the rest of the material. If the rapid breakage is due to a basic inhomogeneity in the material, special analysis is required since in this case harder material is passed on to the lower sizes and all sizes will show a non-first-order behavior. On the other hand, the starting material can contain material which is abnormally weak due to an unusual shape or due to flaws introduced during the preparation. In this case, once the material is broken the fragments become normal and the lower size intervals show no non-first-order behavior. Usually, region a is quite small and occurs only in the larger sizes of a feed prepared from jaw or roll crushing, where it often happens to be associated with some fraction of abnormally flaky material. The influence of this abnormal material is rapidly lost in the size distributions produced after some minutes of grinding. The initial ab effect can be removed by *preconditioning* the sized feed in a ball mill for a short time and resieving to eliminate the broken material.

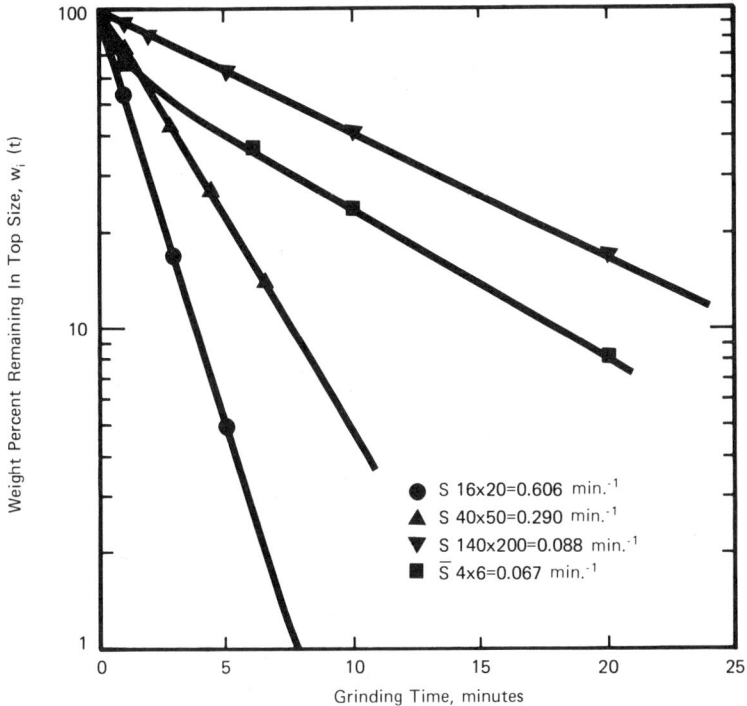

Fig. 3. First-order plots for ●, 1.18 mm × 850 μm (16 × 20 mesh); ▲, 425 × 300 μm (40 × 50 mesh); ■, 4.75 × 3.35 mm (4 × 6 mesh), and ▼, 106 × 75 μm (140 × 200 mesh) cement clinker ground under conditions shown in Table 2 in Chapter 10; Type II cement clinker.

A special case of the above discussion of region a is that of particles which are too large to be crushed normally in the mill [see the 4.75 × 3.35 mm (4 × 6 mesh) results in Fig. 3]. The forces produced in the mill are not sufficient to consistently break particles caught in a breakage action, so weaker particles have a higher probability of breakage than stronger. *Abnormal breakage* of too large particles is associated with a maximum in the S values (see Chapter 5).

It is sometimes found that the b region goes into a c or d region, where again the first-order law does not appear to apply. If a region d is found for a $\sqrt[4]{2}$ or $\sqrt{2}$ size interval, there are a number of possible physical causes of the breakdown of the first-order hypothesis such that breakage is occurring less rapidly than expected.

1. There may be accumulation of stronger fractions in the remaining unbroken material as grinding proceeds.

2. The finer material produced in the milling may lower the rate of breakage of the coarse material by some kind of cushioning action, which has been observed for abnormal breakage of large sizes, for very fine dry grinding, and for wet grinding of thick pastes.

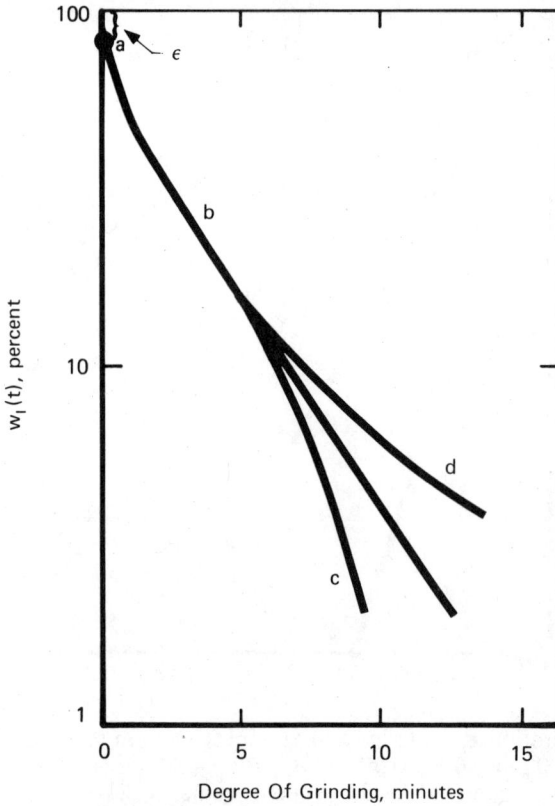

Fig. 4. Illustration of deviations from first-order plot.

3. The production of fines may alter the mechanics of the mill action to give less tumbling, so that the measured energy input to a ball mill test might decrease during the test, indicating decrease in the number of balls being lifted per second, and hence decrease in the specific rates of breakage.

4. Some fine material may be able to agglomerate or *cold-weld* to such an extent that the agglomerates are not broken up by sieving and this might lead to regrowth of large particles as grinding proceeds.

Similarly, region c shows that breakage is occurring more rapidly than expected. This can be due to any one of the following factors.

5. Removal by breakage of larger sizes which were shielding the smaller sizes in the interval may lead to faster overall rates of breakage.

6. The material may become progressively weaker with time due to impacts which weaken, but do not break, the particles.

7. A harder component may be liberated which helps grind the softer material.

8. The mechanics of the grinder may become more efficient (higher power) as fines increase in the mill.

If a non-first-order effect of type c or d is due to change in the properties of the material being ground, causes (1), (5), (6), or (7), this is readily confirmed by running enough tests to collect a fresh starting charge of 100% top size from the material left in region c or d. The initial rate of breakage of this fresh charge will match that in region c or d. If, on the other hand, the effect is due to the change of environment, causes (2), (3), (4), (5), or (8), starting a new test with this 100% top size will restore the environment to the original condition and the material will break at the rate of region b. This type of test unambiguously separates a *material* from an *environment* effect.

If the one-size-fraction method being discussed gives a good first-order plot, this proves only that the accumulation of *finer* material during the grind does not influence the specific rate of breakage of the starting size. The presence of *larger* material might still change the specific rate of breakage. This can only be rigorously tested by tracer experiments in which larger sizes than the traced test material are present. However, it can be *assumed* that breakage is first-order, with the rate measured by the one-size-fraction method being independent of the presence of larger or smaller sizes. The size distributions at various grind times can then be computed on this basis, and if they agree with the experimental values, it can reasonably be inferred that the first-order assumption is indeed correct.

The slowing down of general grinding rate due to the accumulation of fines in the environment (see "Effect of Environment in the Mill" in Chapter 5) is very important in fine grinding. This effect can be detected by the Weibull plot (see "Solution of the Batch/Plug Flow Grinding Equation" and Fig. 3 in Chapter 8), which gives approximate straight lines of slope 1 to 1.3 for first-order grinding of a narrow size feed range. When the grinding becomes non-first-order, the results deviate from the straight line, as shown in Fig. 5. The figure illustrates a *false* grinding time Θ corresponding to the real time t, such that first-order grinding from time Θ gives the same results as non-first-order grinding for time t (see "Effect of Environment in the Mill" in Chapter 5). The ratio Θ/t is the value of K in Eq. 16 in Chapter 5.

Analysis of Data: *B* Values

By definition, the values of B are deduced from the size distributions at short grind times where there is mainly size 1 material breaking and only small amounts of smaller sizes to rebreak. The smaller the amount of material broken out of size 1, the more accurate are the B estimates, especially if the incomplete-sieving correction is also small. However, it is difficult to do accurate sieving and weighing on the material broken below the first size interval if this material is only a small percent of the total, although very careful analysis can give excellent results under these circumstances. In addition, the initial breakage is often abnormal (when this is observed, the feed should be preconditioned as discussed above). Experience suggests that good results are obtained when the time of grinding is chosen to give an amount of material broken out of the top size interval of about 20 to 30%.

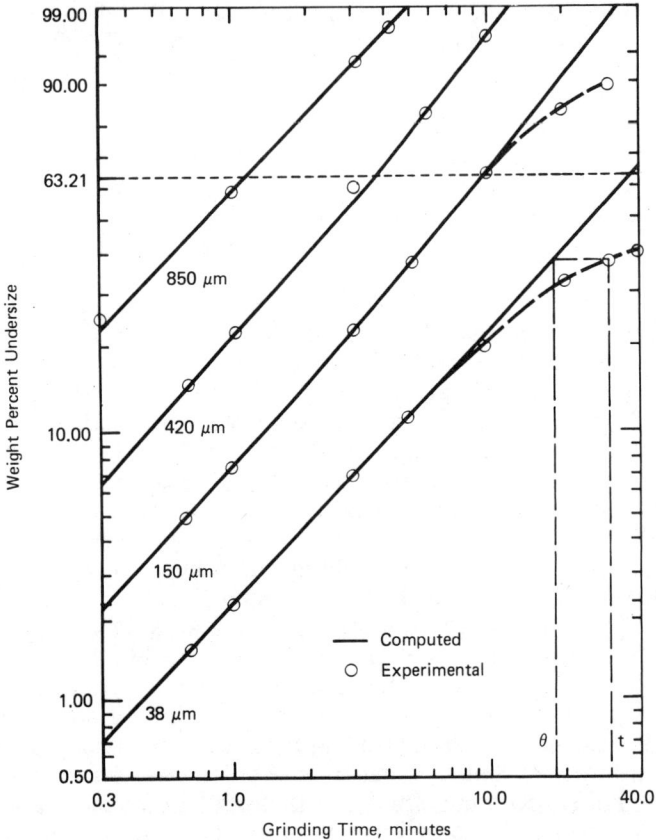

Fig. 5. Weibull plot for 1.18 mm × 850 μm (16 × 20 mesh) cement clinker ground for different times in a laboratory ball mill, showing deviation from first-order grinding at long times.

However, at this level of breakage it is certain that rebreakage of fragments has occurred to some degree and a correction must be made for this. One technique is to measure the size distributions as a function of time and extrapolate back to close to zero time. Then, by definition the b values are calculated from:

$$b_{2,1} = \frac{\text{weight arriving in size 2, } t \to 0}{\text{weight broken out of size 1, } t \to 0}$$

The fractional weight broken out of size 1 is $w_1(0) - w_1(t)$; the fractional weight arriving in size 2 assuming negligible breakage from size 2 is $w_2(t) - w_2(0)$; the fractional weight arriving in size 3 is $w_3(t)$ since $w_3(0)$ is negligible, and so on. Thus

$$b_{2,1} = [w_2(t) - w_2(0)]/[w_1(0) - w_1(t)]$$
$$b_{3,1} = w_3(t)/[w_1(0) - w_1(t)]$$

and

$$b_{i,1} = [w_i(t)]/[w_1(0) - w_1(t)], \quad i > 2$$

In the cumulative form, the fraction broken which falls less than the top size x_1 is clearly 1, hence $B_{1,1} = 1$; the fraction broken which falls less than the upper size of interval 2, x_2, is also clearly 1, hence $B_{2,1} = 1$; the fraction of broken material which falls less than the upper size of interval i is

$$= \frac{P_i(t) - P_i(0)}{[1 - P_2(0)] - [1 - P_2(t)]}$$

or

$$B_{i,1} = \frac{P_i(t) - P_i(0)}{P_2(t) - P_2(0)} t \to 0, \quad i > 1, \quad \text{Method BI} \tag{2}$$

where $P_i(0)$ is normally zero except for the incomplete-sieving error $\varepsilon = P_2(0)$.

Unfortunately, it is difficult to get accurate size distributions at small degrees of breakage, and performing accurate tests at several short times to enable an extrapolation to near-zero times is experimentally tedious. Consequently, we use a procedure (Austin and Luckie, 1972) which corrects for secondary breakage, based on the solution of the batch grinding equation given by the *compensation* condition (see Chapter 8). This solution is

$$1 - P_i(t) = [1 - P_i(0)] \exp[-S_j B_{ij} t], \quad i > j$$

For the top size, assuming that the compensation conditions applied approximately,

$$1 - P_i(t) \simeq [1 - P_i(0)] \exp(-B_{i,1} S_1 t), \quad i > 1$$

For the top size interval, first-order breakage gives

$$1 - P_2(t) = [1 - P_2(0)] \exp(-S_1 t)$$

Then,

$$-S_1 t = \ln[(1 - P_2(t))/(1 - P_2(0))]$$
$$-B_{i,1} S_1 t \simeq \ln[(1 - P_i(t))/(1 - P_i(0))]$$

and

$$B_{i,1} \simeq \frac{\log[(1 - P_i(0))/(1 - P_i(t))]}{\log[(1 - P_2(0))/(1 - P_2(t))]}, \quad i > 1, \quad \text{Method BII} \quad (3)$$

For small degrees of grinding, the value of $P_i(0)$ and $P_i(t)$ will be small for the one-size fraction method [except for $P_1(t) = 1$, of course], and using the expansion of $\log(1 + x)$ for small x,

$$B_{i,1} \simeq [P_i(t) - P_i(0)]/[P_2(t) - P_2(0)], \quad i > 1, > t \to 0$$

Thus the two methods give identical results at sufficiently low degree of grinding.

In general, if the top size is denoted by j, Eq. 3 can also be put as

$$B_{i,j} \simeq \frac{\log[(1 - P_i(0))/(1 - P_i(t))]}{\log[(1 - P_{j+1}(0))/(1 - P_{j+1}(t))]}, \quad i > j \quad (3a)$$

This correction procedure works well for ball milling because the S and B values do indeed often approximately compensate. This has been proved by Austin and Luckie (1972) who used typical values for S and B to compute size distributions for batch grinding at various grind times, and then applied the BI and BII calculation techniques to the simulated data. Since the size distributions are exact, there is no experimental error in the curves and a comparison of the calculated B values with the known *true* values gives a measure of the error in the back-calculation techniques. Table 2 shows this technique applied to the quartz data of Fig. 1 in Chapter 2. The table shows that the BI method only works for a very small degree of grinding, but that the BII method gives reasonable values up to about 30% of the top size broken out, bearing in mind experimental variability in real results (see below). The larger the difference between the descriptive parameters α and γ, the less the compensation condition applies with accuracy to some degree of grinding. Therefore, the analysis may have to be done at 20% or less to get good results; however, the example used has a larger difference between α and γ than usual for many materials. When the BII method is applied to results which do not fit the compensation condition, the BII correction procedure is less accurate. An example is breakage of a size to the right of the maximum in Fig. 2 in Chapter 5. In this case it is recommended to use the BIII method (Austin and Luckie, 1972) (see Appendix 10), which requires an estimate of the S values.

Obtaining good B values experimentally is the most difficult part of the testing for S and B, because it is theoretically necessary (as shown above) to use a small degree of grinding to avoid excessive secondary breakage and necessary to use a single size feed for the same reason. However, the short grinding times are often those where good results are hard to get because of weak starting material or experimental errors. Consequently, the results shown in Fig. 6 are quite common: Replication of the determination gives a

Table 2. Comparison of B Parameters as Calculated by BI and BII
(for $\alpha = 0.8$, $\gamma = 1.28$, $S_1 = 0.52$ min^{-1}).

Size Interval	True B	BI	BII	BI	BII	BI	BII	BI	BII	BI	BII
1	1.0	1.0	1.0	1.0	1.0	1.0	1.0	1.0	1.0	1.0	1.0
2	1.0	1.0	1.0	1.0	1.0	1.0	1.0	1.0	1.0	1.0	1.0
3	0.432	0.485	0.454	0.513	0.464	0.542	0.475	0.615	0.499	0.688	0.522
4	0.250	0.281	0.259	0.301	0.265	0.323	0.273	0.386	0.289	0.454	0.306
5	0.153	0.172	0.157	0.187	0.163	0.202	0.167	0.247	0.178	0.300	0.189
6	0.102	0.114	0.105	0.124	0.107	0.134	0.110	0.165	0.116	0.203	0.123
7	0.063	0.071	0.065	0.078	0.067	0.085	0.069	0.107	0.074	0.133	0.0785
8	0.040	0.045	0.041	0.050	0.0426	0.054	0.0438	0.069	0.047	0.087	0.050
9	0.026	0.029	0.0267	0.032	0.0276	0.035	0.0285	0.045	0.034	0.057	0.0326
10	0.017	0.019	0.0176	0.021	0.018	0.023	0.0188	0.029	0.020	0.037	0.0213
$P_2(t)$		15%		25%		35%		56%		71%	

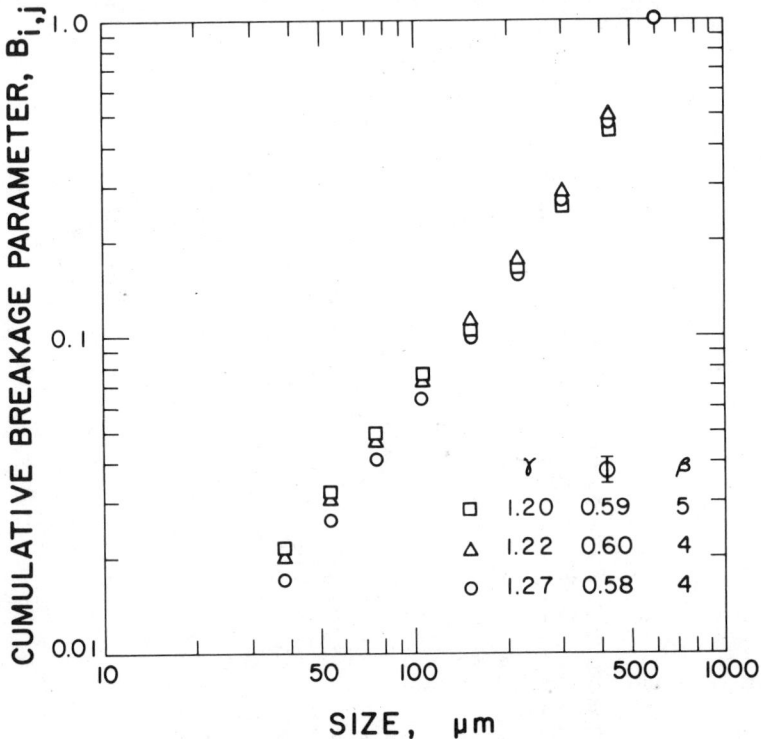

Fig. 6. Triplicate breakage distribution functions ($\sqrt{2}$ intervals) for wet grinding of quartz in a laboratory mill.

band of results and the values of B determined from one test are not closely repeatable in another. This fact must be borne in mind when analyzing B data for different size and/or mill conditions (see "Statistical Analysis" in Chapter 10 for methods of analysis).

Nonnormalized *B* Values

Fig. 7 shows a typical result for a material in which B values are unambiguously *not* normalized. For ball milling, we have always found the nonnormalized B values to exhibit the same pattern as shown in this figure; that is, the smaller the breaking size, the finer the corresponding dimensionless B distribution. Austin and Luckie (1972a) have described a technique for characterizing such distributions, as follows. The shape of the B values can be well described as the sum of two power functions (which appear as two straight lines on a log-log plot; see Fig. 3 in Chapter 5); that is,

$$B_{i,1} = \Phi_1 (x_{i-1}/x_1)^\gamma + (1 - \Phi_1)(x_{i-1}/x_1)^\beta, \quad i > j \qquad (4)$$

where Φ_1 is the intercept shown in Fig. 7 and γ is the slope of the small end

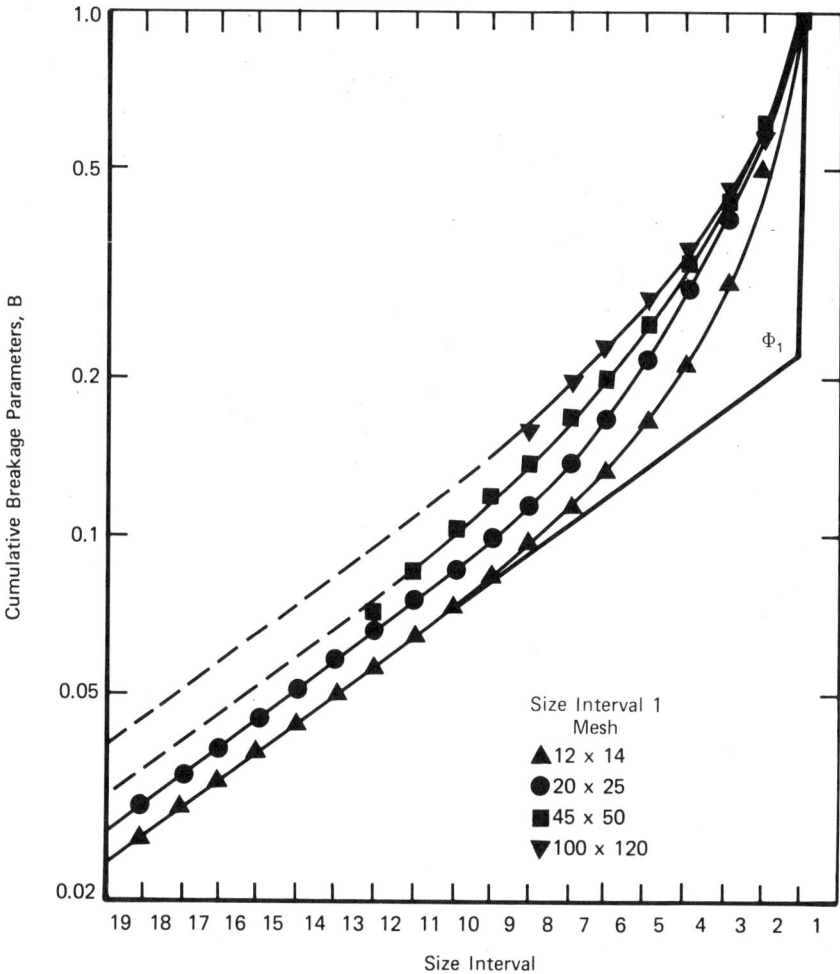

Fig. 7. Experimental values of *B* function of cement clinker Type II for various feed sizes (see Fig. 3).

of the distribution. Since γ and Φ are easily determined from the plot, the final term on the right-hand side of Eq. 4 is calculated and plotted (as shown in Fig. 3 in Chapter 5) to give the value of β.

Since the size intervals are in geometric progression $x_2/x_1 = R$, $x_3/x_2 = R$, and $x_{i-1}/x_j = R^{(i-1-j)}$, Eq. 4 can also be expressed as

$$B_{i,j} = \Phi_j R^{(i-j-1)\gamma} + \left(1 - \Phi_j\right) R^{(i-j-1)\beta}, \quad n \geq i > j \geq 1 \tag{5}$$

with j having values from 1 to $n - 1$.

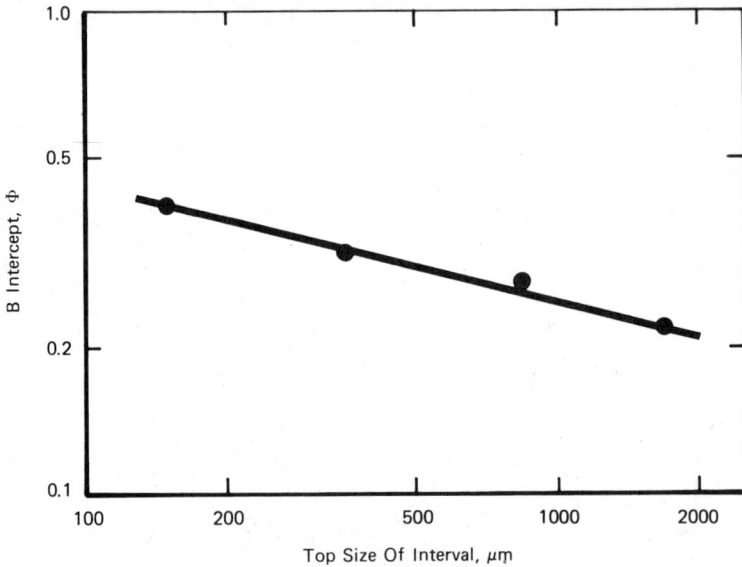

Fig. 8. Intercept of *B* plots as a function of size being broken.

Fig. 8 gives the variation of Φ with size j as

$$\log \Phi_j = -\delta \log(x_j/x_0) + \log \Phi_0 \qquad (6)$$

where x_0 is a standard size of one mm and where $-\delta$ is the slope (negative) of the line in the figure. For geometric sieve intervals of ratio R (lower/upper size)

$$\Phi_j = \Phi_1 R^{(1-j)\delta} \qquad (6a)$$

where Φ_1 is a known intercept for size 1 and δ is positive for ball milling. Thus Eqs. 5 and 6 are an empirical mathematical representation of B values, with the descriptive parameters being γ, β, δ, and Φ_0. For normalized B values, $\delta = 0$. This equation can only be used for extrapolation to $\Phi_j = 1.0$. It is usually sufficient to take $\Phi_j = 1.0$ for small sizes.

The values of γ and β are usually constant for breakage of sizes to the left of the maximum in S values. However, the B values of sizes to the right of the maximum may not fit Eq. 4, or if they do, the γ and β values can be different (see Chapter 16).

Time-Scale Factors for *S* by Inference

The detailed determination of the S and B functions even for one material and one set of milling conditions requires considerable experimental effort. As a consequence, therefore, it is worthwhile to be able to estimate the S and B functions for a new set of conditions without repeating all this work. This can sometimes be done by using the time-scale effect discussed in "Analysis of the Batch Grinding Equation" in Chapter 4. If the B values do not change in going from one set of test conditions to another, and if all the S values are changed by a *constant multiple*, the size distributions as a function of time from a given feed size will be identical between the sets of conditions except for a factor in the time scale. Thus, if S values are increased by a factor of κ, the grinding time taken to reach a given size distribution is reduced by κ (Eqs. 7 and 8 from Chapter 4):

$$S_i' = \kappa S_i, \qquad n \le i \le 1$$

$$P\big(x_i, \tau, \underline{S}, \underline{B}, \underline{w}(0)\big) = P\big(x_i, \tau/\kappa, \underline{S}', \underline{B}, \underline{w}(0)\big)$$

and

$$\tau \propto 1/\kappa$$

This can be applied best by using the Weibull plot (see Fig. 5) to compare two sets of batch grinding data, as illustrated in Fig. 9. If the two sets of data can be reasonably superimposed by shifting the time scale of one of them by κ, then this value of κ is the value in Eq. 7 in Chapter 4. In this case, the dry grinding times have been divided by 1.4. It is then *inferred* that the values of B are the same between the two tests, and the values of S are changed by the constant multiple κ. It is preferable to check this inference by determining one B plot for each of the sets and by comparing S_1 values.

Extension to Subsieve Sizes

Although sieving is the most widely used method of size measurement in the mineral industries, the lowest size available in normal sieves is the 400 mesh of 38 μm. If the size distributions of interest are largely below this size, the sieve size distribution does not contain enough information to describe the bulk of the distribution. In this case, it is necessary to use other methods of size analysis. Some generally applicable comments on subsieve size analysis follow.

First, the definition of size is different from that of sieving so it is usually necessary at least to multiply the measured size by a factor to convert it to sieve sizes, to join the subsieve and sieve size distributions together. Second, if a particle passes through a sieve hole, or remains suspended in a sedimentation test, or is counted as a small volume in the Coulter Counter*, it is undoubtedly a small particle. However, the reverse is not true; a particle counted as large in any of these methods can actually be an agglomerate of fine material. The essence of any measurement of size distribution applied to fine sizes is to obtain a complete *dispersion* of the particles, so that they are examined as separate particles. Thus, methods involving liquids, wetting, dispersing agents, and ultrasonic agitation are normally to be preferred over dry methods. Third, the presence of a small amount of fine material adhering to larger particles will not much affect the weights of sieve fractions of these large particles, but the accumulation

---

*Coulter Electronics, 590 W. 20th Street, Hialeah, FL 33010.

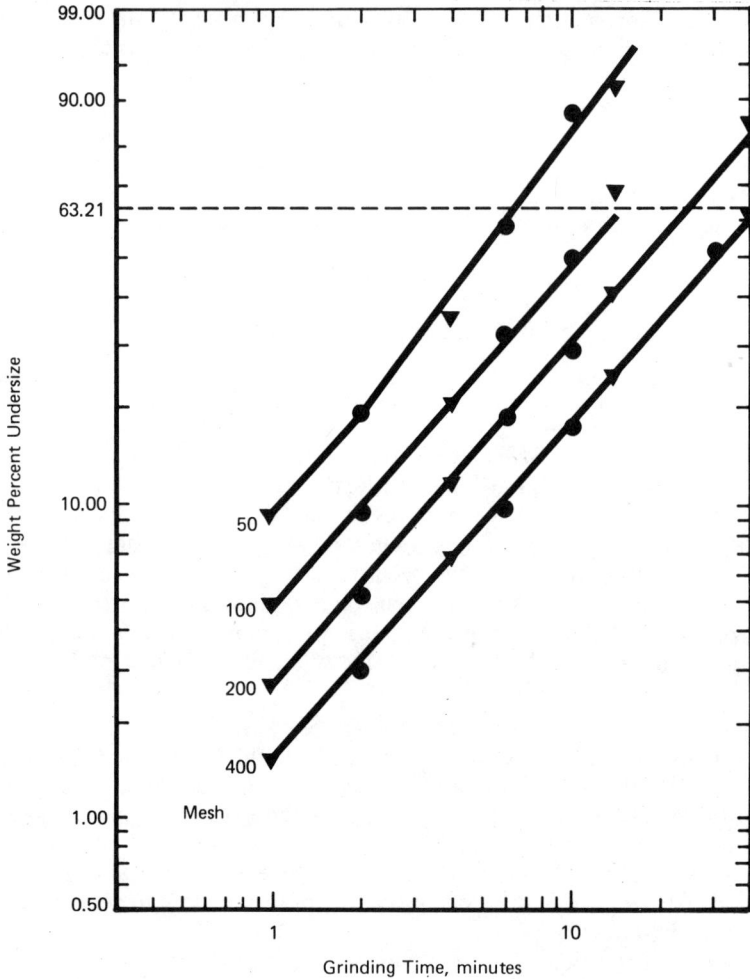

Fig. 9. Weibull plots for grinding of 850 × 600 μm (20 × 30 mesh) petroleum coke: ▼, wet; ●, dry; 1.4 scale factor applies $S_{wet} = 1.4 S_{dry}$, so dry grinding times have been reduced by dividing by 1.4 (see Fig. 1 in Chapter 2 for grinding conditions).

of the *loss* of fines over many sieve intervals can give a serious error in the amount assigned to the smallest size interval, as discussed before (see Fig. 2).

For example, curve *d* in Fig. 2 shows a typical size distribution obtained from dry sieving at 38 μm (400 mesh) and sedimenting the − 38 μm (minus 400 mesh) material. This type of result, seen over and over again, is often erroneously taken to mean that grinding of fine material is not occurring and ultrafines are not being produced. If the sieved material is now carefully washed, it is found that some fine material washes through to the subsieve fraction. After washing, the loss in weight of the upper sizes hardly affects the size distribution in the upper sizes, but the gain in weight of the very fine material straightens out the curve.

For dense materials with particles of *normal* shape, we prefer size analysis by sedimentation using the Andreasen pipette technique or the x-ray Sedigraph (Oliver, Hickin, and Orr, 1971). The latter instrument has the disadvantage that it requires rather a large volume ratio of solid in liquid, but it is rapid and appears to work well for solids as dense or denser than sand in water. It does not work on coal due to the small differences in density between coal and water (or other liquid). Neither does it work on porous material such as petroleum coke, due to the uncertainty of the effective density to be used. The Coulter Counter, being a measure of volume irrespective of particle density or shape, would seem to be ideal, but it *loses* fine material which it cannot see. Again, this gives the effect shown by curve *d* in Fig. 2. The same effect is seen in subsieve size distributions measured in the Microtrac* instrument: Austin and Shah (1983) have given a procedure for correcting for missed material in this instrument which is suitable for materials such as coal.

Micromesh sieves[†] can also give good results; the sieve is dried at 70°C in a continuous air flow, weighed, about 0.1 g of dried sample placed in the screen, and the screen reweighed. It is then mounted in a holder so that it just dips below the level of liquid in a small ultrasonic bath (300 W, 28 kHz). If necessary, a few drops of wetting agent are added to the material on the screen to assist flooding by the bath liquid. The bath is turned on and the screen washed with a fine spray from a wash bottle until no further flow of powder through the screens can be seen (10 to 20 min). The screen and contents are then dried and weighed. Unfortunately, even with the most careful handling, the 5, 10, 15, and 20 μm electroformed screens are so delicate that they rapidly puncture, leading to passage of oversize material.

To date, whenever we have carefully performed extension of the sieve size distributions to subsieve sizes, allowing for shape factor differences, we have concluded that the results are in agreement with predictions from the extension of the power functions of the S and B parameters (Eqs. 1 and 5 in Chapter 5) to small sizes for first-order grinding (see Fig. 1 in Chapter 2). As a general rule, if the shape of a size distribution changes at or close to the size at which the method of size analysis changes, the change in slope must be considered as suspect.

Tracer Techniques

The Kelsall technique for determination of S and B values in a continuous test (see the "Introduction" in this chapter) consists of the determination of the amount of tracer *in each size fraction* in each time collection because this gives the size distribution produced at each residence time. The disappearance of the top traced size j with time then gives S_j, and the

---

*Leeds and Northrup Instruments, 3000 Old Roosevelt Blvd., St. Petersburg, FL 33702.
[†]Buckbea Mears Co., Micro Products Division, 245 E. 6th Street, St. Paul, MN 55101.

distribution of tracer over the size intervals resulting from a single size of traced feed, for a short grind time, gives $B_{i,j}$, by the same calculations as discussed above. Since the data is taken from the mill exit material, there is no guarantee that the extent of grinding can be made small enough to get good B values. Similarly, for batch tests, it is sometimes necessary to determine S values of a fine size breaking in an environment of coarser sizes, which is again conveniently done by tracing the size of interest in the feed material since tracing enables the disappearance of this material to be distinguished from production of this size by breakage of the coarser particles.

Radiotracing techniques are very useful because of their speed and accuracy. A typical procedure is as follows. The apparatus consists of a sodium iodide detector and photomultiplier unit connected to a single channel analyzer with digital output. The sample to be counted is placed in a standard holder on the detector as shown in Fig. 10, *filling a set volume of the sample holder*. A statistically significant number of counts is obtained, corrected for background count rate determined using a blank (nontraced) sample of identical geometry, and the result is always used as a fraction of

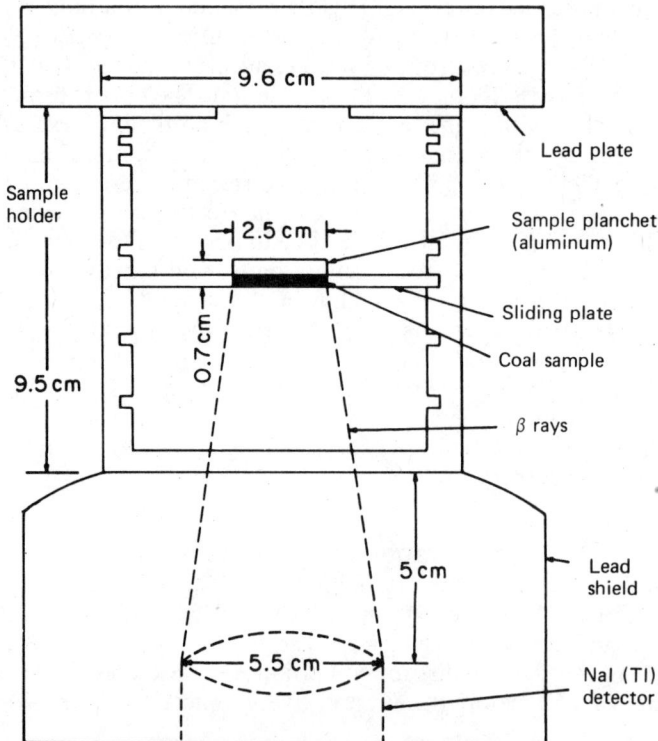

Fig. 10. Detailed features of sample holder and the detector used for radiotracing experiment.

the total count rate determined just before or after (or the arithmetic mean of the before and after measurements).

For example, consider the determination of the fraction of starting traced material of size i still in size i after some time of grinding. A sample is sieved to obtain the fraction by weight of size i, w_i. If the counts per minute per gram on size i and on the total sample are c_i and c_T, the fraction still in size i is clearly

$$w_i^*(t) = w_i c_i / c_T \qquad (7)$$

This ratio technique avoids all problems of calibration, and the use of a standard filling level and sample-detector geometry ensures exactly comparable results between different samples without the necessity for corrections for self-absorption, geometric effects, or other factors.

The radiotracing method is especially convenient when the material under study gives a suitable level of activity, with a suitable half-life, following irradiation by thermal neutrons in a nuclear reactor. With ores and rocks it is frequently a trace component which has a high capture cross-section for thermal neutrons and which gives the radiation suitable for counting. A preliminary x-ray spectrograph will indicate the appropriate channel setting (the radiation energy band to be counted) or it may be convenient to count all radiation with energy above a lower discrimination setting.

Gardner and Austin (1962) have described a number of checks and corrections which must be applied to the results. First, it must be demonstrated that irradiation does not alter the size distributions produced by milling, as they showed with the coals they investigated. The time for counting must be sufficient to get good counting statistics, as evidenced by convergence to a stable count rate (about 10^4 counts). Third, it is essential to check that the count rate of a given size interval is indeed truly proportional to the mass of tracer in that size fraction. This is done by irradiating a single size fraction, grinding in a batch mill, followed by a careful screen analysis. The fractional count rate per gram must be constant for each size interval if the technique is to be valid. This will not occur if the tracer is some mineral fraction which concentrates in a particular size range on grinding. Finally, it is necessary to correct the counts appearing in the size interval *above* the traced feed size interval and counts appearing in the size interval *below* the test interval, without grinding. This is exactly equivalent to the blank incomplete-sieving correction discussed above, and is handled in the same way; traced material found in the blank in the larger size is taken as of size, while traced material in the lower size is taken as being present in this size at the start of grinding.

Summary

Batch tests are the most accurate method for obtaining S and B values for grinding in laboratory scale mills because there are no effects of RTD or variation of hold-up to complicate the analysis. S_j and $B_{i,j}$ values can be determined by batch grinding a top size of j for various times (one-size fraction method), with sampling and screen analysis at each time. A sample of feed should always be screened without grinding to determine the sieving error. Spoon or ladle sampling from the mill is satisfactory for dry grinding but not for wet grinding, where the whole mill contents must be removed, filtered, dried, and mixed before sampling. A suitable screening schedule must be developed to approximate complete screening without excessive abrasion (which allows some near-size particles to pass through the screen). Sample weight must be decreased for fine samples to prevent blinding of screens, e.g., no more than 10 g of fines through a 45 μm (325 mesh) 200-mm (8-in.) diam screen. On the other hand, sample weight must be increased for coarse sizes to ensure the collection of a representative number of large particles. The weight of material lost during screening should be added as subsieve material and should not exceed 0.4%.

If the initial rate of breakage of the top size is higher than the normal rate, the effect may be due to a fraction of flaky material in the feed material. This effect can be removed by *preconditioning* and rescreening. In some cases the rate of breakage increases or decreases toward the later part of the test; this can be due to *material* and/or *environment* effects. An example of a material effect is the remaining material being weaker or stronger than the feed. An example of an environment effect is the accumulation of fines causing a cushioning action. Collection of the remaining fraction of top size from enough tests to perform a fresh test on the collected material enables these two effects to be distinguished. A material effect gives immediately the same rate of breakage as when the material was collected, whereas an environment effect will give initially the same rate of breakage as the original feed followed by the same change with time. The slowing down of grinding rates due to the accumulation of fines in the mill charge can be best seen from Weibull plots.

Best results for B values are obtained for about 20 to 30% by weight of the top size broken. Assuming no rebreakage of fragments

$$B_{i,1} = \frac{P_i(t) - P_i(0)}{P_2(t) - P_2(0)}, \qquad i > 1, \quad \text{BI method} \qquad \text{(S1)}$$

A better method is to use the compensation condition to correct approximately for rebreakage

$$B_{i,1} = \frac{\log\left[(1 - P_i(0))/(1 - P_i(t))\right]}{\log\left[(1 - P_2(0))/(1 - P_2(t))\right]}, \qquad i > 1, \quad \text{BII method} \quad \text{(S2)}$$

If values of S are available, the BIII method is preferred; it is necessary to use this method for breakage of sizes to the right of the maximum in S. It must be remembered that the direct determination of B values from short time grinding data is subject to considerable experimental variability.

It is convenient to use the empirical fitting equations

$$B_{i,j} = \Phi_j\left(x_{i-1}/x_j\right)^{\gamma} + \left(1 - \Phi_j\right)\left(x_{i-1}/x_j\right)^{\beta}, \qquad i > j \qquad \text{(S3)}$$

$$\left.\begin{aligned} \Phi_i &= \Phi_0\left(x_i/x_0\right)^{-\delta} \\ &= 1, \qquad\qquad \Phi_0\left(x_i/x_0\right)^{-\delta} \geq 1 \end{aligned}\right\} \qquad \text{(S4)}$$

Some materials give normalized breakage, that is, $\delta = 0$. Other materials give $\delta \geq 0$ for breakage of sizes smaller than the size where the maximum in S values occurs.

If the B values are approximately constant for one set of test conditions to another, and if the S_j values change by a constant multiple, the sets of batch size distributions are identical except for a shift in the time scale. The Weibull plot is a convenient method of interpolation to a fixed percentage less than size, in order to get the ratio of the time scales involved.

It is not easy to extend sieve size distributions to the subsieve region experimentally because: (a) the definition of size is different for the subsieve analysis technique, requiring a correction factor to size; (b) some subsieve size analysis methods do not measure the quantity of fine particles. Methods using wet dispersion are recommended.

Radiotracing is recommended for the determination of breakage parameters of a traced fraction in the presence of other sizes in the starting feed. The traced material is prepared by thermal neutron irradiation in a reactor and mixed into the mill charge. The technique determines the count rate of a weighed amount of sample (of a given size interval) filling a known standard volume in a sodium iodide detector system. This count rate is converted to count rate for all of the mass in that size interval, and divided by the total count rate for all material, to give the fraction of traced material in each size interval. Enough counts must be taken to get good counting statistics (e.g., 10^4). It is necessary to check that the count rate is proportional to mass of traced material, which is done by irradiating a total feed, grinding, screening, weighing, and counting. The fraction by weight from counting should be the same as the fraction by weight by weighing, that is, the same count rate per gram for all material.

References

Allen, T., 1975, *Particle Size Measurements*, 2nd ed., Wiley, New York.

Aplan, F., 1975, Private communication, Mineral Processing Section, The Pennsylvania State University.

Austin, L. G. and Luckie, P. T., 1972, "Methods for Determination of Breakage Distribution Parameters," *Powder Technology*, Vol. 5, pp. 215–222.

Austin, L. G. and Luckie, P. T., 1972a, "Estimation of Non-normalized Breakage Distribution Parameters from Batch Grinding," *Powder Technology*, Vol. 5, pp. 267–277.

Austin, L. G. and Shah, I., 1983, "A Method for Inter-Conversion of Microtrac and Sieve Size Distributions," *Powder Technology*, Vol. 35, pp. 271–278.

Frühwein, P., 1976, "Algorithm for Estimating the Process Parameters of Continuous Grinding," *Proceedings*, Fourth European Symposium Zerkleinern, H. Rumpf and K. Schonert, eds., Dechema Monographien 79, Nr. 1576-1588, Verlag Chemie, Weinheim, pp. 505–518.

Gardner, R. P. and Austin, L. G., 1962, "A Radioactive Tracer Technique for the Determination of Breakage Functions," *Proceedings*, First European Symposium Zerkleinern, H. Rumpf and D. Behrens, eds., Verlag Chemie, Weinheim, pp. 217–231.

Gardner, R. P. and Sukanjnajtee, K., 1973, "Combined Tracer and Back-Calculation Method for Determining Particulate Breakage Functions in Ball Milling. Part III. Simulation of an Open Circuit Continuous Milling System," *Powder Technology*, Vol. 7, pp. 169–179.

Gardner, R. P., Rogers, R. S. C., and Verghese, K., 1980, "Development and Use of a Short-Lived Radiotracer Method for Applying the Mechanistic Approach to Industrial Continuous Ball Mills," Prep. European Symposium, *Particle Technology*, K. Leschonski, ed., Vol. A, Amsterdam.

Kelsall, D. F. and Reid, K. J., 1965, "The Derivation of a Mathematical Model for Breakage," American Institute of Chemical Engineers—Institution of Chemical Engineers Symposium, Series No. 4, pp. 4–20.

Kelsall, D. F., Reid, K. J., and Restarick, C. J., 1967–68, "Continuous Grinding in a Small Wet Ball Mill. Part I. Influence of Ball Diameter," *Powder Technology*, Vol. 1, pp. 291–300.

Kelsall, D. F., Reid, K. J., and Restarick, C. J., 1968–69, "Continuous Grinding in a Small Wet Ball Mill. Part II. A Study of the Influence of Hold-up Weight," *Powder Technology*, Vol. 2, pp. 162–168.

Olivier, J. P., Hickin, G. K., and Orr, C., Jr., 1971, "Rapid Automatic Particle Size Analysis in the Sub Sieve Range," *Powder Technology*, Vol. 4, pp. 257–263.

Snow, R. H. and Meloy, T. P., 1972, "Measurement of Breakage and Grinding Rate Functions in Ball Milling by Tracer Experiments Using Nuclear Activation Analysis," *Proceedings*, Third European Symposium Zerkleinern, H. Rumpf and K. Schonert, eds., Dechema Monographien 69, Nr. 1292-1326, Verlag Chemie, Weinheim, pp. 535–555.

10

Back-Calculation of Breakage Parameters from Batch and Continuous Mill Data

Introduction

The previous chapter outlined the methods which can be used to determine specific rates of breakage and B values directly by experiment in batch ball mills. It was pointed out that the techniques were not directly applicable to continuous systems because of the added complications of RTD and variable mass hold-up. In this chapter, we outline the techniques for *indirect* determination of S and B values by *back-calculation* from experimental data. The technique uses a computer search program to find the values of characteristic parameters for S and B which make the simulated results of the mill model agree as closely as possible with a set of experimental or plant data. The advantages of a back-calculation method are that: (1) it uses all the available data in the calculation simultaneously, thus spreading error; (2) it can be used on limited data, thus reducing the need for major amounts of experimental work; and (3) it can be applied to continuous full-scale data. The major disadvantage is that it forces the data to fit the assumptions of the proposed model, and it is not always possible to detect when certain assumptions are not valid.

Considering the results shown in Fig. 1 in Chapter 2, it is clear that the computed results agree with the experimental results with considerable accuracy. Thus, solution of the batch grinding equations with an appropriate set of S and B values gives a close match to the experimental data for batch grinding. It should be possible, then, to go backwards: Given the experimental data, what are the correct sets of S and B values necessary to generate this data? Kelsall and Reid (1965) were the first to succeed in this back-calculation. They used an analog computer plus an X-Y recorder, with the analog of the set of batch grinding equations constructed on the computer. The set of values of S_i plus the normalized set of the b_{i-j} values were adjustable via individual potentiometers. The output was $w_i(t)$ vs. t on the plotter or oscilloscope screen, for chosen values of i. Values of S_i and b_{i-j} were adjusted by hand until the computed $w_i(t)$ agreed with the experimental $w_i(t)$ for these values of i, by visual match.

Klimpel and Austin (1970) and Klimpel and Austin (1977) developed a digital program, somewhat equivalent to the analog technique, which included analyses of sensitivity to parameter variations. One very important point should be made right away: The values of S and B inserted into the solution of the batch grinding equation can be substantially in error, yet the *forward computation* will give reasonable match between experimental and computed size distributions. Large errors in S and B, especially if compensatory, will give only small errors in $P(x, t)$, which is fortunate. In the back-calculations, however, it automatically follows that small errors in $P(x, t)$ give large errors in S and B, which is unfortunate. If experimental results do not follow first-order laws exactly, a forward computation assuming first-order gives quite good results, but in a back-calculation, the use of non-first-order experimental data throws all the error into the computed values of S and B. Thus it has proved difficult to construct a digital search program to match the human judgment and experience used in the analog search procedure. Klimpel and Austin did not find it possible to calculate the values of $S_1, S_2, \ldots, S_{n-1}$ and $b_1, b_2, \ldots, b_{n-1}$ independently of one another and get meaningful values which agreed with direct experimental determination.

Consequently, the program developed by Klimpel and Austin used simplifying assumptions concerning the functional forms of S and B with respect to particle size, in order to reduce the number of parameters in the search. In one option, the forms chosen for the S and B values were

$$S_i = A(x_i/x_1)^{\alpha}, \quad n \ge i \ge 1 \tag{1}$$

$$B_{i,1} = \Phi_1 R^{(i-2)\gamma} + (1 - \Phi_1) R^{(i-2)\beta}, \quad n \ge i \ge 1 \tag{2}$$

where R is the geometric sieve ratio (usually $1/\sqrt{2}$), as discussed in "Variation of Breakage with Particle Size" in Chapter 5 and "Analysis of Data: B Values" in Chapter 9, and A is S_1. Then, assuming B to be a normalized set, this reduced the unknown parameters to $\alpha, A, \gamma, \beta, \Phi_1$. For nonnormalized B values (see Chapter 9) they used

$$\Phi_j = \Phi_1(x_j/x_1)^{-\delta}, \quad \delta > 0 \tag{3}$$

which introduced the additional parameter δ. For S values passing through a maximum with size they used

$$S_i = A(x_i/x_1)^{\alpha} Q_i \tag{4}$$

where the correcting factors Q_i were described by a two-parameter log-normal distribution, thus introducing two more parameters.

They demonstrated that, in order to get reliable values for the B parameters (Φ_1, γ, β, and δ) from batch grinding data, it was necessary to include short grinding time data on a one-size-fraction feed. In order to get a good value for the parameter α, it was necessary to have longer grinding time data. Thus the data of Fig. 1 in Chapter 2 is the type of data necessary to get reliable back-calculated values for the complete set of parameters.

Lynch et al. (1977) were among the first to back-calculate breakage parameters from continuous full-scale data. They assumed the mill to be fully mixed and took the B values to be the functional form

$$B_{ij} = \frac{1 - \exp\left[-\left(x_i/x_{j+1} \right)^u \right]}{1 - \exp[-1]} \tag{5}$$

where u is an adjustable parameter, taken as 1.0 for ball milling. Kelsall, Reid, and Restarick (1967–68) assumed a constant vector of normalized B_{ij} values and a mill residence time distribution (RTD) equivalent to one large fully mixed reactor followed by two equal smaller fully mixed reactors (one-large/two-small model), and back-calculated S_i values from the feed and product size distributions to a full-scale mill. This technique has also been used by Hodouin, Berube, and Everell (1979) but using values for the B vector determined by laboratory tests on the same ore. All of these investigators performed an *interval-by-interval* back-calculation of S_i values, where S_1 is determined first and used to calculate S_2, then S_1 and S_2 used to calculate S_3, etc. Herbst, Rajmani, and Kinneberg (1977) have developed a search routine to back-calculate the descriptive breakage parameters of functional forms for S_i and B_{ij}, for an input vector of values representing the experimental residence time distribution for a mill. Fruhwein (1976) has described a back-calculation procedure designed to use size distributions sampled along a stopped dry mill, assuming the convective-diffusion equation to apply; no application to real data was demonstrated.

Each of these methods has advantages and disadvantages and certain conditions which must be met in order for the back-calculated values to have physical validity. However, before discussing these in detail, the back-calculation program options (Klimpel and Austin, 1984) which we have developed by an evolutionary process over the past ten years will be described, since they will be used to illustrate the success and shortcomings of the various techniques. The programs are available on request from Dow Chemical Co., but they must be used within the constraints outlined below if they are to give meaningful values with reliable statistical estimates of variance.

Back-Calculation from Batch Grinding Data

Program A is based on the Reid (1965) solution of the continuous grinding equation set, as modified by Gardner, Verghese, and Rogers (1980) (see "Continuous Milling and Residence Time Distribution" in Chapter 6)

$$p_i = \sum_{j=1}^{i} a_{ij} e_j, \qquad n \leq i \leq 1 \tag{6}$$

where

$$
a_{ij} = \begin{cases} f_i - \displaystyle\sum_{\substack{k=1 \\ i>1}}^{i-1} a_{ik}, & i = j \\[3ex] \dfrac{1}{S_i - S_j} \displaystyle\sum_{k=j}^{i-1} S_k b_{ik} a_{kj}, & i > j \end{cases}
$$

and

$$
e_j = \int_0^{\infty} \exp(-S_j t)\phi(t)\, dt
$$

For batch/plug flow

$$
e_j = \exp(-S_j t)
$$

The values of S_i and B_{ij} are assumed to be of the form of Eqs. 1 through 4, with the values of Q_i represented by

$$
Q_i = 1 \Big/ \left(1 + \frac{x_i}{\mu}\right)^{\Lambda} \tag{7}
$$

Thus the descriptive parameters are A, α, μ, Λ, Φ, γ, β, and δ.

The program input consists of a minimum of three size distributions including the feed, at $t = 0$, t_1, and t_2, in the form of $P(x_i, t)$ values. The program searches for the best set of A, α, μ, Λ, Φ, γ, β, and δ values to minimize the error between computed and experimental $P(x_i, t)$ values, assuming perfect first-order breakage. The objective function used is

$$
\text{minimize } SSQ = \sum_k \sum_{i=1}^{n} w_i (p_i \text{ observed} - p_i \text{ computed})^2 \tag{8}
$$

where the summation over k is for all pairs of times $t = 0$ to t_1, $t = 0$ to t_2, etc. The details of the rest of the program are similar to those for Program B discussed in the next section. The weighing factors w_i depend on the error structure of the data, which is determined by performing replicate batch grinding tests (see "Statistical Analysis" in this chapter).

The technique to determine the descriptive parameters is as follows. First, a size of feed is tested which is certain to give normal breakage, that is, a size to the left of the maximum in S vs. x. A minimum of three size distributions are determined at three grinding times and α, β, γ, Φ_1, and A determined from the printout, as in Table 1, using $\delta = 0$. The times are chosen to give the size distributions corresponding approximately to the size distributions of 1, 2, 15 min in Fig. 1 in Chapter 2, because the short time

Table 1. Summary of Computer Output for Back-Calculation of *S* and *B* Parameters from Batch Grinding Data

```
SIMPLE POWER FUNCTION SEARCH FOR 20 X 30 QUARTZ


    ABSOLUTE LEAST SQUARES CURVE FITTING

    TOL    0.9999999E-03  0.1000000E-05  0.9999999E-03  0.9999999E-03

    CONSTRAINTS WITH VARIABLE NUMBER AND RHS FOR
    PARAMETER GREATER THAN OR EQUAL TO CONSTANT

 1   0.100000E-01   2   0.100000E-01   3   0.100000E-01
 3  -0.990000E+00   4   0.100000E-01   5   0.100000E-01

STEP   A     ALPHA      MU   LAMBDA    PHI  GAMMA   BETA  DELTA      SSQ
 0   0.5000  1.0000   0.0     0.0    0.4500 1.2000  5.0000  0.0   -0.247427
 1   0.5651  0.9769   0.0     0.0    0.4840 1.1845  4.9990  0.0   -0.112945
 2   0.5259  0.9292   0.0     0.0    0.5505 1.1356  4.9988  0.0   -0.075015
 3   0.5084  0.8072   0.0     0.0    0.4924 1.1433  5.0015  0.0   -0.066176
 4   0.5130  0.8063   0.0     0.0    0.4731 1.1283  5.0041  0.0   -0.065703
 5   0.5127  0.8087   0.0     0.0    0.4706 1.1203  5.0056  0.0   -0.065661
 5   0.5127  0.8087   0.0     0.0    0.4706 1.1203  5.0056  0.0   -0.065661

            VARIABLES HELD CONSTANT
    0      0      1      1      0   0      0       1

    RESIDENCE TIME FORM = 0     CONSTANT =    0.0

    SIEVE RATIO =   0.7070     MAX SIZE =   840.0000

    INT    S VALUE      SIZE
     1     0.5127     1.0000
     2     0.3873     0.7070
     3     0.2926     0.4998
     4     0.2211     0.3534
     5     0.1670     0.2498
     6     0.1262     0.1766
     7     0.0953     0.1249
     8     0.0720     0.0883
     9     0.0544     0.0624
    10     0.0         0.0441


    B VALUE
 J =   1
 1.0000 0.4125 0.2329 0.1497 0.1000 0.0676 0.0458 0.0310 0.0210
 J =   2
 1.0000 0.4125 0.2329 0.1497 0.1000 0.0676 0.0458 0.0310
 J =   3
 1.0000 0.4125 0.2329 0.1497 0.1000 0.0676 0.0458
 J =   4
 1.0000 0.4125 0.2329 0.1497 0.1000 0.0676
 J =   5
 1.0000 0.4125 0.2329 0.1497 0.1000
 J =   6
 1.0000 0.4125 0.2329 0.1497
 J =   7
 1.0000 0.4125 0.2329
 J =   8
 1.0000 0.4125
 J =   9
 1.0000
```

Table 1. Continued

```
        PREDICTED CURVES

        T =   0.0                      T =   0.333
******FEED******    ****************PRODUCT*****************
```

EXP CUM	EXP DIF	EXP CUM	EXP DIF	PRED CUM	PRED DIF	RESIDUAL
1.0000		1.0000		1.0000		
0.0327	0.9673	0.1950	0.8050	0.1845	0.8155	-0.0105
0.0	0.0327	0.0771	0.1179	0.0722	0.1123	-0.0049
0.0	0.0	0.0402	0.0369	0.0409	0.0314	0.0007
0.0	0.0	0.0252	0.0150	0.0261	0.0147	0.0009
0.0	0.0	0.0185	0.0067	0.0175	0.0087	-0.0010
0.0	0.0	0.0107	0.0078	0.0118	0.0057	0.0011
0.0	0.0	0.0067	0.0040	0.0080	0.0038	0.0013
0.0	0.0	0.0044	0.0023	0.0054	0.0026	0.0010
0.0	0.0	0.0029	0.0015	0.0037	0.0017	0.0008
0.0	0.0	0.0	0.0029	0.0000	0.0037	0.0000

```
        SPECIFIED SIGMA = 1.0000   MAX KS VALUE = 0.4917   N = 10
            SIGN TEST IN T FORM =       1.26   DEG. FREE. =   9


        T =   0.0                      T =   1.000
******FEED******    ****************PRODUCT*****************
```

EXP CUM	EXP DIF	EXP CUM	EXP DIF	PRED CUM	PRED DIF	RESIDUAL
1.0000		1.0000		1.0000		
0.0327	0.9673	0.4011	0.5989	0.4207	0.5793	0.0196
0.0	0.0327	0.2112	0.1899	0.2126	0.2081	0.0014
0.0	0.0	0.1200	0.0912	0.1253	0.0873	0.0053
0.0	0.0	0.0810	0.0390	0.0807	0.0446	-0.0003
0.0	0.0	0.0561	0.0249	0.0540	0.0267	-0.0021
0.0	0.0	0.0370	0.0191	0.0366	0.0174	-0.0004
0.0	0.0	0.0241	0.0129	0.0249	0.0117	0.0008
0.0	0.0	0.0159	0.0082	0.0169	0.0080	0.0010
0.0	0.0	0.0107	0.0052	0.0115	0.0054	0.0008
0.0	0.0	0.0	0.0107	0.0000	0.0115	0.0000

```
        SPECIFIED SIGMA = 1.0000   MAX KS VALUE = 0.4900   N = 10
            SIGN TEST IN T FORM =       1.26   DEG. FREE. =   9


        T =   0.0                      T =   3.000
******FEED******    ****************PRODUCT*****************
```

EXP CUM	EXP DIF	EXP CUM	EXP DIF	PRED CUM	PRED DIF	RESIDUAL
1.0000		1.0000		1.0000		
0.0327	0.9673	0.7900	0.2100	0.7922	0.2078	0.0022
0.0	0.0327	0.5535	0.2365	0.5541	0.2382	0.0006
0.0	0.0	0.3754	0.1781	0.3733	0.1807	-0.0021
0.0	0.0	0.2610	0.1144	0.2524	0.1209	-0.0086
0.0	0.0	0.1850	0.0760	0.1724	0.0800	-0.0126
0.0	0.0	0.1283	0.0567	0.1183	0.0541	-0.0100
0.0	0.0	0.0810	0.0473	0.0812	0.0371	0.0002
0.0	0.0	0.0549	0.0261	0.0556	0.0256	0.0007
0.0	0.0	0.0312	0.0237	0.0380	0.0176	0.0068
0.0	0.0	0.0	0.0312	-0.0000	0.0380	-0.0000

```
        SPECIFIED SIGMA = 1.0000   MAX KS VALUE = 0.4826   N = 10
            SIGN TEST IN T FORM =       0.0    DEG. FREE. =   9
```

Table 1. Continued

	T =	0.0		T =	5.000	
*****FEED*****			************PRODUCT*****************			

EXP CUM	EXP DIF	EXP CUM	EXP DIF	PRED CUM	PRED DIF	RESIDUAL
1.0000		1.0000		1.0000		
0.0327	0.9673	0.9239	0.0761	0.9255	0.0745	0.0016
0.0	0.0327	0.7428	0.1811	0.7647	0.1608	0.0219
0.0	0.0	0.5715	0.1713	0.5789	0.1858	0.0074
0.0	0.0	0.4218	0.1497	0.4179	0.1610	-0.0039
0.0	0.0	0.2915	0.1303	0.2953	0.1226	0.0038
0.0	0.0	0.2045	0.0870	0.2065	0.0888	0.0020
0.0	0.0	0.1512	0.0533	0.1434	0.0631	-0.0078
0.0	0.0	0.1079	0.0433	0.0990	0.0444	-0.0089
0.0	0.0	0.0676	0.0403	0.0681	0.0309	0.0005
0.0	0.0	0.0	0.0676	-0.0000	0.0681	-0.0000

SPECIFIED SIGMA = 1.0000   MAX KS VALUE = 0.4797   N = 10
SIGN TEST IN T FORM =   0.63   DEG. FREE. =   9

	T =	0.0		T =	10.000	
*****FEED*****			************PRODUCT*****************			

EXP CUM	EXP DIF	EXP CUM	EXP DIF	PRED CUM	PRED DIF	RESIDUAL
1.0000		1.0000		1.0000		
0.0327	0.9673	0.9960	0.0040	0.9943	0.0057	-0.0017
0.0	0.0327	0.9500	0.0460	0.9591	0.0352	0.0091
0.0	0.0	0.8534	0.0966	0.8685	0.0905	0.0151
0.0	0.0	0.7125	0.1409	0.7285	0.1400	0.0160
0.0	0.0	0.5716	0.1409	0.5714	0.1571	-0.0002
0.0	0.0	0.4237	0.1479	0.4270	0.1444	0.0033
0.0	0.0	0.2921	0.1316	0.3090	0.1179	0.0169
0.0	0.0	0.2199	0.0722	0.2191	0.0899	-0.0008
0.0	0.0	0.1553	0.0646	0.1533	0.0658	-0.0020
0.0	0.0	0.0	0.1553	0.0000	0.1533	0.0000

SPECIFIED SIGMA = 1.0000   MAX KS VALUE = 0.4808   N = 10
SIGN TEST IN T FORM =   0.63   DEG. FREE. =   9

	T =	0.0		T =	15.000	
*****FEED*****			************PRODUCT*****************			

EXP CUM	EXP DIF	EXP CUM	EXP DIF	PRED CUM	PRED DIF	RESIDUAL
1.0000		1.0000		1.0000		
0.0327	0.9673	0.9992	0.0008	0.9996	0.0004	0.0004
0.0	0.0327	0.9975	0.0017	0.9936	0.0060	-0.0039
0.0	0.0	0.9638	0.0337	0.9641	0.0294	0.0003
0.0	0.0	0.9085	0.0553	0.8885	0.0756	-0.0200
0.0	0.0	0.7734	0.1351	0.7643	0.1242	-0.0091
0.0	0.0	0.5798	0.1936	0.6147	0.1496	0.0349
0.0	0.0	0.4496	0.1302	0.4685	0.1462	0.0189
0.0	0.0	0.3350	0.1146	0.3437	0.1248	0.0087
0.0	0.0	0.2450	0.0900	0.2458	0.0979	0.0008
0.0	0.0	0.0	0.2450	-0.0000	0.2458	-0.0000

SPECIFIED SIGMA = 1.0000   MAX KS VALUE = 0.4686   N = 10
SIGN TEST IN T FORM =   0.63   DEG. FREE. =   9

data enables accurate estimates of A and the parameters of B, and the longer time data increases the reliability of the estimate of α. B values are most reliable when the feed is a single size interval only. The computation is repeated with δ as a variable, and the sum-of-squares (SSQ) with $\delta = 0$ is compared to the SSQ with optimum δ to see if the addition of δ has produced a statistically significant improvement in fit (see "Statistical Analysis"). If not, δ is taken as zero. Second, a larger feed, chosen to be on the right-hand side of the maximum in S, is also experimentally tested, and (a minimum of) two size distributions determined. The previously determined values of α, β, γ, and δ, plus Φ_1 and A scaled to the new top size, are used as fixed inputs to the back-calculation program, and the search performed on Λ and μ only.

The ability to fix one or more of the parameters during the search has proved very valuable because it enables values which are accurately known from experiment to be held at the correct value, and it enables the sensitivity of any of the parameters to controlled changes in the other parameters to be investigated. In particular, the B_{ij} values for larger sizes may be different from the normal values (see "Ball Size Effects on B Values" in Chapter 16), so it is necessary to input a known matrix of B_{ij} values to determine good values for μ and Λ. It should be noted that this program forces a first-order law into any non-first-order results obtained with larger sizes, so that μ and Λ are effective mean values based on

Table 2. Experimental Conditions for Determination of Breakage Parameters for Cement Clinker

Mill	
Inner diameter	195 mm
Length	175 mm
Volume	5250 cm^3
Balls	
Diameter range	(26.7 − 27.0) mm
Number	63
Volume filling	20% ($J = 0.20$)
Specific gravity	7.8
Average ball weight	77.5 g
Assumed porosity between balls	40% ($\varepsilon = 0.40$)
Quality	alloy steel
Cement	
Bulk specific gravity	1.685
Volume	178 cm^3
Weight	300 g
Void filling	42% ($U = 0.42$)
Speed	
Operational speed	84 rpm (80%c.s.)
Critical speed	105 rpm

Table 3. Results from Grinding Three Sizes of Cement Clinker

Size Interval US Mesh	4 × 6 US mesh* (ε = 0.01 fraction) Weight Percent Less Than Size				16 × 20 US mesh (ε = 0.02 fraction) Weight Percent Less Than Size					40 × 50 US mesh (ε = 0.01 fraction) Weight Percent Less Than Size				
	1.00 min	6.00 min	10.00 min	20.00 min	0.33 min	1.00 min	3.00 min	5.00 min	10.00 min	1.00 min	3.00 min	4.50 min	7.00 min	20.00 min
4	100.00	100.00	100.00	100.00										
6	30.27	66.30	74.05	92.23										
8	14.76	50.36	61.10	86.76										
12	10.00	45.50	56.00	84.26										
16	6.88	39.38	52.86	84.16	100.00	100.00	100.00	100.00	100.00					
20	5.50	37.22	51.50	83.16	19.58	45.85	83.43	94.65	99.92					
30	4.51	35.23	50.28	83.29	9.69	27.50	66.04	85.30	98.56					
40	3.83	33.49	49.15	82.86	6.05	18.36	50.56	72.43	94.82	100.00	100.00	100.00	100.00	100.00
50	3.21	31.38	47.55	82.29	4.30	12.96	38.28	58.75	87.24	26.00	58.52	73.30	87.00	99.72
70	2.74	29.16	45.39	81.29	3.08	9.48	29.06	46.72	76.52	14.65	39.60	53.83	71.01	97.83
100	2.28	26.40	42.16	79.29	2.28	7.07	22.08	36.46	64.48	10.01	28.05	39.71	55.78	92.60
140	1.93	23.32	37.86	75.26	1.69	5.02	16.80	28.26	52.77	7.03	20.47	29.81	43.24	83.81
200	1.51	19.42	31.71	65.86	1.27	3.96	12.75	21.73	42.30	5.10	15.06	22.18	33.30	72.65
270	1.35	16.00	26.00	52.00	0.92	2.95	9.61	16.65	33.38	3.70	11.20	16.67	25.65	60.80
400	1.10	12.95	21.00	44.00	0.70	2.22	7.28	12.72	26.13	2.75	8.33	12.49	19.34	49.64
Weight used for sieving, g	70.08	69.43	52.35	30.01	53.00	45.70	43.58	38.10	24.96	45.63	45.20	39.24	36.49	22.38
Weight loss, g added to −400 mesh	0.02	0.23	0.11	0.12	0.06	0.14	0.23	0.10	0.24	0.04	0.08	0.30	0.33	0.03

*See Table 1 in Chapter 1 for metric conversions.

Table 4. Back-Calculated Parameters and Experimental Parameters for Grinding Cement Clinker *M2*

Parameters	40 × 50 US mesh*		16 × 20 US mesh*	
	Back Calculation Value	Experimental Value	Back Calculation Value	Experimental Value
α	0.89	0.91	0.88	0.91
γ	0.90	0.90	0.88	0.87
β	3.99	4.01	4.00	3.99
δ	0.18	0.20	0.20	0.20
$S_{40 \times 50}$ min^{-1}	0.29	0.29		
$\Phi_{40 \times 50}$	0.61	0.60		
$S_{16 \times 20}$ min^{-1}			0.71	0.71
$\Phi_{16 \times 20}$			0.51	0.54
μ / x_1			1.51	
Λ			3.95	
SSQ	0.00080		0.00104	

*See Table in Chapter 1 for metric conversions.

first-order kinetics. Not surprisingly, the simple forward simulations of the results do not give very exact reproduction of the experimental batch data for this case, because the breakage of the large sizes is not first-order. However, the use of the parameters for continuous mill simulation, where the feed entering the mill has a complete size distribution with small fractions of large sizes, has proved quite successful (see Chapter 12).

It should also be noted that the back-S-and-B batch program is specifically designed to deal with test data using a narrow feed size. When it is applied to data with a wide range of feed size distribution, there is normally too much experimental error to hope to obtain accurate values of the B parameters, because the change of size distribution with time then becomes insensitive to γ values. Put another way, there will be a broad range of combinations of α, γ, A, and Φ which will give almost identical minimization of SSQ. The program will automatically vary one parameter (e.g., A or α) by fixed amounts from the optimum and search for the other parameters. Application of the F-test (see "Statistical Analysis") shows the statistically acceptable range of combinations of values.

The parameters for grinding of a cement clinker are considered as an illustration of the techniques described above. The size distributions for grinding 425×300 μm (40×50 mesh), 1.18 mm $\times$ 850 μm (16×20 mesh), and 4.75×3.35 mm (4×6 mesh) feeds under the conditions shown in Table 2 are given in Table 3. The first-order plots and S values were shown in Chapter 9, Fig. 3. Direct application of the back-S-and-B calcula-

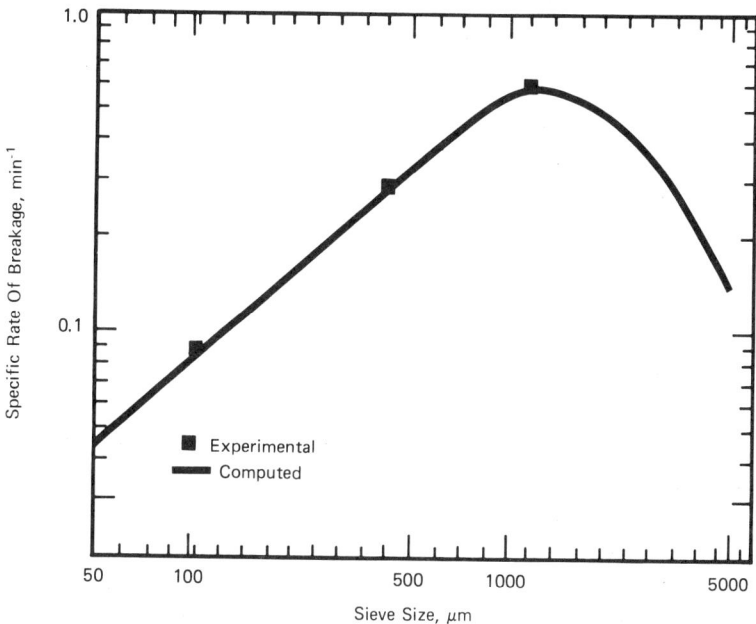

Fig. 1. S values for grinding of cement clinker $M2$ (see Table 2); $\sqrt{2}$ intervals.

Table 5. Final Values of Parameters for Cement Clinker $M2$

Parameter	Values
α	0.88
γ	0.88
β	4.00
δ	0.20
$S_{16 \times 20}$ min^{-1}	0.71
$\Phi_{16 \times 20}$	0.51
Λ	2.5
μ, mm	2

tion to the 425 × 300 μm (40 × 50 mesh) results using the "simple power function/nonnormalized B option" gave the values shown in Table 4. Better fits were obtained on the 1.18 mm × 850 μm (16 × 20 mesh) using the "complete S function/nonnormalized B option," giving the results shown in Table 4. The experimental values for the B parameters were obtained from the BII method and the α value from Fig. 1. Mean values of the parameters

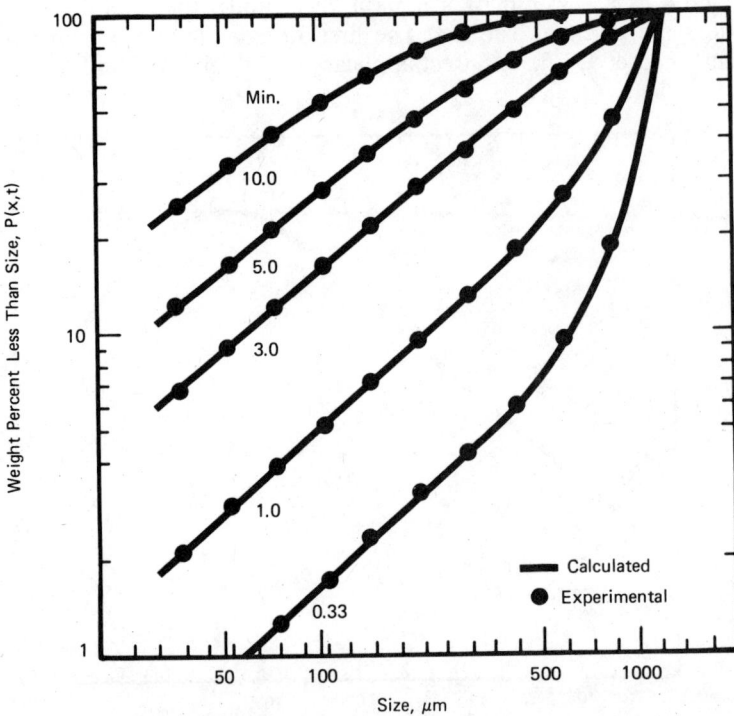

Fig. 2. Size distributions for 1.18 mm × 850 μm (16 × 20 mesh) cement clinker $M2$ (see Tables 2, 3, and 4).

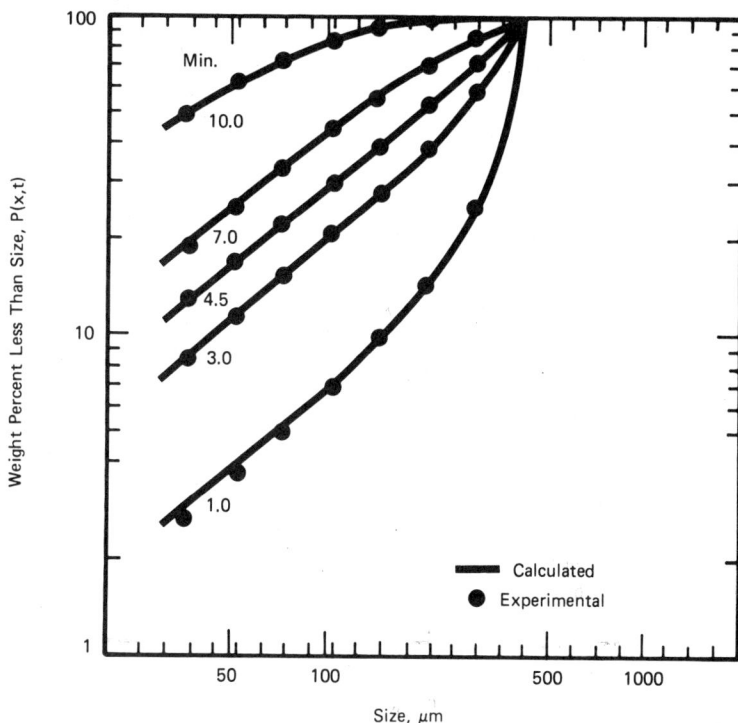

Fig. 3. Size distributions for 425 × 300 μm (40 × 50 mesh) cement clinker $M2$ (see Tables 2, 3, and 4).

were used in the "complete S function/nonnormalized B option" applied to the 4.75 × 3.35 mm (4 × 6 mesh) results, giving $\mu = 2$ mm and $\Lambda = 2.47$. The final value set decided upon is shown in Table 5. Figs. 2 and 3 show the agreement between forward computed and experimental data using these parameters.

Back-Calculation from Continuous Mill Data: Program Description

There are two distinct programs (Klimpel and Austin, 1984), both written in standard Fortran IV with similar input and output, using the same optimal search algorithm.

Program A: Open Circuit / Batch Only

Open Circuit here means that the size distribution of the mill feed and the mill product are the experimental inputs; if the mill is in closed circuit, the circulation ratio is estimated in the usual way and the mill feed size distribution calculated from the makeup feed and recycle size distributions and the circulation ratio. The mill algorithm is Eq. 6 and the objective

function is

$$\text{minimize } SSQ = \sum_{i=1}^{n} w_i (p_i \text{ observed} - p_i \text{ computed})^2 \qquad (8a)$$

The residence time distribution is introduced via the vector of e_j values and six options are available: (1) plug flow, to be used for batch grinding; (2) fully mixed; (3) a series of m equal fully mixed reactors; (4) the Gardner/Rogers RTD model; (5) the Mori semi-infinite solution to the convective-diffusion equation; and (6) the one-large/two-small model. If it is found that the experimentally determined RTD of a mill fits the RTD resulting from any of these models, then that model can be used in the computation. It should be recognized that the hold-up in a mill is not normally determined in a full-scale test, except when an RTD measurement is performed. However, it can be assumed that the form of the dimension-less RTD is similar to that measured on similar mills, and a formal mean residence time calculated assuming that the hold-up corresponds to complete interstitial filling of the ball load ($U = 1.0$). The back-calculated value of α is unchanged by varying τ, and the formal value of τ gives a formal value of A.

Program A has the advantage of flexibility in the choice of residence time distribution. However, the Reid solution (and all forms derived from it) tends to go unstable when there are a number of terms in which $S_i - S_j$ becomes small, as illustrated in the Examples section. This is particularly true when an attempt is made to use all the size distributions around a closed circuit in the objective function. For this reason Program B was developed as a completely stable computation.

Program B: Open and Closed Circuit: Three Reactor Model
The closed circuit mill algorithm used is a reduced form of that developed by Austin, Luckie, and von Seebach (1976) for an RTD of m unequal fully mixed reactors in series (see Chapter 17). The symbolism is given in Fig. 4,

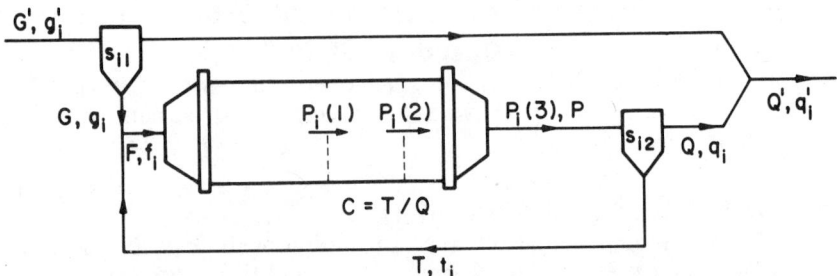

Fig. 4. General circuit nomenclature. The normal closed circuit is $s_{i1} = 1.0$; the reversed closed circuit is $s_{i1} = s_{i2}$; open circuit is $s_{i1} = 1.0$, $s_{i2} = 0$. G', G, F, P, T, Q, Q', represent solid flow rates; g_i', g_i, f_i, p_i, t_i, q_i, q_i' are corresponding size distribution vectors; s_{i1}, s_{i2} are classifier selectivity values.

and a mass-kinetic balance around the mill gives

$$p_i^*(3)$$

$$= \frac{g_i + \sum\limits_{j=1}^{i} b_{ij} S_j \left[\tau_1 p_j^*(1) + \tau_2 (1 + \tau_1 S_i) p_j^*(2) + \tau_3 (1 + \tau_1 S_i)(1 + \tau_2 S_i) p_j^*(3) \right]}{(1 + \tau_1 S_i)(1 + \tau_2 S_i)(1 + \tau_3 S_i) - s_{i2}}$$

(9)

where p^* is the Luckie and Austin (1972) metameter $p(1 + C)$ and

$$p_j^*(2) = p_j^*(3)(1 + \tau_3 S_j) - \tau_3 \sum_{k=1}^{j-1} b_{jk} S_k p_k^*(3)$$

$$p_j^*(1) = p_j^*(2)(1 + \tau_2 S_j) - \tau_2 \sum_{k=1}^{j-1} b_{jk} p_k^*(2)$$

This can be sequentially computed starting with $i = 1$, since g_i is known from g_i' by

$$g_i = g_i' s_{i1} \Big/ \sum_1^n g_i' s_{i1}$$

for set classifier numbers s_{i1}.

The postclassifier product q_i is

$$q_i = (1 - s_{i2}) p_i^*(3)$$

The size distribution of the overall circuit product is

$$q_i' = (1 - s_{i1}) g_i' + q_i \sum_{j=1}^{n} s_{j1} g_j'$$

and expressions for the other size distributions are readily derived.

By appropriate choice of τ_1, τ_2, and τ_3, where the overall mean residence time is $\tau = \tau_1 + \tau_2 + \tau_3$, the RTD model can clearly be one, two, or three equal fully mixed reactors, one-large/two-small reactors, or three unequal fully mixed reactors. Weller (1980) has indicated that the RTDs measured on a number of fully scaled mills can be reasonably approximated by the one-large/two-small model.

The objective function used is:

$$\text{minimize } F = W_1 \sum_i w_{1i} (q_i \text{ observed} - q_i \text{ computed})^2$$

$$+ W_2 \sum_i w_{2i} (f_i \text{ observed} - f_i \text{ computed})^2$$

$$+ W_3 \sum_i w_{3i} (p_i \text{ observed} - p_i \text{ computed})^2$$

$$+ W_4 \sum_i w_{4i} (t_i \text{ observed} - t_i \text{ computed})^2$$

$$+ W_5 \sum_i w_{5i} (q_i' \text{ observed} - q_i' \text{ computed})^2$$

$$+ W_6 (C_{\text{observed}} - C_{\text{computed}})^2 \qquad (10)$$

Normally, the weighting factors W are taken as 1 or 0 for ease of statistical analysis; up to five of them can be set equal to zero if desired.

The search algorithm used is a modified conjugate gradient algorithm (Fletcher and Powell, 1963; Goldfarb and Lapidus, 1968), which is very efficient when supplied with analytical first derivatives of F with respect to the optimization variables (A, α, μ, Λ, Φ, γ, β, and δ in the cases considered here). The back-calculation is normally performed by setting the B parameters to values determined in the laboratory, and calculating A and α. Then the calculation is repeated with μ and Λ as additional variables to see if significantly improved fit is obtained, if large feed sizes are present. Finally, the parameters of Φ and γ are allowed as additional variables and statistical improvement of fit again tested.

As will be discussed in Chapters 12 and 14, there is evidence that some mills can behave as if they had an internal classification occurring as the material leaves the mill. In Chapter 17, the section, "Mill Circuits Assuming First-Order and Simple RTD," shows how this classification is incorporated into the one-large/two-small model. Program B contains the option to incorporate an internal classification action via a set of classifier numbers which return material to the third stage. The classifier numbers are assumed to fit the usual logistic function (Chapter 13, Eq. 9)

$$c_i = \frac{1}{1 + (d_{50}/x_i)^\lambda}$$

where d_{50} and λ are characteristic parameters and c_i is the fraction of material of size i returned to the third stage. The values of d_{50} and λ can be entered as input if they are known from experimental studies. The program also contains the option to search for the optimun d_{50} and λ for the *external* classifier: this option is used when the data around the classifier is not sufficient to determine d_{50} and S.I. independently.

In addition to the above optimal search for the parameters of functional forms for S and B, options A and B were also programmed for the sequential calculation of S_i values (the interval-by-interval calculation) starting with $i = 1$, with B values supplied as input (as a normalized vector, a matrix, or by the descriptive parameters of Φ, γ, β, and δ). This was performed by an iterative binary-type search to find first the value of S_1 to make $|p_1$ observed $- p_1$ computed$| \leq \varepsilon$, where ε is a specified tolerance, then the value of S_2 to make $|p_2$ observed $- p_2$ computed$| \leq \varepsilon$, etc., for the *open-circuit* case where f_i, p_i are the experimental vectors used in the computations. Since S_1 is determined first and then used in the determination of S_2, etc., there is no ability to distribute error over the set of S_i values. If p_1 observed $= 0$, as is often the case in continuous mill data, it is not possible to calculate S_1 directly, so it must be calculated using some small but finite value for p_1 ($= \varepsilon$). All values of S_1 greater than some minimum value will make $|p_1$ observed $- p_1$ computed$| \leq \varepsilon$, since p_1 observed $= 0$. The program chooses the minimum value of S_i to obtain the desired match, with the tolerance taken as 0.1 times the estimated error in data (e.g., if data is accurate to 0.1%, that is, 0.01, ε is taken as 0.001). Because of accumulation of errors in the sequential calculation, it is possible for data with large experimental or model error to lead to a negative S_i value and the computation is stopped when this happens. In general, with real data, the interval-by-interval S_i values vary erratically with x_i, and it is usual to put a smooth curve through the plotted values (log S_i vs. log x_i) and pick off smoothed values for use in circuit simulations.

Statistical Analysis

One statistical test is to compare the variances (SSQ/degree of freedom) of residual error given for different values of the search parameters. For example, the optimal set of parameters can be determined, giving an associated SSQ residual error, SSQ_{optimal}. Then one parameter (e.g., α) is changed by a desired percentage (e.g., $+5\%$) from the optimal value and held constant while the search is repeated on the other parameters, giving an associated SSQ residual error, SSQ_1. Then the F test is applied to

$$F_1 = \frac{SSQ_1 \text{ residual error}/(n - N + 1)}{SSQ \text{ optimal residual error}/(n - N)} \tag{11}$$

The test indicates the significance level of increased error due to the nonoptimum value of α, for example. If the F test does not show a high significance level, this means that the new value of α (e.g., $1.05 \times \alpha_{\text{optimal}}$) plus its associated parameters are as good as the optimal values, within the accuracy of model inadequacy and experimental error. Repeating this procedure for ranges of parameters shows the sensitivity of residual error to variation in a particular parameter.

A more complete method of analysis requires an independent estimate of experimental error. The lack-of-fit SSQ is defined by

$$SSQ \text{ lack-of-fit} = \left[\sum_{i=1}^{n} \sum_{l=1}^{k} w_i (p_{i,l} - \bar{p}_i)^2 \right] / (k-1) \qquad (12)$$

where $\bar{p}_i$ is the arithmetic mean of k observations of $p_{i,l}$, and the degrees of freedom are $n(k-1)$. The F test is then applied to the ratio of variance of residual error to independent experimental error:

$$F_2 = \frac{SSQ \text{ residual error}/(n-N)}{SSQ \text{ lack-of-fit}/(n)} \qquad (13)$$

If the F test on this basis, with degrees of freedom of $n-N$ and $n(k-1)$, does not indicate significant difference ($F_3 < F_{n-N, n(k-1)}$), then the optimal model parameters give a model which does describe the data within the expected accuracy of the experimental data; thus, Eq. 13 is a measure of model adequacy.

As for Eq. 11, the SSQ resulting from parameters changed from optimum is used to investigate sensitivity of model adequacy to the choice of parameters:

$$F_3 = \frac{SSQ_1 \text{ residual error}/(n-N+1)}{SSQ \text{ lack-of-fit}/(n)} \qquad (14)$$

When $F_3 > F_{n-N+1, n}$, the new parameters give a statistically significant worse match of model to experiment.

In cases where several size distributions around a closed circuit are used in the back-calculation, the SSQs and degrees of freedom are appropriately defined for all the size distributions used. This type of calculation can be performed in a number of different ways because of the number of possible combinations of size distributions used to define the SSQ.

The weighting factors appropriate for analysis of plant data were deduced as follows. Table 6 shows triplicate size distributions obtained by sieving analysis of carefully replicated plant data. Each sample was a composite of slurry samples taken at 5-min intervals over a 30 min period of steady operation; the three composite samples were collected at the same time. It can be seen that size intervals containing larger quantities of material give larger values of $(p_i - \bar{p}_i)^2$. A plot of variance vs. $\bar{p}_i$ on log-log scales suggested that $V_i \propto \bar{p}_i$, so that weighting factors of $w_i = 1/\bar{p}_i$ gave random distribution of the weighted errors defined by $(p_i - \bar{p}_i)^2/\bar{p}_i$. A mean variance of these weighted errors defined by

$$V = \sum_k \sum_i \left[(p_i - \bar{p}_i)^2 / \bar{p}_i \right] / n(k-1)$$

was found to be 4×10^{-4}. The same weighting factors are assumed to apply to all the size distributions used in a closed circuit calculation and the mean variance calculated from replicates of all of these size distributions. Other

Table 6. Replicated Data (in triplicate) from a Full-Scale Mill Grinding Copper Ore

Size Interval	Mill Product p_i Values: Fraction			Mean $\bar{p}_i$	Unweighted Variance $\times 10^4$
1	0.010	0.008	0.001	0.0096667	0.023
2	0.008	0.009	0.012	0.0096667	0.043
3	0.026	0.031	0.028	0.0283333	0.063
4	0.049	0.054	0.052	0.0516667	0.063
5	0.076	0.082	0.084	0.0806667	0.173
6	0.084	0.094	0.089	0.0890000	0.250
7	0.095	0.106	0.103	0.1013333	0.323
8	0.101	0.104	0.095	0.1000000	0.210
9	0.092	0.093	0.087	0.0906667	0.103
10	0.083	0.079	0.077	0.0796667	0.093
11	0.084	0.076	0.078	0.0793333	0.173
12	0.062	0.057	0.059	0.0593333	0.063
13	0.041	0.036	0.044	0.0403333	0.163
14	0.045	0.037	0.042	0.0413333	0.163
15	0.032	0.034	0.026	0.0306666	0.173
16	0.113	0.100	0.113	0.1086667	0.563

data sets gave mean weighted variances of 16×10^{-4}, 11×10^{-4}, 8×10^{-4}, and 4.5×10^{-4}, depending on the care taken in sampling, the ease of getting good samples, and the number of size distributions used to calculate the variance.

A similar treatment of a triplicated wet batch grinding test on quartz in a 200-mm i.d. mill at 1, 3, 7, and 15 min of grinding gave the same weighting factors and an overall mean variance of 3×10^{-4}.

Two additional statistical tests (Blau, Klimpel, and Steiner, 1972) are routinely performed. First, the Kolmogorov-Smirnov goodness-of-fit test is applied to confirm that the weighted residual errors are normally distributed with a zero mean and the corrected variance; this checks that the weighting factors $1/p_i$ are appropriate for the data set investigated. Second, the Student t test form of the sign sequence test was used to confirm that a statistical treatment of sum-of-squares was sufficient. In this test, model inadequacy can be detected as a statistically significant sequence of all positive or all negative errors; a sum-of-squares analysis does not detect this type of model error. In all cases it was found that the F tests of Eqs. 13 and 14 were the most sensitive indicators of model inadequacy.

Back-Calculations on Continuous Large-Scale Data

Klimpel and Austin (1984) gave a number of examples to illustrate the problems of back-calculation of values from full-scale data. They concluded:

1. The interval-by-interval back-calculation rarely gives correct S_i values for the top sizes, and small errors in subsieve size measurement give large errors in the S_i interval-by-interval values for these sizes.

2. Back-calculation assuming a ball mill to be fully mixed gives radically incorrect S_i values (too large α) when applied to data from a ball mill with a real RTD.

3. In the presence of typical experimental error, it is not possible to deduce that the fully mixed RTD is incorrect (that is, the model is inadequate) and the incorrect S_i values reproduce the mill product size distribution. The use of the incorrect S_i values *at another mill feed rate* gives radically incorrect predictions of product size distribution.

4. Use of an RTD closer to plug flow than the correct RTD gives too small α, and vice versa.

5. Use of B values with a smaller γ than the correct value gives too large α, and vice versa.

6. Use of only the *circuit product* size distribution in the back-calculation (Herbst, Rajmani, and Kinneberg, 1977) sometimes gave a range of statistically acceptable values of A and α which did not include the experimentally determined α.

7. As many of the size distributions around a closed circuit as possible should be used in the back-calculation, since this gives the smallest range of statistically acceptable values and gives α values more in agreement with those determined by direct experiment.

Table 7 gives the size distributions from a closed-circuit large-scale (3 by 5 m) ball mill test (Klimpel and Austin, 1984) on wet grinding of a copper ore at 100 tph. Laboratory tests had given $\alpha = 0.92$, $\gamma = 0.67$, $\beta = 3.0$, and $\Phi = 0.57$ for the ore. Analysis of variance on triplicates of the circuit size distributions gave the weighted mean variance as 8×10^{-4}. The RTD was assumed to be the one-large/two-small model with relative size 0.70, 0.15, 0.15, and the classification action could be described by $a = 0.28$, $d_{50} = 191$ μm, and S.I. = 0.45. The back-calculation was performed on this data using all the circuit data to spread error, with the laboratory B parameters as fixed input. The optimum α value obtained was 0.96, with $a = 1.99$ per min and the range of $\alpha = 0.78$, $a = 1.4$ per min to $\alpha = 1.11$, $a = 2.7$ per min gave no statistically significant difference in lack-of-fit (95% confidence level). Thus at this level of confidence the value of α is $0.96 \pm 17\%$, which contains the value of α determined in the laboratory.

The method of statistical testing is illustrated in Table 8 and Fig. 5, where the F_3 values are 95% confidence level values. The same data used to perform the back-calculation using only the circuit product size distribution Q gave $\alpha = 0.86$, $A = 1.62$ per min, with an acceptable range of $\alpha = 0.55$, $A = 0.75$ per min to $\alpha = 1.2$, $A = 2.3$ per min, that is, $\alpha = 0.86 \pm 37\%$.

Back-Calculation Using Simple Approximate Mill Models

For some purposes it may be sufficient to use an approximate mill model based on the compensation condition (see Chapter 8), e.g., using the semi-infinite RTD (Chapter 8, Eq. 16)

$$P(x,\tau) = 1 - [1 - F(x)]\exp\left[-\left(\frac{\sqrt{1 + 4D^*\tau S(x)} - 1}{2D^*}\right)\right]$$

Table 7. Experimental and Model Size Distributions for a Closed-Circuit Ball Mill (back-calculations performed using Q, F, P, and T size distribution)

G_i	$Q_i$exp	$Q_i$mod	$F_i$exp	$F_i$mod	$P_i$exp	$P_i$mod	$T_i$exp	$T_i$mod
100	100	100	100	100	100	100	100	100
80	100	100	90	88.9	100	99.6	99	99.2
65	100	100	80	79.9	99	98.6	97	97.0
50	100	100	69	69.6	96	96.3	92	92.1
43	100	99.8	63	62.4	92	92.8	85	84.7
37	99	99.2	54	54.0	87	87.3	74	73.5
32	97	97.0	45	44.7	79	79.5	60	59.3
28	91	91.2	36	36.2	71	70.0	45	45.6
24	82	81.5	28	29.3	61	60.1	36	35.4
20	71	69.8	22	23.9	50	50.5	28	28.3
16	59	57.9	18	19.2	42	41.6	23	22.9
13	48	47.4	15	15.6	34	34.0	19	18.6

Table 8. Statistical Analysis of Data of Table 7 (A varied, search on α: $F_{47,96} = 1.57$ at 95% confidence level)

A	α	SSQ	Degrees of Freedom	F
1.99(optimal)	0.96(optimal)	0.0316	46	0.86
2.18	1.00	0.0348	47	0.92
1.79	0.90	0.0357	47	0.95
2.48	1.07	0.0467	47	1.24
1.49	0.81	0.0505	47	0.34
2.98	1.16	0.0797	47	2.12
0.99	0.63	0.1483	47	3.94

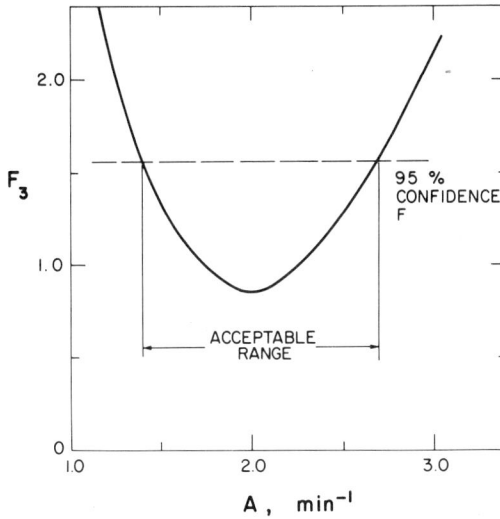

Fig. 5. F test applied to data of Table 7.

The values of $\tau S(x)$ are estimated from the continuous mill test data $P(x, \tau)$ and $F(x)$, for a reasonable value of D^* (for example, $D^* = 0.25$; see Chapter 14). A formal value of hold-up W is estimated from mill dimensions and a formal value of τ calculated from W/F. At any other flow rate, a new value of τ is calculated using the same formal W, and new values of $\tau S(x)$ follow, assuming the same D^*. The model again enables estimates to be made of the effect of changes in the flow rate and/or feed size distribution. These calculations are easily performed by hand.

Closed circuit computations with simple models using a set of classifier selectivity numbers s_i can be performed using the transfer function concept (see "Solution of Grinding Equations" in Chapter 6). The simple solutions of Chapter 8 can be put as

$$1 - P(x_i) = \left[1 - F(x_i)\right] k_i \tag{15}$$

where

$$k_i = \begin{cases} \exp[-S_{i-1}\tau] & \text{plug} \\ 1/(1 + \tau S_{i-1}) & \text{fully mixed} \\ \exp\left[-\left(\dfrac{\sqrt{1 + 4D^*\tau S_{i-1}} - 1}{2D^*}\right)\right] & \text{semi-infinite} \end{cases}$$

The transfer function equation is

$$p_i = \sum_{j=1}^{i} d_{ij} f_j$$

Using $p_i = P(x_i) - P(x_{i+1})$, $f_i = F(x_i) - F(x_{i+1})$ and performing some algebra with Eq. 15 gives

$$d_{ii} = k_{i+1} \qquad (16)$$

$$d_{ij} = d_{ii} - d_{i-1,i-1} = k_{i+1} - k_i \qquad (17)$$

From Eq. 13 in Chapter 6, then,

$$p_i^* = \frac{k_{i+1} g_i + (k_{i+1} - k_i) \sum_{j=1}^{i-1} \left(s_j p_j^* + g_i \right)}{1 - k_{i+1} s_i}, \qquad n \le i \le j \le 1 \quad (18)$$

where $p_i^* = (1 + C) p_i$, $1 + C = \sum_1^n p_i^*$.

Obviously, calculations performed using such a simple mill model must be anticipated to give incorrect results if mill conditions deviate very much from those used to back-calculate the S_i values.

Summary

In principle, it is possible to calculate the vector of S_j values and the matrix of $B_{i,j}$ values by back-calculation from sets of batch grinding data. In practice, quite large errors in individual S and B values will give a forward computation which agrees reasonably well with experimental size distributions. Conversely, however, small errors in the size distributions result in large errors in S_j and $B_{i,j}$ values, even giving negative values in some cases. Thus it has been found desirable to reduce the number of unknown parameters by using functional forms for S as well as B. For values of S to the left of the maximum, and normalized B, the back S-and-B computer program calculates the parameters A, α, Φ, γ, and β. It requires a minimum of three size distributions (at $t = 0$, t_1, and t_2). Short time data on a single size fraction of feed is necessary to give the program enough data to determine good Φ, γ, β estimates, and longer time data enables a good α estimate. The program has options to determine also δ, and using size distributions produced from breakage of large sizes, the values of μ, Λ can be estimated by fixing α, γ, β, Φ, and A.

It is also possible to calculate apparent overall S values from the feed and product size distribution obtained from a steady state, continuous mill test, assuming B values and knowing or assuming the residence time distribution. Digital computation and search procedures are required. This is usually the only method which is feasible for large-scale mills.

It is essential to perform statistical analysis to obtain an estimate of the reliability of back-calculated values. The best method for performing statistical analysis requires replicated data to obtain an estimate of the sum-of-squares (SSQ) lack-of-fit defined by

$$SSQ \text{ lack-of-fit} = \sum_{i=1}^n \sum_{l=1}^k w_i (p_{i,l} - \bar{p}_i)^2 / (k-1) \qquad (S1)$$

where k is the number of independent observations of each p_i value, and $\bar{p}_i$

is the arithmetic mean. The weighting factors w_i were found to be $1/p_i$ for plant data. If more than one size distribution is used in the back-calculation for a closed-circuit system, the SSQ is appropriately defined. Values of weighted variance (SSQ/degree of freedom) from plant trials were 4×10^{-4} to 16×10^{-4}, depending on the accuracy of sampling and analysis.

The back-calculation gives optimal values of the searched parameters (e.g., A and α). One of these is systematically varied and the search repeated on the other parameters. The resulting SSQ residual error is defined by

$$SSQ \text{ residual error} = \sum_{i=1}^{n} w_i (p_i \text{ observed} - p_i \text{ computed})^2 \quad (S2)$$

with degrees of freedom of $n - N + 1$, where N is the number of searched parameters in the optimal search. The F-statistic is then applied to

$$F = \frac{(SSQ \text{ residual error})/(n - N + 1)}{(SSQ \text{ lack-of-fit})/(n - N)} \quad (S3)$$

When F is greater than the F value from statistical tables for the appropriate degrees of freedom, the new (nonoptimal) values of parameters are statistically in error to a significant degree.

Back-calculations are usually performed assuming the form of the RTD (e.g., one-large/two-small reactor model) and using B values determined in the laboratory. Use of an assumed RTD closer to fully mixed than the real value gives too large α and A, and vice versa. Use of too large γ for the B values gives too small α, and vice versa. Narrower ranges of statistically acceptable values are obtained by using all the size distributions round a circuit, e.g., α was $0.96 \pm 15\%$ for one set of plant data.

Back-calculated parameters based on wrong assumptions will usually generate reasonably correct size distributions in comparison to the data set used for the back-calculation, but will give wrong values for other flow rates through the mill.

The use of a realistic RTD model and laboratory B values for back-calculation from well taken replicated plant data usually gives a range of α values which includes the α value determined in the laboratory. The statistically acceptable range of α values enables the mill model to be used to predict the likely *range* of size distributions resulting from simulations of mill behavior under various conditions.

A simple approximate model can be used for hand calculation of apparent S values, using the compensation condition and the semi-infinite RTD,

$$P(x, \tau) = 1 - [1 - F(x)]\exp\left[-\left(\frac{\sqrt{1 + 4D^*\tau S(x)} - 1}{2D^*}\right)\right] \quad (S4)$$

The procedure for closed circuit calculations is given. The model is likely to give valid simulations only for mill conditions close to those used in the back-calculation.

References
Austin, L. G., Luckie, P. T., and von Seebach, H. M., 1976, "Optimization of a Cement Milling Circuit with Respect to Particle Size Distribution and Strength Development by Simulation Models," *Proceedings*, Fourth European Symposium Zerkleinern, H. Rumpf and K. Schonert, eds., Dechema Monographien 79, Nr. 1576–1588, Verlag Chemie, Weinheim, pp. 519–537.

Blau, G., Klimpel, R., and Steiner, E., 1972, "Equilibrium Constant Estimation and Model Distinguishability," *Industrial and Engineering Chemistry, Fundamentals*, Vol. 11, pp. 324–332.

Fletcher, R. and Powell, M., 1963, "A Rapidly Convergent Descent Method for Minimization," *British Comp. Journal*, Vol. 6, pp. 163–168.

Fruhwein, P., 1976, "Algorithms for Estimating the Process Parameters of Continuous Grinding," *Proceedings*, Fourth European Symposium Zerkleinern, H. Rumpf and K. Schonert, eds., Dechema Monographien 79, Nr. 1576–1588, Verlag Chemie, Weinheim, pp. 505–518.

Gardner, R. P. and Sukarijnajtee, K., 1973, "Combined Tracer and Back-Calculation Method, Part III," *Powder Technology*, Vol. 7, pp. 169–179.

Gardner, R. P., Verghese, K., and Rogers, R. S. C., 1980, "The On-Stream Determination of Large Scale Ball Mill Residence Time Distributions with Short-Lived Radioactive Tracers," *Trans. SME-AIME*, Vol. 268, pp. 422–431.

Goldfarb, D. and Lapidus, L., 1968, "Conjugate Gradient Method for Nonlinear Programming Problems with Linear Constraints," *Industrial and Engineering Chemistry, Fundamentals*, Vol. 7, pp. 142–151.

Herbst, J. A., Rajmani, K., and Kinneberg, D. J., 1977, "*Estimill*," *Program Description and User Manual*, Metallurgical Engineering, University of Utah.

Hodouin, D., Berube, M. A., and Everell, M. D., 1979, *Industrie Minerale Mineralurgie*, pp. 29–40.

Kelsall, D. F. and Reid, K. J., 1965, "The Derivation of a Mathematical Model for Breakage," American Institute of Chemical Engineers—Institution of Chemical Engineers Symposium, Series No. 4, pp. 14–20.

Kelsall, D. F., Reid, K. J., and Restarick, C. J., 1967–68, "Continuous Grinding in a Small Wet Ball Mill, Part I. Influence of Ball Diameter," *Powder Technology*, Vol. 1, pp. 291–300.

Klimpel, R. R., Austin, L. G., 1970, "Determination of Selection-for-Breakage Functions in the Batch Grinding Equation by Non-Linear Optimization," *Industrial and Engineering Chemistry, Fundamentals*, Vol. 9, Apr., pp. 230–237.

Klimpel, R. R. and Austin, L. G., 1977, "The Back-Calculation of Specific Rates of Breakage and Non-Normalized Breakage Distribution Parameters from Batch Grinding Data," *International Journal of Mineral Processing*, Vol. 4, pp. 7–32.

Klimpel, R. R. and Austin, L. G., 1984, "The Back-Calculation of Specific Rates of Breakage from Continuous Mill Data," *Powder Technology*, Vol. 38, pp. 77–91.

Luckie, P. T. and Austin, L. G., 1972, "A Review Introduction to the Solution of the Grinding Equations by Digital Computation," *Minerals Science and Engineering*, Vol. 4, Apr., pp. 24–51.

Lynch, A. J., et al., 1977, *Mineral Crushing and Grinding Circuits*, Elsevier.

Mori, Y., Jimbo, G., and Yamazaki, M., 1964, "On the Residence Time Distribution and Mixing Characteristics of Powders in Open-Circuit Ball Mill," *Kagaku Kogaku*, Vol. 28, pp. 204–213.

Reid, K. J., 1965, "A Solution to the Batch Grinding Equation," *Chemical Engineering Science*, Vol. 20, pp. 953–963.

Weller, K. R., 1980, "Hold-Up and Residence Time Characteristics of Full Scale Grinding Circuits," *Proceedings*, Third International Federation of Automatic Control, pp. 303–309.

11

Mill Power

Introduction

In Chapter 3, two principal techniques of sizing ball mills were demonstrated, one of which uses rates of breakage calculated for known mill conditions to determine internal mill dimensions, and the other which uses a global specific energy concept to determine the power required to drive the mill. In the first case, it is necessary to determine the power required to drive the mill from the known mill size and loading conditions; in the second case it is necessary to determine the mill size knowing the mill power. This chapter is a review of the equations of shaft mill power at steady operation as a function of mill size and the loading and speed conditions; appropriate allowance has to be made for starting torque, drive losses, and other factors.

The energy input to mills is obviously linked to the specific energy of grinding (defined from a given feed size distribution to a desired product size distribution). The objective in efficient operation of a grinding device is to make sure it is operating in the region where specific grinding energy is as low as possible (contingent upon high capacity and low wear). As discussed in Chapter 5, there is a range of operating conditions where this is true and the trick is to know how to avoid poor grinding conditions where it is not true. Then the choice of mill conditions within this range becomes dependent on the economics of ball loading and wear (see "Optimization of Mill Power and Ball Loading for Tumbling Ball Mills" in this chapter).

The Theory of Mill Power for Tumbling Ball Mills

There are two main approaches in the derivation of equations (White, 1904; Davis, 1919; Rose and Sullivan, 1958; Hogg and Fuerstenau, 1972) describing the power required to drive tumbling mills. One approach calculates the paths of balls tumbling in the mill and integrates the energy required to raise the balls over all possible paths. The other approach (Hogg and Fuerstenau, 1972) treats the ball-powder bed as shown in Fig. 1, assuming that the turning moment of the *frictional force* must balance the turning moment of the center of gravity of the bed around the mill center. The frictional force presumably arises from friction between the case and the bed plus friction of ball-on-ball as balls move up through the bed. In both cases, the energy used to turn a well-balanced empty mill is small since it consists only of bearing friction.

It is instructive to look into a laboratory tumbling mill fitted with transparent end walls, so that the motion of the charge in the mill can be studied by direct visual observation. An approximate description (see Chapter 5, Fig. 1) is that a ball enters into the bed surface below the halfway mark on the surface and moves round in a circular path without slip with respect to other balls (a locked condition) until it reaches the surface. As it emerges it rolls down the surface, so that there is a stream of rolling balls forming a free-flowing surface, referred to as *cascading*. However, there are some balls, plus some quantity of powder, which *cataract* by ejection from the bed. More cataracting occurs as the rotational speed is increased, while cascading occurs more at lower speeds, but both are always present at the speeds of interest in milling. Note also that *liners* (also called *lifters*) are used to prevent the ball charge slipping as a mass down the mill casing; the liners act as keys which project into the ball charge and keep the layers of balls next to the case moving round with the case.

In principle, a correct analysis of the forces holding the bed in the inclined position against the turning moment due to gravity must lead to the same answer as an analysis of the energy of raising the balls, assuming no slip of ball-over-ball in the bed (a locked bed). However, the turning moment approach does not allow for balls in free flight or for the recovery of energy to turn the mill by falling balls striking the far side of the case. On the other hand, the description of all possible ball paths within the mill is a very difficult task because of the complexity of the forces between balls

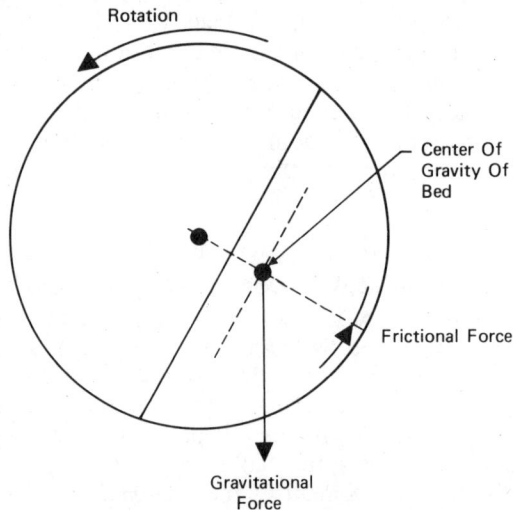

Fig. 1. Illustration of forces of friction and gravity for turning moment about the center of a ball mill.

moving within and outside the bed, especially as a function of the powder trapped between the balls. Energy is necessary to raise the balls against gravity and, perhaps, against frictional forces within the bed. As discussed before ("Kick, Rittinger, and Bond Laws of Batch Grinding" in Chapter 2), however, it is certainly true that the energy input to the mill appears predominantly as heat. Falling balls collide with other balls in the bed and the kinetic energy of downward movement is converted to heat; layers of balls and powder moving over one another produce frictional heat.

Because of the complexity of detailed theoretical analysis of such a system, we will give only an elementary treatment which involves rather simple concepts plus geometrical similarity. Let a hypothetical mean height, $\bar{h}$, of ball raise-drop be defined by

$$\text{mean energy required to raise each ball} \propto \rho_b d^3 \bar{h}$$

where ρ_b is ball density and d ball diameter. In effect, this implies that energy is used to raise balls rather than to overcome frictional forces, and it is assumed that $\bar{h}$ is independent of ball diameter, density, surface hardness, or friction coefficient. The second major assumption is that for otherwise constant conditions

$$\bar{h} \propto D$$

D being mill diameter. This assumes geometrically similar tumbling actions independent of the scale of the mill, providing ball diameter is much less than mill diameter and the mill length is such as to make end effects negligible.

For a given percentage mill filling by balls, that is, a prescribed value of J,

$$\text{number of balls} \propto D^2 L/d^3$$

it can be assumed that the number of balls raised per minute is proportional to the number which are present times the revolutions per minute. But the rpm at a fraction ϕ_c of the critical speed is

$$\text{rpm} = \phi_c c/D^{1/2}$$

(c is 42.3 for D in m, 76.6 for D in ft). Therefore, the energy rate (power) required to turn the mill is

$$m_p \approx K\rho_b(D^2L/d^3)(d^3D)/D^{1/2}$$

and the ball size cancels out, giving

$$m_p \approx K\rho_b LD^{2.5} \tag{1}$$

where the value of K is constant only for given conditions within the mill.

Because the model used in the derivation is oversimplified, it is better to introduce a degree of latitude by stating the equation in the more general form

$$m_p = K\rho_b LD^{2+n_2} \tag{1a}$$

where n_2 is expected to be close to 0.5. Eq. 1 assumes that effects from the end walls of the cylinder are negligible, so that doubling the length of the mill will double the power requirement. Rose and Sullivan (1958) demonstrated in a laboratory mill that power was proportional to ball density, using lead, steel, and glass balls, holding $L/D = 1$, $D/d = 11$, and $J = 0.5$. They found the hardness and coefficient of restitution of balls to have no effect. There is, however, less agreement on the variation of K with mill conditions, as will be seen in the following sections.

The value of K will vary with the ball loading in the mill. At very low values of loading, it is reasonable to expect that power to the mill will increase in proportion to the ball loading. On the other hand, at high loadings where the volume of balls almost completely fills the mill, the load will tend to centrifuge around in a mass without tumbling, which would require only low power. Thus it is expected that the power will pass through a maximum as ball load is increased. This is demonstrated in the next section.

Power Equations for Tumbling Ball Mills

The relation given by Rose and Sullivan (1958) for net mill power for dry grinding is

$$m_p = (2.8)(10^{-4})(LD^{2.5}\rho_b)(\phi_c)(1 + 0.4\sigma U/\rho_b)F(J), \text{kW} \tag{2}$$

where the empirically determined constant $(2.8)(10^{-4})$ is for units of ball density ρ_b in lb/ft^3 and L and D in feet. For L and D in meters and ρ_b in kg/m^3, the constant is $(1.12)(10^{-3})$. In this equation it is assumed that power to the mill is proportional to ϕ_c, from $\phi_c = 0$ to 0.8, so the equation should not be extrapolated beyond $\phi_c = 0.8$. The term $1 + 0.4JU/\rho_b$ arises as follows. Assume that the powder tumbles with the balls, so that

$$\text{power} \propto (\text{weight ball charge} + \text{weight powder charge})$$

Compared to balls alone, therefore, there is a correction factor

$$\xi = \frac{\text{weight balls + powder}}{\text{balls alone}}$$

Assume the formal porosity of ball bed and powder to be 0.4. Let σ be the density of the powder. Let the fraction of the voids in the ball bed filled with powder be U. Then, since 0.6 of the bed volume is solid balls and 0.4 is porosity, the volume of powder is $0.4U$. This powder volume contains 0.4 as porosity and 0.6 as solid. Thus,

$$\xi = \frac{(0.6)(\rho_b) + (0.4)(0.6)(\sigma U)}{(0.6\rho_b)} = 1 + 0.4\sigma U/\rho_b$$

Note that in the term $0.4\sigma U/\rho_b$ the units of σ and ρ_b are not important providing they are consistent with each other, so the ratio of specific gravities can be used.

The function $F(J)$ allows for the effect of ball filling and was empirically determined by measuring power at known values of J in a small laboratory mill; by fitting a polynomial to the curve given by Rose and Sullivan the following expression is obtained,

$$F(J) = 3.045J + 4.55J^2 - 20.4J^3 + 12.9J^4, J < 0.5 \qquad (3)$$

Fig. 2 shows the values, with a maximum in power at $J \approx 0.40$.

For comparison purposes, it is more convenient to express Eq. 2 in terms of the mass of grinding media, where the media is usually steel balls. The mass of media in the mill is $M = (L\pi D^2/4)(J)(1 - 0.4)\rho_b$ or

$$M = 0.47JLD^2\rho_b \qquad (4)$$

and

$$m_p/M = C_{Ro}D^{0.5}\phi_c(1 + 0.4\sigma U/\rho_b)[F(J)/J], \text{kW/t}, \phi_c \leq 0.8 \qquad (5)$$

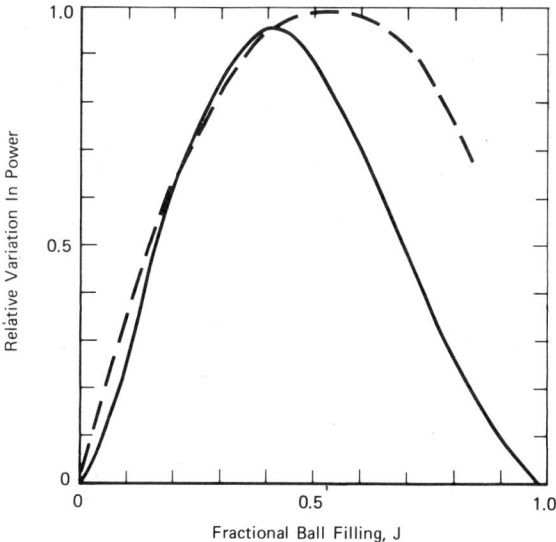

Fig. 2. Relative variation of power with ball loading:
— function $F(J)$ of Rose and Sullivan equation; -----
Bond equation.

where C_{Ro} is 1.17 for M in short tons (2000 lb), D in feet, and 2.38 for M in metric tons (1000 kg), D in meters. Bond (1960) gives the empirical equation, for *shaft* power of wet grinding in overflow ball mills,

$$m_p/M = C_{Bo}D^{0.3}\phi_c(1 - 0.937J)(1 - 0.1/2^{9-10\phi_c}), \text{kW/t} \qquad (6)$$

where C_{Bo} is 9.91 for D in feet, M in short tons, and 15.6 for D in meters, M in metric tons (diameters are inside mill linings). He states the results should be multiplied by 1.08 for dry grinding in grate discharge mills. He also gives a *slump* correction to be applied when the maximum ball diameter (d_m) is less than 45.72 mm (1.8 in.) and the mill diameter greater than 2.4 m (8 ft):

$$S_s = (1.8 - d_m)/2, \text{kW/st} \qquad (6a)$$

where S_s is to be subtracted from the values given by Eq. 6 for wet overflow mills; d is the ball diameter in inches. For a ball of 12.7 mm (0.5 in.), the correction is 0.7 kW/t. Rowland (1974) has modified this to

$$S_s = \left(\frac{3D}{20} - d_m\right)\Big/2, \text{kW/st} \qquad (6b)$$

for mills greater than 3.6 m (12 ft) in inside diameter, D in feet. The relative variation of power with J is shown in Fig. 2, and it is clear that the variation does not agree with the Rose and Sullivan equation.

The empirical equation of Beeck (1970), for dry grinding of cement is

$$m_p/M = 42.3C_{Be}D^{0.5}\phi_c, \text{kW/t} \qquad (7)$$

where M is in metric tons, D is in meters, and the value of C_{Be} varies with J as shown in Fig. 3. Also shown are values of C_{Be} calculated from the values of ball loading, fraction of critical speed, and mill power for US cement finish mills collected by Hackman, Pitney, and Hagemeier (1973). US cement practice is substantially different from German practice. The filling levels are consistently higher, with an average of 36%; that is, $J = 0.36$; the percentage of critical speed is normally between 70 and 80%, with an average of 75%. The wider scatter in the measured values of C_{Be} is shown in Fig. 3; it must be recognized that the measurements of mill power per ton of media may not be accurate in large operating mills and the measured power will also vary with the motor and drive efficiency. The spread of data is of the magnitude of $\pm 15\%$ of the mean, and although the original data show that larger diameter mills have a significantly higher m_p/M, the data is too scattered to get an accurate value of the exponent of D.

Measurements of Mill Power

Because of the apparent divergence in the equations given above, additional tests and analyses of mill power were performed. Shaft power was measured with a torque device for mills of about 200 mm i.d., and

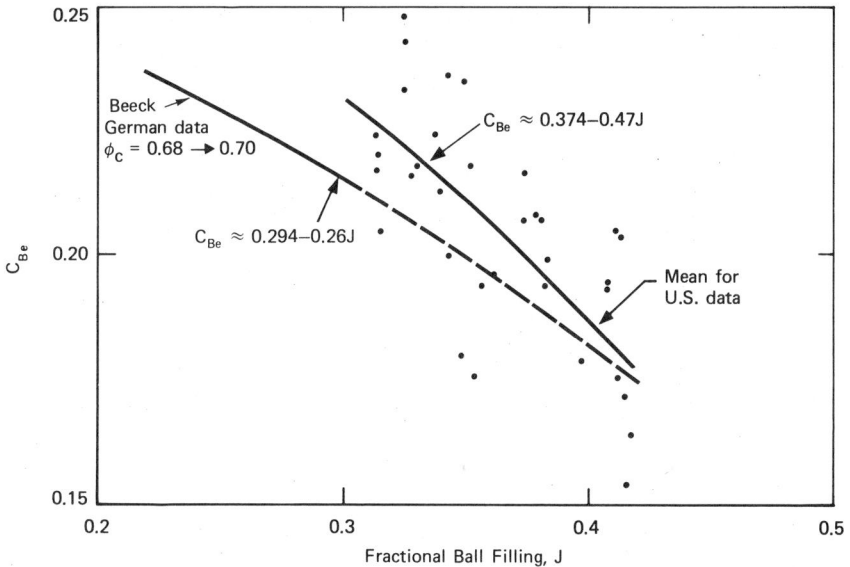

Fig. 3. Coefficient for mill power using the Beeck equation for cement grinding, US data (Hackman, Pitney, and Hagemeier, 1973). $D = 2.9$ to 4.5 m (9.5 to 15 ft), $L / D = 2.7$ to 3.7, $\phi_c = 0.70$ to 0.80.

corrected for no-load (no balls) power due to bearing friction. For larger mills the gross power was measured by a wattmeter on the motor or the shaft power was measured on a hydraulic drive. The no-load loss was measured with the mill empty and with the mill completely filled with sand (no tumbling), with linear interpolation and extrapolation for other weights in the mills. The no-load loss for a given weight was subtracted from the measured power to eliminate drive losses and bearing friction.

One of the problems with power measurement in the laboratory is shown in Fig. 4; the power varies with the particle size of the material being ground. The 200-mm diam by 175-mm long mill used in this test was a commercially supplied test mill with welded bead lifters less than 2 mm high. It gave a variation of net power with rotational speed as shown in Fig. 5. For a relatively coarse powder charge (measured in the horizontal region of Fig. 4 at 10 to 30 min), the mill power passed through a maximum at greater than critical speed for low ball loading (Rose and Sullivan, 1958; Hukki, 1957), with a shift of the maximum to 70% c.s. for $J = 0.4$. For a fine charge (produced after 60 min of grinding), the power was lower for lower values of ϕ_c but became higher at higher ϕ_c, again producing a maximum at greater than critical speed.

It is concluded that the small lifters allowed slip to occur as rotational speed was increased, at low ball load with fine particles. Coarser particles help to key the charge to the case and give more tumbling, as does a higher ball load, which gives more force on the case. A high value of rotational speed is required to give maximum power under slipping conditions (Rose

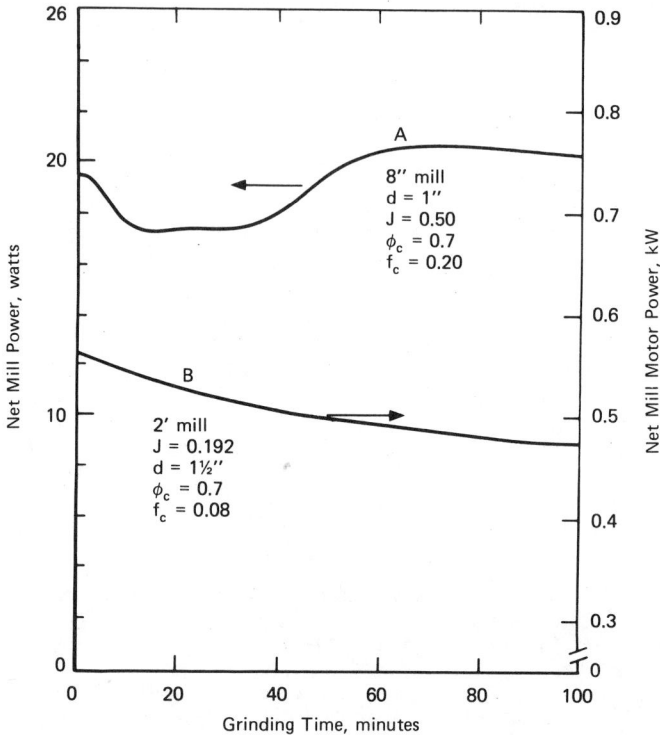

Fig. 4. Variation of mill power with grinding time (size of powder): (A) $D = 200$ mm, $L = 175$ mm; (B) $D = 0.6$ m, $L = 0.3$ m.

and Sullivan, 1958). Note that the derivation of the centrifuging condition for critical speed assumes that *no slip* occurs, with balls traveling round on the case. The maximum power is higher in the presence of slip because energy is used both for raising the balls and to produce frictional heat at the case. It is well known (Rose and Sullivan, 1958) that power measurements in the absence of any powder give anomalous results due to slip and our experience confirms this.

The mill was fitted with six lifters of 19-mm diam semicircular cross-section. In this case, the mill power went through a maximum of 13 W at 75 to 80% critical speed with coarse sand, and 11 W at 70 to 75% critical speed with fine sand, for a ball load of 20%. Thus slip at high rotational speeds was almost eliminated by these lifters. It is interesting that the B values with and without the lifters at a standard speed of 70% of critical and $J = 0.2$ were essentially identical, but the S values were increased by a factor of 1.25 with lifters even though the power draw was virtually identical.

It was found that variation of f_c from 0.04 to 0.2 gave no significant variation in power, showing that the Rose and Sullivan treatment for powder tumbling (see Eq. 2) is not generally valid.

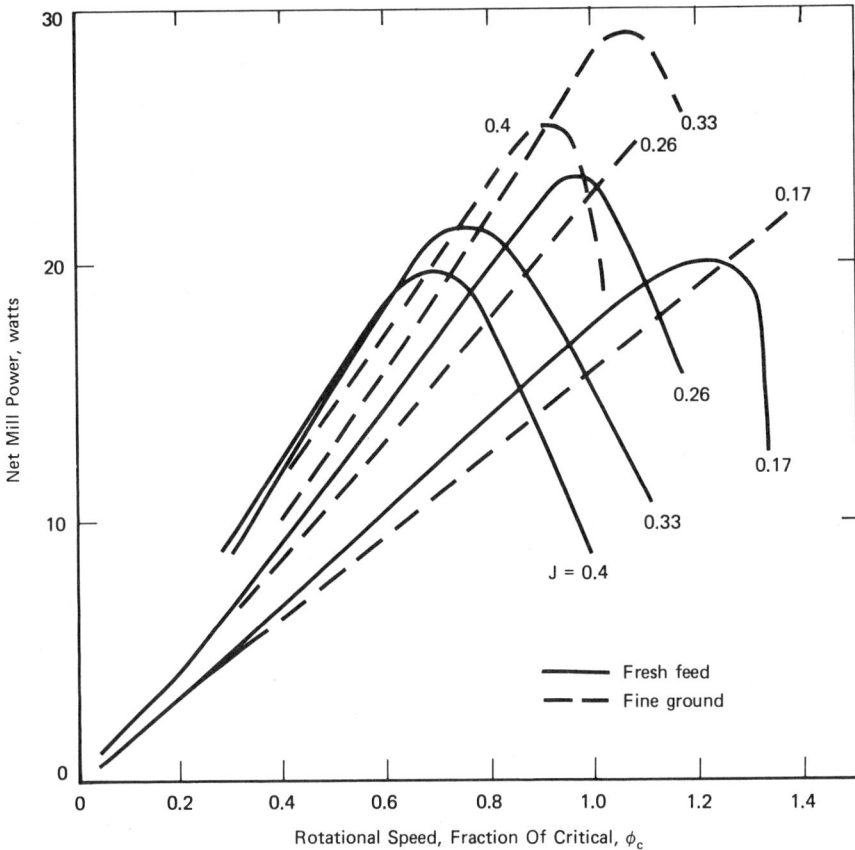

Fig. 5. Abnormal variation of mill power with rotational speed for a mill with small lifters (D = 200 mm, d = 25.4 mm).

Tests were also performed in a mill of 0.60-m internal diameter by 0.31-m long mill fitted with 20 lifters of 25-mm diam semicircular cross-section. This test mill gave a no-load power of 0.244 kW for all values of speed and for the mill empty or filled with 130 kg of sand, showing the no-load power to be due entirely to drive losses. As with the 200-mm diam mill, grinding of coarse feed particles gave significantly higher power than finer particles (see Fig. 4). However, the variation of power with rotational speed was of normal form, with a maximum in power at 75 to 80% of critical speed.

A low powder load gave significantly higher power than normal powder loads at speeds below maximum power, but over the normal powder load range there was negligible effect of powder load. There was no effect of ball size (19 to 38 mm) on power below 75% of critical speed for J = 0.2, but the rotational speed and magnitude of the maximum power was highest for 25.4-mm (1-in.) balls. Fig. 6 shows a typical result at higher ball filling. Using a small percentage (10%) of smaller or larger balls mixed in the ball

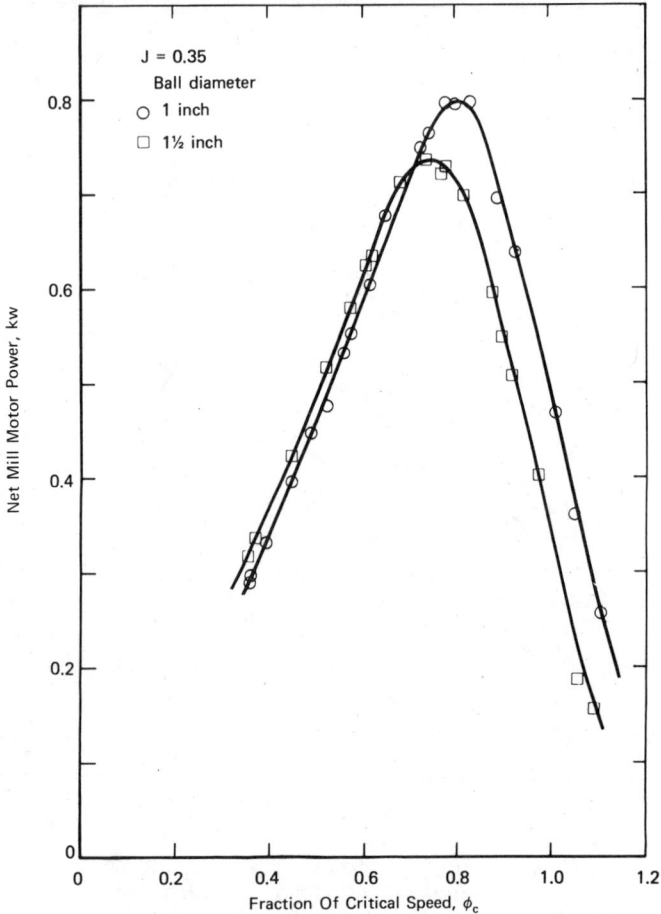

Fig. 6. Variation of mill power with mill speed, as a function
of ball size (0.6 m mill, $J = 0.35$, $f_c = 0.14$).

charge eliminated the differences, so it can be concluded that a practical ball
charge containing a mixture of balls draws the same power irrespective of
ball diameter for normal ranges of ball sizes.

Fig. 7 shows typical variation of power with ball load, for small mills, at
various fractions of critical speed. Maximum power draw occurred at about
45% ball loading for each rotational speed. Calculations show that the
expression $\phi(1 - 0.1/2^{9-10\phi_c})$ in Eq. 6 (Bond) did not give the right form
of the variation with rotational speed for this mill. Empirical fitting of the
results of Fig. 6 for 38-mm (1 1/2-in.) balls gave

$$m_p \propto (\phi_c - 0.1)\frac{1}{1 + \exp[15.7(\phi_c - 0.94)]}, \quad 0.4 < \phi_c < 0.9 \quad (8a)$$

Fig. 8 shows the data plotted in a form to test the Bond prediction that

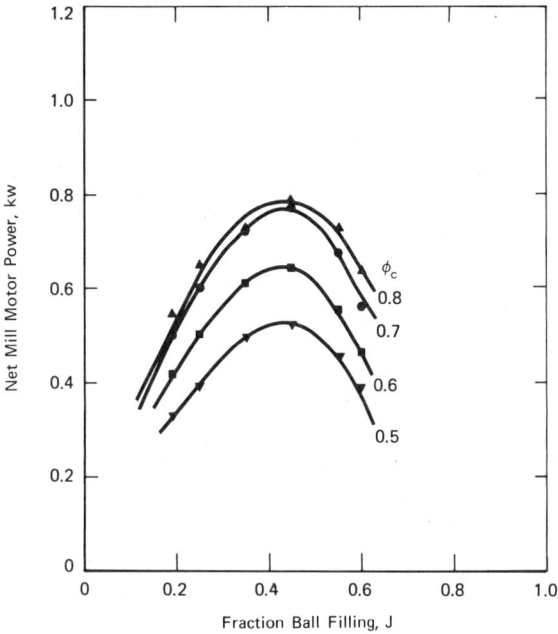

Fig. 7. Mill power as a function of ball loading and fraction of critical speed; $D = 0.6$ m, $d = 27$ mm.

$m_p/M \propto 1 - 0.937\,J$. It was found that it was necessary to modify the expression for small mills to

$$m_p/M \propto \frac{1 - 0.937J}{1 + 5.95J^5} \tag{8b}$$

Discussion of Mill Power Expressions

Fig. 9 suggests that the disagreement between the formulae proposed may be due to a change in the exponent of diameter as mill diameter increases beyond 2.4 m (8 ft). However, in the region of 2.4 to 4.6 m (8 to 15 ft) diam, the Bond, Beeck, and Rose and Sullivan equations for $\phi_c = 0.7$, $J = 0.3$ to 0.4 give answers within about $\pm 10\%$. It is proposed that the Bond equation (Eq. 6) be used for large mills [$D \geq 2.4$ m (8 ft)] but that the following expression be used for the *net* mill power of smaller mills run in the dry batch mode (see Figs. 8 and 9),

$$m_p/M = C_A D^{0.5}(\phi_c - 0.1)\frac{1}{1 + \exp[15.7(\phi_c - 0.94)]}$$

$$\times \left(\frac{1 - 0.937J}{1 + 5.95J^5}\right), \text{kW/t} \tag{9}$$

where $C_A = 13.0$, for D in meters, M in metric tons.

This equation is for net power for batch dry ball milling whereas the Bond equation is for pinion shaft power for wet continuous overflow mills.

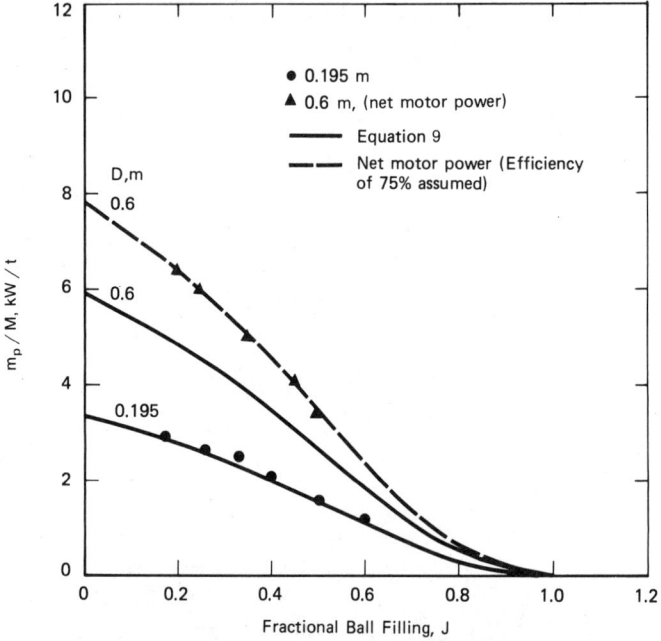

Fig. 8. Mill power per metric ton of grinding media as a function of ball charge at 70% critical speed.

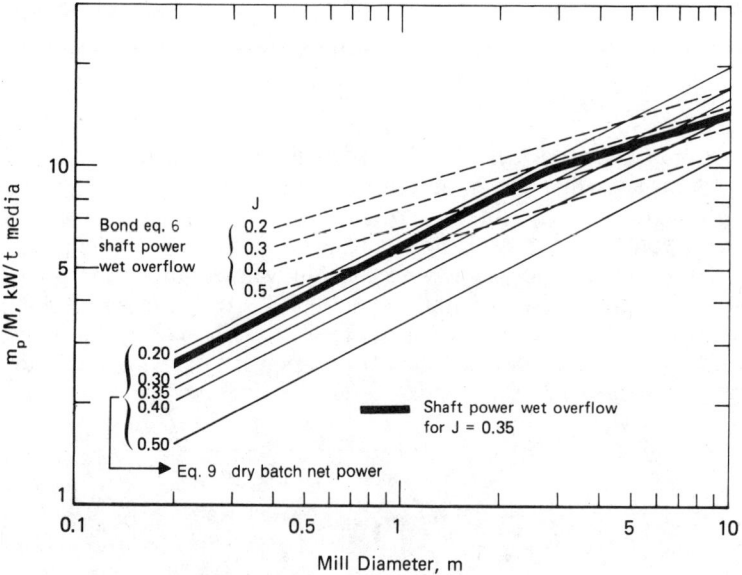

Fig. 9. Mill power per metric ton of media as a function of mill diameter ($\phi_c = 0.7$).

A 0.82 m i.d. by 1.53 m long mill with hydraulic bearings was run as a batch dry mill and as a continuous wet overflow mill, under identical J, ϕ_c conditions. Continuous wet operation gave higher power, with a conversion factor of 1.07 to go from net power for dry batch to net power for wet continuous overflow, and a further factor of 1.10 to go from net power to pinion shaft power: the total conversion factor is then 1.18. This gives an intersection of the equivalent result for small mills (the heavy solid line in Fig. 9) with the Bond equation results at $D = 2.5$ m (≈ 8 ft) for $J = 0.35$.

It should be noted that the 10% increase of net power to give shaft power is to be used only to give a shaft power for comparison with large mills with efficient bearings and drives. The actual shaft power for a small mill can be much higher due to high friction of nonhydraulic bearings. This can have the effect of making the pinion power measured on a pilot-scale mill agree *fortuitously* with the Bond expression (Eq. 6), but replacement of simple bearings with good quality bearings gives results which fit Eq. 9. Clearly the true specific grinding energy must be based on true net power excluding excessive bearing friction.

The way in which power varies as a function of ball loading and rotational speed implies the following physical picture. At low ball loadings, where the balls travel around close to the mill case, they are raised quite effectively in the mill, requiring a large power per unit mass of balls. As J is increased, a smaller fraction of the balls are close to the case, and instead of being lifted high, some fraction of them roll down the bed surface before a full lift is obtained. Thus, the power per unit mass of balls decreases, but because the ball load is simultaneously increasing, the total power increases. Eventually, when the mill is filled to between 0.4 and 0.5 bed volume, the cascading action applies to more and more of the balls and the total power passes through a maximum. Increase in rotational speed tends to tumble more balls around per unit time and, hence, increases the power (energy per unit time) required to run the mill. However, at rotation speeds above 70 to 80% of critical, this is counteracted by the increased centrifugal force. Thus, the power required to run the tumbling ball mill is a complex function of the frequency and height of tumbling, and these act against each other to give a maximum in power in the region of $J = 0.4$ to 0.5 and $\phi_c = 0.7$ to 0.8.

The mill power equations contain no allowance for lifter design, although it is certain that some lifter designs give more cataracting than others at the same fraction of critical speed and ball loading, and should therefore give maximum power at different J and ϕ_c values. There appears to be few quantitative relations in the literature for the effect on mill power of different lifter designs. For example, Rowland (1974) shows a maximum of power at $J = 0.42$ for a 5.5-m (18-ft) i.d. mill with 76.2-mm (3-in.) makeup balls. New single wave liners give 10% more power than the Bond equation, while new double wave liners give 10% less power than the Bond calculation. Rogers et al. (1982) have reported on differences in normal breakage (dominated by cascading) and breakage of oversize feed (dominated by cataracting) produced by three different liner designs in a 0.9-m diam mill, at a fixed fraction of critical speed. Figs. 10a, b, and c give their results. At 70% of critical speed, the corrugated and angular spiral liners required

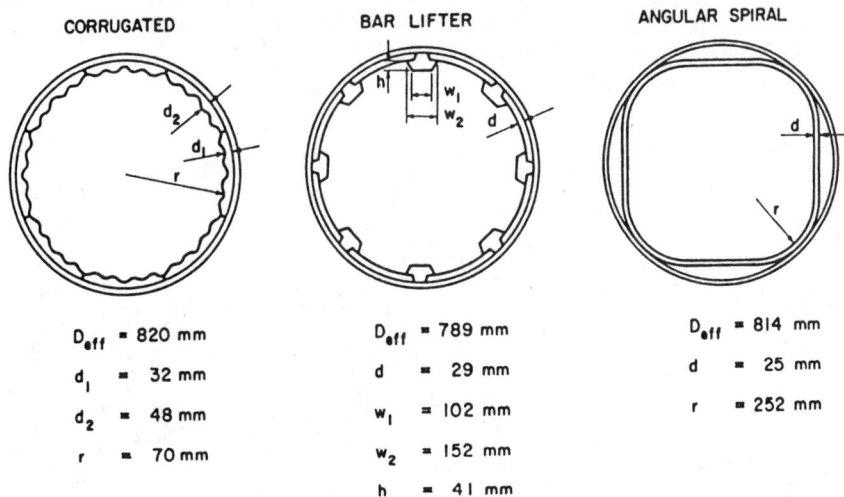

Fig. 10a. Cross-sectional view and dimensions for the corrugated, bar lifter, and angular spiral liners.

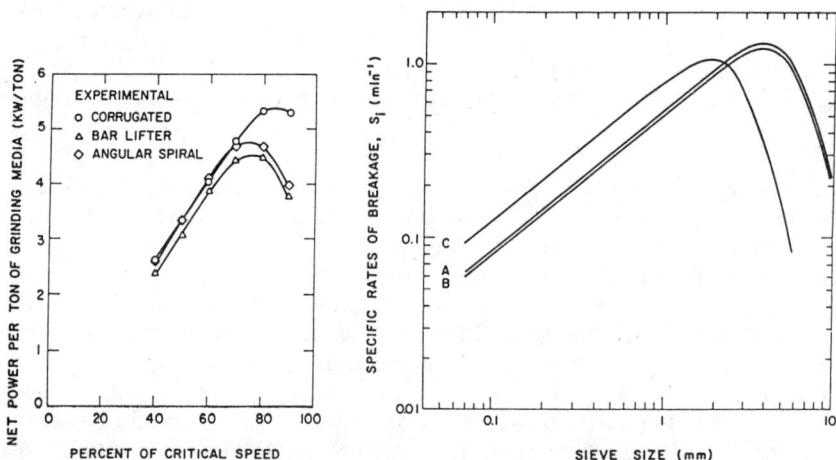

Fig. 10b. Left. Net mill power data for the 0.91 × 1.52 m ball mill fitted with different liners and operated with a 35% (by volume) ball loading.

Fig. 10c. Right. Specific rates of breakage for quartz ground in the 0.91 × 1.52 m ball mill fitted with corrugated (C), bar lifter (B), and angular spiral (A) liners.

almost the same power, but the corrugated liner gave higher rates of normal breakage (more cascading) and lower rates of breakage of larger sizes (less cataracting). Bar lifters gave lower power draw but equally effective breakage as the angular spiral liners, at this speed, for all sizes.

Results reported in Chapter 5 (see Fig. 5) indicate that the maximum mill capacity occurs at a ball load of $J = 0.4$, which is not quite at the maximum power input ($J = 0.45$), presumably because cascading and cataracting are not equally effective in giving breakage. However, this conclusion is based on results from a laboratory mill of 0.6 m diam fitted with semicircular lifters and may not be applicable to any given large-scale mill with more or less effective keying of the charge to the mill case.

Optimization of Mill Power and Ball Loading for Tumbling Ball Mills

From the arguments given in the previous section, it might be concluded that ball mills should be run at values of J and ϕ_c close to maximum power input, since this gives the maximum tonnage output from the mill. However, the cost of grinding per ton of product is dependent on a number of fixed and variable costs. These include the capital cost of the mill (C_M, \$), the capital cost of the ball charge and liners in the mill (C_B, \$), the cost of energy used per ton ground (c_E, \$/t), the cost of steel (balls and liners) worn away per ton ground (c_B, \$/t), maintenance (repair) costs per ton ground (c_R, \$/t) and finally, other fixed costs such as supervisory labor and control systems (C_L, \$). It is the net cost per ton of solid processed which has to be minimized.

Consider an investment and tax situation where the conversion of fixed charges to the basis of dollars per hour is at a rate of R dollars per hour per dollar of fixed cost. Then,

$$\text{grinding cost, } \$/t = \left(\frac{R}{Q}\right)(C_M + C_B + C_L) + (c_E + c_R + c_B) \quad (10)$$

where the mean output from the grinding circuit is Q tph. At a specific grinding energy of E kWh/t, this will require a machine power of $m_p = QE$. This power can be used to drive a mill of volume V_1 with a value of J_m which gives the maximum power input, but it can also be used to drive a *larger* mill, volume V_2, with a *smaller* value of J_2; that is, $V_1 J_m > V_2 J_2$ (see Fig. 8). In the second case, C_M will increase but C_B will decrease because fewer tons of balls are used; in addition, the value of c_E will be lower. Obviously, the minimum in grinding costs depends on the relative costs of the mill volume and ball charge, how cost increases with mill volume, and the value of c_E at various ball loads.

The qualitative application of these concepts can be illustrated by comparing the practice of the United States vs. that of Europe for the dry grinding of cement clinker to the finished product of cement powder. In the US, conventional steel grinding balls have been cheap, so that C_B is relatively small in Eq. 10. This steel wears away fairly rapidly, so that the term c_B is substantial. In Europe, usual technology is to use high-cost alloy steel with a very slow wear rate, so that the term c_B is small but C_B is high.

It is then economically advantageous to use a *longer* mill, giving a higher value of C_M, combined with a lower J giving a smaller value of C_B. Because the power per ton of balls is higher (see Fig. 8) for lower values of J, the required tonnage of expensive balls is reduced in the longer mill. This is one reason (see Fig. 3) why Beeck gives values of J for West German practice of 0.22 to 0.3, whereas in US practice, the mean value of J is 0.36. Fig. 6 in Chapter 5 suggests that specific grinding energy is reduced for lower ball loadings.

The value of c_B depends on the rotational speed of the mill, as well as the abrasiveness of the material, ball mix, and other factors. A larger mill diameter means higher peripheral speed, since peripheral speed = $\pi D \phi_c (42.3/\sqrt{D}) = 132\phi_c\sqrt{D}$ m/min. Slip at the case will give increased wear for higher peripheral speed, as will impact from cataracting balls, especially large balls. Rowland and Kjos (1980) recommend that the fraction of critical speed be reduced for larger mills, as shown in Table 1, presumably to reduce the level of cataracting and compensate for the higher peripheral speed. A similar conclusion for rubber lifters has been reported by Trelleborgs Gummifabriks AB (Anon, undated). A reduction in lifter service life of four to one was found when rotational speed was increased from 75% of critical, with life approximately proportional to the reciprocal of ball diameter at a given rotational speed.

Summary

In order to complete a mill design and calculate the specific energy of grinding, it is necessary to have a mechanical law which relates mill power and mill size and operating conditions. There is a range of feasible and efficient mill conditions where grinding efficiency and mill capacity vary somewhat with ball load, and the choice of operating conditions within this range must be made on economic considerations of capital and operating cost.

An elementary treatment of the net energy required to raise balls in a mill, using geometric similarity of the mean ball path for mills of different diameter, gives

$$m_p = K\rho_b LD^{2.5} \tag{S1}$$

Table 1. Recommended Rotational Speed for Ball Mills*

Mill i.d. m	Recommended % Critical Speed
0 − 1.8	80 − 78
1.8 − 2.7	78 − 75
2.7 − 3.7	75 − 72
3.7 − 4.6	72 − 69
> 4.6	69 − 66

*After Rowland and Kjos, 1980.

where ρ_b is ball density and K depends on ball loading and rotational speed as a fraction of critical speed. In practice

$$m_p = K\rho_b LD^{2+n_2} \tag{S2}$$

where n_2 is close to 0.5. K is determined by experiment as a function of J and ϕ_c.

For comparison purposes, mill power equations are expressed as power per metric ton of media, $M = (L\pi D^2/4)(J)(1 - 0.4)\rho_b$. The Rose and Sullivan equation for net power is

$$m_p/M = 2.38D^{0.5}\phi_c(1 + 0.4\sigma U/\rho_b)(3.045 + 4.55J - 20.4J^2 + 12.9J^3) \tag{S3}$$

where D is in meters, m_p/M is in kW/t media. The Bond equation for shaft power is

$$m_p/M = 15.6D^{0.3}\phi_c(1 - 0.937J)(1 - 1/2^{9-10\phi_c}) \tag{S4}$$

for wet overflow mills, multiplied by 1.08 for dry, grate mills. The Beeck equation (cement finish grinding) for total power is

$$m_p/M = 42.3C_{Be}D^{0.5}\phi_c \tag{S5}$$

where

$$C_{Be} = 0.294(1 - 0.884J), 0.2 < J < 0.3 \quad (German\ data)$$

$$C_{Be} = 0.374(1 - 1.257J), 0.3 < J < 0.4 \quad (US\ data)$$

A laboratory mill of 200 mm diam with small lifters gave slippage of the charge at higher rotational speeds and low ball charge, leading to maximum power at rotational speeds greater than critical. The effect was reduced with coarser particles and higher J. Power is abnormally high in the presence of slippage since energy is used both to raise balls and to generate frictional heat at the case. Using larger lifters eliminated the abnormality in power vs. rotational speed, giving a maximum in power at 70 to 80% critical speed, depending on particle fineness. The quantity of powder had negligible influence on mill power over the range $f_c = 0.04$ to 0.2.

Tests in laboratory mills (with reasonably effective lifters) up to 1 m diam gave the equation of net mill power for dry batch milling as

$$m_p/M = 13.0D^{0.5}(\phi_c - 0.1)\frac{1}{1 + \exp[15.7(\phi_c - 0.94)]}$$

$$\times \left(\frac{1 - 0.937J}{1 + 5.95J^5}\right), kW/t \tag{S6}$$

which gives a maximum in power at $J = 0.45$ and $\phi_c = 75$ to 80% of critical speed. There were only small influences of powder load or ball diameter. The data gave an exponent of $D^{0.5}$ for the scale-up of power per ton of media. A factor of times 1.18 is used to convert this power to the shaft power for wet continuous overflow ball mills at the same conditions, for comparison with the Bond equation. It is recommended that Eq. S6 be used for the net power of laboratory and pilot-scale mills ($D < 2.5$ m) with effective lifters and the Bond Eq. S4 for the shaft power for larger mills. Pilot-scale mills often have high bearing friction losses, so the shaft power is much higher than the net power.

It seems likely that lifter design influences the power equations, so that maximum power may occur at different ball loading and fraction of critical speed for different lifters.

The mill power per ton of media decreases as ball load increases, that is, a smaller ball load pulls proportionally more power and gives a higher mill capacity per ton of media. If high-cost alloy steel balls are used, it may be economically advantageous to have a larger mill shell and a smaller ball load to produce a desired capacity. Small-scale data suggests that the specific energy of grinding is a minimum at about 25% ball loading. Larger mill diameters are run at lower fractions of critical speed to reduce liner wear.

References

Anon., undated, "Trelleborg Mill Lining," Trelleborgs Gummifabriks AB, Trelleborg, Sweden.

Beeck, R., 1970, *Zement-Kalk-Gips*, Vol. 23, pp. 413–416.

Bond, F. C., 1960, *British Chemical Engineering*, Vol. 6, pp. 378–391, 543–548.

Davis, E. W., 1919, "Fine Crushing in Ball Mills," *Trans. AIME*, Vol. 61, pp. 250–296.

Gaudin, A. M., 1939, *Principles of Mineral Dressing*, McGraw-Hill, New York.

Hackman, A. H., Pitney, R. J., and Hagemeier, D. F., 1973, "Survey of U.S. Cement Finish Mills," *Pit and Quarry*, Vol. 66, pp. 112–122.

Hogg, R. and Fuerstenau, D. W., 1972, "Power Relations for Tumbling Mills," *Trans. SME-AIME*, Vol. 252, pp. 418–432.

Hukki, R. T., 1957, "Grinding at Supercritical Speeds in Rod and Ball Mills," *Progress in Mineral Dressing, Proceedings*, Fourth International Mineral Processing Congress, E. Ohman, ed., Jernkontoret, Stockholm, pp. 85–122.

Rogers, R. S. C., et al., 1982, "The Effect of Liner Design on the Performance of a Continuous Wet Ball Mill," prepared for XIV International Mineral Processing Congress, Canadian Institute of Mining and Metallurgy, Toronto, pp. I5.1–19.

Rose, R. H. and Sullivan, R. M., 1958, *Ball, Tube and Rod Mills*, Chemical Publishing Co., New York, pp. 69–108.

Rowland, C. A., Jr., 1974, "Comparison of Work Indices Calculated from Operating Data with Those from Laboratory Test Data," *Proceedings*, Tenth International Mineral Processing Congress, M. J. Jones, ed., Institution of Mining and Metallurgy, London, pp. 47–61.

Rowland, C. A., Jr. and Kjos, D. M., 1980, "Rod and Ball Mills," *Mineral Processing Plant Design*, A. L. Mular and R. B. Bhappu, eds., 2nd ed., AIME, New York, pp. 239–278.

White, H. A., 1904, "Theory of the Tube Mill," *Journal of the Chemical, Metallurgical and Mining Society of South Africa*, Vol. 5, pp. 290–305.

12

The Real World: Analysis of Full-Scale Mills

Introduction

This chapter has two main functions: (1) to show how the information developed previously is applied to the analysis of pilot and full-scale operating plants, and (2) to point out some of the uncertainties involved which require further investigation. It must be realized that there is no *a priori* reason why breakage in a full-scale mill should be the same as in a laboratory mill, so it is necessary to validate the scale-up laws by comparing predictions from models with actual plant performance.

It is useful at this point to summarize the major features of ball mill operation in qualitative statements. There are two regions of breakage of particles by tumbling balls. Normal first-order breakage occurs when particles are small compared to the ball diameter; the primary progeny fragment distributions in this region have a constant slope γ and are *normal*, and the specific rates of breakage decrease as particles become smaller. *Abnormal* breakage occurs when the particles are too large for the ball diameter; non-first-order breakage occurs, with an initial faster rate followed by a slower rate, and the primary progeny fragment distributions usually contain a higher proportion of fine material. The values of the effective mean specific rates of breakage decrease as particles become bigger.

There are two types of grinding inefficiency. The first type, which we call *indirect* inefficiency, occurs when the mill circuit produces excess fines by overgrinding of material which is already fine enough for the following process. This occurs when the mill is closer to fully mixed flow, the classifier is inefficient, and the circulating load is low. If the circuit and classifier are adjusted to produce a feasible two-point matched size distribution (i.e., a specific product from a given feed) then the mill capacity and specific grinding energy are essentially independent of RTD and classifier efficiency.

On the other hand, there are causes of *direct* inefficiency, in which some energy is consumed without producing size reduction. These include: (1) underfilling the mill; (2) overfilling the mill; (3) use of wrong ball diameters, either too small for breakage of larger sizes or too large for efficient breakage of small sizes; (4) slip in the mill due to poor liner design, incorrect rotational speed; and (5) use of wrong slurry density and viscosity. If two mill conditions are compared when producing the same product size

249

distribution from the same feed, these factors will give a lower capacity and higher specific energy for incorrect as compared to correct conditions.

Overfilling at high flow rates through the mill is especially important (see Chapter 7). If the reduction ratio required of the mill is relatively small, material must be passed at high rates and the mass transfer law (see Eq. 30 in Chapter 14) states that the mill will overfill. Breakage efficiency decreases as the interstitial filling of the balls increases above 1, leading to direct inefficiency. An optimum value of circulating load exists where the decrease of indirect inefficiency is balanced by the increase of direct inefficiency.

The tests on mill power show that there are probably two types of power consumption in a mill: efficient power consumption used to tumble balls and less efficient power consumption used to supply frictional heat in the presence of slip. Balls tumble in a *cascading* action, where there are many ball-ball collisions with powder nipped between the balls, and in a *cataracting* action where balls fall through a large distance before striking the toe of the charge. Under good conditions the normal breakage rates will tend to match the power input to the cascading action, since higher power means more tumbling action. Under improper conditions it is possible to get lower breakage rates at higher power consumption, a direct inefficiency. Power used to produce cataracting is likely to be efficient for abnormal breakage of large sizes, but less efficient for normal breakage.

Very fine dry grinding or fine wet grinding at high slurry density gives a *slowing-down* of breakage rates in a mill, due to powder cohesion and slurry viscosity effects. This is again a direct inefficiency.

Producing a Simulation Model of a Large-Scale Plant: Fitted and Real Models

A number of workers have fitted simulation models to data from large mills (Freech, et al., 1970; Lynch, A. J., et al., 1977; Herbst and Rajamani, 1982; and Gelpe, Flament, and Hodouin, 1983). There are two basic approaches. The first consists of using data from a full-scale plant to back-calculate effective S and B values assuming a first-order mill model, starting either with a known or assumed residence time distribution or even back-calculating the RTD in addition to S and B values. The models produced by this technique can be called *fitted* models because there is no guarantee that the calculated values of S and B are real. They are just values which force-fit the mill data to the model, and if the wrong RTD is used or if the breakage is in actuality non-first-order, then errors in the assumed model will appear as non-real S and B values. The second approach is to start with a completely detailed knowledge of the breakage processes of the material involved under the mill conditions involved, usually from small-scale tests, and scale up this information to create a *real* simulation model of the full-scale plant. In practice, there is often insufficient information to apply this technique, so a compromise between the two approaches is used. The fewer assumptions made in the back-calculation, the more likely are the back-calculated values to be closer to real values.

Scaling from Laboratory Data: Wet Grinding of Copper Ore

The procedure followed by Klimpel and Austin (1984b) is given as a first example. It assumes: (1) first-order breakage; (2) the same B values for the mixture of balls in the mill as measured for 25 mm balls in a 200 mm i.d. laboratory mill; (3) the same α value (in $S_i = ax_i^\alpha$) in the large mill as in the small; (4) actual smoothed classifier s_i values to calculate mill feed in closed circuit; (5) an estimated RTD based on reported literature values for similar mills (one-large/two-small model of sizes 0.7, 0.15, 0.15). The circuit tested was a normal closed circuit with an overflow mill of 3.05 m i.d. by 4.7 m long, wet grinding copper ore at 80% of critical speed, giving 50 t/h of product with a flow through the mill of 97.5 t/h. Ball loading was estimated as 38% filling, with makeup of 25.4 and 50.8 mm balls. The breakage parameters for the ore as measured in a laboratory batch mill (Klimpel, 1982, 1983) are given as Ore 1 in Table 1: no values were reported for μ and Λ, since tests were not performed for large sizes in the abnormal breakage region. The values for the size distributions around the full-scale mill are given in Table 2. Samples were analyzed in triplicate, and statistical analysis indicated that the size distributions were accurate only to two significant figures.

Table 1. Laboratory Tests on Copper Ore

	Ore 1	Ore 2
Mill diameter, mm	200	200
Mill volume, cm^3	5800	5800
Ball diameter, mm	25.4	25.4
Mill speed, rpm	60	60
Fraction of critical	0.64	0.64
Volume load balls J, % (based on 0.4 porosity)	32.5	32.5
Density of ore, kg / m^3	2.65×10^3	2.65×10^3
Solids by weight, %	64	72
Weight of solid, kg	1.11	1.36
Interstitial filling U (based on 0.4 porosity)	0.93	1.14
$S_{18 \times 25} = a$, per min	0.34	0.30
α	0.93	0.91
γ	0.67	0.61
Φ	0.57	0.63
β	3.0	2.9
δ	0.0	0.0
Bond Work Index, kWh / t	12.7	15.0*
		13.8*

Source: Klimpel, 1982, 1983.
*Determined on different samples than those used for the breakage tests.

Table 2. Experimental and Predicted Size Distributions for a Closed-Circuit Ball Mill at 50 t / h, Mill Power 700 kW

Size	Makeup feed	Circuit product		Mill feed		Mill product		Recycle	
μm	G_i	Q_i exp	Q_i sim	F_i exp	F_i sim	P_i exp	P_i sim	T_i exp	T_i sim
				Cumulative % less than size					
1680	100	100	100	100	100	100	100	100	100
1180	80	100	100	90	88.8	100	99.6	99	99.2
850	65	100	100	80	79.6	99	98.6	97	97.0
600	50	100	100	69	69.3	96	96.4	92	92.1
425	43	100	99.8	63	62.1	92	92.9	85	84.8
300	37	99	99.3	54	53.9	87	87.6	74	73.8
212	32	97	97.2	45	44.8	79	80.1	60	59.9
150	28	91	91.7	36	36.4	71	71.0	45	46.4
106	24	82	82.3	28	29.6	61	61.2	36	36.3
75	20	71	70.8	22	24.2	50	51.7	28	29.1
53	16	59	59.0	18	19.5	42	42.8	23	23.6
38	13	48	48.6	15	15.8	34	35.1	19	19.2

An analysis around the classifier [610 mm (24 in.) hydrocyclone] as described in Chapter 13 gave a reasonable fit of classifier selectivity numbers by a constant bypass and the logistic function in log x for classification (Chapter 13, Eq. 9):

$$s_i = a + (1 - a)c_i$$

$$c_i = \frac{1}{1 + (d_{50}/x_i)^\lambda}$$

where λ is related to Sharpness Index by $\lambda = -\ln(9)/\ln(S.I.)$. The characteristic parameters were $a = 0.25$, $d_{50} = 193$ μm, and S.I. = 0.50; these values were used as input to the closed-circuit simulation program. The value of circulation ratio based on the plant data was 0.95.

To proceed further, it was necessary to make estimates of those factors not directly determined. Since the feed to the mill contained only a small fraction of sizes above 1 mm, it was not expected that abnormal breakage would be a significant factor, so the absence of values for μ and Λ was not important. The ball size distribution in the mill was assumed to be a Bond ball mix (see Chapter 16, $\Delta = 0$) with an upper size of 50.8 mm. The residence time distribution was assumed to be equivalent to the one-large/two-small model (see Chapter 14) with reactors in the ratio 0.7, 0.15, 0.15. The major unknown was the hold-up in the mill. The mass transfer law of Eq. 30 in Chapter 14 predicted a hold-up of $U \approx 1.0$, just on the borderline of the start of overfilling, showing that the circuit had been well optimized with respect to circulating load. This value gave a formal hold-up of 8.34 tons of ore.

The laboratory values for a_T, α, γ, Φ and β were used as input to the simulation model, without allowance for any variation of B values with ball diameter. The simulation was performed to give a circuit product passing through the one-point specification of 59% minus 270 mesh (53 μm), corresponding to the plant performance. The simulated result is also shown in Table 2: the value of τ was 5.7 min, the predicted circulation ratio was $C = 0.85$, and the value of Q/W was 0.095, giving a circuit capacity of 47 t/h, as compared to the measured value of 50 t/h. However, it must be noted that there was some addition of 25.4 mm makeup balls, which would make the simulated capacity increase by leading to a finer mix of balls in the mill.

The Bond calculation using the feed and product 80%-passing sizes gave a predicted capacity of 59 t/h. However, the Bond calculation is based on a circulation ratio of 2.5, whereas the plant had a circulation ratio of $C = 0.95$.

As a second example, this data was also treated by the *fitting* procedure, using the back-calculation technique of Klimpel and Austin (1984a) (see Chapter 10). The same assumptions were made concerning the residence time distribution and the classification action, and the laboratory B values were assumed to apply to the full-scale mill. However, this method back-calculates the set of $\overline{S}_i$ values necessary to get best least-squares fits to the

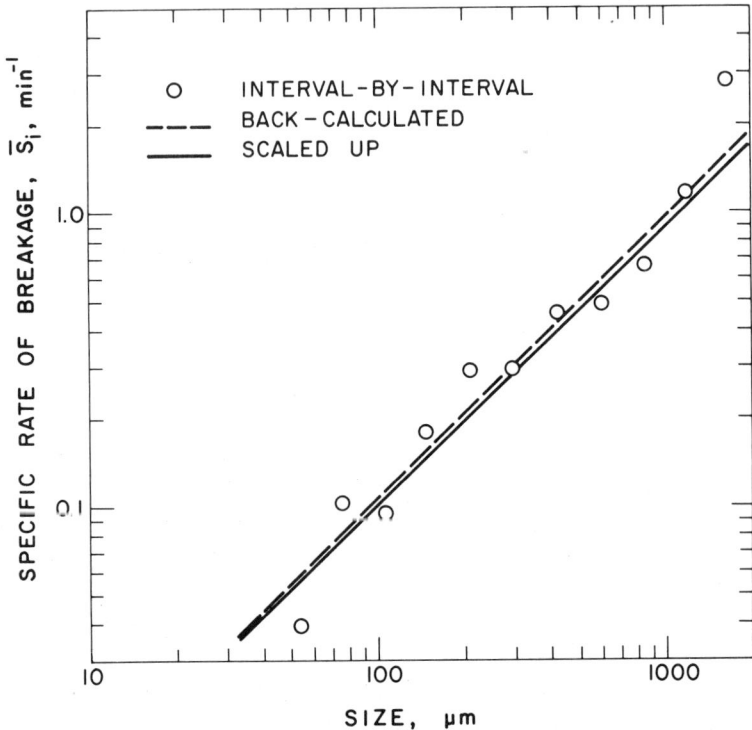

Fig. 1. Comparison of scaled-up $\overline{S}_i$ values with back-calculated values for 3-m diam mill ($\sqrt{2}$ intervals, plotted at upper size of interval).

experimental data, so no assumptions concerning the mixture of balls in the mill or mill conditions were necessary. A comparison of the back-calculated $\overline{S}_i$ values with the scaled-up values is shown in Fig. 1. The scaled-up values and the back-calculated values using $\overline{S}_i = ax_i^\alpha$ clearly have almost the same α value. Also shown are the interval-by-interval back-calculated $\overline{S}_i$ values, which indicate that the choice of a correct line through these points is difficult (see Chapter 10).

Fitted vs. Scaled-up Models for Another Wet Ball Mill

The simple analysis reported in the previous section is not always possible. Klimpel and Austin (1984b) have given results for the reversed closed circuit shown in Fig. 2, with the size distributions given in Table 3. The tests were performed by holding the water flow constant and increasing the solid rate, to increase pulp density. The tests were not performed with the objective of constructing a simulation model and the experimental size distributions were incomplete, and it was clearly not generally possible to perform accurate extrapolations to larger and smaller sizes. The makeup

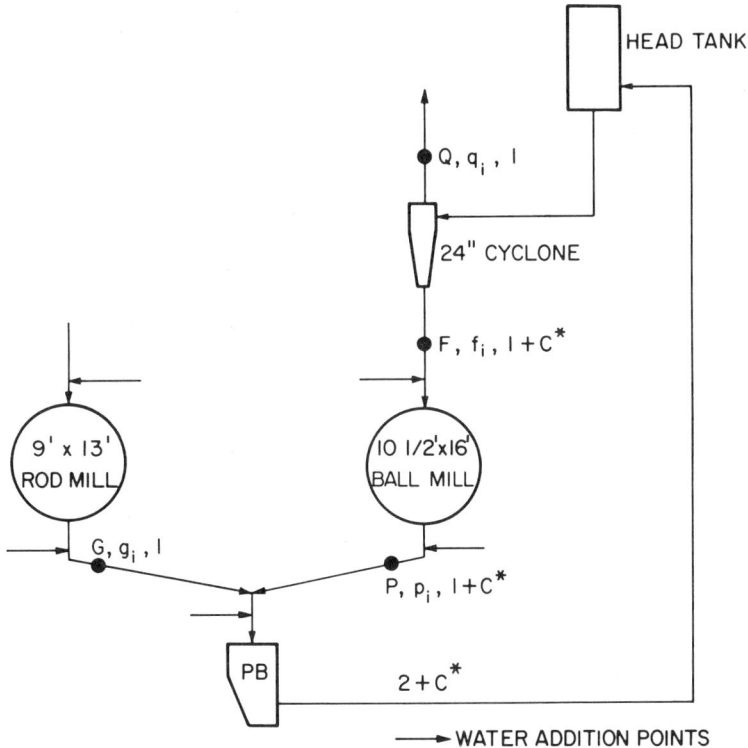

Fig. 2. Grinding circuit tested, sample points.

feed was a rod mill discharge; plotting the experimental size distribution data gave a straight line on a Schuhmann plot, so that reasonable estimates could be made of the smaller and larger makeup feed sizes. It was not possible to perform accurate classifier analysis, but the circulating loads (P/Q) could be calculated with reasonable accuracy using Eq. 1a in Chapter 13, a best estimate of Sharpness Index was 0.45, and the bypass was within the range 0.26 to 0.30.

Unfortunately, a sample of ore for S and B analysis was not taken during the tests. Testing on three samples taken at later times gave the breakage properties given as Ore 2 in Table 1, and values of Bond Work Index (determined at 200 mesh) of 13.8 kWh/t for one sample and 15 kWh/t for another. Comparing the breakage properties of Ore 2 with Ore 1 for the same filling level shows that Ore 2 should have a lower Work Index than Ore 1, not a higher one. Thus it was accepted that the breakage properties of Ore 2 given in the table were probably for a sample of weaker material.

During the full-scale tests the hold-up in the mill was measured by dumping the mill, at the three different flow rates. On this basis, the value of

Table 3. Comparison of Computed and Experimental Results for a Wet Overflow Ball Mill in Reverse Closed Circuit (3 m diam)

Capacity, t/h	Size, mesh	Int. No.	G Rod mill discharge	F Cyclone underflow to ball mill Exp.	F Sim.	P Ball mill discharge Exp.	P Sim.	Q Cyclone overflow Exp.	Q Sim.	Circulating load P/Q
87	6	1	100*		100		100			Expt. = 1.75
	8	2	99.5*		99.7		99.9			
	12	3	96*		97.3		99.3		100	Sim. = 1.57
	16	4	88*		91.6		99.7		99.9	wt% solids
	20	5	79*		84.4		94.6	100	99.7	mill exit
	30	6	69*		75.1	91.2	89.2	99.8	99.2	73.5
	40	7	59.5	68.2	63.9		80.8		97.6	
	50	8	51*		51.3	74	70.4	95.7	93.8	
	70	9	44.8	44.6	39.3	60.6	58.8	86.8	86.0	
	100	10	38.3	32.4	28.5	47.6	48.3	73.5	75.3	
	140	11	32.6	22.4	21.2	38.9	40.2	65.0	65.8	
	200	12	29.7	19.3	17.3					
102	6	1	100		100		100			Expt. = 1.95
	8	2	95.0*		97.3		99.8			
	12	3	88.0*		93.1		99.2		100	Sim. = 1.96
	16	4	79.0*		87.1		97.8		99.9	wt% solids
	20	5	69.0*		79.0		94.8		99.7	mill exit 76
	30	6	60.0*		69.5		89.7		98.8	
	40	7	49.7	57.3	56.1	84.1	81.1	99.7		

Table 3. Continued

Capacity, t/h	Size, mesh	Int. No.	G Rod mill discharge	F Cyclone underflow to ball mill		P Ball mill discharge		Q Cyclone overflow		Circulating load P/Q
				Exp.	Sim.	Exp.	Sim.	Exp.	Sim.	
	50	8	45.0*		44.3		70.7		96.9	
	70	9	38.4	36.2	30.6	64.4	57.5	92.0	91.3	
	100	10	33.1	26.1	20.5	50.8	45.3	81.2	81.8	
	140	11	28.6	17.9	14.5	39.3	35.7	70.2	70.2	
	200	12	27.4	13.1	11.8	30.9	29.2	59.6	61.6	Expt. = 2.80
108	6	1	100		100		100			
	8	2	96*		98.2		99.8			
	12	3	87*		93.7		98.7			
	16	4	75*		86.4		96.1		100	
	20	5	64*		77.5		91.4		99.9	
	30	6	55*		66.9		84.1		99.5	Sim. = 2.59
	40	7	46.3	48.1	53.0	66.7	73.1	99.0	98.4	wt% solids
	50	8	41.5*		39.4		60.3		95.5	mill exit
	70	9	35.9	27.3	25.7	44.9	46.1	87.0	88.5	80.5
	100	10	31.0	18.9	16.5	34.3	34.4	76.1	77.4	
	140	11	26.8	12.7	11.5	25.4	26.4	65.4	65.4	
	200	12	23.2	9.3	8.7	21.4	20.9	56.3	54.6	

*Extrapolated or interpolated.

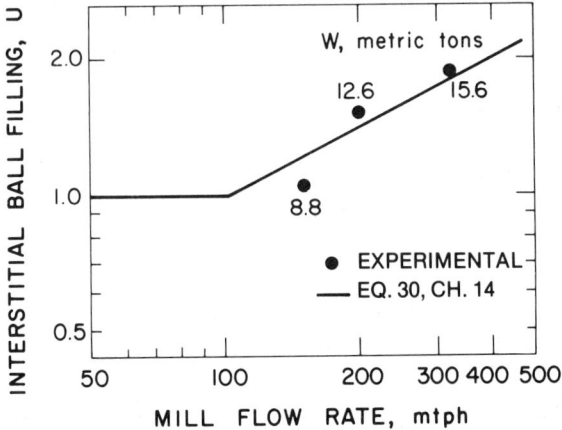

Fig. 3. Variation of hold-up with solid flow rate through a wet overflow mill, $D = 3.05$ m.

interstitial filling of the balls by solid hold-up is shown in Fig. 3. It is quite evident that the high flow rates through the mill gave substantial overfilling of the mill.

The data for the ore was scaled to the full-scale mill using an estimate of ball load of 38% mill filling, the mill speed of 70% of critical speed, and assuming an equilibrium Bond ball mix for 50.8 mm makeup balls. The scaled values of the breakage parameters were used in closed-circuit simulations (RTD = 0.7, 0.15, 0.15, as before) and the mill capacity varied to give a one-point match on the circuit product: 75.3% < 106 μm for $Q = 87$ t/h; 70.2% < 150 μm for $Q = 102$ t/h; and 65.4% < 75 μm for $Q = 108$ t/h. At the same time, the d_{50} of the cyclone was varied, assuming bypass to be 0.28, until the simulated mill feed and mill product analyses were in best agreement with the experimental size distributions at 75 μm. It was found that the three simulated data sets required a d_{50} of about 190 μm to give reasonable match to the circulating loads and size distributions around the circuit. Note that the makeup feed to the ball mill was different for each test, being coarser as the capacity was increased since the residence time in the preceding rod mill was also decreased.

The size distributions in Table 3 and Fig. 4 show three major features. First, the relatively small changes in overall capacity produce large changes in the mass flow of solid through the mill, which were approximately 150, 200, and 300 t/h. Second, it is clear that the simulated size distributions do not agree very closely with the measured plant values. There is a consistent variation between simulated and plant data: the experimental size distributions have a steeper slope than the simulated values for the lowest feed rate and a lower slope for the highest feed rate. The explanation is probably that the high filling level and higher slurry density in the mill at high flow rates give breakage with a lower value of γ (see Fig. 13 in Chapter 5), producing size distributions with lower slopes and proportionally more fines.

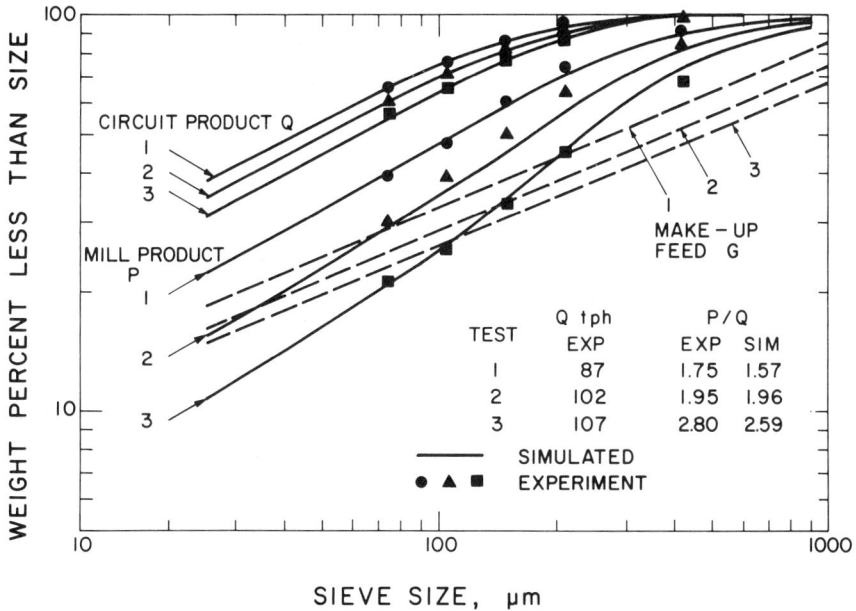

Fig. 4. Comparison of model and experimental results for wet grinding of copper ore at three flow rates (see Table 3).

The values of breakage parameters for Ore 2 given in Table 1 in this chapter gave too high a circuit capacity, as expected. Therefore the value of a was adjusted to give the correct capacity at the lowest feed rate. It was found that it had to be multiplied by a factor of 0.87, that is, it was too high by about 15%. The values of U and hence the overfilling factors were calculated from the mean line of U vs. P of Fig. 3. The simulated values were $U = 1.2$, $K_o = 0.92$ at $Q = 87$ t/h and $P = 150$ t/h; $U = 1.45$, $K_o = 0.80$ at $Q = 103$ t/h and $P = 202$ t/h; $U = 1.66$, $K_o = 0.68$ at $Q = 108$ t/h and $P = 277$ t/h. It is seen, therefore, that adjusting the value of a to make the capacity correct at one flow rate gave good predictions of the other two mill capacities.

Using the experimental 80%-passing sizes and a Bond Work Index of 15 kWh/t gave values of Q of 79.5, 82, and 89 t/h. A Work Index of 13.8 kWh/t gave values of 86, 89, and 96 t/h for the three tests.

Even though the simulation model is not perfect, it is concluded that it exhibits all the major features of the experimental data. Expressed in another way, making the model fit the data for one set of conditions (a *fitted* model) enabled reasonable predictions to be made for the other conditions. The evidence for overfilling of the mill at high flow rates through the mill, and consequent decreased breakage rates, is quite clear. Normally, increase of flow rate through the hydrocyclone would decrease d_{50} (e.g., see Eq. 22 in Chapter 13), but the increase of slurry density in these tests appeared to compensate to give essentially constant d_{50}.

A Simulation Model for an Air-Swept Ball Mill
Grinding Coal*

The characteristic feature of this type of mill is that the ground coal is removed entirely in the air stream. The stream can be used immediately for direct firing of a furnace or it can be passed through a cyclone and filter system to recover the coal. Air sweeping is used to dry the coal (the entering air is heated) and to keep the size distribution in the mill relatively free of fines to prevent the slowing-down effect (see "Effect of Environment in the Mill" in Chapter 5). Such mills have large feed and exit trunnions to reduce pressure drop through the mill and are usually operated at comparatively low ball loadings, e.g., $J = 0.25$. The air-flow rate is adjusted to give an optimum hold-up in the mill for the coal feed rate being used, as determined by maximum capacity. The specification of the size distribution of the product is obtained by adjustment of an air classifier, with return of oversize to the mill feed.

The pilot-scale mill tested was an air-swept mill of 0.98 m (3.24 ft) mean internal diameter by 1.5 m (5 ft) long, with inlet and discharge trunnions of approximately 0.30 m (1 ft) diameter. The ball load corresponded to 24% mill filling ($J = 0.24$), with a ball mix of 25% of 25.4 mm (1 in.) diameter, 37.5% of 31.8 mm (1 1/4 in.), and 37.5% of 38.1 mm (1 1/2 in.). The mill was run in closed circuit at steady state at 80% of critical speed, with the feed rate adjusted to give the correct sound level by an experienced operator. The mass flow rate of recycle from the classifier was measured by diverting the flow to a collecting bin for a short time interval, and the makeup feed rate was measured by a calibrated variable speed belt feeder. The exit air temperature was close to 65°C, ensuring coal of low free moisture content. Fig. 5 shows the test circuit (Luckie and Austin, 1980) and ancillary equipment. Classification was by a 0.7 m diameter twin cone air separator, as shown in Fig. 6, adjusted by varying the setting of the entrance vanes.

The breakage parameters of the test coal (subbituminous, Belle Ayre-Wyoming, Hardgrove Grindability Index of 58) were determined in a laboratory mill in the usual way, for an air-dried sample (12 to 15 wt% moisture), giving $\alpha = 0.80$, $\gamma = 0.90$, $\Phi = 0.374$, $\beta = 2.8$, and $a = 1.80/\text{min}$ under the laboratory test conditions (see Table 1 in Chapter 5). Breakage was found to be normalized, that is, $\delta = 0$, and the same B values were found for breakage of larger sizes to the right of the maximum in S_i values. The breakage of these larger sizes was approximately first-order and gave $\Lambda = 3.0$ and $\mu = 3.70$ mm for 25.4 mm balls.

The scale-up of these breakage values from the laboratory test mill of $D = 195$ mm to the pilot-scale mill of $D = 980$ mm was tested by performing batch tests in the pilot-scale mill (without air-sweeping, of course) under the loading conditions expected for continuous operation. Fig. 7 shows that the scaled-up batch simulation gave good prediction of the experimental data. It was noted that the slowing-down effect seen in the laboratory batch

---

*Austin et al.

Fig. 6. Air separator used in coal mill test circuit.

LEGEND

→	MATERIAL FLOW
⇢	AIR FLOW
⊸	AIR & MATERIAL FLOW
—·—	INSTRUMENTATION FLOW
	PRESSURE CONNECTION
Ⓣ	TACHOMETER
SCR	D. C. VARIABLE SPEED CONTROLLER
	THERMOCOUPLE
⟓	DAMPER
⊽	SONIC EAR

DAMPER POSITION INDICATOR

CLASSIFIER

AIR LOCK

FEED

FEEDER

SCR

AIR LOCK

AMBIENT AIR

HOT AIR SOURCE

TEMPERATURE INDICATOR W/HIGH LIMIT ALARM

MILL

COARSE

MILL LEVEL CONTROLLER

MILL EXHAUST FAN

FINE

CLASSIFIER DIFFERENTIAL PRESSURE INDICATOR

MILL OUTLET TEMPERATURE RECORDER CONTROLLER

MILL DIFFERENTIAL PRESSURE INDICATOR

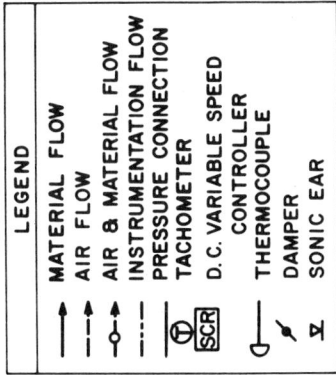

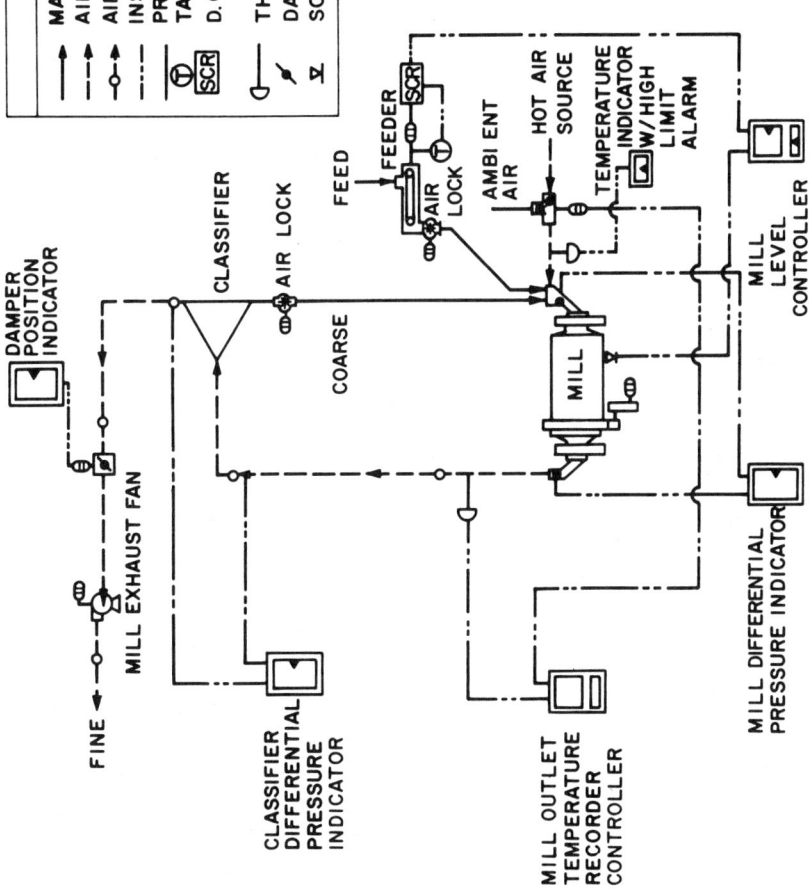

Fig. 5. Test circuit for coal grinding in an air-swept ball mill.

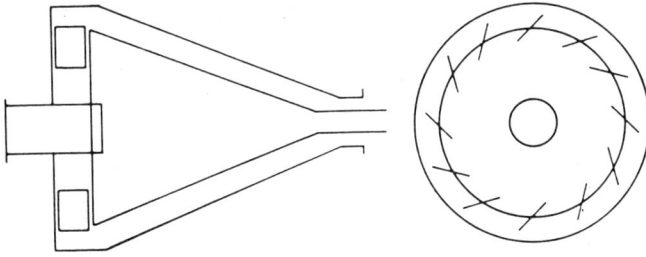

261

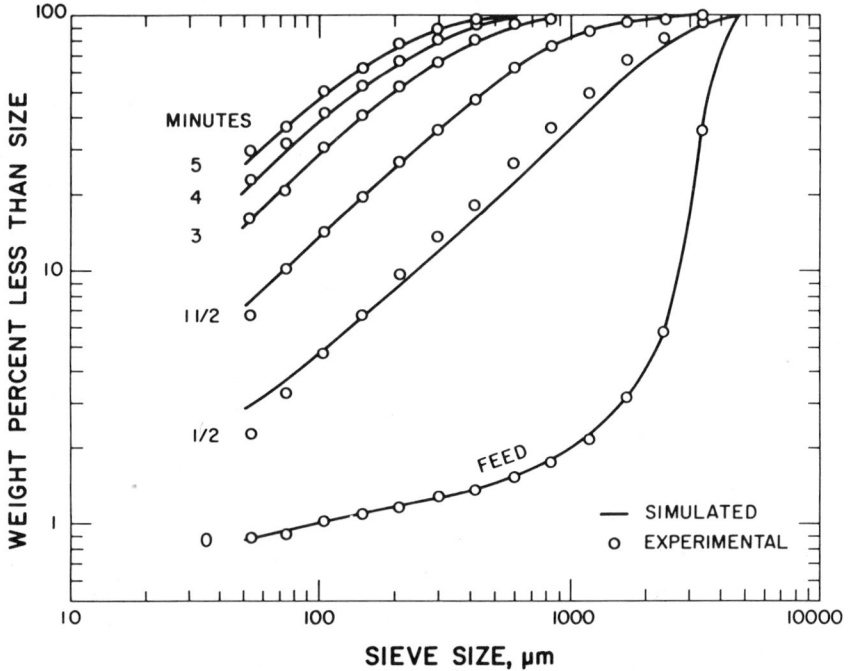

Fig. 7. Results of batch grinding 75 kg Belle Ayre coal ($J = 0.25$, $f_c = 0.083$, moisture content of coal = 15%) in pilot-scale mill.

test as fines accumulated was also seen in this batch test at grinding times greater than five minutes.

The model for this type of mill (see "Air-Swept Ball Mills for Coal or Cement Finish Grinding" in Chapter 17) involves parameters additional to those for non-swept mills. These extra parameters describe the internal classification action as the air sweeps up material and allows larger sizes to fall back into the bed. This was investigated as follows. The mill was operated at steady state and samples taken for feed, recycle, and product size analysis. The mill power, feed, and air-flow rate were then suddenly and simultaneously stopped; the mill came to rest in less than one third of a revolution; and the contents were tipped out, weighed, mixed, and samples taken for size analysis. Four tests at various coal flow rates gave mill hold-ups of 71.3, 72.6, 71.3, and 79.4 kg of coal (dry basis), that is, a mean value of about 73 kg. This is a fractional formal mill filling of $f_c = 0.081$, which is a fractional filling of the ball interstices of $U = 0.83$. This correlates well with the optimum value of 0.83 predicted from Eq. 7 in Chapter 5 for dry grinding.

Knowing the steady mass flow rate to the classifier, the mass flow rate of recycle, and the recycle and product size distribution, the size distribution of the mill discharge, p_i', was readily calculated. τ was calculated from the mill feed rate and hold-up (both dry basis). The internal classification action was described by the balance

$$Fp_i' = \eta \omega w_i (1 - c_i) W \tag{1}$$

where F is the solid flow rate leaving the mill; p_i' is the weight fraction of F which is of size i; η is the fraction of hold-up W which is exposed to the air-sweeping action per mill revolution; ω is the mill revolutions per unit time; w_i is the fraction of W which is of size i; c_i is the fraction of swept-up size i which falls back into the bed. Sampling of powder along the axis of the stopped mill showed that the contents were almost fully mixed, so w_i was taken as the mean value over the total hold-up.

It was found that p_i'/w_i was constant for small sizes since $c_i = 0$, that is, from Eq. 1,

$$\eta \omega = p_i'/\tau w_i, \text{ small sizes} \tag{2}$$

where $\tau = W/F$. This enabled the calculation of η. Fig. 8 shows the variation of η as a function of air-flow rate. This was expressed as

$$\eta = 1.2 \times 10^{-2} u^{0.8} \tag{3}$$

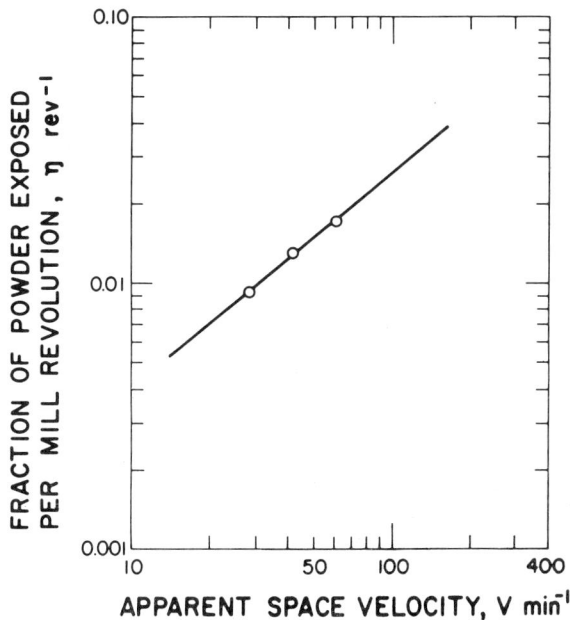

Fig. 8. Value of η for different flow rates of air through a 0.98 m I.D. by 1.5 m long air-swept ball mill (apparent space velocity is volume changes per minute).

UPPER SIEVE SIZE, μm

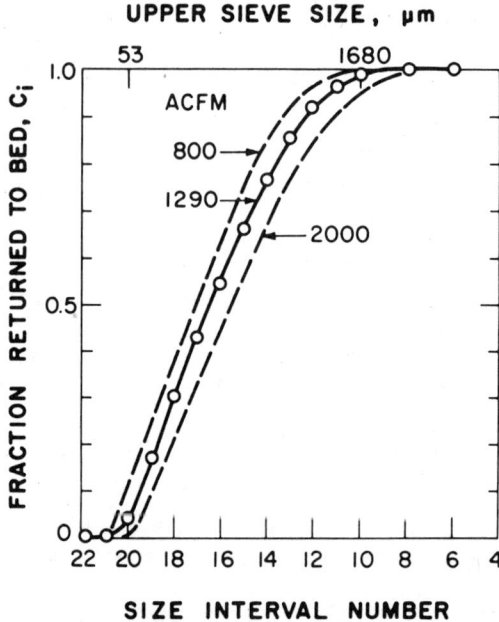

Fig. 9. Smoothed values of internal classification numbers c_i, for $\sqrt{2}$ intervals, plotted at upper size; 1290 Actual Cubic Feet per Minute = 0.61 m^3 / s.

where u is the velocity of air flow through the mill in m/s based on a cross-section of $1 - J$. The value of η for a particular test was used to calculate c_i values from Eq. 1, giving the result shown in Fig. 9. Material greater than about 2000 μm did not leave the mill, that is, $c_i = 1.0$ for large sizes.

The scaled-up breakage parameters and the values for η and c_i were used in the simulation model (see Appendix 7 for computational procedure) and gave the result shown in Fig. 10 for a simulation using the actual mill feed (f_i) and the air-flow rate. Also shown is the size distribution of the circuit product, using the model in closed circuit with classifier selectivity numbers for the twin cone classifier.

Finally, the model was used with appropriate selectivity numbers for the external classifier to predict the capacity and size distribution of the circuit product for a range of air-flow rates. The result is shown in Fig. 11. The model gives correct predictions within the experimental variability of the data.

Several points are of interest. First, because coal is a friable material, it is possible to use larger makeup feed sizes than usual with hard rock. Fig. 10 shows the feed to contain substantial amounts of material greater than 10 mm. This leads to the typical plateau in the size distribution in the mill, representing persistence of large sizes. However, the internal classification

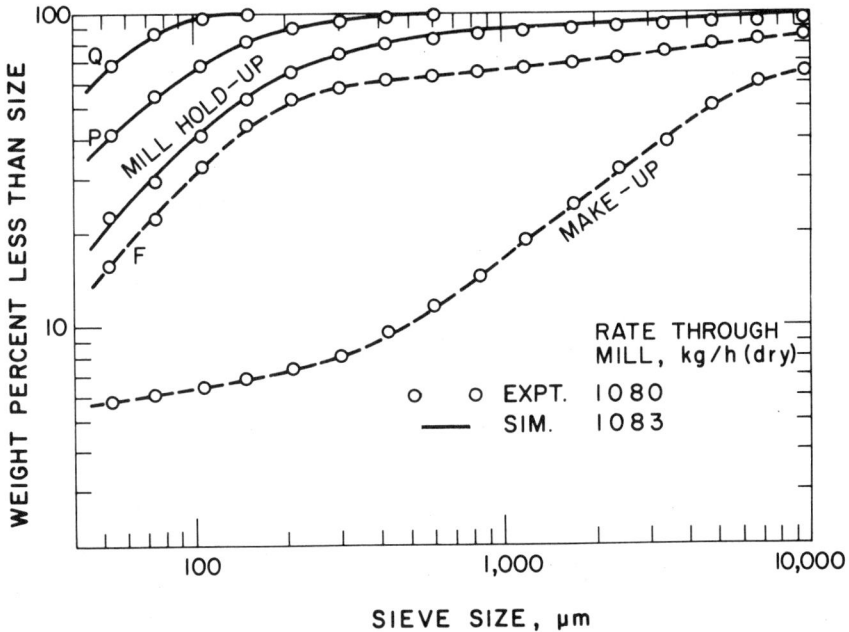

Fig. 10. Comparison of experimental results and computed results; mill hold-up 73 kg, air-flow rate, 0.61 m³ / s (1290 ACFM) at 65°C.

action prevents these larger sizes from leaving the mill, so that the size distribution of the mill product is a normal type of size distribution. Second, the removal of fines by the air stream means that the size distribution of the mill hold-up does not contain the same extent of fines as the mill product. This avoids the direct inefficiency which would otherwise be caused by the slowing-down effect if the mill charge contained fine material. Third, if the external classifier is adjusted, it is possible to get a range of circuit product size distributions at the same air-flow rate. However, keeping the mill hold-up at optimum by using mill sound as a guide requires that the coal flow rate be correspondingly adjusted, thus altering the coal-air ratio. The model can be used to predict the air flow and classifier conditions necessary to get a desired product specification at a desired air-coal ratio, and it will predict the unique capacity for those specifications.

In this particular pilot mill, the drive and bearing friction losses were 5.9 kW and the mill power was 7.5 kW, giving a total motor power of 16.2 kW since the motor efficiency was 82%. The fan power at 0.71 m³/s (1500 ACFM) was 2.6 kW and at 0.95 m³/s (2000 ACFM) was 6 kW. Thus the use of too high air velocities carries a penalty of energy loss in excess fan power.

The models for this type of mill cannot be considered as complete because scale-up laws for η and c_i have not been developed. In addition, breakage characteristics are affected by the moisture content of coals and no systematic study of this effect was performed.

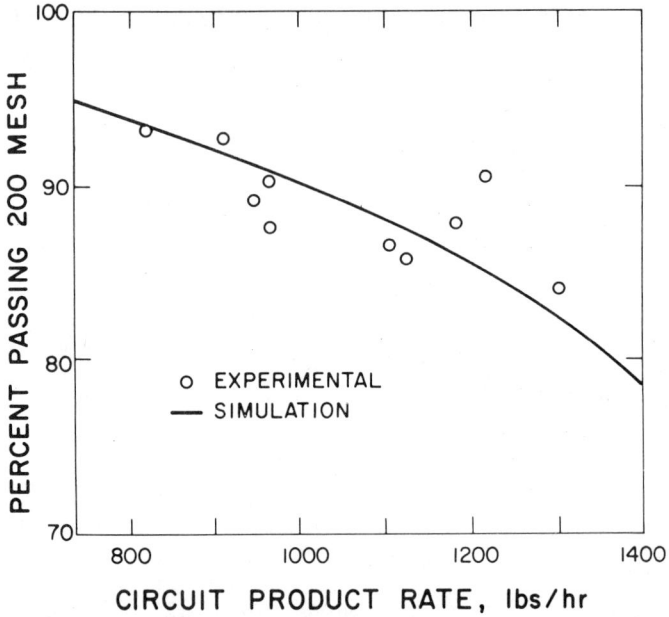

Fig. 11a. Comparison of experimental and predicted fineness of grinding for Belle Ayre coal ground closed circuit in
the pilot-scale mill.

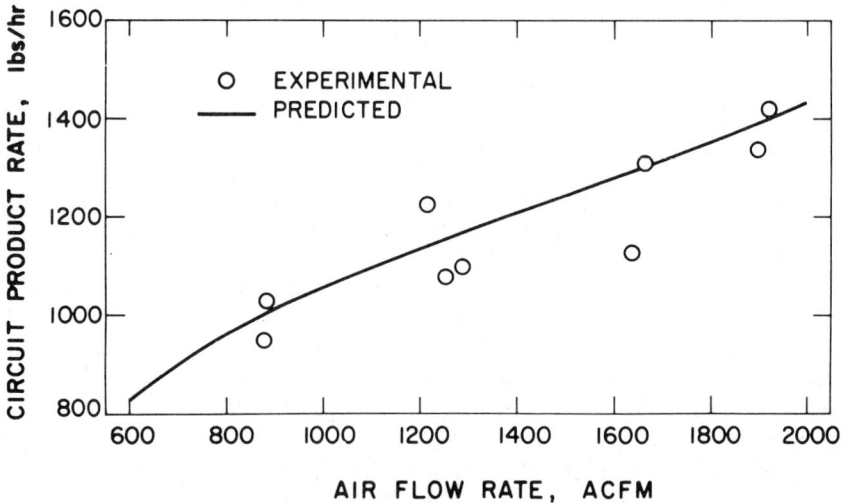

Fig. 11b. Circuit capacity vs. air-flow rate for Belle Ayre coal ground in the
pilot-scale mill.

A Simulation Model for a Cement Mill

Austin, Luckie, and Wightman (1975), Austin, Luckie, and von Seebach (1976) and Austin, Weymont, and Knobloch (1980) have presented partial simulation models for dry milling of cement clinker in two compartment cement tube mills. The most recent model (Austin, Weymont, and Knobloch) incorporates the effect of air sweeping, with an internal classification action. The mathematical details of the model are given in "Air-Swept Cement Mill" in Chapter 17; it treats the mill as a series of equal fully mixed reactors (see Chapter 14), with appropriate mixtures of balls in each reactor, and with material swept up and falling back in each reactor. The concept of internal classification by air-sweeping is identical to that discussed in the previous section; the values measured in those tests were used as a guide to the action in a cement mill. Each reactor was taken to be equivalent to $D/2$ in length, with internal classification values of the form

$$c_i = \frac{1}{1 + (d_{50}/x_i)^{1.5}}$$

where d_{50} is the particle size at which half drops back into the powder hold-up and half continues in the air stream out of the reactor, x_i is particle size, and c_i is the fraction returned to the hold-up. The value of d_{50} for coal of specific gravity 1.3 was 320 μm at an air velocity of about 1 m/s in the mill.

Table 4 gives the breakage parameters for the test clinker, measured in the usual way from batch tests in a laboratory mill. It was found that the B values for large sizes, to the right of the maximum in S_i values, were different from the normal values. As a reasonable approximation, breakage to the right of the maximum for a given ball size was assumed to have the same B parameters, $\gamma = 0.32$, $\beta = 3$, $\Phi = 0.14$. The B values to the left of

Table 4. Characteristic Breakage Parameters for Leimzem Cement Clinker in a Laboratory Test Mill with $J = 0.20$, $U = 0.42$, $d = 27.0$ mm, $D = 195$ mm, $\phi_c = 80\%$ critical

Parameter	Value
α	0.91
a, per min	0.79
Λ	2.5
μ, mm	1.75
Normal breakage, small sizes	
β	4.00
γ	0.75
$\Phi_{16\times20}$	0.34
δ	0.23
Abnormal breakage, large sizes	
β	4.0
γ	0.32
$\Phi_{4\times6}$	0.14

the maximum were non-normalized, with $\delta = 0.23$. The minimum ball size, d_{min}, which gives normal (to the left of the maximum) breakage of size j is given by

$$d_{min} = \left[\left(\frac{x_j}{\mu} \right) \left(\frac{\Lambda - \alpha}{\alpha} \right)^{1/\Lambda} \right]^{1/2} \qquad (4)$$

and therefore the fraction of normal breakage of size j by the mixture of balls is given by

$$m'_j = \sum_{k=1}^{k_{min}} m_k S_{j,k} / \bar{S}_j \qquad (5)$$

where k_{min} corresponds to d_{min}. This leads to a mean $B_{i,j}$ value for the breakage of each size given by

$$\bar{B}_{i,j} = m'_j B_{i,j} + \left(1 - m'_j \right) B'_{i,j} \qquad (6)$$

where $B_{i,j}$ is the normal value for size j and $B'_{i,j}$ is the abnormal (right of maximum in S) value. It was assumed that B values do not depend on ball diameter except for this balance between normal and abnormal breakage.

Fig. 12 shows the circuit which was analyzed. The airborne powder (62,000 m³/hr of air) was passed through a 3.5-m diam static air separator (*Statopol*) and the coarse stream joined with the powder flow as feed to a pair of 6-m diam mechanical air separators (*Turbopols*). The size distributions around the circuit were determined by sieving and by using the Cilas laser diffraction subsieve measurement apparatus (using methyl alcohol as the liquid). The size distributions were used to determine the classifier selectivity values as discussed in Chapter 13.

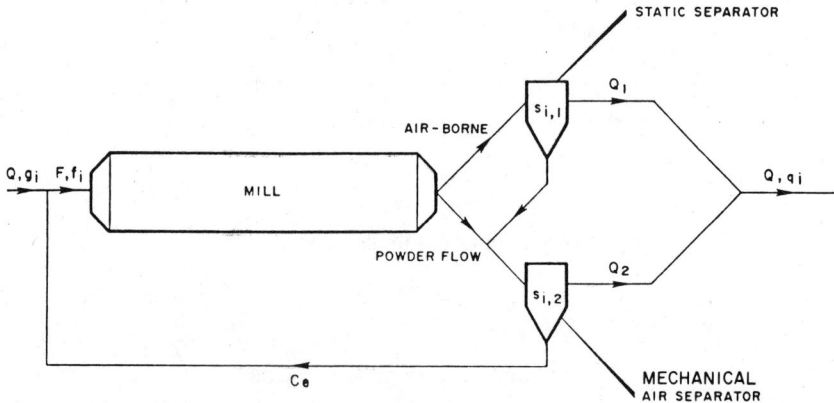

Fig. 12. The closed circuit mill with two external classifiers used for the large-scale mill test.

The mill tested was a two compartment 4.2 m internal diameter mill with a 5.0 m long first compartment and a 10.25 m long second compartment, giving a total mill volume of 210 m³. The mass of balls in the first compartment was 84 t and in the second 164 t, consisting of the mixture of sizes shown in Table 5. The fractional ball loading was 0.26 and 0.25 respectively, based on a bulk density of 4700 kg/m³. The mill was operated at 16 rpm, that is, 73% of critical speed, at a power of 3520 kW and the final product was 117 t/h of cement with a Blaine specific surface area of 3230 cm²/g, giving a specific grinding energy of 30 kWh/t.

Fig. 13 shows the scaled-up values of breakage rates, assuming the ball charges to be fully mixed in each compartment and assuming the optimum powder loading of $U = 0.84$. These $\bar{S}_i$ values were used in two reactors in the first compartment and five in the second, corresponding approximately to reactor lengths of $D/2$. This number of equal fully mixed reactors in series gives an equivalent RTD corresponding approximately to those measured for tube mills (see Chapter 14).

Fig. 14 shows the selectivity curve for the mechanical separator, calculated from the size distributions around the separator. The increase in the fraction returned to the coarse stream at sizes below 20 μm is typical of air separators used for cement (see "Mechanical Air Separators" in Chapter 13). Complete size distributions were not available around the static separator, but previous experience had indicated that the amount of airborne material was approximately 12% of the total flow through the mill at this air-flow rate, and had an 80%-passing size of about 70 μm. The fines from the static separator were about 15% of the total production of finished cement.

Since size analyses around the static separator were not available, it was not possible to calculate directly the selectivity curve. Therefore, the simulation was first performed on the mill alone. Since the makeup feed, recycle size distributions, and circuit capacity were known, calculation of the circulation ratio around the mechanical separator enabled the feed to the mill

Table 5. Ball Size Distributions in the 4.2-m Diam Cement Mill (Compartment 1, 5.0 m long; Compartment 2, 10.25 m long)

Compartment	d_k, mm Range	Mean	m_k Mass Fraction
1	100 – 95	97.5	0.17
	95 – 85	90.0	0.33
	85 – 75	80.0	0.31
	75 – 65	70.0	0.14
	65 – 55	60.0	0.05
2	60 – 50	55.0	0.09
	50 – 40	45.0	0.13
	40 – 30	35.0	0.13
	30 – 25	27.0	0.18
	25 – 23	24.0	0.33
	23 – (19)	21.0	0.14

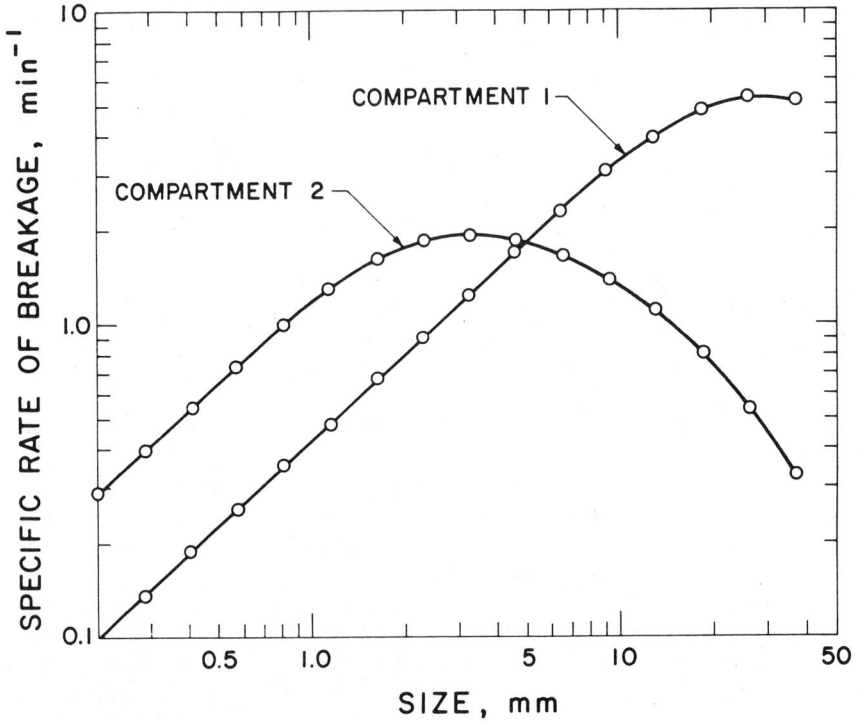

Fig. 13. The scaled-up overall specific rates of breakage $\bar{S}_i$ used for the simulations (values for $\sqrt{2}$ intervals, plotted at upper size of interval).

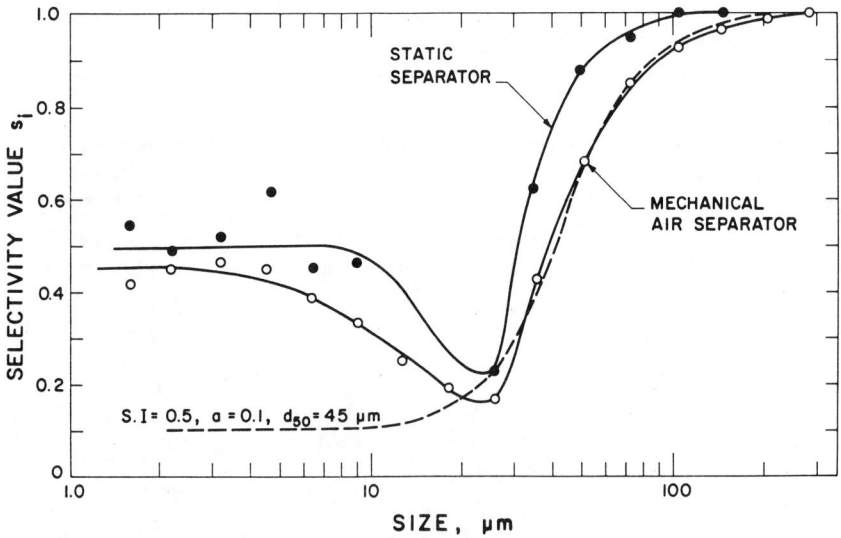

Fig. 14. Tromp curves for the mechanical and static separators (values for $\sqrt{2}$ intervals plotted at upper size of interval).

to be estimated. The mill simulation was performed with this feed and the values of scaled-up breakage parameters. The flow rate through the mill was adjusted to give a mill product to match the known feed to the mechanical separator. At the same time, the values of d_{50} in Eq. 4 and η in Eq. 1 were varied to obtain the correct quantity and size of the airborne material. It was found that the size of the airborne material was most sensitive to d_{50}, whereas the quantity of airborne material was most sensitive to η. Values of $d_{50} = 200$ μm and $\eta = 0.003$ fraction per revolution were found to be reasonable estimates. It was known that the static separator product was 15% (18 t/h) of the total circuit production (117 t/h), and the model simulation showed the airborne material also to be 15% of the flow through the mill. This gave the feed, rejects, and fine product around the separator to be in the ratio 3 : 2 : 1, and enabled the estimation of the Tromp curve for the static separator, as shown in Fig. 14. Since the fraction of airborne material to powder flow was relatively small, the circuit capacity is not very sensitive to errors in the selectivity values for the static separator.

It was now possible to perform a complete circuit simulation, to make the circuit produce a final product of 49.4% minus 18 μm to correspond with the experimental value. The simulation gave a circuit capacity of 150 t/h compared to the experimental value of 117 t/h, which was clearly too high. This was expected because cement of this fineness is in the region where slowing-down of breakage rates is expected (see "Effect of Environment in the Mill" in Chapter 5). Examination of the size distributions predicted in the final stages of the mill showed that the fineness of the charge would lead to slowing-down. Consequently, the simulation was rerun iteratively (Austin, Klima, and Knobloch) with correction of the S_i values in each section using the set of slowing-down factors shown in Table 6.

Incorporating the slowing-down effect gave a simulated circuit capacity of 114 t/h compared to the actual capacity of 117 t/h; the flow through the mill was 328 t/h compared to an actual of 346 t/h. The predicted fraction

Table 6. Slowing-down Factors κ as a Function of Fineness of Size Distribution*

80%-Passing Size, μm	κ
> 595	1.0
595	1.0
420	0.99
297	0.94
210	0.78
149	0.53
105	0.28
74	0.07
53	0.02
37	0
26.3	0
18.5	0

*See Fig. 18 in Chapter 5.

Table 7. Size Distributions as Cumulative Weight % Minus Size

Size, μm	Int. No.	Fresh feed	Mill feed Expt.	Mill feed Model	Mill product: Powder; Model	Mill product: Airborne; Model	Mechanical separator Feed Expt.	Recycle Expt.	Recycle Model	Fines Expt.	Fines Model	Circuit Product Expt.	Circuit Product Model
36900	1	100	100	100	100	100							
26090	2	96	98.6	98.6	100	100							
18440	3	87.5	95.8	95.7	100	100							
13040	4	66	88.5	88.2	100	100							
9219	5	31.5	76.8	76.2	100	100							
6518	6	12.5	70.4	69.6	99.9	100							
4608	7	-5.0	67.9	66.9	99.9	100							
3258	8	2.0	66.9	65.8	99.8	100							
2303	9	0.7	66.4	65.3	99.7	100							
1629	10	0.4	66.0	65.1	99.6	100			99.5				
1151	11	0.2	65.6	64.8	99.3	100			99.2				
814	12		64.9	64.4	98.9	100			98.6				
575	13		64.2	63.6	97.9	99.9			97.4				
407	14		62.9	62.1	96.0	99.8			95.2				
288	15		60.9	59.6	92.9	99.4	94	92	91.3	99.9	100		
203	16		57.6	55.5	88.1	98.5	91	87	85.1	99.8	100		
144	17		51.0	49.7	81.1	96.2	84	77	76.1	99.5	99.6	99.6	99.6
102	18		41.0	41.5	71.7	91.3	75	62	63.6	98.5	98.7	98.7	98.7
72	19		30.1	30.9	59.8	82.0	62	45.5	47.4	95.5	95.9	96.1	96.9
51	20		19.5	19.5	46.4	67.5	47.5	29.5	29.9	89	88.7	90.3	91.0
36	21		11.6	11.3	34.8	52.1	35	17.5	17.3	76	75.0	78.4	79.1
25.4	22		7.9	7.9	26.8	41.2	27.5	12	12.2	59	59.4	62.0	64.1
18.0	23		7.3	7.3	21.7	34.7	23	11	11.1	47	46.5	49.4	50.5
12.7	24		6.7	6.6	17.7	29.2	19.5	10.1	10.1	38.5	36.7	39.8	40.0
9.0	25		5.8	5.8	14.4	24.4	15.5	8.7	8.9	29	28.8	29.9	31.7
6.4	26		4.9	5.0	11.7	20.0	12.5	7.4	7.6	23	22.6	23.8	25.1
4.5	27		3.9	4.1	9.3	16.2	9.8	5.9	6.3	17.5	17.7	18.1	19.9
3.2	28		3.0	3.2	7.4	12.8	9.7	4.6	4.9	14	14.0	14.5	15.7
2.2	29		2.3	2.5	5.8	10.0	5.6	3.4	3.9	10.7	11.0	11.1	12.3
1.6	30		1.7	2.0	4.5	7.9	4.2	2.5	3.0	8.0	8.6	8.3	9.7

272

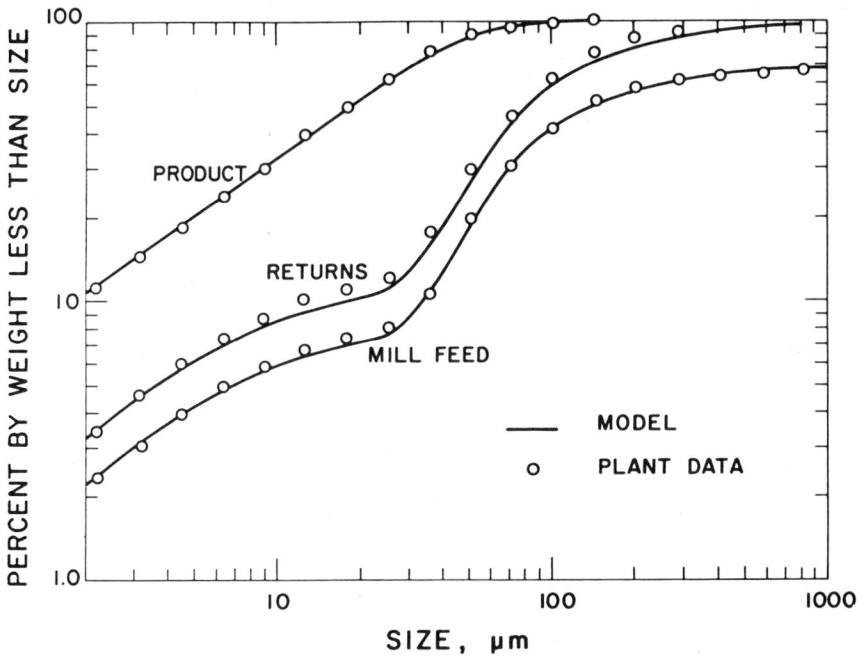

Fig. 15. Comparison of predicted and plant size distributions for the 4.2 m by 15 m long cement mill.

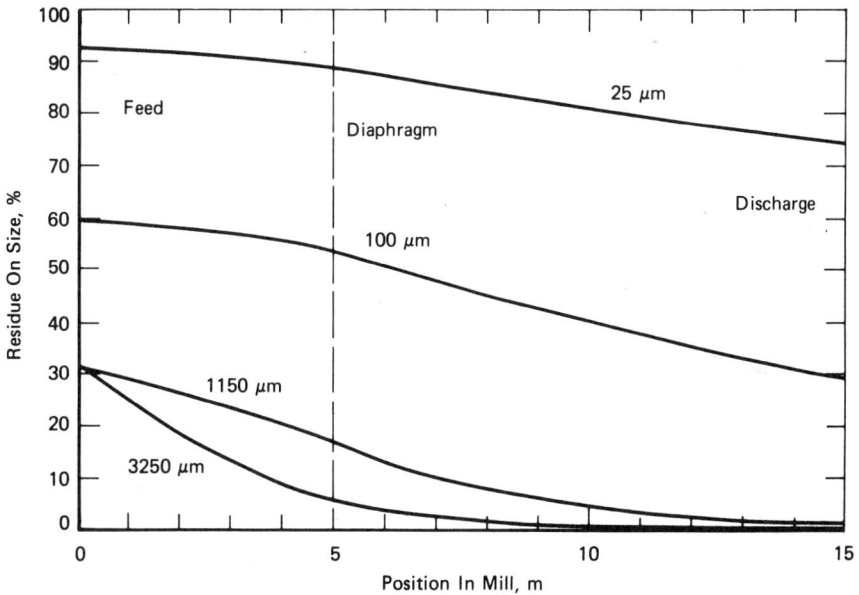

Fig. 16. Simulated change of powder size distribution along the mill.

of total product as fines from the static separator was 19%; the airborne material leaving the mill had a median size of about 35 μm and was almost entirely less than 200 μm. The complete size distributions around the circuit are given in Table 7 and Fig. 15. The simulated size distributions are in reasonable agreement with the plant data, bearing in mind: (1) the actual residence time distribution of the mill was not measured; (2) the actual hold-up in the mill was not known; (3) the Cilas instrument approximates the amount of very fine material; (4) the data on airborne material was incomplete; and (5) the slowing-down factors for very fine sizes involved substantial extrapolation.

Fig. 16 shows the decrease of residue on given sieves along the mill given by the reactors-in-series simulation, plotted at the midpoints of each reactor and joined with smooth lines. This is only an approximate representation, since the mixing inside the mill is not in a series of equal reactors, but the results are roughly in agreement with those reported from sampling along stopped cement mills.

Ball Milling of Multicomponent Ores

The most complex ball mill design situation is the design of mills and classification for ores containing significant volumes of several recoverable components. The most complete investigations of this type of system are the works of Kelsall et al. (1972); Kelsall, Weller, and Heyes; Stewart and Restarick (1971); and Cameron et al. (1971). Austin and Weller (1982) have examined the same type of data, for grinding of a three component Australian ore consisting of gangue, marmatite (Zn), and galena (Pb), as shown in Table 8. In rich ores of this kind, the entering feed has a substantial volume of liberated valuable ores, marmatite and galena in this case, and these will be ground finer with their own specific rates of breakage as the material passes through the mill.

The technique used by Kelsall et al. (1972) was to grind the ore in a laboratory scale, continuous, closed-circuit ball mill, using a makeup feed as close as possible to that expected for the full-scale circuit, to give a product as close as possible to that desired. The streams around the circuit were split into $\sqrt{2}$ sieve fractions and the assay of Zn and Pb performed on each

Table 8. Details of Samples from Continuous Mill Tests on a Multicomponent Ore

	Full-Scale				Small-Scale			
			Head assay*				Head assay*	
	Flow,	Wt%	Wt%		Flow	Wt%	Wt%	
Sample	t / h	solids	Pb	Zn	kg / h	solids	Pb	Zn
Fresh Feed	51.5	61.1	9.1	7.4	30.0	–	9.3	7.1
Mill Disch.	159.2	72.1	14.0	7.4	127.2	70.2	12.7	6.9
Classif. U / F	107.7	82.2	17.4	7.2	97.2	70.4	14.0	7.4
Classif. O / F	51.5	55.9	8.5	7.0	30.0	48.5	8.6	7.1

Source: Austin and Weller, 1982.
*Calculated from size analyses.

fraction. The test mill used (Kelsall, Stewart, and Reid, 1968) is described in Table 9. It was closed with a small spiral classifier and oversize in the makeup ore was crushed to pass 2362 μm. Microscopic examination of the makeup feed showed that there was substantial liberation with respect to galena and marmatite, thus size distributions of these components calculated from the sizing and assay values were *assumed* to represent the size distribution of free particles of these components. Gangue minerals were treated as a single component in terms of grinding and classification. The resulting apparent size distributions for the three components in these laboratory scale continuous tests are shown in Fig. 17.

The data was used to perform the analysis round the classifier, giving the result shown in Fig. 18. Two important conclusions can be drawn from this figure. First, the shape of the classifier partition curve was very abnormal for Zn and the top portion was abnormal for gangue, suggesting that the assumption of complete liberation of material leaving the mill exit was not completely valid or that the classifier was behaving abnormally. Second, the effective d_{50} of the Pb ore was much smaller than for the other components, certainly smaller than expected from the ratio of $\sqrt{\text{s.g.} - 1}$ (see Eq. 22 in Chapter 13, for example). This means that the galena had the highest degree

Table 9. Test Mills, Normal Closed Circuit

	Laboratory Mills		Full-Scale Mill
	Batch	**Continuous**	**Full-Scale Mill**
Mill i.d., D mm	200	305	2130
Length, L, mm	175	305	3050
Ball load, M, kg	5.22	41.0	n.a.
J	0.2	0.4	n.a.
Ball size, mm	25	25	50 (max.)
Speed, ϕ_c	0.89	0.75	0.73
Pulp density:			
% wt solids	86 (galena)	70.2 (overall)	72.1 (overall)
% vol solids	45	39.6	41.8
Raw feed rate	–	30 kg / h	51.5 t / h
Circulation ratio, C	–	3.23	2.09
Mill feed rate	–	127 kg / h	159.2 t / h
τ, min	–	2.31	1.78
Hold-up W	0.900 kg	4.9 kg	4.730 t
galena	0.900	0.72 kg	0.79 t
marmatite	–	0.64 kg	0.60 t
gangue	–	3.54 kg	3.34 t
Interstitial filling, U	0.45	0.65	
Specific gravity:			
galena	7.5		
marmatite	4.0		
gangue	3.2		
overall	3.5		

Source: Austin and Weller, 1982.

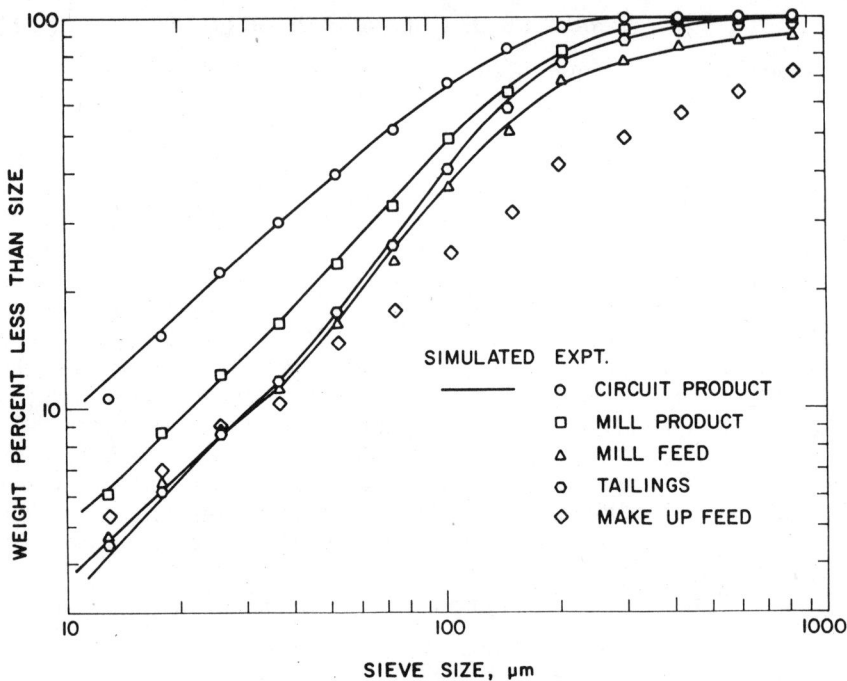

(a) Gangue

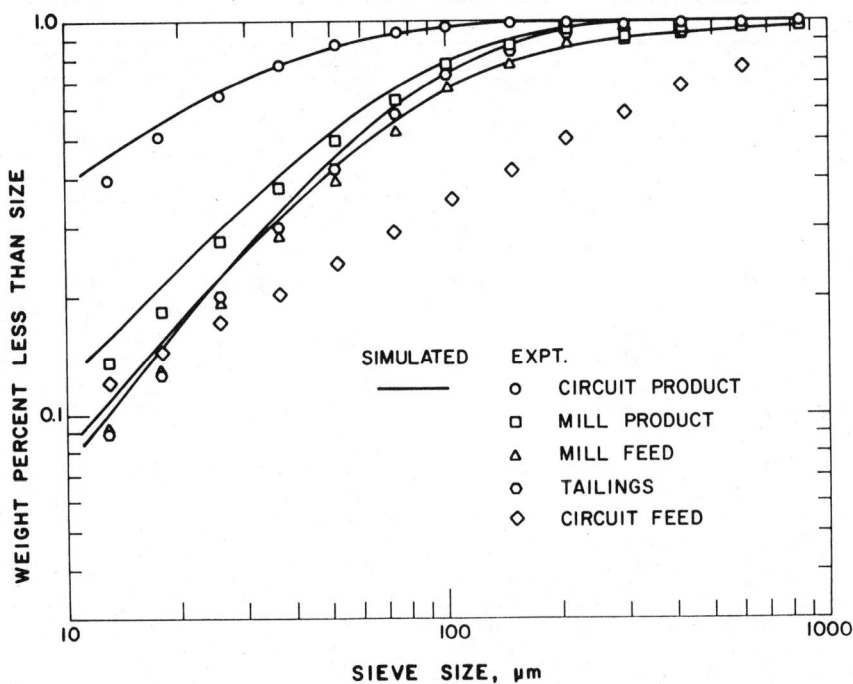

(b) Galena

Fig. 17. Experimental and simulated size distributions around small-scale circuit.

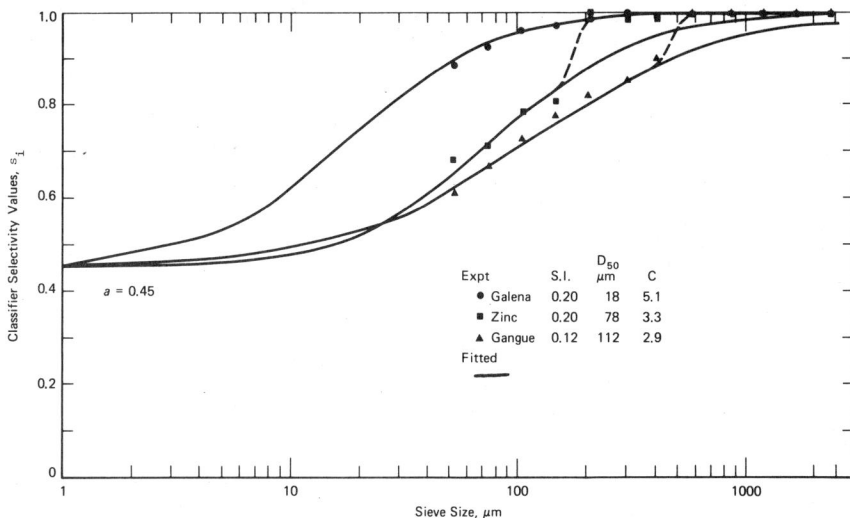

Fig. 18. Experimental and fitted classifier partition curves for the small-scale test.

of recirculation in spite of the fact that it is the most easily ground of the three components and leaves the mill in finest form. Assuming that the residence time in the mill is the same for all feed material, which seems reasonable, this means that the relative proportions of Pb to Zn to gangue in the mill are increased above those in the makeup feed:

$$\tau = \frac{Ww_{Pb}}{(1 + C_{Pb})Gm_{Pb}} = \frac{Ww_{Zn}}{(1 + C_{Zn})Gm_{Zn}} = \frac{Ww_g}{(1 + C_g)Gm_g}$$

where w_{Pb} is the mass fraction of hold-up W which is of galena and m_{Pb} is the mass fraction of the makeup feed rate G which is galena, etc. Thus, $W_{Pb}/m_{Pb} > W_{Zn}/m_{Zn} > W_g/m_g$, since $C_{Pb} > C_{Zn} > C_g$. This means that the breakage action in the mill is preferentially concentrated on the galena at some expense of the gangue. The values expressed as % Pb, Zn, and gangue for comparison with the assay of the makeup feed of Table 8 are Pb = 13.7%, Zn = 7.4%, and gangue = 78.9%.

Table 9 also gives details of the full-scale mill of 2.31 m diameter by 3 m long; Fig. 19 gives the size distributions around the mill, and Fig. 20 gives the classifier analysis around the 510 mm hydrocyclone. In this case, the partition curves were normal and could be fitted with reasonable accuracy by a logistic function. However, the same high degree of recycle of galena was apparent. The ratio of d_{50} between the small-scale and full-scale tests was about 0.45 for each component.

It is clear that the laboratory test gives the same basic trends as seen in the full scale, and would thus be a valuable guide to design of the full-scale system. However, the smaller system has a different top size of feed, the

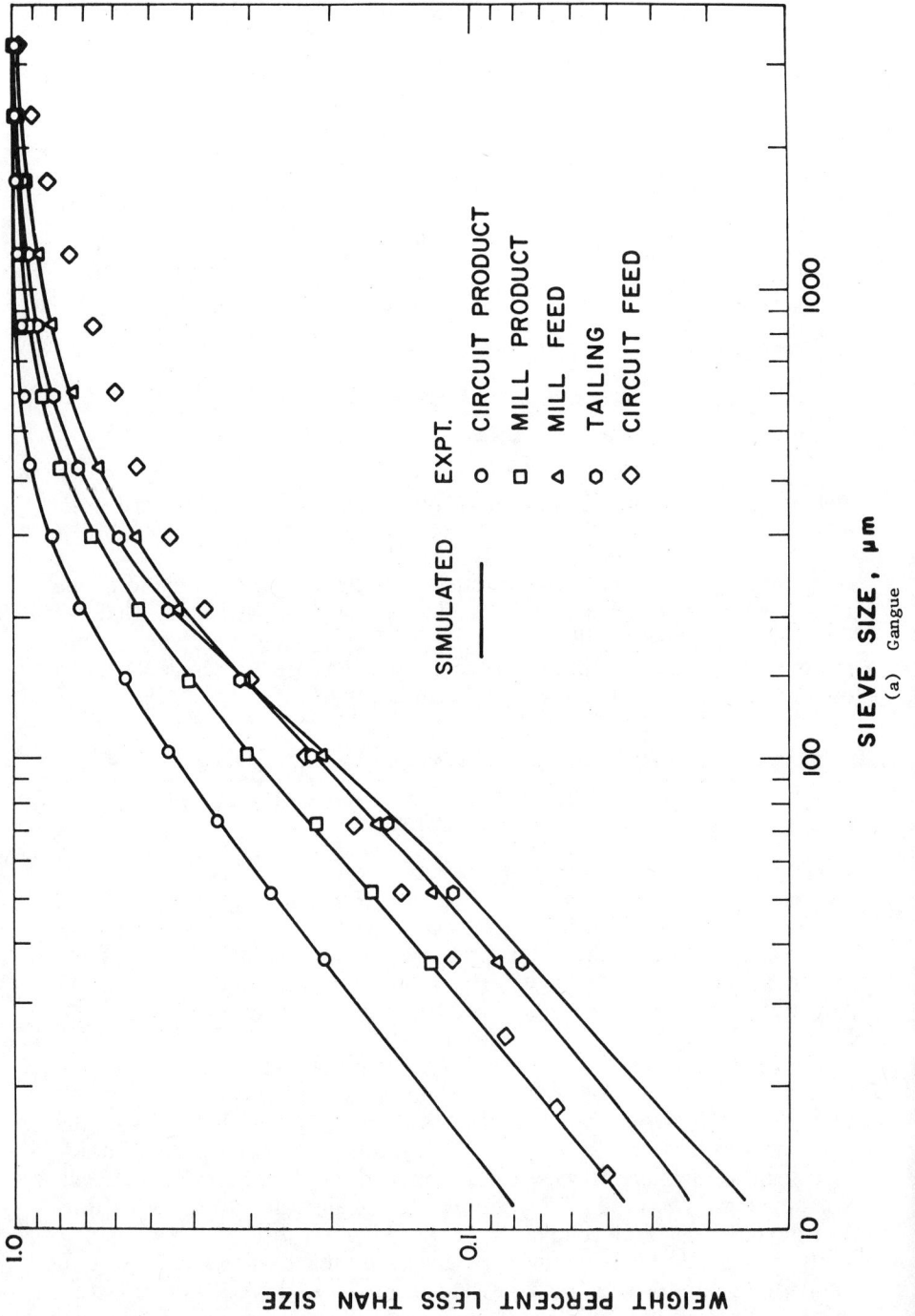

SIEVE SIZE, μm
(a) Gangue

WEIGHT PERCENT LESS THAN SIZE

SIMULATED ——— EXPT.

○ CIRCUIT PRODUCT
□ MILL PRODUCT
△ MILL FEED
◇ TAILING
◇ CIRCUIT FEED

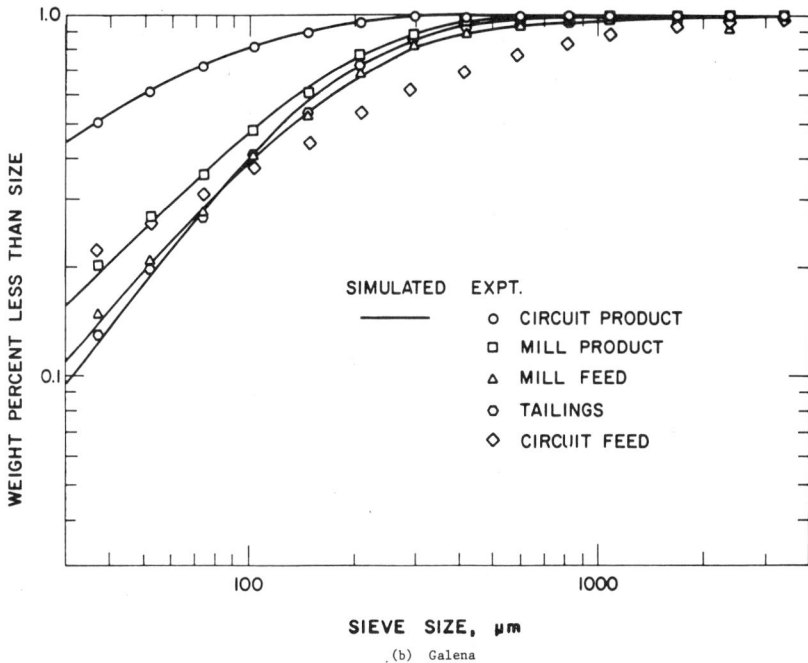

(b) Galena

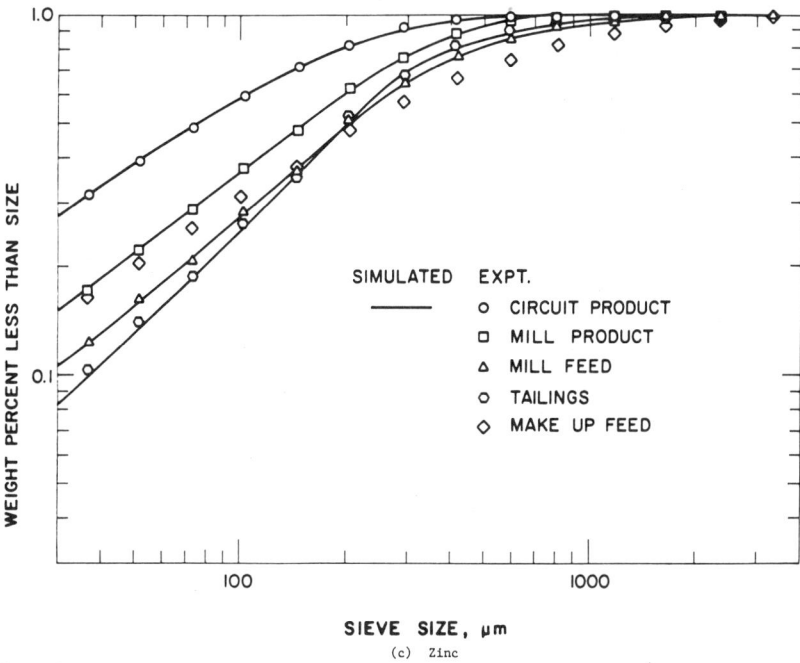

(c) Zinc

Fig. 19. Experimental and simulated size distributions around full-scale circuit.

Table 10. Breakage Parameters for Galena*

Mill diameter, mm	200
Mill length, mm	175
Ball diameter, mm	25
Mill speed, % critical	89
Volume load balls, % (based on 0.4 porosity)	20
Density of ore, kg / m^3	7.5
Solids by volume, %	45
Weight of solid, kg	0.9
Interstitial filling, U (based on 0.4 porosity)	0.45
$S_{18 \times 25} = a$, per min	1.26
α	0.87
γ	0.84
Φ	0.68
β	3.0
δ	0

*See Table 9.

classifier behaved differently, and the product size distributions were different. The construction of a simulation model would enable the predictions of the small-scale system to be varied by simulation, to predict the effect of different feed size and classification in the full-scale system.

Unfortunately, the specific rates of breakage and B values were not determined directly, except for galena (see Table 10). Instead they were back-calculated from the small-scale continuous test results using the inter-

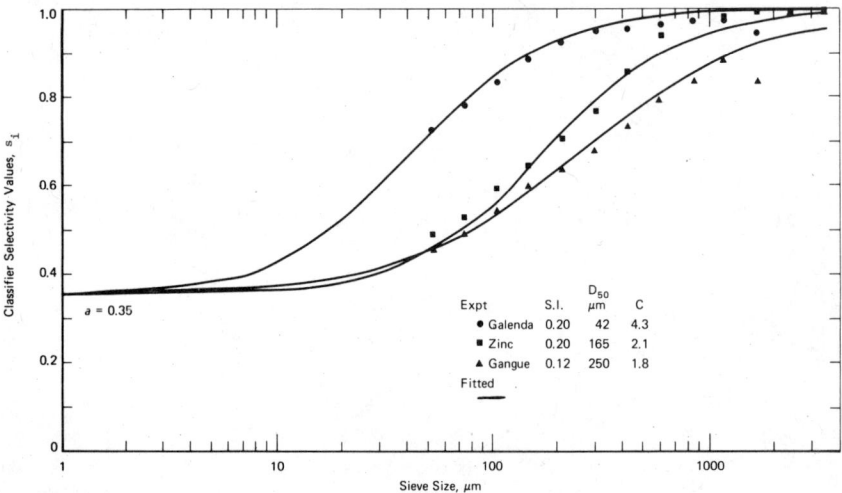

Expt	S.I.	D_{50} μm	C
● Galenda	0.20	42	4.3
■ Zinc	0.20	165	2.1
▲ Gangue	0.12	250	1.8
Fitted ▬			

a = 0.35

Fig. 20. Experimental and fitted classifier partition curves for the full-scale circuit.

val-by-interval method on the mill feed and product (see Chapter 10), and an experimental residence time distribution. The dimensionless residence time distributions for the small and full-scale mills were determined from liquid tracer impulse tests (Weller, 1980). The results could be fitted to a reasonable approximation by the equivalent RTD of one large fully mixed reactor followed by two smaller equal fully mixed reactors (see Chapter 14). For the full-scale mill $\tau_1/\tau = 0.73$ and for the small-scale mill, 0.80; the values of τ are given in Table 9. The B values used were experimental values for galena (see Table 10), and the Kelsall, Reid, and Restarick (1968) standard mean values ($\Phi = 0.72$, $\gamma = 0.96$, $\beta = 3.5$) for marmatite and gangue. Fig. 21 shows the results.

Also shown are the back-calculated values similarly determined from the full-scale mill test data. It is apparent that the small-scale results are not in similitude with the full-scale data, that is, they are not of the same shape. A fit of $S_i = ax_i^\alpha$ to the mid-range of S_i values for galena, for example, gives an α value of at least 1.5, which is much higher than the value of 0.87 determined from batch breakage tests. It is not possible, therefore, to apply scale-up laws to the small-scale values back-calculated from continuous tests, since the different shapes of S_i values produce different shapes of product size distributions.

Figs. 17 and 19 show that these back-calculated S_i values give reasonable agreement between simulations and experiment. However, this is largely a circular calculation and it does not prove that the back-calculated values have physical meaning or that they will apply under any other set of conditions. Note that the actual degree of size reduction from mill feed to mill product is low, which makes it difficult, using this technique, to get good back-calculation of S_i values in the presence of experimental error.

Hogg and Rogovin (see Chapter 14) have recently concluded that different particle sizes move through small continuous mills at different velocities. Larger particles stay in the mill for longer residence times due to preferential settling in the liquid pool. If S_i values are back-calculated assuming the same RTD for all sizes, then larger sizes will have high false S values. For a large-scale mill this effect becomes much less because the velocities of flow are higher and the settling distances much bigger, yet the hindered settling velocity of a given particle in a given slurry is constant. This is one possible explanation for the lack of similitude between the small-scale and full-scale apparent S_i values.

The characteristic breakage parameters for galena determined in the laboratory batch test (see Table 10) were scaled to the full-scale mill using the equations given in Chapter 5. The mixture of balls was assumed to be that of the Bond equilibrium mixture for two-inch makeup balls, with a ball loading of 38%, giving a U of 1.3. These parameters were used in a simulation with the galena feed, RTD, and classifier selectivity values for the full-scale circuit, and the simulation performed to give 50.8% minus 38 μm corresponding to the experimental full-scale value. The result is given in Table 11. The predicted size distributions were significantly closer to the experimental full-scale values than those given by the small-scale back-calculated values of Fig. 21. The predicted circulation ratio was 4.6 for

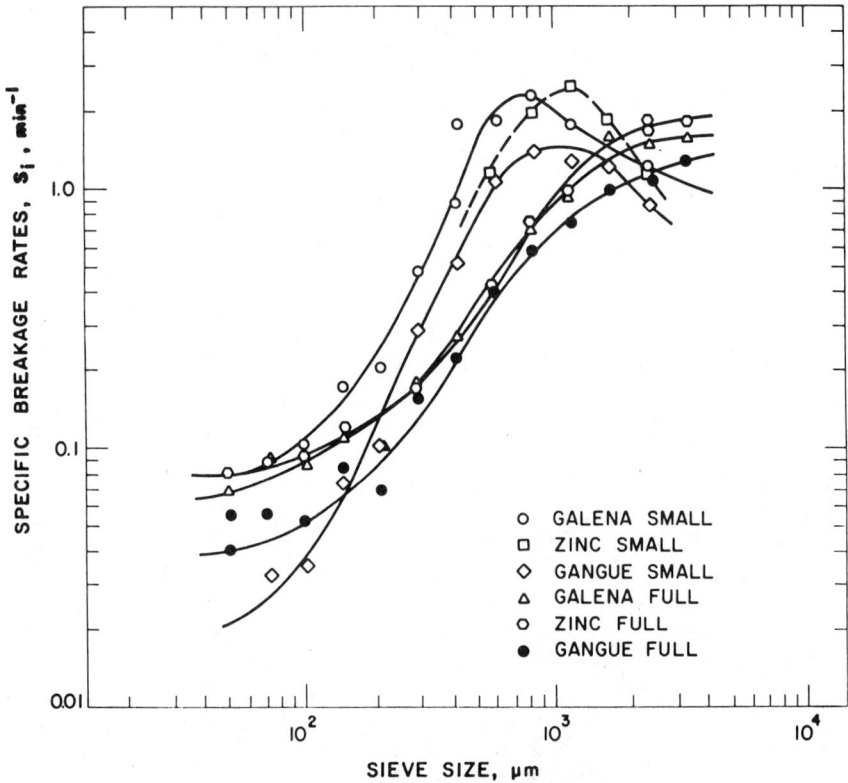

Fig. 21. Interval-by-interval back-calculated specific rates of breakage from small and full-scale tests.

comparison with the experimental value of 4.3. However, the value of τ was 1.43 min instead of the real value of 1.78 min, which would lead to an overestimate of mill capacity of 18%. However, Table 11 shows that the product size distribution given by the simulation was substantially steeper than the real value, and a simulated size distribution predicting less fines would automatically predict higher capacity.

It seems likely that there exists a degree of interlocking of gangue and galena, which has the effect of making the gangue S values back-calculated from the large-scale values (assuming no interlocking) larger than for gangue alone. Similarly the galena values would be smaller than for the liberated galena used in the laboratory breakage test. In order to refine scale-up and simulation models, it would be necessary to investigate the degree of liberation and the breakage characteristics of particles containing two or more components in each particle. This could perhaps be done by float-and-sink specific gravity separation into fractions, with subsequent chemical analysis to determine the proportions of galena to gangue, for example. Applied to fractions from the size distributions around the circuit, this would show quantitatively the change in liberation at various stages.

Table 11. Comparison of Size Distribution of Galena for Full-Scale Circuit, Predicted by Scale-up Laws

Size, μm	Circuit Feed	Ball Mill Product		Recycle		Circuit Product	
		Exp.	Sim.	Exp.	Sim.	Exp.	Sim.
3340	100.00	100.00	100.00	100.00	100.00	100.00	100.00
2362	96.10	99.84	99.79	99.80	99.74	100.00	100.00
1670	92.20	99.59	99.37	99.50	99.24	100.00	99.99
1181	88.80	99.25	98.71	99.10	98.43	99.90	99.97
834	83.40	98.42	97.36	98.10	96.80	99.80	99.92
590	77.10	96.85	94.97	96.20	93.92	99.60	99.78
418	69.80	94.00	90.90	92.80	89.05	99.10	99.39
295	61.90	87.98	84.46	85.70	81.41	97.70	98.42
209	57.30	77.74	76.72	73.60	72.38	95.40	96.59
148	44.20	60.38	62.15	53.70	55.79	99.80	91.27
104	37.80	47.84	48.66	40.00	40.98	81.30	83.82
74	31.60	35.80	35.82	27.60	27.60	70.80	73.39
52	26.70	27.30	25.85	19.40	17.96	61.00	61.96
37	22.30	20.09	18.64	12.90	11.60	50.80	50.81
26	18.20		13.53		7.58		40.74
18	15.20		10.18		5.24		32.76
13	12.80		7.85		3.80		26.40

Circulation ratio C: exp. 4.3; sim. 4.6; residence time τ, min: exp. 1.8; sim. 1.43.

The collection of sufficient sample for laboratory batch tests would enable the breakage characteristics to be determined. It is also possible that breakage of a mixture of strong and weak components in a ball mill is not the simple weighted sum of their breakage alone, due to interaction effects.

There is a great need for an understanding of the effect of mixtures on hydrocyclone performance (see Chapter 13), since performance seems to be much poorer for mixtures than for each component alone. The low effective d_{50} values of galena as compared to gangue, much lower than expected from density difference, mean that the circulating load of galena is high. Consequently, in order to grind fine enough to liberate the ore from gangue, the liberated galena is overground. A promising field of work might be to evaluate selectively active chemicals which influence the classification behavior of one component with respect to another.

Uncertainties and Future Work

It is extremely important to remember the function of a mill circuit simulation model. It is to be used to predict the best operating design. The purpose of comparing the results of model simulations with actual plant data is to validate the models, so that they can be used with confidence to predict the correct recycle conditions, ball mixture, etc. An *incomplete* model which accurately predicts the behavior of a real plant under one set of conditions is of little value because it cannot show the mill behavior under other conditions and cannot, therefore, be used for optimal design. A good example of this is the scale-up procedure proposed by Herbst and Rajamani (1982); Herbst, Grandy, and Fuerstenau (1974); and Herbst and Fuerstenau (1980). Their procedure is to determine the specific rates of breakage and B values in a batch laboratory mill under good hold-up conditions and also to measure the net power input to the mill. The specific rates of breakage are then replaced in the steady-state, continuous grinding equations by

$$S_i \tau = \left(S_i W / m_p \right) \left(m_p / F \right) \tag{7}$$

where the values of $S_i W / m_p$ have been determined. It is then *assumed* that the values of $S_i W / m_p$ are independent of mill size and operating conditions, which is not always true. The simulation is performed for the full-scale plant conditions of mill power (calculated from a power equation) and F value through the mill, and the size distributions compared to the actual values.

It is implicit in this treatment that the filling condition is always in the correct range where the overfilling factor K_o is 1. For this requirement, the results reported in Chapter 5 show that $S_{i1} W_1 = S_{i2} W_2$, that $S_i \propto D^{0.5}$ and $W_1 \propto D^2 L$. The mill power laws also give $m_p \propto D^{3.5} L$. Thus, if the experimental determinations are carried out with the same ball load, fraction of critical speed, ball mix, and slurry density (rheology) as in the full-scale mill (or if $S_i W$ and m_p each vary in the same way with these conditions), then $S_i W / m_p$ is constant with mill diameter. The scale-up procedure is then *algebraically identical* with that proposed in Chapter 5 and applied in the *algebraically identical* with that proposed in Chapter 5 and applied in the example in the section titled, "The Behavior of Different Designs of Mill Circuits" in Chapter 7 for the condition of no mill overfilling. Variation of F to obtain a given one-point product specification would be identical to varying τ.

This algebraic manipulation appears to have no particular advantage. In fact, it has the great disadvantage of being an incomplete model for two reasons: (1) the assumption that $S_i W/m_p$ is constant breaks down as the mill overfills and, (2) it does not include the information necessary for ball charge optimization. Use of the incomplete model in simulations will always predict increased capacity as circulation ratio is increased, whereas the simulations in Chapter 7, and actual plant experience, show that low circulation ratios give highest capacity under many conditions. The circulation ratio of a well-operated plant producing a fine product will be adjusted to give the optimum capacity, which will correspond to the filling conditions of $U \approx 1.0$, and this method of simulation will then agree with the plant data just like the example given in Chapter 7. It will not work for any circuit where the reduction ratio is low enough to produce overfilling.

Similarly, the model proposed by Lynch et al. (1977) is also incomplete. It assumes that the mill is fully mixed and that the B values are the same for different materials. The *apparent* S values are then back-calculated from plant data. To use this approximate model with any degree of validity would require a mass transfer law, which is not incorporated in the Lynch model. It must, therefore, give incorrect conclusions as to plant optimization, just as does the Herbst model.

It is also important to note that a mill simulation must be compared with plant data by running the simulation to match a point on the experimental size distribution, and then comparing predicted and actual capacities. If the simulation is performed at the plant capacity, the predicted 80%-passing size of the product may be within, say, 2% of the experimental value, which at first sight seems a good match. However, a deviation of 2% in size distribution can represent a large error in mill capacity, typically 10%.

There can be no doubt that the basic concepts of size-mass balance models and similitude between *breakage* (not mass transfer) in small mills and large mills are broadly correct. In the cases considered in this chapter, the detailed analysis of test data using these concepts gives a far deeper insight into the operation of the mills and circuits than any analysis based solely on kWh/t. However, the breakage of materials by impact forces is not a process which follows the precise physicochemical laws which can be applied to chemical reactions, heat transfer, etc. In addition, the amount of investigation of breakage in mills by such detailed analysis is minor compared to the efforts which have been expended on analysis of other unit operations in chemical process engineering. It is not surprising then, that the simulation models are by no means perfect.

For example, in spite of the hundreds of reports of investigation of ball milling available in the literature, the laws allowing the calculation of the effect of different mixtures of balls are imprecisely understood. Investigations of breakage rates and primary progeny fragment distributions in relatively small mills are easy to perform and use simple concepts and techniques, yet what is still missing is systematic *experimental* study of the effect of variables, performed by different workers on different materials.

The general problem of predicting the distribution of components with respect to classes of size and component concentration remains to be solved for complex ores. If the valuable component is a small volume fraction of the total, the problem of liberation is easier, and can probably be handled by development of the models formulated by King (1979), Klimpel and Austin (1983), and Klimpel (1984). For cases where, for example, component B in an AB composite forms a substantial fraction of the volume, then a full model must allow for the breakage properties of a range of materials from pure A to pure B, with all intermediate combinations of AB. It is not difficult to formulate and program these models (Austin and Klimpel), but insufficient data exist at present to test their validity.

Tests on full-scale mills are much more difficult to perform over a range of operating variables. Accurate sampling is always a problem, especially for wet systems. It is recommended that triplicate samples be taken and analyzed to enable standard analysis of variance to be performed. However, there are undoubtedly cases where a systematic sampling or analytical error can lead to reasonable standard deviations, yet the results still contain substantial error. Especially, in validating simulation models, it is important that all test data and materials be collected at the same time to avoid uncertainties caused by variable ore, coal, or clinker properties.

Harris and Arbiter (1982) have discussed the problems experienced with large diameter mills at the Bougainville plant, with comparison to results from similar mills at Pinto Valley, as shown in Table 12. The use of Eq. 30 in Chapter 14 for these conditions gives a critical flow rate of about 450 t/h, a predicted U value of 2.2, and a K_o of 0.45 for Bougainville, as compared to $U = 1.4$, $K_o = 0.8$ for the Pinto Valley operation. Harris and Arbiter feel that the problem at Bougainville is due to inadequate residence time for radial mixing, but they use a formal mean residence time based on $U = 1.0$, which is almost certainly incorrect. If the predictions of our mass transfer law are valid for these mills, the mean residence time for each case would be 2.4 and 3.7 min, respectively, at a mill speed of 12.5 rpm. This is quite sufficient for good radial mixing.

The major unusual characteristic of the Bougainville plant is that the makeup feed to the mill circuit is somewhat large, 80%-passing 8.2 mm. To compensate for this, the ball sizes used in the design were also large. The concepts given in this book imply that ball mills are not very efficient crushers, that large balls are inefficient for breakage of small particles, and that the Bond sizing method does not always work well for large feed sizes. Kavetsky, Whiten, and Narayanan (1982) believe that breakage of large sizes, greater than 2 mm, becomes proportionally less efficient for larger mill diameters. In the terms we use, this means that the Λ value increases with mill diameter. It is difficult to see the physical reason for this phenomenon, which is based on inference from back-calculated apparent S values, using

Table 12. Data on Large Diameter Mills

	Bougainville	Pinto Valley
D, m	5.5	5.5
L, m	6.4	6.4
rpm	12.5	12.3
ϕ_c, %	68	67
J, %	40	37
m_p, kW	3300	3000
Circulation ratio	4	1.5
Q, t / h	440	330
F, t / h	2200	830
80%-passing size, mm	8.2	7.3

Source: Harris and Arbiter, 1982.

incorrect assumptions concerning both residence time distribution and B values for breakage of large sizes. Klimpel has shown that poor breakage of large sizes can also be observed in small wet mills; he ascribes this to poor (dilatant) rheological properties of the slurry in the mill. If a mill cannot efficiently remove large particles or break small particles, a high circulating load automatically results. In turn, this leads to a relatively coarse size distribution in the mill and to overfilling. The coarse size distribution gives a dilatant rheological character to the slurry, further decreasing the breakage efficiency of large sizes. Presumably, this is due to rapid settling of large particles, which perhaps then cannot be so readily moved into the tumbling charge.

The conventional solution to the problem would be to reduce the makeup feed size, with a decrease in the makeup ball size, but this requires costly additions to the crushing line. Another solution is to increase the slurry density to improve the rheological properties. However, this might aggravate the overfilling effect. The use of a chemical fluidity modifier under these circumstances has two advantages: it improves the rheological properties (Klimpel) and it reduces the overfilling effect by allowing slurry to flow more easily through the mill. This is a clear case where the determination of residence time distribution and hold-up with and without chemical additive would be very informative. It is also clear that future research involving detailed investigation of abnormal breakage of large particle sizes as a function of mill conditions will become very important as mill feeds become coarser.

It is possible that the ore at the Bougainville plant has an unusually low μ value compared to Pinto Valley, thus leading to persistence of large sizes.

As far as scale-up is concerned, it must be recognized that there is no *theoretical* reason to assume the same breakage characteristics in a large mill as in a smaller mill. It is always possible that correcting factors will be necessary, that is, the α or γ values may have to be changed. It is possible that scale-up laws deduced from small- and medium-size mills will not extrapolate to very large mills. However, it is to be expected that comparison of carefully taken plant data with model predictions based on accurately determined laboratory data will lead to the correct adjustments to the models. The scale-up simulations are already sufficiently close to reality to show the promise of accurate design and optimization. Further theoretical and laboratory studies of the mechanisms of mass transfer should provide a proper base for analysis and interpretation of large-scale results and design for optimal performance. A sound theoretical basis for the way balls move in the tumbling bed would enable extrapolation of breakage and power laws to big mills.

In summary, although the basic concepts have been formulated, there is still substantial laboratory and full-scale experimental work to be done and theoretical analysis to be performed, to accomplish the objective of precise computer-aided design of ball mills and mill circuits.

Summary

Five examples are reported of model predictions, based on breakage parameters measured in laboratory tests scaled up to full size, compared to actual plant data. In increasing order of complexity they are:

1. Prediction of capacity and size distributions for a 3×5 m, (700 kW) wet overflow mill in normal closed circuit, grinding a rod mill product to 70% minus 75 μm (200 mesh).

There was good prediction of plant performance, and the mill was not overfilled at the operating circulating load of 1.95 and the production rate of 50 t/h.

2. Prediction of changes in capacity and size distributions for a 3×5 m (700 kW) wet overflow mill in reverse closed circuit, grinding a rod mill product to 65 to 55% minus 200 mesh, at three different solid flow rates through the circuit.

Predictions of changes in capacity and circulating load were good, and predictions of circuit product size distributions were satisfactory, but the Schuhmann slope of the predicted size distributions were somewhat lower than those of the plant for the lowest feed rate and somewhat higher for the highest feed rate, probably due to the changes in slurry density. Changing the makeup feed from 87 to 103 to 108 t/h caused the flow through the mill to change from 150 to 200 to 280 t/h, leading to increased overfilling of the mill. This was confirmed by measurements of actual mill hold-up, which showed them to correspond to $U = 1.1$, 1.5, and 1.9, respectively. The coarsening of product size distribution was far greater than would be predicted if the effect of overfilling, which decreases breakage rates, were not incorporated into the model.

3. Prediction of size distributions and capacities for an air-swept 1×1.5 m (16 kW) ball mill grinding coal, from a feed of minus 38 mm to products of about 85 to 95% minus 200 mesh, at varying air-flow rates.

The model for this system is more complex than for wet overflow ball mills because the coal is removed from the mill entirely in the air stream. However, the air flow to the mill is always adjusted to give optimum hold-up, using measurements of mill sound, corresponding to $U = 0.84$. The air-sweeping action was treated as an internal classification and the classification parameters measured as a function of air-flow rate by tests involving stopping the mill to measure the size distribution of the mill contents. The prediction of mill capacities and product size distributions as a function of air-flow rate, again by scaling up from laboratory breakage data, was within the scatter of experimental results for ten closed circuit tests at varying flow rates.

4. Prediction of size distributions and capacities for a two-compartment cement tube mill, 4.2 × 5 and 4.2 × 10 m compartments (3520 kW), grinding minus 38 mm clinker to a product of 3250 Blaine cm^2/g, in normal closed circuit, at a rate of 115 t/h.

This model is more complex still because the high level of air-sweeping carried out a substantial fraction of material, there were two separate classifiers (a static and a mechanical air separator), and the cement was ground fine enough to cause the slowing-down effect to come into play toward the discharge end of the mill. The data was not complete for all streams in the circuit, but reasonable estimates were made of the balance around the static air separator. Scaling up of laboratory data in the same way as the previous examples gave good predictions of capacity and reasonable predictions of size distributions, measured experimentally to 1.6 μm using the Cilas laser diffraction instrument. The model indicates that only about 0.3% of the mill charge is exposed to the air sweeping action per mill revolution.

5. Analysis of the recycle and size distributions of a three-component rich ore (galena, marmatite, gangue) in a normal closed circuit wet overflow mill, 2 × 3 m, at 50 t/h.

This system is again complex because of the interaction effects of the major volume components in the hydrocyclone, and the different degrees of recycle of each component. Breakage properties were only available for one component and use of this data *assuming* that all feed to the mill was completely liberated gave unacceptable over-prediction of capacity and incorrect size distributions. Analysis of data from the Kelsall laboratory-scale closed-circuit mill did not give good similitude between the S values back-calculated from the test and those back-calculated from the full-scale test. This may be due to different mass transport effects in the small-scale system. Of particular note is the high recirculation of galena, due to a much lower hydrocyclone d_{50} for this heavy ore than expected from its specific gravity. Thus although this weak material ground finer in the mill, it was still recycled excessively, leading to overgrinding of this component. It was concluded that this type of system cannot be accurately modeled without model and test development to eliminate the assumption of complete liberation.

It is important to realize that a model must be sufficiently complete to predict correctly the effect of varying operating conditions. If it is not, it gives erroneous conclusions (e.g., the Herbst et al. and Lynch et al. models) concerning the optimization of design, even though it may correctly predict a single set of plant data for a simple case such as Example 1 above. Applying the concepts of this book to large wet overflow mills suggests that the problems experienced with some large mills are not due to the breakdown of scale-up laws at large mill diameters. For some materials large sizes are not readily broken even by large balls, so that the slurry in the mill tends to have dilatant rheology. This leads to low rates of breakage, high circulating loads, and overfilling of the mill.

References

Austin, L. G. and Weller, K. A., 1982, "Simulation and Scale-up of Wet Ball Milling," Preprint, 14th International Minerals Processing Congress, Canadian Institute of Mining and Metallurgy, pp. I-8.1–8.24.

Austin, L. G., Luckie, P. T., and Wightman, D., 1975, "Steady-State Simulation of a Cement-Milling Circuit," *International Journal of Mineral Processing*, Vol. 2, pp. 127–150.

Austin, L. G., Luckie, P. T., and von Seebach, H. M., 1976, "Optimization of a Cement Milling Circuit with Respect to Particle Size Distribution and Strength Development by Simulation Models," *Proceedings*, Fourth European Symposium Zerkleinein, H. Rumpf and K. Schonert, eds., Dechema Monographien 79, Nr. 1576–1588, Verlag Chemie, Weinheim, pp. 519–537.

Austin, L. G., Weymont, N. P., and Knobloch, O., 1980, "The Simulation of Air-Swept Cement Mills, Part I. The Simulation Model," *European Symposium on Particle Technology, B*, K. Schonert, W. Gregor, and F. Hofmann, eds., Amsterdam, pp. 640–655.

Austin, L. G. and Klimpel, R. R., "A Ball Mill Simulation Model Incorporating a Model for Liberation," not available.

Austin, L. G., Weymont, N. P., and Knobloch, O., "The Simulation of Air-Swept Cement Mills. Part II. Model Validation," not available.

Austin, L. G., Klima, M., and Knobloch, O., "The Simulation of Air-Swept Cement Mills. Part III. Incorporating the Slowing-Down Effect," not available.

Austin, L. G., et al., "A Simulation Model for an Air-Swept Ball Mill Grinding Coal," accepted by *Powder Technology*.

Cameron, A. W., et al., 1971, "A Detailed Assessment of Concentrator Performance at Broken Hill South, Ltd.," *Proceedings*, Australian Institute of Mining and Metallurgy, Vol. 240, pp. 53–67.

Freech, E. J., et al., 1970, "Mathematical Model Applied to Analysis and Control of Grinding Circuits. Part II. Simulation of Closed-Circuit Grinding," SME Preprint No. 70B28, SME-AIME Annual Meeting, Denver, CO, Feb.

Gelpe, T., Flament, F., and Hodouin, D., 1983, "Computer Design of Grinding Circuit Flowsheets--Application to Cement and Ore Processing," SME Preprint No. 83-191, SME-AIME Annual Meeting, Atlanta, GA, Mar., 22 pp.

Harris, C. C. and Arbiter, N., 1982, "Grinding Mill Scale-up Problems," *Mining Engineering*, Vol. 34, No. 1, pp. 43–46.

Herbst, J. A. and Fuerstenau, D. W., 1980, "Scale-up Procedure for Continuous Grinding Mill Design Using Population Balance Models," *International Journal of Mineral Processing*, Vol. 7, pp. 1–31.

Herbst, J. A. and Rajamani, K., 1982, "Developing a Simulator for Ball Mill Scale-up--A Case Study," *Design and Installation of Comminution Circuits*, A. L. Mular and G. V. Jergensen, eds., AIME, New York, pp. 325–342.

Herbst, J. A., Grandy, G. A., and Fuerstenau, D. W., 1974, "Population Balance Models for the Design of Continuous Grinding Mills," *Proceedings*, Tenth International Minerals Processing Congress, M. C. Jones, ed., Institution of Mining and Metallurgy, London, pp. 23–45.

Kavetsky, A., Whiten, W. J., and Narayanan, S., 1982, "Studies on the Scale-up of Ball Mills," *Proceedings*, Mill Operators Conference, Australian Institute of Mining and Metallurgy, pp. 113–122.

Kelsall, D. F., Reid, K. J., and Restarick, C. J., 1968, "Continuous Grinding in a Small Wet Ball Mill. Part I. A Study of the Influence of Ball Diameter," *Powder Technology*, Vol. 1, pp. 291–300.

Kelsall, D. F., Stewart, P. S. B., and Reid, K. J., 1968, "Confirmation of a Dynamic Model of Closed-Circuit Grinding with a Wet Ball Mill," *Trans. IMM*, Vol. 77, pp. C120–C127.

Kelsall, D. F., Weller, K. R., and Heyes, G. W., "A Comparison of Small and Large Scale Grinding Models for a Three-Component Ore," in preparation.

Kelsall, D. F., et al., 1972, "The Effects of a Change from Parallel to Series Grinding at Broken Hill," *Australian Journal of Mining and Metallurgy* Conference, New Castle, pp. 337–347.

King, R. P., 1979, "A Model for the Quantitative Estimation of Mineral Liberation by Grinding," *International Journal of Mineral Processing*, Vol. 6, pp. 207–220.

Klimpel, R. R., 1982, "Slurry Rheology Influence on the Performance of Mineral/Coal Grinding Circuits. Part I," *Mining Engineering*, Vol. 34, No. 12, pp. 1665–1668.

Klimpel, R. R., 1983, "Slurry Rheology Influence on the Performance of Mineral/Coal Grinding Circuits. Part II," *Mining Engineering*, Vol. 35, No. 1, pp. 21–26.

Klimpel, R. R., 1984, "Some Practical Mathematical Approaches to Predicting Liberation in Binary Systems," *Process Mineralogy III*, W. Petruk, ed., AIME, New York.

Klimpel, R. R., "The Influence of Slurry Rheology on the Breakage in Large Mills with Coarse Feeds," in preparation.

Klimpel, R. R. and Austin, L. G., 1983, "A Preliminary Model of Liberation from a Binary System," *Powder Technology*, Vol. 34, pp. 121–130.

Klimpel, R. R. and Austin, L. G., 1984a, "The Back-Calculation of Specific Rates of Breakage from Continuous Mill Data," *Powder Technology*, Vol. 38, pp. 77–91.

Klimpel, R. R. and Austin, L. G., 1984b, "Simulation of a Mill Grinding Copper Ore by Scale-up of Laboratory Data," *Proceedings of Mintek 50*, Johannesburg, South Africa.

Luckie, P. T. and Austin, L. G., 1980, *Coal Grinding Technology: A Manual for Process Engineers*, NTIS, Springfield, VA.

Lynch, A. J., et al., 1977, Chapter 4, *Mineral Crushing and Grinding Circuits*, Elsevier Scientific Publishing Co., New York, pp. 45–85.

Stewart, P. S. B. and Restarick, C. J., 1971, "A Comparison of the Mechanisms of Breakage in Full Scale and Laboratory Scale Grinding Mills," *Proceedings*, Australian Institute of Mining and Metallurgy, Vol. 239, pp. 81–92.

Weller, K. R., 1980, "Hold-up and Residence Time Characteristics of Full Scale Grinding Circuits," *Proceedings*, Third IFAC Symposium, J. O'Shea and H. Polis, eds., Pergammon Press, pp. 303–309.

13

Classification and Hydrocyclones

Introduction

The concepts of classification and the combining of classifiers with grinding mills to form closed mill circuits were introduced in Chapter 6. Classification consists of splitting the stream of particles leaving the mill (flow rate P) into two streams primarily by the criterion of size: giving a circuit product stream (flow rate Q) containing more fine material per unit of mass, and a recycle stream (flow rate T) containing more coarse material; the latter stream is returned to the mill feed (see Figs. 3 and 4 in Chapter 6). The circulation ratio, C, has been defined as the ratio of the mass flow rates, T/Q, and the circulating load, $1 + C$, by the ratio P/Q.

There are several advantages of closed-circuit operation over the simpler open-circuit grinding. The problem of overgrinding (the creation of excessive fine material) associated with meeting a product specification (e.g., a desired percent passing a certain mesh size) has been mentioned in earlier chapters. A major advantage of closed-circuit operation is that a significant portion of the material already fine enough can be removed from the circuit and, hence, prevented from regrinding. The circuit produces a cumulative size distribution which is steeper (less unnecessary fine material) as the degree of recirculation increases, see Fig. 9 in Chapter 7. The partial elimination of fine material may also reduce the *slowing-down* effect caused by high pulp viscosity or cushioning as fines accumulate along the mill (see Chapters 5 and 15). On the other hand, there is evidence that too much recycle causes overfilling of the mill, so an optimum amount of recirculation exists for wet grinding (see Chapter 7).

In general, the energy required for grinding to a specified percent less than some size is reduced for closed circuit as compared to open-circuit grinding. The use of recycle also provides better operating stability in comparison to open-circuit operation, especially for grate discharge mills. In this case, recycle is especially important to help maintain optimum powder or slurry hold-up in the mill and to avoid cycling of hold-up (and, hence, cycling in the mill power). Finally, the use of classification adds an additional degree of freedom in the control or periodic adjustment of the mill circuit to help correct for changes in the makeup feed size or feed particle strength to the circuit. The major disadvantage of a closed-circuit system is, of course, the additional capital outlay for the classifier and recycle system.

Ideal classification can be defined as the separation of a particle stream into a stream of only fine particles and another of only coarse particles. Practical classifiers, however, are not ideal; even particles of exactly the same size and with the same physical characteristics of shape and density will receive different classification actions within the same classifier, due to a finite width of the entering stream and edge effects, and to dispersive-mixing actions in the fluid. Some of these identical particles will be sent to the coarse particle stream while some will be sent to the fine particle stream. Even for very simple classification actions, the presence of statistical dispersion makes it difficult to obtain adequate theoretical models for the split of sizes. Other complicating factors arise from the influence of other variable physical properties of the feed material, such as shape and specific gravity, which also affect the separation process. In ore grinding there are often several mineralogical species involved, and material being presented to classification can consist of a range of specific gravities due to locking of different minerals together. Very frequently, the separation forces act on particles in a dense suspension, and current theoretical treatments of particle movement in dense suspensions (hindered settling, for example) are not completely satisfactory. Theoretical treatments assume that the classifiers are operating in a smooth manner without unusual or unsteady state fluid flow in the classifier because it is very difficult to treat such conditions, yet real classifiers frequently operate in a pulsing manner.

For these reasons, theoretical models are usually too simplified to give quantitative predictions. For example, although the term *cut size* is often used in the literature to mean the size (deduced from a theoretical treatment) such that all larger particle sizes go to the coarse stream and all smaller sizes to the fine stream, this definition is not usually applicable for real classifiers. It is normally found that there is a size such that all *larger* sizes are sent to the coarse stream, but nearly always some of the smaller sizes also are found in the coarse stream.

For a complete description of a grinding circuit with a particular classifier, it is necessary to calculate sets of classifier selectivity values s_i to use in the simulation model, as discussed previously (see "The Effect of Design Classifier Efficiency" and "Classifier Design: The Hydrocyclone" in Chapter 7). This is not a simple task because the classifier action will change, hence the set of s_i values will change, depending on how the grinding device and the classifier are operated. When changes are made in the mill to produce changes in factors such as product feed rate, product particle size, slurry density, then selectivity values usually will also change. Therefore, sufficiently accurate mill simulations require, in many modeling situations, a quantitative knowledge of the variation of classifier parameters with different operating conditions and design changes. This type of quantitative information is usually not available. For example, even with the most widely used type of classifier in mineral processing, the hydrocyclone, such data is not completely available, and there is disagreement among researchers about the quantitative relations involved.

Thus the approach of this chapter is to demonstrate how data from various types of classifiers is quantified in a manner appropriate for mill circuit simulations. The methodology developed is then applied primarily to the hydrocyclone, which has received more quantitative investigation than other types of classifier.

Principles of Classifier Action

The various types of classifiers fall into two general categories: those that separate by forces of fluid dynamics and those that involve screening. Fluid dynamic classification takes advantage of the differences in rates of travel of particles in a fluid due to differences in particle size. Common classifiers of this type include the simple mechanical agitation devices used in wet grinding circuits, such as spiral and rake classifiers (see Fig. 1), centrifugal force devices such as the cyclone (see Fig. 2), and vane classifiers (see Fig. 3). Cyclones are termed gas cyclones when the conveying fluid is air or some other gas, and hydrocyclones when the conveying fluid is water or some other liquid.

Spiral and rake classifiers work on the principle that the slower settling finer material stays in suspension while the larger (or heavier) particles settle due to gravity. The devices are essentially settling tanks with mild agitation of the settled coarse material to resuspend trapped fines; the coarse particles are transported (raked) away from the rest of the slurry.

The separation action is produced in centrifugal devices by the rotation of the fluid, which develops high centrifugal forces. These forces tend to direct the larger (heavier) particles to the wall where they flow down to the underflow exit. At the same time, the smaller (lighter) particles are left nearer the center of the cyclone with the bulk of the fluid, because the higher drag effect (relative to inertial forces) of the fluid on smaller particles causes slower movement of particles with respect to fluid. The major part of the fluid, carrying the finer particles, reverses direction at the bottom of the tapered end of the cyclone (apex or spigot) and flows upward to the cyclone overflow (see Fig. 2). In a gas cyclone, the coarser solid leaving the tapered

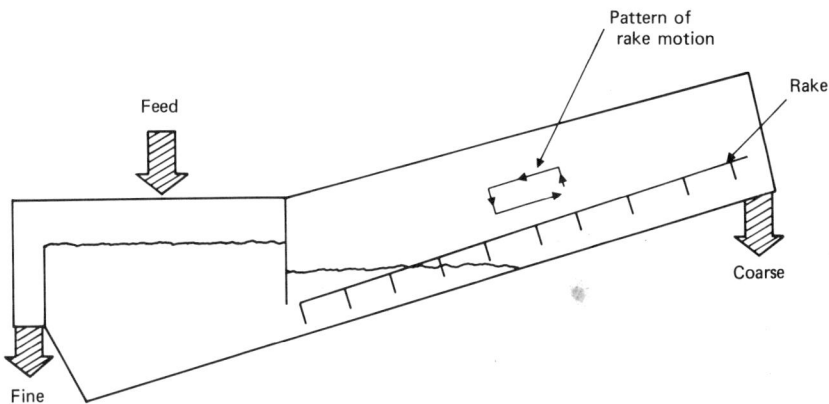

Fig. 1. Diagram of rake classifier.

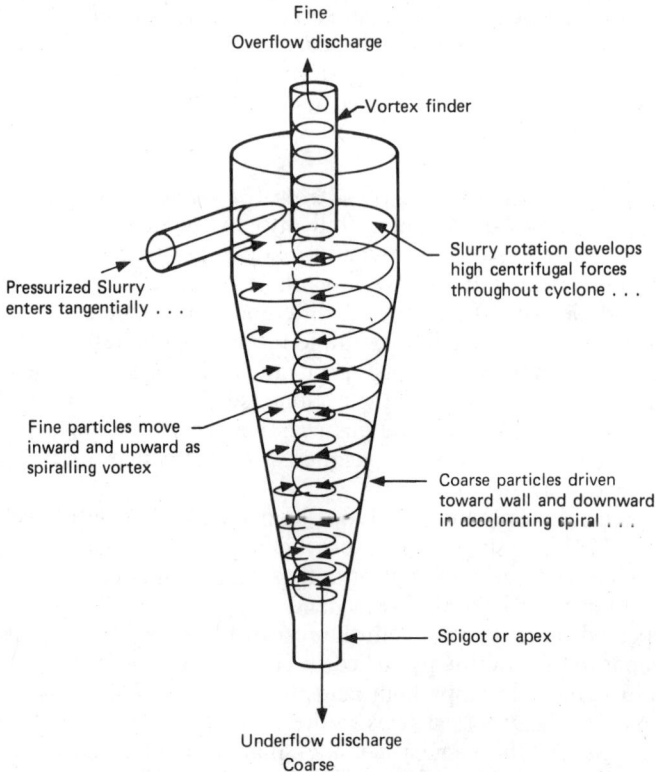

Fig. 2. The mode of operation of a cyclone.

end of the cyclone falls by gravity into a container or chute. In the hydrocyclone, sufficient liquid must emerge with the underflow particles to maintain a fluid slurry, and a cone of discharge spray is formed which falls into a sump. Cyclones are cheap, easy to use, flexible and reliable; the major problem in their use is the balancing of pressure drop across each cyclone in a bank of cyclones in parallel, in order to maintain proper operation of each cyclone.

The principle of the vane classification action is that the vane causes the gas to change direction: larger particles tend to continue in a straight line due to inertial momentum whereas smaller particles tend to follow the fluid path. The larger particles strike the vane, lose velocity, and fall back under the force of gravity. Vanes can be used in conjunction with a cyclone action. The vane classifier shown in Fig. 3 is used typically on fine particles suspended in air and contains two fans driven in opposite directions. The primary fan pulls air around the classifier as shown, whereas the secondary fan acts in the opposite direction. The faster the rotation speed of the secondary fan, the less chance a slow moving large particle has of sweeping between the blades before it is struck by the vanes and rejected from the airstream.

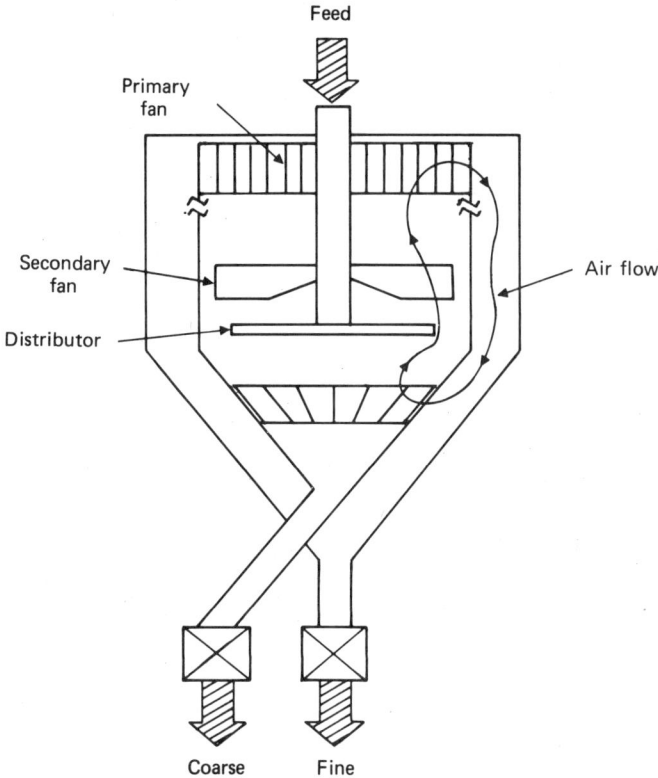

Fig. 3. Diagram of vane classifier.

The second category of classification equipment is based on the repeated presentations of particles to uniformly sized apertures in screens. Two major types of screens are used, vibrating screens which involve a bouncing particle action on a slightly tilted screen, and wedge-wire screens or sieve bends. The latter are used on wet slurries, with the particles flowing across a steeply pitched stationary wedge bar screen at right angles to the slots (see Fig. 4). The slotted surface tends to remove slices of the flowing slurry and the size of the particles removed is much smaller (1/2 to 2/3) than the actual slot width. The finer particles and liquid pass through the slots while the coarser particles overflow at the end of the screen.

A complete or even partial physical characterization of each type of classifier is clearly beyond the scope of this book. A number of references are available in this area (Lynch et al., 1977; Roberts and Fitch, 1956; Plitt, 1976; Austin and Luckie, 1976; Bradley, 1965; Dahlstrom, 1949; Kelsall, 1953; Dekok, 1975; Lynch and Rao, 1975; and Mular and Jull, 1980). For our purposes it is only necessary to describe in a simple, and if necessary empirical, manner how the various types of classifier vary in terms of the effect on a grinding circuit. Ideally, models are needed which describe changes in the mass split of fluid and solid to coarse and fine streams, along

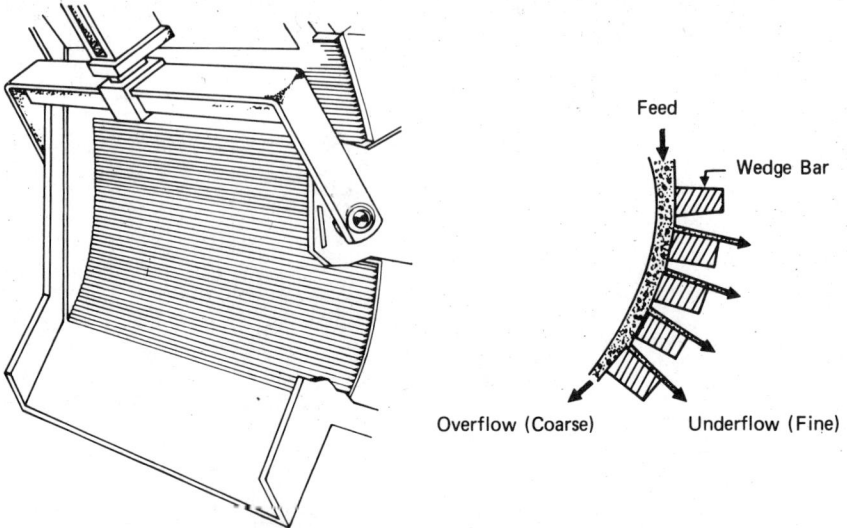

Fig. 4. Diagram of sieve bend.

with the corresponding size distributions, as a function of the operating conditions of each type of classifier.

Calculation of the Circulation Ratio for an Operating Classifier

The estimation of the circulation ratio C is important in the description of classification equipment and a brief review will be given of the various methods which can be used. C is not usually measured directly by mass flow of the streams, but is calculated from the three size distributions around the classifier. Klimpel (1980) has given the criteria by which a formula for consistent calculation of circulation ratio can be chosen, based on the error structure of the measured size distribution data p_i, q_i, and t_i for a given classifier (see Fig. 3 in Chapter 6). Three possible error structures are shown in Fig. 5. Each type of error structure gives rise to a different calculation formula for C which is most likely, from statistical theory, to give the most consistent and reliable estimate of C; Table 1 summarizes the various formulae.

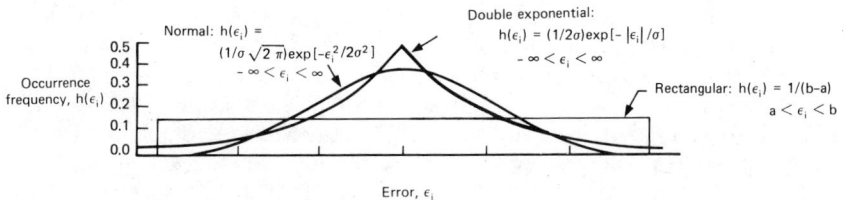

Fig. 5. Illustration of error distribution functions (area under each distribution normalized to 1.0 using an appropriate σ or $b - a$).

Table 1. Circulation Ratio Formulae for Classification
(All Σ Refer to the Sum $\Sigma_{i=1}^{n}$)

Error Structure Criterion	Normal Least squares	Double Exponential Least absolute sum	Rectangular Minimax	Equation residuals for all distributions
$C = T/Q$, Circulation ratio	$\dfrac{\Sigma(p_i - q_i)(t_i - p_i)}{\Sigma(t_i - p_i)^2}$	$\dfrac{\Sigma|p_i - q_i|}{\Sigma|t_i - p_i|}$	$(p_{i'} - q_{i'})/(t_{i'} - p_{i'})$	$\varepsilon_i = (p_i - q_i) - (T/Q)(t_i - p_i)$
$1 + C = P/Q$, Circulating load	$\dfrac{\Sigma(q_i - t_i)(p_i - t_i)}{\Sigma(p_i - t_i)^2}$	$\dfrac{\Sigma|q_i - t_i|}{\Sigma|p_i - t_i|}$	$(q_{i'} - t_{i'})/(p_{i'} - t_{i'})$	$\varepsilon_i = (q_i - t_i) - (P/Q)(p_i - t_i)$

The identification of the error structure of a particular set of data can be carried out in two ways. The first is to try all three formulae for T/Q (or P/Q) on a given set of size data (p_i, q_i, and t_i) to calculate ε_i, where *error* for an interval i is defined in the last column of Table 1. Then, for each formula, a plot can be made of the number frequency of ε_i, $h(\varepsilon_i)$, against ε_i for the n size intervals; $h(\varepsilon_i)\, d\varepsilon_i$ is the number fraction of ε_i which occurs between ε_i and $\varepsilon_i + d\varepsilon_i$. The curve which most closely matches one of the theoretical error structures is the appropriate error structure. This approach may not lead to a clear-cut choice for one data set and the choice becomes clearer for several sets of data, plotted on a normalized basis as in Fig. 5. A second approach consists of taking k duplicate samples of each stream, and then plotting the residuals for each size of each stream, where the residuals are now defined by $\varepsilon_i = (q_i - \bar{q}_i)$, $\bar{q}_i$ being the average q_i value over the k samples, using the circuit product stream as the example. The shapes of these plots matched to those of Fig. 5 will indicate which error distribution is appropriate.

The calculation of C or $1 + C$ is self-evident for the normal (Gaussian) and double exponential error structures; for rectangularly distributed error, it is necessary to calculate the residuals for all other elements j, for each element i, from

$$\text{residual} = |(p_j - q_j) - [(p_i - q_i)/(t_i - p_i)](t_j - p_j)|, \qquad j \neq i.$$

In effect, this uses a C value calculated from one interval, i, to determine the errors for all size intervals j. For each value of i, the maximum residual is then chosen, and the minimum of these maximum residuals gives the i value to be used in the equation in Table 1. This is the best choice of i, that is, i'. In other words, if C had to be calculated from one interval alone, this would be the best one to choose. The circulating load is similarly calculated. For example, in Table 2 the maximum residual for $i = 1$ from the set, -0.00535, 0.00206, -0.1394, -0.00529, 0.02188, 0.037, 0.03764, is 0.03764. The final column shows these maximum residuals for all i values; the minimum value is 0.0235. Therefore, C is calculated for $i' = 5$, giving 1.50. Note that the three error structures would give $C = 1.40$, $C = 1.46$, and $C = 1.50$, respectively.

The least absolute sum formula

$$C = \frac{\Sigma_i |p_i - q_i|}{\Sigma_i |t_i - p_i|} \tag{1}$$

was found to be generally satisfactory for several sets of hydrocyclone data (Klimpel, 1980). Working down from $i = 1$, it is readily shown that if $p_i - q_i$ and $t_i - p_i$ change sign in intervals i^* and i', respectively, and stay changed in sign for smaller sizes, then Eq. 1 can also be put as

$$C = (Q_{i^*} - P_{i^*})/(P_{i'} - T_{i'}) \tag{1a}$$

Table 2. Sample Calculations of Circulation Ratio Involving One Set of Classifier Streams

Sieve Interval i	Experimental data p_i	t_i	q_i	$(p_i - q_i)$	$|p_i - q_i|$	$(t_i - p_i)$	$|t_i - p_i|$	$(p_i - q_i) \times (t_i - p_i)$	$(t_i - p_i)^2$	$\dfrac{(p_i - q_i)}{(t_i - p_i)}$	Maximum residual
1	0.028	0.045	0.000	0.028	0.028	0.017	0.017	0.000476	0.000289	1.647	0.0376
2	0.021	0.037	0.000	0.021	0.021	0.016	0.016	0.000336	0.000256	1.313	0.0256
3	0.039	0.059	0.004	0.035	0.035	0.020	0.020	0.000700	0.000400	1.750	0.0535
4	0.060	0.097	0.013	0.047	0.047	0.037	0.037	0.001739	0.001369	1.270	0.0260
5	0.092	0.128	0.038	0.054	0.054	0.036	0.036	0.001944	0.001296	1.500	0.0235
6	0.103	0.114	0.063	0.040	0.040	0.011	0.011	0.000440	0.000121	3.636	0.0344
7	0.128	0.145	0.137	−0.009	0.009	0.017	0.017	−0.000153	0.000289	0.529	0.0349
8 (sink interval)	0.529	0.375	0.745	−0.216	0.216	−0.154	0.154	0.033264	0.023716	1.403	0.0246
$\sum_{i=1}^{n}$	1.000	1.000	1.000	0.000	0.450	0.000	0.308	0.038746	0.027736		

Criterion   Calculated $C = T/Q$

Least squares $(0.038746 / 0.027736) = 1.40$

Least absolute sum $(0.450 / 0.308) = 1.46$

Minimax $C \equiv$ minimum residual $= 1.50$

which is a convenient method of calculation. If $i^* = i'$ it is especially appropriate when the size distribution data is limited, since it uses only one data point from each distribution. The least squares equation

$$C = \frac{\Sigma_i(p_i - q_i)(t_i - p_i)}{\Sigma(t_i - p_i)^2} \qquad (2)$$

was also found to be satisfactory. Formulae based on the *sums* of cumulative size distribution data were found to be unsatisfactory. Hence, only formula 1 will be used in this book.

Classifier Partition Curves

In principle, the fraction $s(x)$ of size x to $x + dx$ sent to the coarse stream can be calculated from

$$s(x) = \frac{T(dT/dx)}{P(dP/dx)}$$

leading to a continuous function of $s(x)$ vs. x. However, because the mill simulations are performed on finite size elements (usually a $\sqrt{2}$ screen sequence) it is convenient to define the *selectivity* number for size interval i as the fraction (or percent) of size i in feed sent to the coarse stream. The separation of a feed stream into two output streams can then be characterized by the set of numbers s_i. Thus, from Chapter 6, Eq. 11,

$$s_i = \frac{t_i T}{p_i' P} = \frac{Ct_i}{(1 + C)p_i'}$$

where C has been determined under the appropriate test conditions using the equations of the previous section and p_i' is the reconstituted feed,

$$p_i' = \frac{q_i}{1 + C} + \frac{t_i C}{1 + C} \qquad (3)$$

Reconstitution of the stream having the greatest error (usually the feed stream, due to sampling difficulties) is often necessary to provide consistency in the calculated mass balances. The fraction of the feed sent to the finer stream is, of course, $1 - s_i$.

The set of s_i values calculated from a given set of data describe how the masses in the various size intervals are being split. If these same numbers are used for a different feed size distribution to the classifier, it is implicitly assumed that the process of size selection of each size interval in the classifier is first-order. In this context, first-order means that the amount of some size sent to the coarse stream is proportional to the amount of that size presented to the classifier, so that the fraction, s_i, sent to the coarse stream is a constant for that size interval i, even if the overall distribution of the feed to the classifier changes. As in the case of the experimental determination of the specific rates of breakage using sieve size data (see Chapter 7), the size interval must be chosen so that the first-order assumption is reasonably valid. Experience has shown that the geometric sieve ratio of $\sqrt[4]{2}$ is a small enough interval and that the ratio of $\sqrt{2}$ is in most cases also small enough.

A plot of the selectivity numbers vs. size is called a *Tromp* curve, *partition* or *selectivity* curve; it is usually convenient to have size on a log scale. The x_i value can be the upper size of the interval i, or the lower sieve size, or some other size such as the geometric mean size of the interval. Much of the data in the literature is difficult to interpret because it is not stated whether the $s(x)$ values are for differential size intervals, for $\sqrt[4]{2}$ or $\sqrt{2}$ size intervals, or for irregular size intervals, nor is it stated whether x_i is the lower, upper, or geometric mean of an interval. In this book, the s_i values will always be for $\sqrt{2}$ intervals and are always plotted against the upper size of the interval. In computer simulations, of course, only the vector of discrete s_i values is used.

An ideal classification action would be where all sizes smaller than a *cut-size* reported to the fine stream and all larger sizes to the coarse stream (see Fig. 6). In practice, since real classifiers do not behave in this ideal way, additional descriptive parameters are required to characterize measured classifier data. The first type of non-ideal behavior is apparent bypassing. If some of the particles of each size in the feed stream end up in the coarse

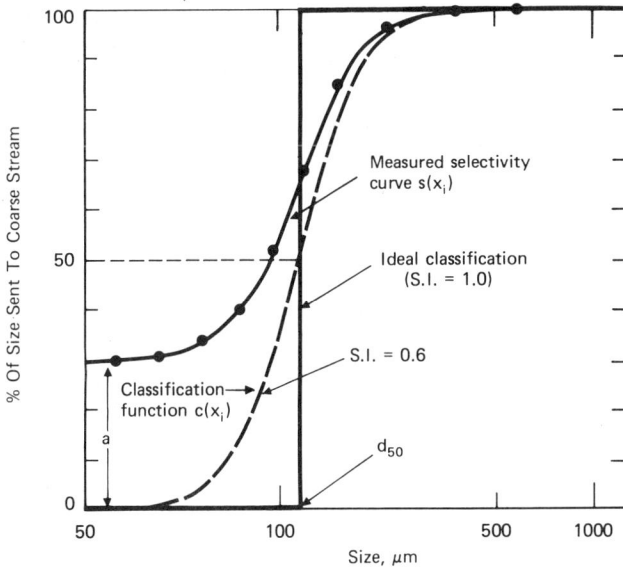

Fig. 6. Illustration of classifier selectivity partition curve; $\sqrt{2}$ size intervals, values plotted at upper size of interval. Data is typical of hydrocyclone data.

stream, then a certain portion of the coarse stream will have the same size distribution as the classifier feed stream. This has the same effect as if a fraction a of the classifier feed stream bypasses the classifier and reports directly to the coarse stream. This occurs with most classifiers, to a greater or lesser extent. On the other hand, apparent bypassing to the fine stream is not a normal condition and indicates an operating malfunction or an incorrect design condition. Apparent bypassing to the coarse stream is always seen in wet classifiers, where the bypassed finer particles are considered to arrive by being trapped in the dense slurry of coarse particles traveling to the underflow. The apparent bypass fraction is marked as a in Fig. 6.

The remaining fraction of the feed, $1 - a$, is subjected to the classification action and for any size interval a fraction c_i is sent to the coarse stream and a fraction $1 - c_i$ to the fine stream. Thus the definition of the *classification numbers* c_i is

$$c_i = \frac{s_i - a}{(1 - a)} \tag{4}$$

In the same way as s_i values, the values of c_i can be plotted vs. the upper size of the interval x_i. The size at which $c(x) = 0.5$ is called the d_{50} value and the curve $c(x_i)$ is called the *classification function* (see Fig. 6). The steepness of the $c(x_i)$ function is conveniently indexed by the *Sharpness Index* S.I., defined as the size d_{25} at which $c(d_{25}) = 0.25$ divided by the size d_{75} at which $c(d_{75}) = 0.75$, that is, S.I. $= d_{25}/d_{75}$. S.I. $= 1$ for ideal classification and S.I. $= 0$ for no size classification at all, that is, the device acts purely as a sample splitter. The variation of selectivity numbers as a function of S.I. and bypass is illustrated in Fig. 12 in Chapter 7. The variation of S.I. over the typical range of 0.4 to 0.7 gives very profound changes in selectivity values.

Eq. 4 shows that the values of c_i can be calculated from a knowledge of the s_i values and a. It has proved very useful to calculate c_i values because the same classifier operated at different conditions will often give the same shape of $c(x_i)$ vs. a dimensionless size defined by x_i/d_{50}. The curve of $c(x/d_{50})$ vs. x/d_{50} is called the *reduced classification* curve.

Because of the scatter in s_i values usually found in industrial classifier data, it is desirable to fit experimental s_i data to a functional form of $c(x_i)$ using Eq. 4. This may be done graphically or by using a parameter estimation technique based on a least squares criterion

$$\underset{a, \text{S.I.}, d_{50}}{\text{minimize}} \sum_{i=1}^{n} w(x_i)[s_i(\text{observed}) - s(x_i)]^2 \tag{5}$$

where the $w(x_i)$ are weighting factors for each of the size intervals, s_i (observed) is given by Eq. 11 in Chapter 6, $s(x_i) = a + (1 - a)c(x_i)$, and $c(x_i)$ is a preselected functional form having parameters S.I. and d_{50}. Techniques for performing this type of calculation are well documented

(Blau, Klimpel, and Steiner, 1972; Wilde and Beightler, 1967), as are procedures for selecting appropriate weighting factors, which are most often taken to be either 1.0 or the fractional weight of feed, p_i, divided by the estimated variance of the error of measurement of p_i (Klimpel, 1980; Luckie and Austin, 1974).

Four equations have been most used for the function $c(x_i)$:
1. Rosin-Rammler (Plitt, 1971; Reid, 1971)

$$c(x_i) = 1 - \exp\left[-(x_i/x_o)^\lambda\right] \tag{6}$$

where

$$d_{50} = x_o(0.693)^{1/\lambda} \tag{6a}$$

and

$$\text{S.I.} = \exp\left[-1.5725/\lambda\right] \tag{6b}$$

2. Log-normal (Aso, 1957)

$$c(x_i) = \frac{1}{\sqrt{2\pi}} \int_{-\infty}^{(\ln[x_i]-\ln[d_{50}]/\lambda)} \exp\left[-u^2/2\right] du \tag{7}$$

where

$$\text{S.I.} = \exp\left[-1.349\lambda\right] \tag{7a}$$

3. Exponential sum, as used by Lynch et al. (1977)

$$c(x_i) = \frac{\exp\left[\lambda(x_i/d_{50})\right] - 1}{\exp\left[\lambda(x_i/d_{50})\right] + \exp\left[\lambda\right] - 2} \tag{8}$$

where

$$\text{S.I.} = \frac{\ln\left[(\exp\left[\lambda\right] + 2)/3\right]}{\ln\left[3\exp\left[\lambda\right] - 2\right]} \tag{8a}$$

(Note that determination of λ from S.I. using this equation requires a trial-and-error calculation.)
4. Logistic (Molerus, 1967; Finney, 1964) in log x_i

$$c(x_i) = \frac{1}{1 + (x_i/d_{50})^{-\lambda}} \tag{9}$$

where

$$\text{S.I.} = \exp\left[-2.1972/\lambda\right] \tag{9a}$$

Other functional forms have been suggested (Luckie and Austin, 1974), but usually at least one form from these four will have sufficient flexibility to fit a particular data set with reasonable accuracy. A computer program for curve fitting classifier data with any of the above models is available (Klimpel, 1982).

A problem which is often encountered occurs when the s_i values are calculated from experimentally measured size distributions of the three streams which do not extend to small enough sizes to get the complete set of s_i values down to and including the bypass fraction. Austin and Klimpel (1981) have developed a procedure for obtaining the complete selectivity curve based on straight line extrapolation of the feed size distribution on a "log size vs. log fraction-less-than-size" plot (*Schuhmann plot*). It is often possible to perform this extrapolation accurately because the product size distribution from a mill gives a straight line on this type of plot over much of the size range, whereas it is usually not possible to extrapolate accurately the T and Q size distributions (see Fig. 7, for example).

In this technique the parameters a, d_{50}, S.I. are determined by the calculation

$$\underset{a,\,\text{S.I.},\,d_{50}}{\text{minimize}} \sum_{k=1}^{n} w(x_i)\left[\gamma_k(\text{observed}) - \gamma_k(\text{calculated})\right]^2 \qquad (10)$$

where $\gamma_k(\text{observed})$ for each size interval k is given by

$$\gamma_{k(\text{observed})} = \frac{C}{(1+C)}\left(\frac{T_k}{P_k}\right), \qquad 1 \le k \le n \qquad (11)$$

where T_k and P_k are the cumulative weight fractions less than size k. $\gamma_k(\text{calculated})$ is given by

$$\gamma_k(\text{calculated}) = a + (1-a)\sum_{i=N}^{k} p_i c(x_i)/P_k, \quad 1 \le k \le n, \quad N > n$$

$$(12)$$

where N is the number of the smallest *extrapolated* size interval, n is the number of the last size interval for which all three particle streams were *measured*, and $c(x_i)$ is a suitable function containing S.I. and d_{50} as parameters. The value of N must be chosen to be large enough so that $c(x_N)$ approaches zero.

This technique uses the information contained in the nth interval (via the use of P_n and T_n) which would otherwise not be used because $s_n = CT_n/(1+C)P_n$ is not a valid calculation unless $s_n = a$. A program to use any of the four equations, 6 to 9, for $c(x_i)$ is available (Austin and Klimpel, 1981). Note that the use of the technique with these functions is valid only for classifier data which has a bypass a, and it will obviously not work for classification curves of the form of Fig. 19, since they do not fit any of the four equations.

Figs. 7 and 8 demonstrate the use of this technique on a set of hydrocyclone data, using the logistic model for $c(x_i)$. Data from sizes

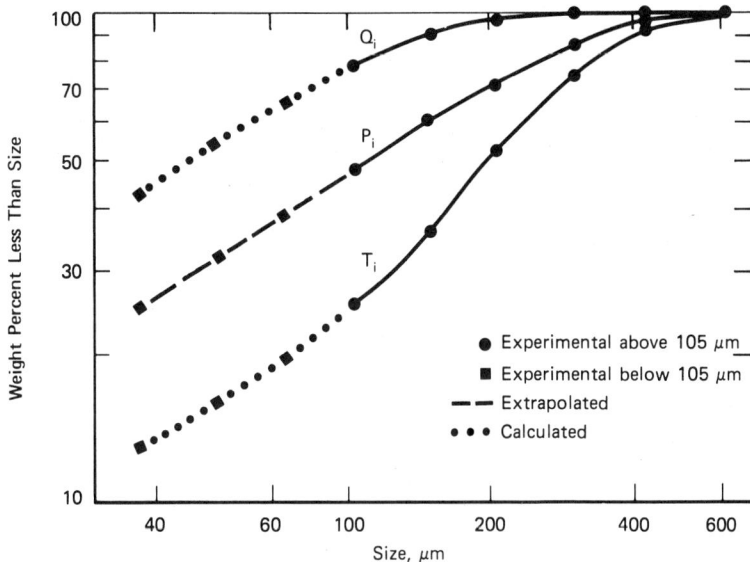

Fig. 7. Size distribution of three streams around a 600 mm hydrocyclone classifier operating on a feed slurry of 33% by volume of copper ore of specific gravity of 2.65.

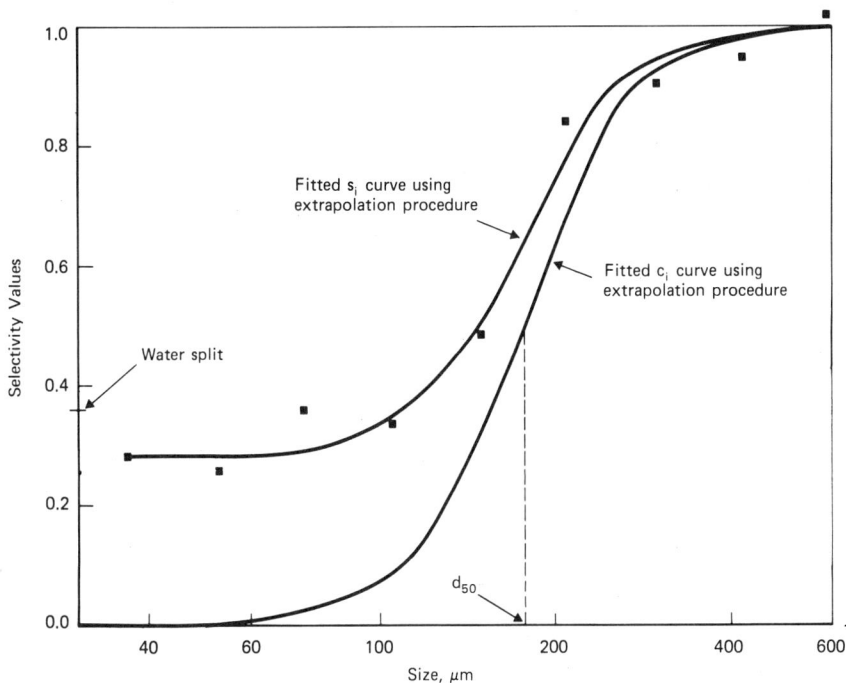

Fig. 8. Selectivity values for data of Fig. 7: ■ individual s_i values from experimental data.

greater than 105 μm ($n = 6$) were used to fit the selectivity curve using Eq. 5, giving the parameter set $a = 0.42$, $d_{50} = 190$ μm, and S.I. = 0.62. When the feed size is extrapolated as shown in Fig. 7 and the data analyzed using Eq. 10 ($n = 6$, $N = 9$), the resulting parameter set was $a = 0.27$, $d_{50} = 170$ μm, and S.I. = 0.62. This new set of classifier parameters was used to predict the size distribution for values less than 105 μm. As can be seen from Fig. 8, the match between experimental and predicted values is good. The set $a = 0.42$, $d_{50} = 190$ μm, S.I. = 0.62 does not give good predictions below 105 μm.

Hydrocyclones: Variables Affecting Hydrocyclone Performance

Several good reviews of hydrocyclone principles are available (Lynch et al., 1977; Bradley, 1965; Dahlstrom, 1949; Kelsall, 1953; Dekok, 1975; Lynch and Rao, 1975; and Mular and Jull, 1980). For example, the book by Bradley gives an excellent review of the behavior of hydrocyclones used on dilute slurries. A given cyclone will only perform properly over a range of flow rates and associated pressure drops; at too low flow rate there is insufficient fluid rotation to create the air core and give separation; at too high flow rates the air core fills the discharge opening at the apex, too little fluid leaves, and separated solid is not carried out but is redistributed into the overflow stream. A simple approximate hydrodynamic analysis gives the relation between volume flow rate Q, pressure drop ΔP, and cyclone diameter D_c as

$$\Delta P = (\text{constant})(Q/D_c^2)^2 \tag{13}$$

where the constant varies with cyclone geometry. Operating pressure drops are in the range of 35 to 350 kPa, and a correct operating range of pressure is $\Delta P \max/\Delta P \min \approx 4$, so that the range of flow rate is no more than about a one to two range for a given cyclone.

By a very simple analysis of the forces acting upon single particles in the rotating fields observed in cyclones, Bradley derived the following equations relating d_{50} to the operating characteristics of a hydrocyclone:

$$d_{50} = (\text{constant})(D_i^2/D_c^{0.5})\left[\frac{\eta(1 - R_f)}{Q(\rho_s - \rho_l)}\tan(\Theta/2)\right]^{1/2} \tag{14}$$

where η is the slurry viscosity, D_i the cyclone inlet diameter, ρ_s and ρ_l the solid and liquid densities, and Θ the included cone angle. R_f is the ratio of underflow volume to feed volume: R_f normally decreases as Q increases. Again, the constant depends on cyclone geometry. This equation applies only to very dilute slurries, in hydrocyclones operating in the correct range of pressure drop. Since the volume flow rate Q scales according to D_c^2, the variation of d_{50} with cyclone diameter, for a fixed cyclone geometry (D_i/D_c = constant) is

$$d_{50} \propto D_c^{1/2} \tag{15}$$

Unfortunately, these simple equations cannot be used to calculate the d_{50} values for cyclones used on dense slurries because the complex rheological behavior of such slurries destroys the applicability of the simple derivations. As indicated earlier there are not, at the present time, generally accepted complete quantifications of hydrocyclone performance on dense slurries as a function of operating conditions and design parameters. The problem arises because there are a number of interacting variables which significantly affect hydrocyclone performance, and to control or determine all of the variables in any given experimental situation is extremely difficult. Thus the data collected in a given study usually results from measurement of certain variables only, while the remainder are *uncontrolled*: in statistical terminology, there is confounding of the test variables. This has the effect that seemingly clear conclusions from one study may not be applicable quantitatively, or even qualitatively, to another study, due to differences in the uncontrolled variable settings. The following paragraphs summarize the qualitative information available.

Cyclone Size

It is well established that smaller diameter cyclones separate at a finer size (smaller d_{50}) than do larger hydrocyclones. In addition, the data collected by Lynch et al. (1977) on geometrically similar but different diameter hydrocyclones showed that hydrocyclones ranging from 100 to 380 mm (4 to 15 in.) diameter gave essentially the same reduced classification function vs. x_i/d_{50}. Fig. 9 from Klimpel (1982) shows data collected on geometrically

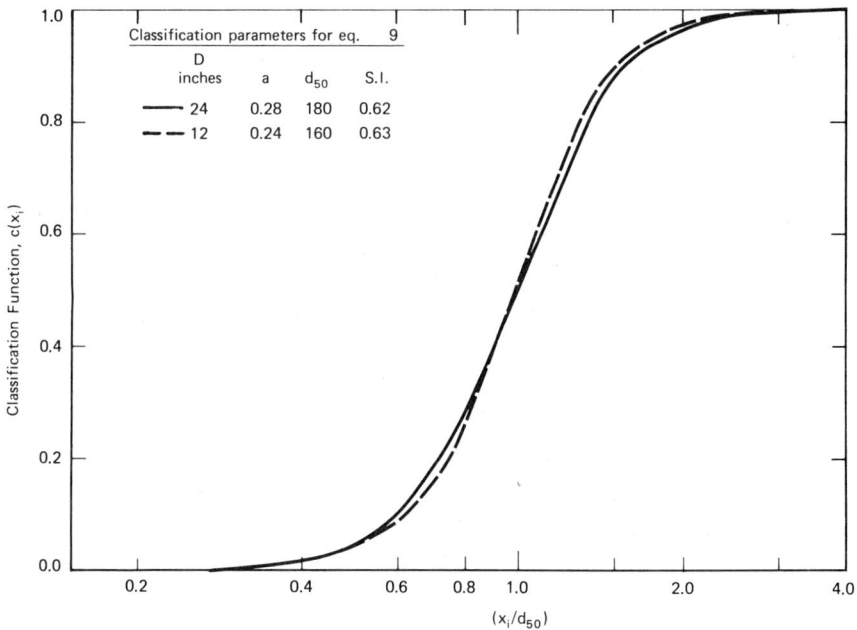

Classification parameters for eq.   9

D inches	a	d_{50}	S.I.
24	0.28	180	0.62
12	0.24	160	0.63

Fig. 9. Reduced classification curves for geometrically similar hydrocyclones of 305 and 610 mm (12 and 24 in.) diameter operating on a copper ore slurry under similar conditions.

similar hydrocyclones of 305 mm (12 in.) and 610 mm (24 in.) diameter, operating on copper ore slurry of constant feed size, pulp density, etc. The S.I. (the shape) for both cyclones is essentially the same, but there are, of course, significant differences in the d_{50} values.

Feed Rate

It has been shown by numerous workers that an increase in the volumetric feed rate causes a decrease in the d_{50} value, providing other factors are held constant (Lynch et al., 1977; Klimpel 1982; Bradley, 1965; Dahlstrom, 1949; Kelsall, 1953; Dekok, 1975; Lynch and Rao, 1975; and Mular and Jull, 1980).

Material

It has been found that hydrocyclone parameters, including S.I. values (Lynch et al., 1977), are different for different minerals. This is not surprising considering the wide range of shape factors, specific gravity, etc., for different minerals and mineral combinations. It is currently assumed (Mular and Jull, 1980) that increasing the specific gravity of feed particles causes the d_{50} value to decrease. However, there is at present insufficient information to enable the precise calculation of this effect using measured material properties.

Even more difficult is the prediction of the behavior of mineral mixtures. Klimpel (1982) has reported analysis of the performance of a 610 mm (24 in.) hydrocyclone operating on a mixture of a nonmagnetic, copper-containing ore (s.g. = 2.81) and a magnetic iron ore (s.g. = 4.39). Each component had a similar size distribution and the feed mass was made up of

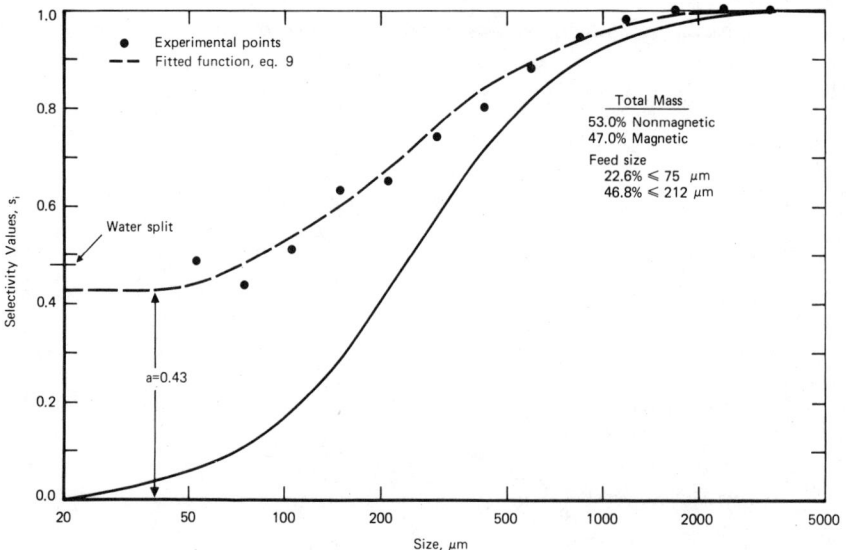

Fig. 10. Overall selectivity values for the total mass of a mixture of copper and magnetic iron ores in a 610 mm (24 in.) hydrocyclone near roping conditions (values plotted vs. top size of $\sqrt{2}$ screen interval).

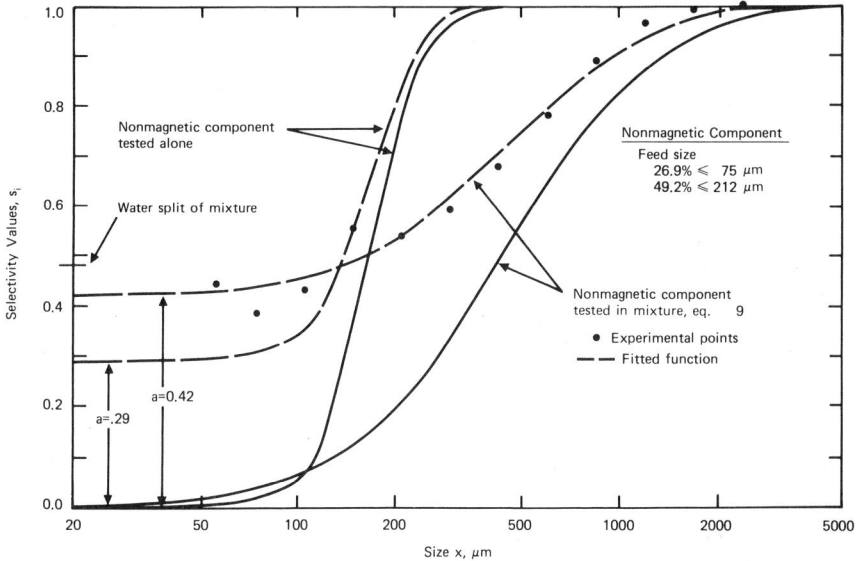

Fig. 11. Selectivity values of the nonmagnetic copper ore component of a mixture of copper and magnetic iron ores and of copper ore tested alone, in a 610 mm (24 in.) hydrocyclone near roping conditions (values plotted vs. top size of $\sqrt{2}$ screen interval).

an equal weight of each component. The three streams around the cyclone were analyzed for the fraction of each ore in each size interval. Fig. 10 shows the overall selectivity values of the mixture ($a = 0.43$, $d_{50} = 250$ μm, S.I. = 0.31). Fig. 11 shows the selectivity values of the nonmagnetic ore in this mixture ($a = 0.42$, $d_{50} = 440$ μm, S.I. = 0.31), and the values for the same ore tested in a separate trial with the ore alone ($a = 0.29$, $d_{50} = 170$ μm, S.I. = 0.64). Similarly, Fig. 12 shows the results for the magnetic ore in the mixture ($a = 0.54$, $d_{50} = 220$ μm, S.I. = 0.37) and tested separately ($a = 0.34$, $d_{50} = 145$ μm, S.I. = 0.55). Both ores in the mixture show larger a values, much lower S.I. values, and much higher d_{50} values, that is, very inefficient classification. It is quite clear that the mixed slurry gives selectivity values which are radically different from the separate components tested in the same cyclone, presumably due to different rheological properties of the mixed slurry.

The pattern of these results contradicts the analysis of hydrocyclone classification of mineral mixtures by Lynch et al., 1977. The implications of Klimpel's results are that the reduced classification curve for each mineral changes as a function of the proportion of that mineral in the mixture. Obviously, additional investigation is needed into this important aspect of industrial classification.

Vortex Finder Diameter, Apex Diameter, Inlet Diameter

It is generally acknowledged (Lynch et al., 1977; Dekok, 1975; Mular and Jull, 1980) that d_{50} will increase with an increase in vortex finder diameter.

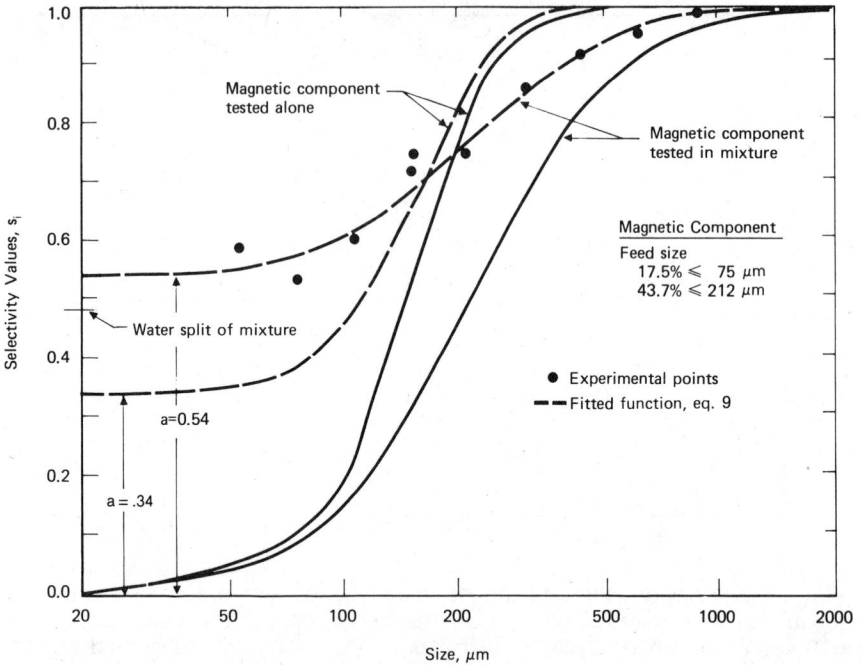

Fig. 12. Selectivity values of iron ore component of a mixture of copper and magnetic iron ores and of iron ore tested alone, in a 610 mm (24 in.) hydrocyclone near roping conditions (values plotted vs. top size of $\sqrt{2}$ screen interval).

It has been shown that d_{50} increases with a decrease in spigot (apex) diameter (Lynch et al., 1977).

The influence of increasing the inlet diameter appears from most studies to cause an increase in the d_{50} value (Lynch et al., 1977; Klimpel, 1982; Bradley, 1965; Dahlstrom, 1949; Kelsall, 1953; Dekok, 1975; Lynch and Rao, 1975; Mular and Jull, 1980).

Slurry Percent Solids
There is considerable confusion concerning the influence of slurry density on hydrocyclone performance. Some workers (Mular and Jull, 1980) state that an increase in percent solids will cause a consistent increase in the d_{50} value while other studies (Lynch et al., 1977; Klimpel, 1982) indicate that the d_{50} will increase for sufficiently low percent solids, pass through a maximum, and then decrease at relatively high percent solids.

Feed Particle Size Distribution
This is again a difficult area to quantify but it is reasonable to assume that the finer the feed (if there are not excessive amounts of coarse particles) the smaller the d_{50} value will be, due to the increase in effective slurry viscosity

caused by finer particles (see below). This has been demonstrated in several studies (Lynch et al., 1977; Klimpel, 1982).

Slurry Viscosity

Until recently almost no work had been done on the effect of slurry viscosity at conditions approximating industrial dense slurry operating conditions. In many cases the effect has not been separated from that of slurry percent solids, since changing the percent solids (hence varying the water and solid volume flow rates) has been the major technique used to control viscosity. Klimpel (1982) has shown that the d_{50} and S.I. values increase when viscosity is lowered by addition of a chemical fluidity modifier without change in the total slurry volume. This is one of the few methods by which the shape (S.I.) of the classification function can be significantly influenced under operating conditions: Fig. 13 gives an example. The characteristic parameters of the fit of Eq. 9 for $c(x_i)$ were $a = 0.29$, $d_{50} = 170$ μm, S.I. $= 0.64$ without chemical and $a = 0.23$, $d_{50} = 190$ μm, S.I. $= 0.70$ with chemical.

Water Flow and Bypass a

It is generally assumed (Lynch et al., 1977; Bradley, 1965; Dahlstrom, 1949; Kelsall, 1953; Dekok, 1975; Lynch and Rao, 1975; and Mular and Jull,

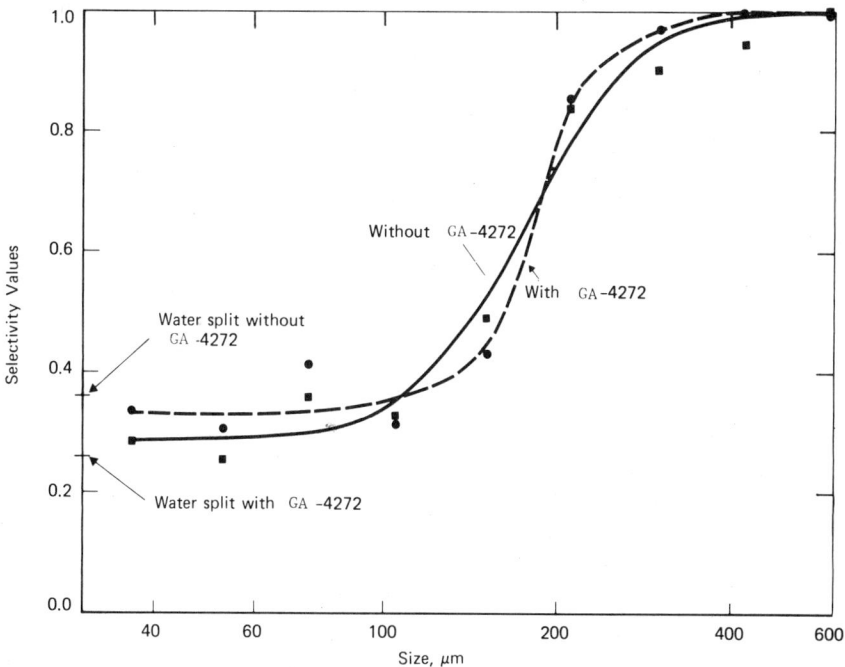

Fig. 13. Effect of fluidity modifier on classifier selectivity curve for a 610 mm (24 in.) diameter hydrocyclone operating on a copper ore slurry without chemical; with 0.09 kg GA-4272 per dry ton feed (see Chapter 15 for description of GA-4272).

1980) that the apparent bypass fraction *a* is equal to the fraction of feed water reporting to the underflow. However, there are some cases (Klimpel, 1982; Austin and Klimpel, 1981) where this assumption has been shown not to apply, if it is assumed that the apparent bypass obtained from analysis of sieve data is valid through the subsieve region. Fig. 13 shows a typical result where the water split is not equal to the bypass fraction; when a fluidity modifier was added, the bypass fraction increased, whereas the water split to underflow decreased. Studies by Lynch et al. (1977) have shown that the spigot diameter, water flow rate, and particle feed size have significant influences on the water split, with an increase in spigot diameter causing a decrease in the fraction of water in the overflow.

Quantitative Models for Hydrocyclones

There are three major models used to predict the variation of d_{50} with operating variables for dense slurry cyclones. Lynch et al. (1977) found that the following relationship held for their data:

$$\log d_{50} = K_1(\text{vortex finder diameter}) - K_2(\text{apex diameter})$$

$$+ K_3(\text{inlet diameter}) + K_4(\text{percent solids by weight in feed slurry})$$

$$- K_5(\text{volume flow rate of feed pulp}) + K_6 \qquad (16)$$

where $K_1, K_2, \ldots, K_6$ are empirical constants greater than zero. These were determined by curve-fitting the results of tests using the same feed material, with $\log d_{50}$ as the fitted result. For a limestone slurry in a Krebs hydrocyclone of 508 mm (20 in.) diameter, the parameters were $K_1 = 0.0418$, $K_2 = 0.0576$, $K_3 = 0.0366$, $K_4 = 0.0299$, $K_5 = 0.00005$, and $K_6 = 0.0806$, for diameters in centimeters, flow in liters per minute, and d_{50} in micrometers.

Lynch et al. (1977) suggest that when a large hydrocyclone is being evaluated, it is convenient to determine the factors K_1 to K_5 with tests on a geometrically similar small cyclone, and to determine d_{50} from a test on the large cyclone. Since K_1 to K_5 do not depend on cyclone diameter, the value of K_6 for the large cyclone can then be determined from Eq. 16. They also found that the feed size distribution had a significant influence on the empirical constants, so that three different sets of constants were required to describe accurately the d_{50} values for three different limestone feeds (the set of K constants given previously is the average set for the three feed size distributions studied).

Lynch also developed an expression for predicting the water split. If the *a* value is approximated by the water split, then $a = R_f/100$, where

$$R_f = K_o + \frac{[(\text{apex diameter}) - K_1]}{(\text{flow rate of water})} \qquad (17)$$

As before, they developed individual expressions for three different feed size distributions, coarse, medium and fine, as well as a mean expression for all three. The coefficients for the mean of all sizes are $K_o = -1.61$ and

$K_1 = 1.41$, for diameters in centimeters and flow rates in tons per hour. Lynch states that the λ in Eq. 8 should be determined for each particular material.

A check similar to the one given by Mular and Jull (1980) should be used to be certain that the spigot discharge is not operating in a roping condition. Their graphical data defining the probable roping condition regime for ores of known specific gravity, percent solids by weight in the underflow and percent solids by weight in the overflow were replotted as percent solids by volume, and gave a single approximation for the probable non-roping regime (see "Classifier Design: The Hydrocyclone" in Chapter 7):

(volume fractions solids in underflow)

$$\leq 0.5385(\text{volume fractions solids in overflow}) + 0.4931$$

In practice, plant engineers have found that each hydrocyclone operating in a given circuit on a given material requires a different set of parameters in Eq. 16, and quite often one or more extra terms such as slurry percent solids, viscosity, temperature, are required to predict satisfactorily the d_{50} values being observed in the plant.

The second quantitative model is from Plitt (1971),

$$d_{50} = K_0 \frac{(\text{cyclone diam})^{K_1}(\text{inlet diam})^{K_2}(\text{vortex finder diam})^{K_3}}{(\text{apex diam})^{K_5}(\text{free vortex height})^{K_6}(\text{volume flow rate})^{K_7}}$$

$$\times \frac{\exp[K_4(\text{volume percent solids})]}{(\rho_s - \rho_l)^{0.5}} \quad (18)$$

Curve fitting was used to determine the empirical constants K_0 through K_7 for particular sets of data. Plitt gives $K_0 = 35$, $K_1 = 0.46$, $K_2 = 0.6$, $K_3 = 1.21$, $K_4 = 0.063$, $K_5 = 0.71$, $K_6 = 0.38$, and $K_7 = 0.45$ for diameters and heights in inches, flow rates in cubic feet per minute, and densities in grams per cubic centimeters. As was the case with the Lynch model, industrial practice has shown that different K values and/or additional terms are required to use the Plitt model to predict accurately d_{50} values for a given application.

Plitt developed an expression for predicting the ratio of the underflow pulp rate to the overflow pulp rate on a volumetric basis:

$$S = K_0(\text{apex diam})^{K_1}(\text{free vortex height})^{K_2}$$

$$\times \left[(\text{apex diam})^2 + (\text{vortex finder diam})^2\right]^{K_3}$$

$$\times \exp[K_4(\text{volume percent solids})]/(\text{inlet pressure})^{K_5}(\text{cyclone diam})^{K_6}$$

$$(19)$$

where $K_0 = 2.9$, $K_1 = 3.31$, $K_2 = 0.54$, $K_3 = 0.36$, $K_4 = 0.0054$, $K_5 = 0.24$, and $K_6 = 1.11$ for diameters and heights in inches and pressures in feet. As before, a check similar to that given by Mular and Jull (1980) should be used to be certain that the underflow discharge is not operating in a roping condition.

A mass balance on water and solid volumes gives fractional water split to underflow as

$$a' = \frac{\left(\dfrac{\text{underflow pulp volume rate}}{\text{feed pulp volume rate}}\right) - (\text{volume fractions solids in feed})\dfrac{C}{1+C}}{1 - \text{volume fractions solids in feed}}$$

(20)

The value of circulation ratio C is related to the size distribution entering the classifier by

$$1 + C = \frac{1}{(1-a)(1 - \Sigma_i c_i p_i)}$$

Inserting into Eq. 20 and assuming that the apparent bypass is approximately equal to the water split, then the apparent bypass can be extracted as

$$a = $$

$$\frac{\left(\dfrac{\text{underflow pulp volume rate}}{\text{feed pulp volume rate}}\right) - (\text{volume fractions solids in feed})\Sigma_i c(x_i) p(x_i)}{1 - (\text{volume fractions solids in feed})\Sigma_i c(x_i) p(x_i)}$$

(20a)

The term (underflow pulp volume rate/feed pulp volume rate) can be calculated using Eq. 19, as $S/(1 + S)$.

Plitt also used $c(x_i)$ defined by Eq. 6, with the λ values in Eq. 6 varying with hydrocyclone geometry and operating conditions according to

$$\lambda = \left[(\text{cyclone diam})^2(\text{free vortex height})/(\text{total pulp volume rate})\right]^{K_1}$$

$$\times \exp[0.58\text{–}1.58(\text{pulp underflow volume rate})/$$

$$(\text{total pulp volume rate})]$$

(21)

where $K_1 = 0.15$ and diameters and heights are in inches and flow rates in cubic feet per minute.

The third quantitative model is from Arterburn (1982). He assumes a *typical* cyclone (Krebs) where the geometry is:

$$\text{inlet area} = (0.015 \text{ to } 0.02)\,\pi\,(\text{cyclone diam})^2$$

$$\text{vortex finder diam} = 0.4(\text{cyclone diam})$$

$$\text{apex diam} > 0.1(\text{cyclone diam})$$

$$\frac{\text{included}}{\text{angle}} = \begin{cases} 20°, & \text{cyclone diam} > (250 \text{ mm})(10 \text{ in.}) \\ 12°, & \text{cyclone diam} < (250 \text{ mm})(10 \text{ in.}) \end{cases}$$

Then,

$$d_{50}, \mu m$$

$$= 16.9 \frac{(\text{cyclone diam cm})^{0.67}}{(\text{pressure drop, kPa})^{0.28}(\rho_s - \rho_l)^{0.5}(1\text{–}1.9 \text{ volume fraction feed solids})^{1.43}}$$

$$(22)$$

where ρ_s is the specific gravity of the solid and ρ_l is the specific gravity of the liquid.

Again, if the apparent bypass is assumed to be equal to the water split then it can be calculated using Eq. 20a. The feed pulp volumetric flow rate and the underflow pulp volumetric rate can be determined from the approximation of Arterburn's graphical data:

feed pulp volumetric flow rate, l/s

$$= 0.0025\sqrt{(\text{pressure drop, kPa})}\,(\text{cyclone diam, cm})^2 \qquad (23a)$$

$$\text{underflow pulp volumetric rate, } l/s = K_0(\text{apex diam, cm})^2 \qquad (23b)$$

where K_0 has the range of 0.244 to 0.098.

No relation was given for K_0 vs. pressure drop for a given cyclone diameter, so we have assumed

$$K_0 = 0.00215\,\Delta P + 0.02665 \qquad (23c)$$

which gives a value of 0.171 in cyclone calculations at 69 kPa (10 psi) of pressure; it must be remembered that an appropriate degree of uncertainty is thus introduced. Arterburn's graphs are strictly for water through a cyclone and flow rates are somewhat higher for slurry. However, it is the *ratio* of flow rates which is required in the calculation of bypass, so that the error introduced by using Eq. 23 for slurry is largely canceled out:

$$\text{pulp split} = \frac{\text{underflow volumetric rate}}{\text{feed volumetric flow rate}} = \frac{400K_0\left[\dfrac{(\text{spigot diam})}{(\text{cyclone diam})}\right]^2}{\sqrt{(\text{pressure drop, kPa})}}$$

$$(24)$$

Arterburn assumes a standard reduced classification curve, with a λ value in Eq. 8 of 4.1. Again, a check similar to the one given by Mular and Jull (1980) should be used to be certain that the spigot discharge is not operating in a roping condition.

The Arterburn equations are used as an illustration of hydrocyclone design in association with a mill model in Chapter 7.

It is convenient to express the interrelations of the mass balances of solid and water around a hydrocyclone in terms of the overflow density, since this is normally specified to give a good feed to flotation. The basic equations are:

$$c_u = \frac{1}{1 + a'\left(\dfrac{1 + C}{C}\right)\left(\dfrac{1 - c_f}{c_f}\right)} \qquad (25a)$$

$$c_o = \frac{1}{1 + (1 - a')(1 + C)\left(\dfrac{1 - c_f}{c_f}\right)} \qquad (25b)$$

where c_u, c_o, and c_f are volume fractions of solid in slurry in the cyclone underflow, overflow, and feed, respectively.

These rearrange to

$$a' = 1 - \left(\dfrac{1 - c_o}{c_o}\right)\left(\dfrac{c_f}{1 - c_f}\right)\left(\dfrac{1}{1 + C}\right) \qquad (25c)$$

$$c_u = \frac{1}{1 + \left[\left(\dfrac{1 + C}{C}\right)\left(\dfrac{1 - c_f}{c_f}\right) - \left(\dfrac{1 - c_o}{c_o}\right)\left(\dfrac{1}{C}\right)\right]} \qquad (25d)$$

$$c_f = \frac{(1 - a')(1 + C)\left(\dfrac{c_o}{1 - c_o}\right)}{1 + (1 - a')(1 + C)\left(\dfrac{c_o}{1 - c_o}\right)} \qquad (25e)$$

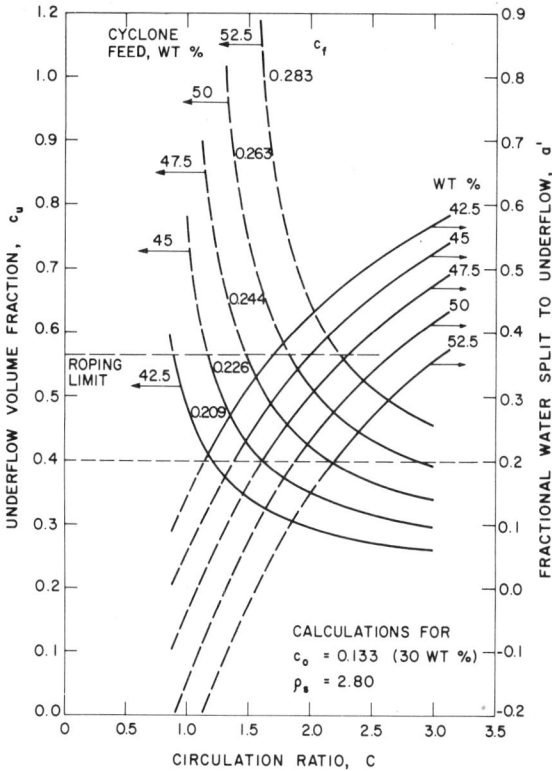

Fig. 14. Mass balances around hydrocyclone for overflow slurry density of 30 wt %, for a solid of specific gravity 2.80.

Fig. 14 gives an example of the mass balance relations, and also indicates the Mular roping limit. In practice, a water split to underflow (equal to solid bypass a) of less than 0.2 is not likely to be achievable under normal operating conditions, so the solid lines in the figure define the feasible operating region.

Other Types of Classifiers

Rake Classifiers

This type of classifier consists of an inclined tank, equipped with an overflow weir and box to collect the overflow product of fine particles and water (see Fig. 1). The coarse particles settle to the bottom and are discharged at the upper end of the tank by the to-and-fro movement of the rakes. After the feed slurry is introduced into the pool, the distance that a particle settles depends upon its hindered settling rate and the length of time it is in the pool. The length of time it stays in the pool is determined by the distance from the inlet to the overflow weir. Thus the feed must be located so that the velocity of the pulp toward the weir, together with the distance, allows sufficient time for the finer particles to be carried out over the weir while the coarser particles settle out below. The coarser particles are agitated and washed as they are conveyed by the rakes, thus reducing the amount of undersize particles carried out in the coarse stream.

The control of liquid and slurry density is important, but not as critical as for hydrocyclones. Increasing the solid concentration (by reducing the water flow) decreases the settling rates but also reduces the velocity toward the weir, giving longer settling times which act to compensate for the reduced settling rates. However too high water flow rate can decrease the classification efficiency because of the reduced time for separation.

Lynch et al. (1967) attempted to develop a model for a Dorr DSF rake classifier, similar to the model developed for the hydrocyclone. However, they concluded that "it would not be possible with the present data to develop a conventional mathematical model of rake classifiers in the sense that the products could be calculated directly from the feed through one series of equations as was done in the case of cyclone classifiers."

Roberts and Fitch (1956) give results for a 1.83 m × 7 m (6 ft × 23 ft) DSF Dorr Classifier operating at 19 strokes/min with a weir depth of about 1 m (40 in.) and a slope of 0.21 m/m. The data were

Size, μm	Cumulative percent less than size		
	Feed	Coarse	Fine
590	86.3	79.5	100.0
420	76.0	64.2	99.7
297	57.7	38.4	96.2
210	43.5	23.3	83.8
149	32.8	14.8	68.7
105	26.7	11.0	58.0
74	22.1	8.4	49.0
Relative mass flow rate	1	0.67	0.33
Percent solids	52	74	32.6

This set of data gives selectivity values of

Size, μm	s_i	c_i
590 × 420	99.0	98.7
420 × 297	93.6	91.5
297 × 210	70.9	61.0
210 × 149	53.0	36.9
149 × 105	41.5	21.6
105 × 74	38.0	16.8
74 × 0	25.3	0

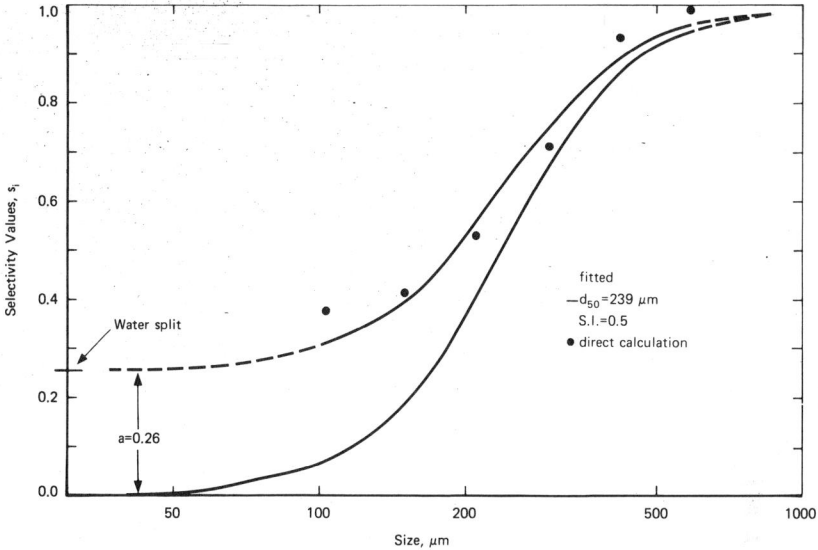

Fig. 15. Selectivity values for a rake classifier.

The split of water to the coarse stream was 25.4%, but the data, although typical of reported data, is not sufficiently complete for accurate analysis for bypass *a*. Consequently, these data were fitted to the logistic function, Eq. 9, using the extrapolation technique presented in the section titled, "Classifier Partition Curves." The characteristic parameters developed were $d_{50} = 240$ μm, S.I. = 0.50, *a* = 0.26, see Fig. 15. A more detailed analysis would require: (a) extended size distribution data since half of the fine product is in the sink interval and (b) specific gravity data, since the unusual shape of the plotted data points could be due either to experimental error or the presence of different specific gravity materials.

Sieve Bends

The sieve bend is a member of the family of profile wire screens for wet screening, which also includes the cross flow screens and the Vor-Siv (Leonard, 1979). The "screen" is a slotted deck made of stainless steel wedge bar oriented at right angles to the direction at which the solid-liquid slurry streams cross the screen. The feed slurry is fed evenly across the entire width of the deck, tangentially to the screen. The full stream of slurry flowing over the sieve bend decreases in depth in increments of about one-quarter of the slot width each time it passes a slot, giving a separation of the feed solids at a size considerably smaller than the opening in the sieve bend: specific gravity of the particles has no influence on the fineness at which they are screened. The curved surface (see Fig. 4) ensures that the slurry layer stays in contact with the screen surface. The slurry of water and fines is collected in the effluent chamber, from where it is piped for further processing; the cake, which contains some portion of the fine particles as "void-filling" material, is discharged over the lip of the sieve bend and returned to the mill. Periodic reversal of the sieve bend is practiced to reduce the rate of wear of the bevel edge of each bar as it slices off the lower layer of slurry.

Data (Leonard and Mitchel, 1968) from a sieve bend preparing minus 65 mesh coal to use as feed to froth flotation cells gave

Size, μm	Cumulative percent less than size		
	Feed	Coarse	Fine
840	91.5	86.0	100.0
590	78.7	64.8	99.8
420	65.5	44.5	96.0
297	53.6	29.4	90.2
210	45.0	20.5	79.5
149	37.4	15.6	68.5
105	31.2	12.1	58.5
74	25.8	9.6	49.8
Relative mass flow rate	1	0.63	0.37
Percent solids	37.7	58.9	23.5

The selectivity values from this set of data are

Size, μm	s_i	c_i
590 × 420	90.0	86.5
420 × 297	81.4	75.0
297 × 210	58.3	44.1
210 × 149	42.8	23.2
149 × 105	37.0	15.5
105 × 74	32.6	9.5
74 × 0	24.5	0

The split of water to the coarse stream was 26.5%, and again this data is not complete enough for accurate analysis for *a*. The data fitted to the logistic function, (Eq. 9), using the extrapolation technique presented in "Classifier Partition Curves," gave characteristic parameters of $d_{50} = 300$ μm, S.I. = 0.46, *a* = 0.25 (see Fig. 16).

Lynch et al. (1977) have developed an approximate expression for predicting the d_{50} of a sieve bend based upon operating data:

$$\log[d_{50}] = K_0 + K_1(\text{water underflow rate}) + K_2(\text{wt \% solids})$$

$$+ K_3\log[\text{slot width}] \qquad (26)$$

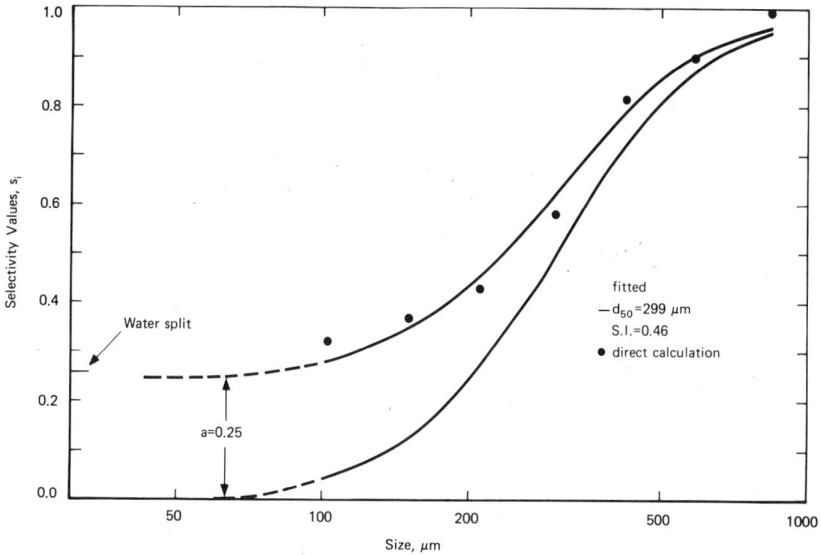

Fig. 16. Selectivity values for a sieve bend.

where $K_0 = 2.45$, $K_1 = 0.001372$, $K_2 = 0.0029$, and $K_3 = 1.1718$ for widths in millimeters and flow rates in metric tons per hour. The water underflow rate is calculated from the approximate expression

$$\text{water underflow rate} = K_1(\text{slot width}) - K_2(\text{wt \% solids})$$

$$+ K_3(\text{slot width})^{1/3}(\text{water feed rate}) \quad (27)$$

where $K_1 = 2$, $K_2 = 0.06$, and $K_3 = 0.98$ for widths in millimeters and flow rates in tons per hour. The **a** value can be estimated as the water split,

$$a \simeq \frac{(\text{water feed rate}) - (\text{water under flow rate})}{(\text{water feed rate})}$$

Lynch found that a λ value of 4.0 in Eq. 8 was appropriate for sieve bends.

Vibrating Screens
There are several different types of screens: fixed screens called grizzlies that protect machinery from oversize lumps; rotating screens called trommels; and shaking screens that are used to size, dewater and transport particles. However, the most common type of classifying screen is the vibrating screen.

The screen area required to separate a stream of particles on a vibrating screen is determined from a basic screen capacity given by the manufacturer of the screen. Depending upon the particular manufacturer, either the total feed rate to the screen, the feed rate to the screen of material larger than the screen aperture, or the feed rate to the screen of material smaller than the screen aperture is divided by the basic screen capacity, which is obtained from an empirical plot of tph/unit area vs. the nominal (square) aperture size, applicable for a material of bulk density of 1600 kg/m³ (100 lb/ft³). Then this screen area is corrected to allow for: (1) differences in bulk density; (2) quantity of fine particles in the feed; (3) geometry of the aperture; (4) position of the screen in a multiple deck design; (5) open area of the screen deck; (6) whether the screening is wet or dry; (7) efficiency of the screening. After the screening area is determined, the width is established by the depth of oversize material at the discharge end of the screen, which must be less than four times the nominal aperture size. The length, which determines the efficiency of the sizing, must be at least twice the width.

The size distributions of the oversize and undersize streams are not established by this procedure. The *screening efficiency*, which does not usually exceed 95% in industrial practice, is defined as the undersize stream recovery evaluated at the screen size. It can be readily shown that this definition is consistent with a partially ideal size selectivity curve with bypass *a*.

$$s_i = \begin{cases} 1, & \text{size} \geq \text{nominal size} \\ a = 1 - \text{screening efficiency}, & \text{size} < \text{nominal size} \end{cases} \quad (28)$$

Data (Batterham et al., 1980) for dry screening on a 20 mm (0.75 in.) aperture vibrating screen gave

Size, mm	Cumulative percent less than size		
	Feed	Coarse	Fine
65	100.0	100.0	100.0
46	96.7	89.9	100.0
32.5	89.2	67.3	100.0
23	79.4	37.5	100.0
16.25	66.8	6.0	96.7
11.5	51.7	1.3	76.4
8	43.3	1.2	63.9
5.75	35.4	1.1	52.2
4	31.5	0.9	46.5
3	27.4	0.85	40.5
2	24.0	0.8	35.3
Relative mass flow rate	1	0.33	0.67

The selectivity values for this data are:

Size, mm	s_i	c_i
65 × 46	100.0	100.0
46 × 32.5	100.0	100.0
32.5 × 23	100.0	100.0
23 × 16.25	82.2	82.1
16.25 × 11.5	10.2	9.75
11.5 × 8	0.5	0.0
8 × 5.25	0.5	0.0
5.25 × 4	1.4	0
4 × 3	0.6	0
3 × 2	0.3	0
2 × 0	1.1	0

The size selectivity values indicate that the screen is relatively efficient, since the apparent bypass is about 0.5%. The higher value for the sink interval would indicate that there were fines clinging to the larger particles when the coarse stream was sampled. The steep shape of the classification curve shown in Fig. 17 is typical of coarse vibrating screens.

However, as the screen size becomes smaller, the classification curve becomes flatter and the bypass increases. It becomes more and more difficult to prevent screens from *blinding*, which is the term for blocking of

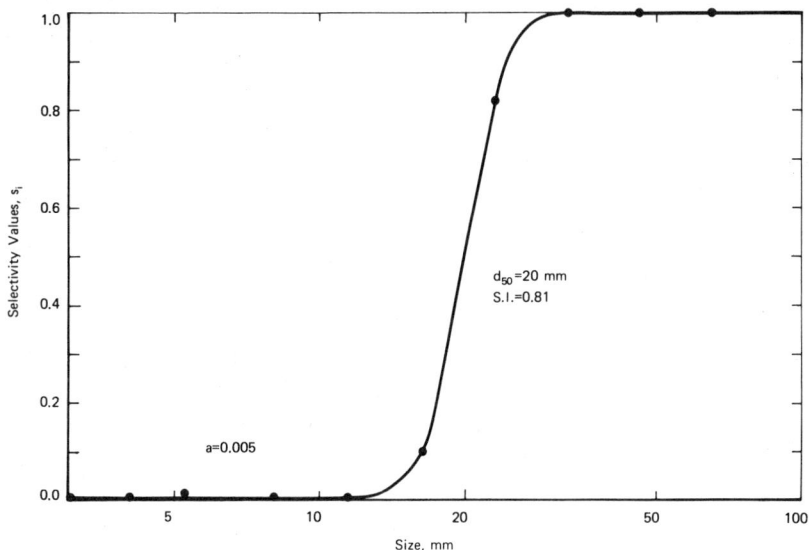

Fig. 17. Selectivity values for a 20 mm aperture vibrating screen.

the holes by near-size particles which become locked in the hole. For example, variability in the feed material may introduce material which does not screen well due to particle shape. As a general rule, large-scale screening below 4 mesh requires careful analysis and testing. It appears that a systematic presentation of partition curves for various screen types and materials as a function of screen loading is not currently available.

Mechanical Air Separators

A mechanical air separator is constructed with an inner shell and an outer shell (see Fig. 3). Material fed to a rotating plate is dispersed within the inner shell and air, pulled in from the outer shell, passes upward through the descending curtain of dispersed feed, elutriating intermediate and fine material out of the feed. The elutriated particles enter a section that contains rotating blades, which separate the intermediate material from the fine particles, returning the intermediate material to the coarse stream. The fine particles exit from the inner chamber into the outer shell suspended in the airstream, passing through the turbine of the blower which maintains the forced circulation. The rotation of particulates and air in the outer shell helps to produce a separation between the fluid and the solid, the solids discharge, and the air reenters the inner shell.

Data for a 4.5 m (15 ft) diameter mechanical air separator classifying cement clinker gave:

Size, μm	Cumulative percent less than size		
	Feed	Coarse	Fine
125	100.0	100.0	100.0
88	86.5	81.3	99.3
63	78.8	68.7	97.6
44	55.3	36.6	84.4
31.5	50.0	32.2	80.3
22	36.2	21.6	58.3
15.5	27.3	15.7	42.2
11	23.3	13.1	36.5
7.75	16.7	9.7	24.7
5.5	13.0	8.2	20.1
4	11.8	7.0	16.7
3	10.8	6.3	14.5
2	8.9	5.4	11.7
1.5	7.1	4.4	8.7
Relative mass flow rate	1	0.63	0.37

The selectivity values for this data are

Size, μm	s_i	c_i
125 × 88	97.8	96.6
88 × 63	92.7	88.7
63 × 44	80.5	69.8
44 × 31.5	64.3	44.6
31.5 × 22	45.2	15.0
22 × 15.5	38.4	4.5
15.5 × 11	43.8	12.9
11 × 7.75	32.7	0
7.75 × 5.5	36.9	2.2
5.5 × 4	35.8	0.5
4 × 3	35.0	0
3 × 2	35.6	0.15
2 × 1.5	36.0	0.8
1.5 × 0	46.4	16.9

Examination of this data would indicate that the *a* value should be around 35.5%. The high selectivity value for the very finest material might be attributed to those fine particles sticking to the larger particles. However, analysis by Luckie and Austin (1975) of this type of classifier, in which the classifying air is recirculated, indicated that a *fishhook* selectivity curve might be expected because the outer shell does not give complete separation of solid from gas, and material is hence returned to the main classification

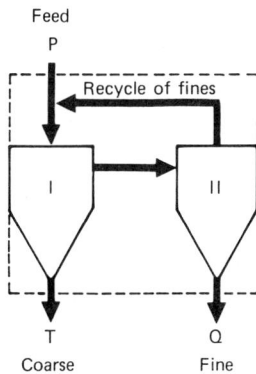

Fig. 18. Proposed model of classification action for mechanical air separator.

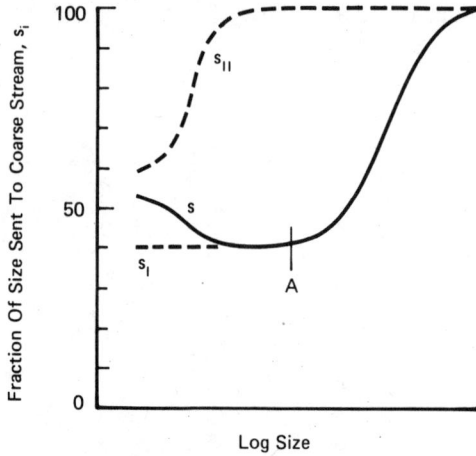

Fig. 19. Illustration of fishhook selectivity curve based on two classification actions in series as shown in Fig. 18.

action. When the classification action is analyzed as this special two-stage classifier configuration (see Fig. 18), the resulting overall size selectivity curve takes on a fishhook shape as depicted in Fig. 19.

Another data set for the classification (Austin and Luckie, 1976) of cement clinker demonstrates this behavior more clearly:

Size, μm	Feed	Coarse	Fine
177	98.9	98.5	100.0
125	92.5	89.9	100.0
88	78.4	71.1	100.0
63	62.1	49.3	99.8
44	51.9	36.1	98.4
31.5	36.9	20.5	84.8
22	27.5	14.2	66.6
15.7	19.9	10.0	48.9
11	14.0	7.6	32.9
7.75	9.4	5.4	21.3
5.5	5.0	3.0	12.5
4	2.9	1.8	7.5
Relative mass flow rate	1	0.75	0.25

The size selectivity values calculated from this data are

Size, μm	s_i
177 × 125	100.0
125 × 88	100.0
88 × 63	100.0
63 × 44	97.0
44 × 31.5	78.0
31.5 × 22	50.3
22 × 15.5	41.4
15.5 × 11	30.5
11 × 7.75	35.9
7.75 × 5.5	40.0
5.5 × 4	42.8
4 × 0	46.5

The fishhook pattern is quite evident. It should be noted that the estimation procedure of the section, "Classifier Partition Curves" does not work for this type of data because the overall $s(x_i)$ function is not of the simple three parameter form.

The variation of selectivity values with operating conditions is quite complex: the most detailed analysis available is of data from a laboratory separator of 1 m (3 ft) diameter (Austin and Luckie, 1976) and the applicability of the quantitative results to industrial separators is questionable.

Two-Stage Classification

Two-stage classification is the process of combining two classifiers in a series arrangement in order to improve the overall classification. Since the arrangements result in the process still producing a fine and a coarse stream from a feed stream, the two stages can be modeled as an overall single set of selectivity values if the selectivity values of both classifiers are known for the conditions of the series configuration.

For example, the arrangement shown in Fig. 20a reclassifies the coarse stream from the first stage and combines both fine streams. In principle, this arrangement removes additional fine material from the coarse stream in order to improve the recovery of product. This arrangement can be used when the combined fine streams give a stream that meets the product specification, which is not always possible. The overall selectivity value is readily derived as

$$s(x_i) = s_1(x_i)s_2(x_i) \tag{29}$$

Another arrangement, shown in Fig. 20b, reclassifies the fine stream from the first stage and combines both coarse streams. This arrangement is used when the feed stream contains a relatively small quantity of potential fine product. A rule-of-thumb of classification is that to produce a powder containing 95% finer than a particular size, the feed stream should contain at least 50% finer than that same size. However, with this two-stage arrangement, a lower percentage in the feed can be tolerated. The overall selectivity value can be calculated from

$$[1 - s(x_i)] = [1 - s_1(x_i)][1 - s_2(x_i)] \qquad (30)$$

Two other arrangements involve feeding one of the streams from the second stage back into the feed to the first stage. These arrangements are used to decrease product loss while producing a fine product meeting specification. For example, in the arrangement depicted in Fig. 20c, the coarse stream from the first stage is reclassified and the fine stream from the second stage returned as feed to the first stage. In this case, the overall selectivity value is given by

$$[1 - s(x_i)] = \frac{1}{1 + \dfrac{s_1(x_i)s_2(x_i)}{[1 - s_1(x_i)]}} \qquad (31)$$

In the arrangement depicted in Fig. 20d, the fine stream from the first stage is reclassified and the coarse stream from the second stage returned as feed to the first stage. In this case, the overall selectivity value is given by

$$s(x_i) = \frac{s_2(x_i)}{1 - s_2(x_i)[1 - s_1(x_i)]} \qquad (32)$$

The special two-stage arrangement for classifiers in which recirculation of the conveying fluid is required, as shown in Fig. 18, has been discussed

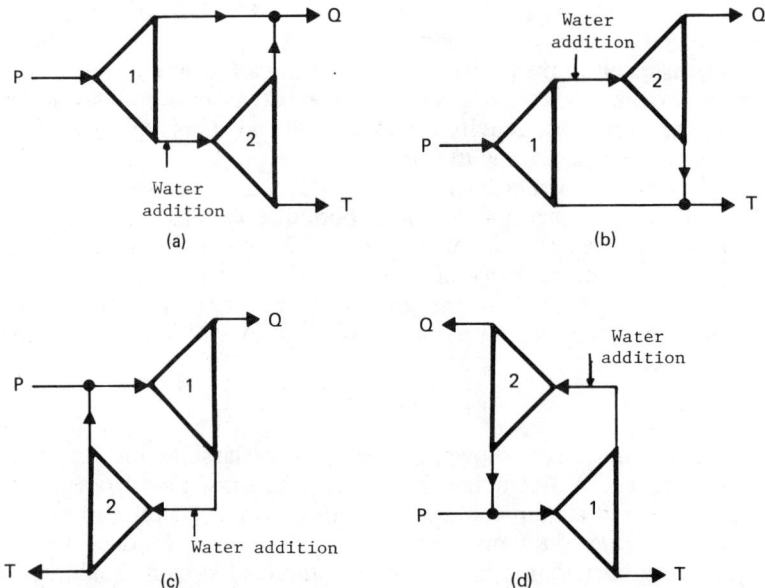

Fig. 20. Two-stage classification arrangements.

already (see "Mechanical Air Separators"). The overall selectivity value is given by

$$s(x_i) = \frac{s_1(x_i)}{1 - [1 - s_1(x_i)][1 - s_2(x_i)]} \tag{33}$$

The mass balances around each cyclone must satisfy Eq. 25, with the water addition between the cyclones as a design variable. For illustrative purposes we will consider the case where water addition is used to make the values of $s_1(x_i)$ and $s_2(x_i)$ the same. Eqs. 29 to 32 then go to simpler forms and letting $s_1(x_i) = s_2(x_i) = a_1$ for smaller x_i gives the overall apparent bypass a at fine sizes, as shown in Table 3. An overall classification function $c(x_i)$ can then be defined by $s(x_i) = (1 - a)c(x_i) + a$, and substituting for a gives the equations for $s(x_i)$ shown in Table 3. Setting $c(x_i) = 0.5$ gives the $s(d_{50})$ values for each arrangement.

The size, d_{50}, at which this occurs is obtained by rearranging Eqs. 29 to 32 to give the inverse functions (see Table 3)

$$s_1(x_i) = g(s(x_i))$$

hence,

$$c_1(d_{50}) = \frac{g(s(d_{50})) - a_1}{1 - a_1} = \text{known}$$

therefore d_{50} follows. For example, if Eq. 9 is used for $c_1(x_i)$,

$$x_i/d_{50_1} = \left[\frac{g(s(x_i)) - a_1}{1 - g(s(x_i))} \right]^{-\ln(S.I.)/2.197}$$

and $x_i = d_{50}$ when $s(x_i) = s(d_{50})$. Similarly, expressions can be obtained for d_{75}, d_{25} for $c(x_i) = 0.75$ and 0.25, respectively, and the overall S.I. value calculated.

For example, if the individual classifier selectivity curve is taken as Eq. 9 with characteristic parameters of $a_1 = 0.3$, $(S.I.)_1 = 0.6$ and $d_{50_1} = 100\ \mu m$, then the overall selectivity curve parameters are:

Arrangements of Fig. 20

		a	b	c	d
a	=	0.09	0.51	0.11	0.38
$d_{50}, \mu m$	=	113	81.5	101	83
S.I.	=	0.62	0.65	0.64	0.67

In all four arrangements, the S.I. values have increased, one of the advantages of such arrangements. However, the other characteristic parameters vary. It would appear that arrangement 20c is better than 20a, particularly if the individual classifiers have high apparent bypass values. Similarly, arrangement 20d appears to be better than 20b.

Table 3. Functional Forms for Calculating the Characteristic Parameters for the Overall Selectivity Curves of a Two-Stage Classifying Arrangement

Arrangements of Fig. 20

	a	b	c	d
$a =$	a_1^2	$2a_1 - a_1^2$	$\dfrac{a_1^2}{1 - a_1 + a_1^2}$	$\dfrac{a_1}{1 - a_1 + a_1^2}$
$s(x_i) =$	$(1 - a_1^2)c(x_i) + a_1^2$	$(1 - a_1)^2(1 - c(x_i)) + 1$	$\dfrac{(1 - a_1)^2 c(x_i) + a_1^2}{1 - a_1 + a_1^2}$	$\dfrac{(1 - a_1)^2 c(x_i) + a_1}{1 - a_1 + a_1^2}$
$g(s(x_i)) =$	$\sqrt{s(x_i)}$	$1 - \sqrt{1 - s(x_i)}$	$\dfrac{\sqrt{s(x_i)(4 - 3s(x_i))} - s(x_i)}{2(1 - s(x_i))}$	$\dfrac{1 + s(x_i) - \sqrt{(1 - s(x_i))(1 + 3s(x_i))}}{2s(x_i)}$

The work of Rogers et al. (1981) can be examined as a more detailed example of the analysis of two-stage classifier circuits. The system tested was wet grinding of limestone in a 0.9 m diameter by 1.54 m long (2.9 by 5 ft) pilot scale mill, closed with Krebs (Arterburn, 1982) cyclones of 76 mm diameter. Separate tests on the cyclone gave the empirical relation

$$d_{50} \propto \frac{(\text{vortex finder diam})^{0.52}\exp(0.08c_f)}{(\text{volumetric feed rate})^{1.3}(\text{apex diam})^{0.35}} \tag{34}$$

where c_f is the fraction by volume of solids in the cyclone feed. The fractional water split to the underflow was found experimentally to be

$$a' = \frac{0.25(\text{apex diam})^{1.66}(0.006P_{75} + 0.55)}{(\text{vortex finder diam})^2} \tag{35}$$

where P_{75} is the percentage minus 75 μm in the cyclone feed. It was found that the apparent bypass could be approximated by $a = a'$. (Note that the occurrence of P_{75} in the calculation of a values means that the overall classification is not truly first order.) The sharpness index was S.I. = 0.58.

Eqs. 29 to 32 enable the calculation of the water splits for two of these cyclones staged together in any of the arrangements, by writing appropriate water mass balances for a given entering flow rate and feed size distribution. Instead of testing the cyclone stages with a given feed size distribution as in the example above, Rogers et al. (1981) recycled the coarse material to the mill feed and adjusted makeup feed rate to give a desired circuit product. Experimental results were compared to a complete mill/two-stage cyclone model, and Eqs. 29 to 32 plus mass balance equations formed the two-stage classification models. Good agreement was found between experiment and model results, thus validating the model, e.g.:

Hydrocyclone dimensions	Apex diam, mm	Vortex finder diam, mm
Primary	15.9	19.1
Secondary	15.9	19.1
	Measured	Simulated
Product rate (kg/h)	765	761
Recycle rate (kg/h)	1575	1581
Product wt % solids	18.4	20.0
Recycle wt % solids	72.0	73.5
Percent passing stated size in the circuit product:		
Size, μm		
297	100.00	100.00
149	99.7	99.96
74	89.6	91.24
53	71.2	71.74

The model was then used to compare the capacities of the mill-cyclone arrangements for the production of desired product specifications. Fig. 21 shows their results, for arrangements a and c. As expected, two stage arrangements a and c gave higher output than a single stage, for a one-point product specification. Arrangement c was slightly better than a, but more important, the circuit product of arrangement c gave a substantially higher percent solids content in the circuit product. Of course, the size distributions were somewhat steeper (less fines) for the a and c arrangements.

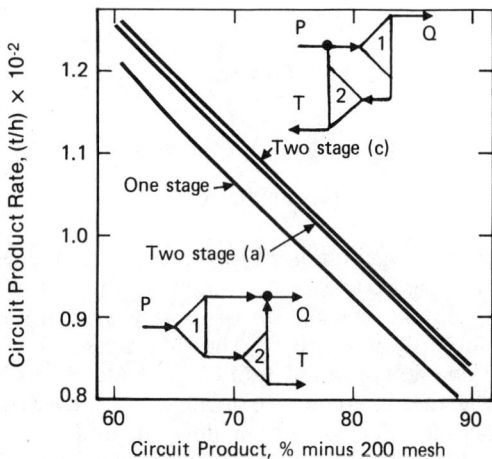

(a)

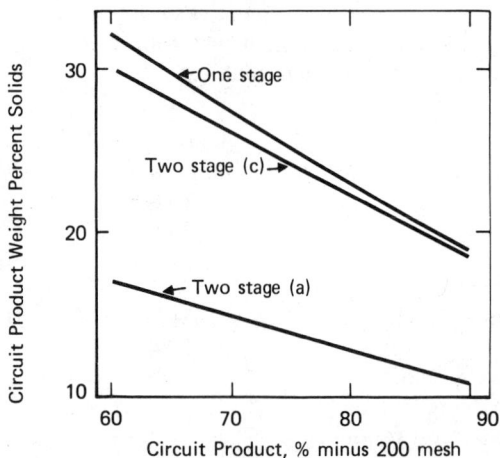

(b)

Fig. 21. Circuit product (a) rates (b) percent solids (by weight) as a function of two-stage classification and circuit product fineness.

These examples show the utility of detailed analysis (via simulation) of two-stage classification before the installation of a trial arrangement.

Summary

The advantages of closed-circuit grinding are that efficient classification and high recycle rates: (1) reduce overgrinding of fines and reduce slowing-down effects due to accumulation of fines; (2) give more stable operating conditions, especially for grate discharge mills; and (3) give added flexibility in mill control.

Even for particles of identical size and physical properties, it is predicted that there will be a statistical dispersion in a classifier such that some of them will go to the coarse particle stream and some to the fine particle stream. In addition, an assembly of real particles will have a variety of shapes and, in ore systems, a range of specific gravities, which will affect the trajectories through the classifier. The classifiers frequently operate with dense suspensions, and there is no satisfactory theory of particle-fluid behavior in such suspensions. It is difficult to treat unsteady state pulsing behavior. Thus, theoretical models cannot at present be used for quantitative predictive purposes.

Mill circuit simulations require a knowledge of classifier selectivity values, and how they vary with changing conditions. However, this type of information is not generally available, even for hydrocyclones. Thus emphasis is laid on the methodology of obtaining such data, and the methodology is illustrated in the hydrocyclone section.

There are two general categories of classifiers, fluid dynamic classifiers and screen classifiers. The first category includes spiral and rake classifiers, cyclones, and vane classifiers. Spiral and rake classifiers keep fines suspended by mechanical stirring actions while larger particles settle and are transported away from the rest of the slurry. The cyclone is a simple device with a cylindrical-conical shape, where tangential entry of a fluid stream establishes a rotating path. This throws larger particles to the wall by centrifugal action, where they flow to the underflow exit. Finer particles stay suspended due to proportionally higher drag forces, and are carried out in a fluid stream (overflow) leaving at the top of the cyclone. In vane classifiers, the vanes change the gas direction and larger particles have a greater chance of striking a vane and falling out of the gas due to their greater linear momentum. The two major types of screen classifiers are vibrating screens where particles in a thin layer are repeatedly presented to screen holes by a jogging action (dry or wet), and the sieve bend type of screen used on wet slurries. In this second type, the slurry flows over a curved section of plate with slots and smaller particles pass through the slots with some fraction of the water.

Analysis of the error structure of several sets of hydrocyclone data showed that a satisfactory equation for calculating circulation ratio from the three size distributions around the classifier (feed = p_i, underflow = t_i, overflow = q_i) was

$$C = \frac{\sum_i |p_i - q_i|}{\sum_i |t_i - p_i|} \tag{S1}$$

In the absence of detailed data this can be used in the form

$$C = (P_{i*} - Q_{i*})/(T_{i'} - P_{i'})$$
(S1a)

where $i*$ and i' are the size intervals where $p_i - q_i$ and $t_i - p_i$, respectively, change sign. The selectivity numbers for $\sqrt{2}$ size intervals, s_i, are calculated from

$$s_i = \left(\frac{C}{1 + C}\right)\left(\frac{t_i}{p_i'}\right)$$
(S2)

where p_i' is the reconstituted feed,

$$p_i' = \frac{q_i}{1 + C} + \frac{Ct_i}{1 + C}$$
(S3)

A plot of s_i vs. size is called a *Tromp, partition* or *selectivity* curve.

Real classifiers are non-ideal, and behave as if a fraction *a* of all feed sizes bypass to the coarse stream. If it is assumed that a fraction $1 - a$ is subject to the actual classification action, then a set of *classification* numbers c_i are defined by

$$s_i = a + (1 - a)c_i$$
(S4)

These numbers plotted against particle size enable d_{50} to be obtained, where $c(d_{50}) = 0.5$. The *Sharpness Index* is defined as d_{25}/d_{75} and ranges typically from 0.4 to 0.7. The curve of $c(x/d_{50})$ vs. x/d_{50} is often approximately constant for different operating conditions or classifier sizes, and is called the *reduced classification* curve.

The four most usual functions for fitting this curve are

1. $$c(x_i) = 1 - \exp\left[-(x_i/x_o)^\lambda\right]$$
(S5)

where $$d_{50} = x_o(0.693)^{1/\lambda}$$
(S5a)

and $$S.I. = \exp[-1.5725/\lambda]$$
(S5b)

or $$c(x_i) = 1 - 2^{-\exp[-1.5725\log[x_i/d_{50}]/\log[S.I.]]}$$
(S5c)

2. $$c(x_i) = \frac{1}{\sqrt{2\pi}} \int_{-\infty}^{(\ln[x_i] - \ln[d_{50}]/\lambda)} \exp\left[-u^2/2\right]\, du$$
(S6)

where $$S.I. = \exp[-1.349\lambda]$$
(S6a)

or $$c(x_i) = \frac{1}{\sqrt{2\pi}} \int_{-\infty}^{(-1.349\log[x_i/d_{50}]/\log[S.I.])} \exp\left[-u^2/2\right]\, du$$
(S6b)

3.
$$c(x_i) = \frac{\exp[\lambda(x_i/d_{50})] - 1}{\exp[\lambda(x_i/d_{50})] + \exp[\lambda] - 2} \qquad \text{(S7)}$$

where
$$\text{S.I.} = \frac{\ln[(\exp[\lambda] + 2)/3]}{\ln[3\exp[\lambda] - 2]} \qquad \text{(S7a)}$$

4.
$$c(x_i) = \frac{1}{1 + (x_i/d_{50})^{-\lambda}} \qquad \text{(S8)}$$

where
$$\text{S.I.} = \exp[-2.1972/\lambda] \qquad \text{(S8a)}$$

or
$$c(x_i) = \frac{1}{1 + \exp[-2.1972\log[x_i/d_{50}]/\log[\text{S.I.}]]} \qquad \text{(S8b)}$$

It is sometimes difficult to estimate a from data because the data does not extend to fine enough size distributions to give $s_n = a$. In this case a technique is available based on extrapolation of the feed to the classifier as a Schuhmann distribution; the technique uses the information contained in the P_n, T_n values which would otherwise not be used.

Hydrocyclones operate properly with flow rates between a minimum and a maximum, due to the fluid dynamics of the devices. The semitheoretical operating-design equations for hydrocyclones dealing with dilute slurries in this correct operating range are

$$\Delta P = (\text{constant})(Q/D_c^2)^2 \qquad \text{(S9)}$$

$$d_{50} = (\text{constant})(D_i^2/D_c^{0.5})\left[\frac{\eta(1 - R_f)}{Q(\rho_s - \rho_l)}\tan(\Theta/2)\right]^{1/2} \qquad \text{(S10)}$$

Unfortunately these equations cannot be used for hydrocyclones operating on dense slurries because of the complex rheological behavior of such slurries.

For dense slurries there are qualitative rules for the variation of d_{50} with the size and operating conditions of a hydrocyclone. Smaller diameter cyclones give smaller d_{50} values. Increased flow rate gives decreased d_{50} values. Reduced classification curves are the same for geometrically similar cyclones. Mixtures of minerals do not behave as the weighted sum of each mineral tested alone in the same cyclone. An increase in vortex finder diameter increases d_{50}; a decrease in apex diameter increases d_{50}; an increase of inlet diameter increases d_{50}. The effects of slurry density, particle size distribution, and slurry viscosity are complex and not fully explored to date. Lowering viscosity with a chemical fluidity modifier increases the sharpness of classification and increases d_{50} also. It is generally assumed that the water split is equal to the bypass a, but this is not true for all cases.

Three forms are given for the variation of d_{50} with hydrocyclone dimensions and conditions:

Lynch:

$$\log d_{50} = K_1(\text{vortex finder diam}) - K_2(\text{apex diam})$$
$$+ K_3(\text{inlet diam}) + K_4(\text{percent solids by weight in feed pulp})$$
$$- K_5(\text{volume flow rate of feed pulp}) + K_6 \qquad \text{(S11)}$$

Plitt:

$$d_{50} = K_0 \frac{(\text{cyclone diam})^{K_1}(\text{inlet diam})^{K_2}(\text{vortex finder diam})^{K_3}}{(\text{apex diam})^{K_5}(\text{free vortex height})^{K_6}(\text{volume flow rate})^{K_7}}$$
$$\times \frac{\exp\left[K_4(\text{volume percent solids})\right]}{(\rho_s - \rho_l)^{0.5}} \qquad \text{(S12)}$$

Arterburn:

$$d_{50} = K_0 \frac{(\text{cyclone diam})^{K_1}}{(\text{pressure drop})^{K_2}(\rho_s - \rho_l)^{0.5}\left[1-1.9(\text{volume fraction solid})\right]^{K_3}} \qquad \text{(S13)}$$

Equations are also given for the fractional water split to the underflow, and these can be used to estimate apparent bypass **a**. The values of λ in equations S5 to 8 are to be determined for each material. The mass balance around a hydrocyclone gives

$$c_u = \frac{1}{1 + a'\left(\dfrac{1+C}{C}\right)\left(\dfrac{1-c_f}{d_f}\right)} \qquad \text{(S14a)}$$

$$c_o = \frac{1}{1 + (1-a')(1+C)\left(\dfrac{1-c_f}{c_f}\right)} \qquad \text{(S14b)}$$

The Mular roping limit is

$$c_u \leq 0.5385 c_o + 0.4931 \qquad \text{(S15)}$$

Typical selectivity values are given for rake classifiers and sieve bends, but quantitative models for the variation with operating conditions are not available. Selectivity values for vibrating screens with large apertures show sharp classification with small bypass, but again quantitative models do not exist for smaller screen apertures where classification can be very inefficient.

Mechanical (air) separators used on cement or pulverized coal frequently show a *fishhook* in the selectivity curve, with higher fractions of very fine sizes in the coarse stream. This is possibly due to very fine material being carried into the coarse stream by adherence to larger particles. However, in those systems where air is recirculated, the recirculation behaves like a two-stage classification carrying fine powder back to the feed to the main classifier. This can also give a fishhook in the selectivity curve. Again, quantitative rules for the variation of selectivity values with operating conditions are not generally available for mechanical classifiers.

Four two-stage classification arrangements are shown in Fig. 20. The equations for the overall classification action of these arrangements are:

a) $$s(x_i) = s_1(x_i)s_2(x_i) \tag{S16}$$

b) $$1 - s(x_i) = \left[1 - s_1(x_i)\right]\left[1 - s_2(x_i)\right] \tag{S17}$$

c) $$1 - s(x_i) = 1 \Big/ \left[1 + \frac{s_1(x_i)s_2(x_i)}{1 - s_1(x_i)}\right] \tag{S18}$$

d) $$s(x_i) = \frac{s_2(x_i)}{1 - \left[1 - s_1(x_i)\right]\left[1 - s_2(x_i)\right]} \tag{S19}$$

In principle, these enable the overall advantage of these arrangements to be calculated, but the values of $s_2(x_i)$ will depend on the feed conditions to the second cyclone, which depend on water addition.

To estimate the effect of two-stage classification on a mill circuit, it is thus necessary to have a combined milling-water split-classification model. One example of this type of analysis using hydrocyclones shows that arrangement c gives substantially increased output over a single stage, with only a slight reduction in solids content of the circuit product; arrangement a also gave increased output but at the expense of lower solids content.

References

Arterburn, R. A., 1982, "The Sizing and Selection of Hydrocyclones," *Design and Installation of Comminution Circuits*, A. L. Mular and G. Jergensen, eds., AIME, New York, pp. 592–607.

Aso, K., 1957, "On the Theory of Partition Curve and Its Application to Coal Preparation or Mineral Dressing," *Memoirs of the Faculty of Engineering, Kyushu Imperial University*, Vol. 17, pp. 18–83.

Austin, L. G. and Luckie, P. T., 1976, "An Empirical Model for Air Separator Data," *Zement-Kalk-Gips*, Vol. 29, pp. 452–457.

Austin, L. G. and Klimpel, R. R., 1981, "An Improved Method for Analyzing Classifier Data," *Powder Technology*, Vol. 29, pp. 277–281.

Batterham, R. J., et al., 1980, "Screen Performance and Modelling with Special Reference to Iron Ore Crushing Plants," *Particle Technology*, Amsterdam.

Blau, G., Klimpel, R. R., and Steiner, E., 1972, "Nonlinear Parameters Estimation and Model Distinguishability of Physiochemical Models at Chemical Equilibrium," *Canadian Journal of Chemical Engineering*, Vol. 50, pp. 399–410.

Bradley, D., 1965, *The Hydrocyclone*, Pergammon Press, London.

Dahlstrom, D., 1949, "Cyclone Operating Factors and Capacities in Coal and Refuse Slimes," *Trans. AIME*, Vol. 184, pp. 331–344.

Dekok, S., 1975, "The Design, Operation and Performance of Cyclone Classifiers in Grinding," *Proceedings of Winter School of South African Institute of Mining and Metallurgy*, Johannesburg.

Draper, N. and Smith, H., 1966, *Applied Regression Analysis*, John Wiley and Sons, Inc., New York.

Finney, D., 1964, *Probit Analysis*, University Press, Cambridge.

Kelsall, D. F., 1953, "A Further Study of the Hydraulic Cyclone," *Chemical Engineering Science*, Vol. 2, pp. 254–272.

Klimpel, R. R., 1980, "Estimation of Weight Ratios Given Component Make-up Analyses of Streams," *Trans. SME-AIME*, Vol. 266, pp. 1882–1886.

Klimpel, R. R., 1982, "The Influence of a Chemical Dispersant on the Sizing Performance of a 24-inch Hydrocyclone," *Powder Technology*, Vol. 31, pp. 255–262.

Leonard, J. W., ed., 1979, *Coal Preparation*, 4th ed., AIME, New York.

Leonard, J. W. and Mitchel, D. R., eds., 1968, *Coal Preparation*, 3rd ed., AIME, New York.

Luckie, P. T. and Austin, L. G., 1974, "Technique for Derivation of Selectivity Functions from Experimental Data," *Proceedings*, Tenth International Minerals Processing Congress, M. J. Jones, ed., Institution of Mining and Metallurgy, London, pp. 773–790.

Luckie, P. T. and Austin, L. G., 1975, "Mathematical Analysis of Mechanical Air Separator Selectivity Curves," *Trans. IMM*, Vol. 84, pp. C253–C255.

Lynch, A. J. and Rao, T., 1975, "Modelling and Scale-up of Hydrocyclone Classifiers," *Proceedings*, Eleventh International Minerals Processing Congress, Universita di Cagliari, Italy, pp. 245–269.

Lynch, A. J., et al., 1967, "An Analysis of the Performance of a Ball Mill-Rake Classifier Comminution Circuit," Australian Institution of Mining and Metallurgy, pp. 9–18.

Lynch, A. J., et al., 1977, Chaps. 5 and 6, *Mineral Crushing and Grinding Circuits*, Elsevier Scientific Publishing Co., New York, pp. 87–126.

Molerus, O., 1967, "Stochastisches Model der Gleichgewichssicktung," *Chemie-Ingenieur-Technik*, Vol. 39, pp. 792–796.

Mular, A. and Jull, N., 1980, "The Selection of Cyclone Classifiers, Pumps and Pump Bases for Grinding Circuits," *Mineral Processing Plant Design*, A. L. Mular and R. B Bhappu, eds., 2nd ed., AIME, New York, pp. 376–403.

Plitt, L., 1971, "The Analysis of Solid-Solid Separation in Classifiers," *CIM Bulletin*, Vol. 64, pp. 42–47.

Plitt, L., 1976, "A Mathematical Model of the Hydrocyclone Classifier in Classification," *CIM Bulletin*, Vol. 69, pp. 114–123.

Reid, K. J., 1971, "Derivation of an Equation for Classifier-Reduced Performance Curves," *Canadian Metallurgical Quarterly*, Vol. 10, pp. 253–254.

Roberts, E. and Fitch, E., 1956, "Predicting Size Distribution in Classifier Products," *Mining Engineering*, Vol. 8, pp. 1113–1118.

Rogers, R. S. C., et al., 1981, "An Evaluation of the Use of Two vs. One Stage of Hydrocyclones in a Pilot Scale Ball Mill Circuit," SME Preprint No. 81–125, SME-AIME Annual Meeting, Chicago, IL.

Wilde, D. and Beightler, C., 1967, *Foundations of Optimization*, Prentice-Hall, Inc., Englewood Cliffs, NJ.

14

Residence Time Distributions and Mass Transfer

Introduction

As discussed in previous chapters, especially Chapter 6, a knowledge of the residence time distribution (RTD) of a mill is essential to constructing an accurate simulation of the grinding behavior of the mill. However, the simulations in Chapters 7 and 10 show that it is possible to get the same grinding performance from a mill with values of RTD varying from plug flow to fully mixed. In Chapter 7, in the section titled, "The Behavior of Different Designs of Mill Circuits," for example, it was shown that producing a two-point matched size distribution (from a given feed) requires almost the same specific grinding energy for a plug-flow RTD as that for a fully mixed RTD, providing the mill is not in the overfilled region. Of course, the circulating load and the classifier settings are different for each RTD considered, with the value of circulation ratio being lower for a residence time distribution closer to the plug-flow limit. Thus, as far as the detailed design of a grinding circuit is concerned, it is only necessary to be able to estimate the RTD of a given mill with sufficient accuracy to get reasonably close to actual operating conditions. The grinding circuit can then be fine-tuned by the operator to obtain the desired steady-state conditions, since some degree of adjustment is always possible with any classifier-circulating load system. It is concluded, then, that it is not necessary to be able to describe the residence time distributions of mills with great accuracy.

A complete solution to the problem of optimal control of a grinding circuit must include a fairly detailed knowledge of the mass transfer laws of movement of material through the grinding device. The RTD of a grinding device is inherently bound up with the way mass moves in the device, and the two problems, mill control and mill design using RTD, are complementary. If the mass transfer laws are known in detail, the variation of RTD with mill conditions would also be known, in the same way as a knowledge of the laminar velocity profile in fluid flowing in a pipe enables the RTD of a laminar tubular reactor to be calculated. Unfortunately, there have been virtually no detailed studies of the mass transfer laws in mills. Consequently, it is not possible at present to calculate from first principles the residence time distributions of grinding devices as a function of design or operating conditions, either for steady-state design or for dynamic mill control models.

In this chapter, then, we are limited to reporting the existing data on residence time distributions from ball mills, and to placing the data in a framework convenient for interpolation and extension to similar mills. It will be clear that much work remains to be done on this area of grinding theory. On the other hand, there is certainly enough empirical information to go ahead with *reasonable* design decisions, at least for tumbling ball mills.

Definitions and Experimental Measurements

A working definition of residence time distribution is as follows: Consider a steady-state feed into a reactor and let some fraction of the feed admitted over a very short time interval Δt be marked with a tracer of some kind. If the exit stream from the reactor is tested for the appearance of traced material, then at time t after the pulse of traced feed entered, a fraction $1 - \Phi(t)$ will still be inside. The curve of $\Phi(t)$ vs. t is the cumulative RTD. Fig. 6 in Chapter 1 shows the limiting values for a plug flow and a fully mixed reactor:

$$\left.\begin{array}{ll} \Phi(t) = 0, & 0 \leq t \leq \tau \\ \Phi(t) = 1, & t \geq \tau \end{array}\right\} \text{ plug flow} \tag{1}$$

$$\Phi(t) = 1 - \exp(-t/\tau), \qquad 0 \leq \tau, \text{ fully mixed} \tag{2}$$

where τ is the mean residence time defined by (amount in the reactor)/(flow rate into the reactor). For ball mills, $\tau = W/F$, W being the mass of powder in the mill and F the feed rate of powder.

This definition is directly applicable when all the powder in the device mixes, disperses, and flows independently of particle size (and shape and density, if particles of different shape and density are present). This appears to be true as a reasonable first approximation for most mechanical grinding machines, since they tend to be efficient mixers as a consequence of the mechanics of the grinding action.

In practice, the experimental measurement of RTDs in a mill usually gives a set of values of $c(t)$, where $c(t)$ is the proportional concentration of tracer in the exit stream at time t, as shown in Fig. 1. A sample of traced material is usually collected over a very short sampling time, and samples are taken once every 15 or 30 sec for example, by simple scoop or bucket sampling. A known weight of each sample is analyzed for the *comparative* content of tracer, and the amount is expressed on a fixed sample weight basis, so that the $c(t)$ scale in Fig. 1 is usually quite arbitrary. Thus, the concentration of tracer is rarely found as "g tracer per g powder" because of the analytical difficulties involved. It is much more likely to be in a form similar to "counts per minute per g powder," to use radiotracing as an example. If the system is at a steady-state mass flow rate of F, the amount of tracer which has left the device at time t after admission is *proportional* to $\int_0^t c(t) F \, dt$. In an open-circuit system there is no recycle of tracer back to the mill feed, and providing samples are taken frequently enough and for long enough times, the values of $c(t)$ enable a smooth curve to be drawn, coming down to $c(t) = 0$ at long times. Then the total amount of tracer which has left the machine is *proportional* to $\int_0^\infty c(t) F \, dt$ and, hence, since F is constant,

$$\Phi(t) = \int_0^t c(\theta) \, d\theta \Big/ \int_0^\infty c(\theta) \, d\theta \qquad (3)$$

The value of infinity in the limit of the integral is, of course, just a sufficient time to give $c(t)$ tending to zero. The integration is performed by any numerical technique consistent with the accuracy of the data; counting of squares under the curve of $c(t)$ vs. t plotted on good quality, fine-ruled graph paper is often sufficient.

As defined above, $\Phi(t)$ is the cumulative RTD, and the differential RTD is defined by $\phi(t) = d\Phi(t)/dt$, that is, fraction per unit time. Here, $\phi(t)$ has the physical meaning that $\phi(t) \, dt$ is the fraction of tracer which leaves the mill between time t and $t + dt$ after admission. A plot of $\phi(t)$ vs. t is

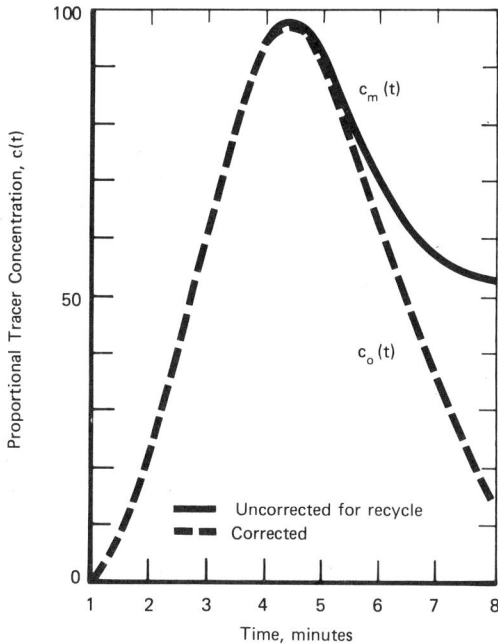

Fig. 1. Tracer concentrations at mill exit at time t after a pulse of tracer addition to mill feed.

similar to that of $c(t)$ vs. t, but the area under the curve is 1. From Eq. 3

$$\frac{d\Phi(t)}{dt} = c(t)/M_o = \phi(t) \tag{4}$$

where $M_o = \int_0^\infty c(t)\, dt$, with units of concentration multiplied by time. It can be shown that the mean residence time τ is related to $\phi(t)$ by

$$\tau = \int_0^\infty t\phi(t)\, dt \tag{5}$$

Thus τ is the first moment of the differential RTD function $\phi(t)$. The variance of the distribution is the second moment about the mean, $\sigma^2 = \int_0^\infty (t - \tau)^2 \phi(t)\, dt$.

Different systems will require different experimental techniques for the measurement of the primary RTD. In wet grinding it is often assumed that the pulp density in the mill is the same as the pulp density entering and leaving the mill, so that the residence time of powder is the same as the residence time of water. The values of RTD can then be determined by adding a soluble tracer to the water, providing the tracer is not adsorbed on the solid in the mill. If the mineral and water being used have a low background electrical conductivity, then salt (NaCl) can be used as a tracer, with its proportional concentration determined using a conductivity cell. Copper sulfate has been used, with the copper ion determined in several different ways. A soluble dye with colorimetric determination can be used. It is better, of course, to trace the solid particles since this avoids the assumptions concerning pulp density and absorption of tracer. The earliest method was the fluorescein tracer technique used for the analysis of dry grinding of cement clinker, as follows (Mardulier and Wightman, 1971).

The mill is tested at steady state, grinding normally, and the total feed rate into the mill is estimated from mass flow measurements or circulating load determination (see Chapter 13). The quantity of fluorescein to be added is about one gram per t/h (0.03 oz. per 1.1 stph) of material and this is made into a solution with about one and one-half times the weight of water. About ten times this weight of clinker is placed in a plastic bag in a container which is then evacuated to about one torr of pressure. The fluorescein solution is admitted onto the evacuated material, the bag sealed and shaken, and the solution is absorbed into the microporous structure of the cement clinker, where the water reacts with the cement leaving the dye as a coating on the internal pores. The bag is thrown into the mill feed, a stop-watch started, and a scoop sample of the fine powder leaving the mill is put in a labelled plastic envelope. The scoop is wiped clean and another sample taken after 30 sec, and so on for 5 min. Samples are taken every minute for another 10 min and every two minutes for another 20 min. The comparative fluorescein content is determined by mixing 2 g (0.07 oz.) of sample with 50 cm^3 (3 cu in.) of water in a beaker, stirring with a glass rod for 30 sec, settling for two minutes, and decanting the supernatant liquid through a dry medium-fast filter paper. The liquid is placed in a Nessler tube and the relative intensity determined visually by comparison with standard solutions. A photofluorometer, which enables the contents to be measured with an accuracy of one part per 10^8 parts of sample, can also be used.

Radiotracing by neutron irradiation of a sample of the solid to make it radioactive, with count rate measurement by Geiger-Muller tube or NaI photomultiplier detectors, has been used successfully on large scale mills (Gardner, Verghese, and Rogers, 1980; Rogers and Gardner, 1979; Gardner, Rogers, and Verghese, 1977). Samples can be collected for counting (see Chapter 9) in a similar fashion to the fluorescein tracer method or a detector placed at the mill exit to pick up emitted radiation from the issuing stream. Rogers and Gardner (1979) have also used a detector in the discharge sump in wet systems, with appropriate allowance for the RTD of the sump itself. It is frequently necessary to correct for decay by counting a blank sample of irradiated solid identical to that added to the mill feed and subtracting the background radiation count. If $Y(t)$ is the fractional count rate (based on the count rate at zero time as 1) of the blank and if c_b is the background count rate

$$c_m(t) = Y(t)c(t) + c_b$$

where $c(t)$ is the desired corrected value and $c_m(t)$ is the measured value. The counting or sampling should be carried out for a total time of four or five mean residence times after admission of tracer, to ensure the tail of the distribution is obtained for accurate calculation of τ from Eq. 5.

Unfortunately, the manipulation of tracer data to obtain RTD is rarely as simple as described above by Eq. 3, because of the recycle of tracer to the feed, via a classifier. In principle, it is possible to take traced samples from the recycle stream at the same time as from the mill exit, but in practice this is often difficult or impossible to do. Even if this complication is not present, the tail of the $c(t)$ curve at long times is often difficult to determine accurately, hence $M_o = \int_0^\infty c(t)\, dt$ cannot be estimated accurately. It is convenient, therefore, to have mathematical models for the RTD which at least enable internal consistency to be obtained for a given set of data. Before these are discussed, however, the procedure is outlined for correcting for recycle to obtain the equivalent of an open-circuit or *primary* RTD.

The Effect of Recycle

When recycle is present, the feed entering the mill between time θ and $\theta + d\theta$ will contain a proportional quantity of tracer $q(\theta)$ in addition to tracer in the makeup feed; $q(\theta)$ is given by $q(\theta) = Tc_R(\theta)\, d\theta$, where $c_R(\theta)$ is proportional concentration of tracer in the recycle (T) measured at the point of mixing into the mill feed. The *fractional* concentration of tracer due to the *primary* differential RTD of a feed pulse is $\phi(t)$; let the measured proportional concentration of tracer at the mill exit (corrected for decay and background) be $c(t)$. Then $c(t)$ will contain not only the contributions from the original test pulse, but also the sum of contributions from recycled tracer $q(\theta)$ admitted at θ which is leaving after a residence of $t - \theta$, summed over all values of θ from zero to t. The fraction of $q(\theta)$ which leaves at time t to $t + dt$ is $\phi(t - \theta)\,dt$. Thus a mass balance on tracer leaving over time t to $t + dt$ gives

$$Fc(t)\,dt = Fc_o(t)\,dt + \phi(t - \theta_1)q(\theta_1)\,dt + \phi(t - \theta_2)q(\theta_2)\,dt + \cdots$$

$$= T\left[\int_{\theta=0}^{t} \phi(t - \theta)c_R(\theta)\,d\theta\right]dt + Fc_o(t)\,dt$$

where $c_o(t)$ is the proportional concentration of tracer due only to the original test pulse. Using the circulation ratio C as defined before $F/T = (1 + C)/C$,

$$c(t) - c_o(t) = \left(\frac{C}{1 + C}\right)\int_{\theta=0}^{t}\phi(t - \theta)c_R(\theta)\,d\theta \qquad (6)$$

Knowing $c(t)$, $c_R(\theta)$, and C, it is possible to estimate $c_o(t)$ and, hence, $\phi(t)$ by deconvoluting the equation (see Appendix 4). Thus, in order to correct for recycle of tracer back to the feed, it is necessary to measure the concentration of tracer in the recycle, using the same basis as the measurement of tracer in the material leaving the grinding machine. It is also necessary to calculate or measure the circulating load (see Appendix 4).

It is implicit in the measurement of RTDs that the grinding device mixes the contents so well that all particles move through the mill at the same speed. In principle, this can be checked by tracing sized fractions to add to the feed. However, the traced fraction grinds finer as it goes through the mill so that the tracer concentration measured at the mill exit is spread over a range of smaller particle sizes. The quantity of tracer recycled will vary with the size traced because this will alter the particle sizes carrying tracer into the classifier. Let the ratio of tracer in the recycled (classifier coarse exit) material to the tracer in the mill exit (classifier feed) material be $r(\theta)$. At times below the mean residence time, the tracer coming out has been in the mill less than the average time and the tracer will be concentrated toward the coarse fractions. On classification, a bigger proportion of the tracer will end up in the tailings, less in the fines. Thus $r(\theta)$ will be greater than one. At time greater than the mean residence time, the tracer is ground finer and $r(\theta)$ is less than one. Adding tracer to different size fractions of the feed will clearly give a different apparent overall RTD than adding it uniformly to the mill feed, because the shift of tracer toward the fines or coarse is changed.

This problem is avoided if traced material can be detected in the mill feed (combined makeup and recycle stream) as well as at the mill exit. The measured tracer in the mill feed is in effect treated as a series of impulses. The total effective input of tracer is $\int_o^\infty c_F(t)\,dt$, where $c_F(t)$ is the proportional concentration of tracer in the mill feed. The fraction of this which enters at time θ to $\theta + d\theta$ is $c_F(\theta)\,d\theta/\int_o^\infty c_F(t)\,dt, = \phi_F(\theta)\,d\theta$, say. The fraction of this which leaves the mill at time t to $t + dt$ is $\phi(t - \theta)\,dt$, so that

$$\phi_P(t) = \int_0^t \phi(t - \theta)\phi_F(\theta)\,d\theta \qquad (6a)$$

where $\phi_p(t) = c(t)/\int_0^\infty c(t)\,dt$; the value of $\phi(t)$ is again obtained by deconvolution.

Rogers, Bell, and Hukki (1982) have developed a method for correcting for recycle which does not involve measurement of recycled tracer concentrations. The method is applicable for any classifier which returns a constant fraction a (the apparent bypass fraction, see Chapter 13) of small sizes to the recycle stream. It consists of using material of sufficiently small particle size as the traced sample, because then a known constant fraction a will be recycled when the traced material leaves the mill after grinding. In the symbolism used above $Tc_R(\theta) = aPc(\theta)$, $c_R(\theta) = a[(1 + C)/C]c(\theta)$ and Eq. 6 becomes

$$c(t) - c_o(t) = a\int_0^t \phi(t - \theta)c(\theta)\,d\theta \tag{6b}$$

from which $c_o(t)$, $\phi(t)$ are obtained by trial-and-error computation.

In order to appreciate the overall effect of recycle, the $c_R(\theta)$ values in Eq. 6 can be considered as arising from once-passed, twice-passed, thrice-passed, etc. tracer. The total returned tracer is then

$$q(\theta) = M_o\left[\left(\frac{C}{1 + C}\right)\phi(t)r_1(t) + \left(\frac{C}{1 + C}\right)^2 r_2(t)\right.$$

$$\times \int_0^t \phi(t - \theta_1 - t_d)\phi(\theta_1)r_1(\theta_1)\,d\theta_1$$

$$+ \left(\frac{C}{1 + C}\right)^3 r_3(t)\left(\int_{t_d}^t \phi(t - \theta_2 - t_d)\phi(\theta_2)r_2(\theta_2)\right.$$

$$\left.\left.\times \int_0^{\theta_2}\phi(\theta_2 - \theta_1 - t_d)\phi(\theta_1)r_1(\theta)\,d\theta_1\,d\theta_2\right) + \cdots\right]dt \tag{7}$$

where t_d is the transport delay time from mill exit to mill feed, and the second term in square brackets on the RHS is only present for $t > t_d$, the third term for $t > 2t_d$, etc. The values of $r_1(\theta)$ are the classifier split ratios for once-passed tracer, $r_2(\theta)$ for twice-passed, etc. Using the Rogers method of tracing the particle sizes in the bypass region, the values of $r = a$. Eq. 7 can then be analytically integrated for suitably simple assumed forms of $\phi(t)$, and the parameters describing $\phi(t)$ obtained by a search procedure to match the left-hand side of the equation to the right-hand side.

It is clear that recycle may in principle produce a series of peaks in the values of $c(t)$, and this can be illustrated by using the approximate RTD obtained from the analytical solution of an infinite diffusion model (see Eq. 11) assuming that $r(t) = 1$ for all recycle (which would be correct if the classifier acted purely as a sample splitter). Substituting Eq. 11 into Eq. 7

and performing the integrations gives

$$
\left.
\begin{aligned}
\phi^*(t/\tau) = {} & \frac{1}{2\sqrt{\pi D^* t^*}} \exp\left[-\frac{(1 - t^*)^2}{4D^* t^*} \right] \\
& + \frac{[C/(1 + C)]}{2\sqrt{\pi D^*(t^* - t_d^*)}} \exp\left[-\frac{(2 - t^* - t_d^*)^2}{4D^*(t^* - t_d^*)} \right] \\
& + \cdots\cdots\cdots\cdots\cdots \\
& + \frac{[C/(1 + C)]^{n-1}}{2\sqrt{\pi D^*(t^* - t_d^*)}} \exp\left[-\frac{\left(n - (t^* - \overline{n - 1}t_d^*)\right)^2}{4D^*(t^* - \overline{n - 1}t_d^*)} \right]
\end{aligned}
\right\}
\tag{8}
$$

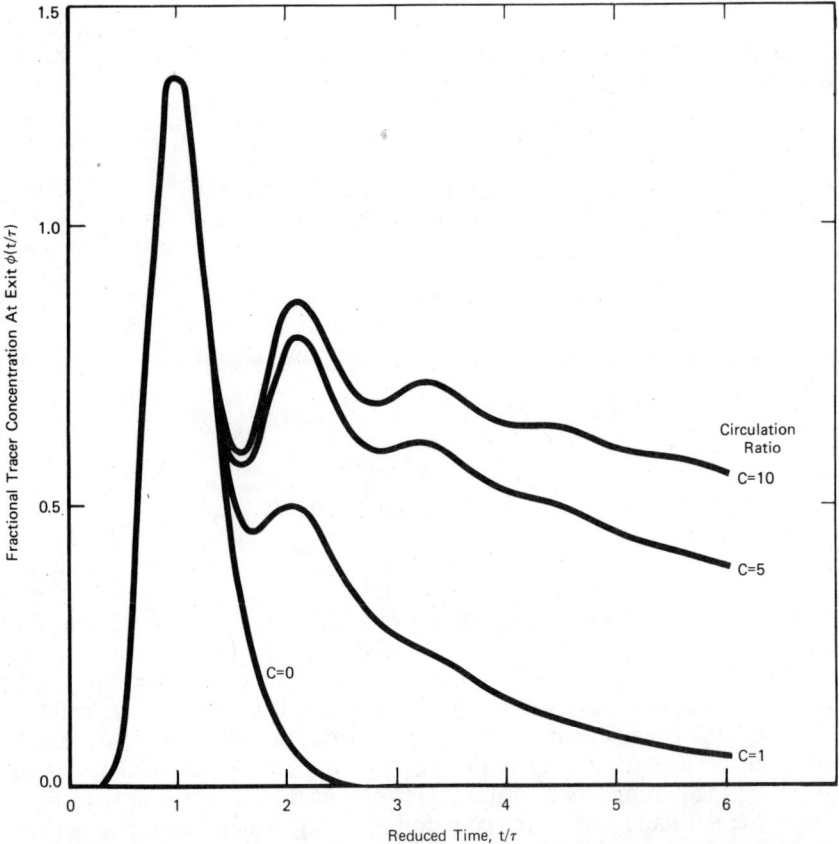

Fig. 2. Tracer concentration at the mill exit predicted by the idealized recycle model of Eq. 8 for a tracer pulse: $t_d^* = 0.2$, $D^* = 0.05$, that is, a system close to plug flow.

where t^* is reduced time t/τ and t_d^* is reduced delay time t_d/τ. Fig. 2 shows a typical result for a low value of D^*, toward the plug flow regime. The influence of the first recycled material is not apparent until after the first peak concentration is passed. For larger values of D^*, that is, systems closer to fully mixed flow, and for t_d^* smaller, the peaks move closer together and in the presence of experimental scatter, they are not usually detectable.

The *Diffusion* Model for Tumbling Ball Mills

In the introductory section it was suggested that it is convenient to have models for the RTD. A true model would create the RTD from a knowledge of the flow and mixing processes in the reactor, but a partial model based on empirical measurements is also of considerable utility. Considering a tumbling ball mill, it requires no great stretch of the imagination to think of traced powder or slurry tumbling down and spreading or *dispersing* backward and forward along the mill axis as it falls. If the mill is running at steady state under conditions where the filling level along the mill is almost constant, the movement of tracer along the mill might be described by two terms: (1) the flow of traced material at a constant bulk flow velocity and (2) the superimposed dispersion or *diffusion* effect due to the tumbling, mixing action.

Thus, if the traced powder is considered to *diffuse* at a rate dependent on its partial concentration in the bulk of the powder, the rate of diffusion would be expected to follow Fick's law,

$$\text{mass flow along axis in } l \text{ direction} = -D\,\partial c/\partial l$$

where c is the concentration of tracer (in g tracer per g solid, for example) along the mill and is a function of length and time, $c = c(l, t)$. l can be defined as starting from the mill feed end, as in Fig. 2 in Chapter 6, so that the velocity of mass flow is in the l direction. An equivalent linear velocity is defined by $u = L/\tau$, where L is the total mill length and τ the mean residence time defined by W/F. In the usual way, then, a mass balance on an element at l to $l + dl$ gives the well-known expression (Fokker-Planck or convective-diffusion equation)

$$\frac{\partial c}{\partial t} = \frac{\partial\left(D\frac{\partial c}{\partial l}\right)}{\partial l} - u\frac{\partial c}{\partial l} \tag{9}$$

The flux through a cross-sectional plane perpendicular to the axis is

$$\text{flux} = -D\frac{\partial c}{\partial l} + uc$$

In most treatments it will be assumed that D does not vary along l, giving

$$\frac{\partial c}{\partial t} = D\frac{\partial^2 c}{\partial l^2} - u\frac{\partial c}{\partial l} \tag{9a}$$

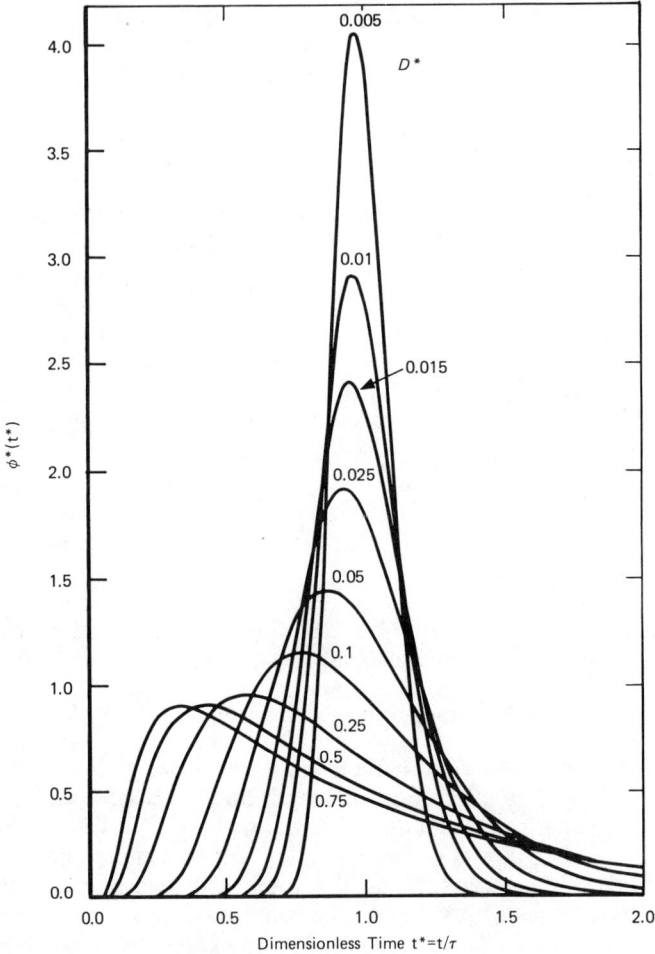

Fig. 3a. Dimensionless solution of diffusion model: complete boundary conditions.

As far as tumbling mills are concerned, the wall at the feed end prevents diffusion through the end, so the boundary condition at $l = 0$ is "flow of tracer through $l = 0$ equals tracer carried in the feed." The same condition applies at the exit end, where tracer flow through $l = L$ is only by the exiting stream. For a pulse of tracer added over a very short time period at $t = 0$, these two boundary conditions are

$$\text{At } l = 0, \text{ flux of tracer} = 0, \qquad t > 0$$

$$\text{At } l = L, \text{ flux of tracer} = uc(L), \qquad t \geq 0$$

Applying these boundary conditions to mass flux of tracer through the

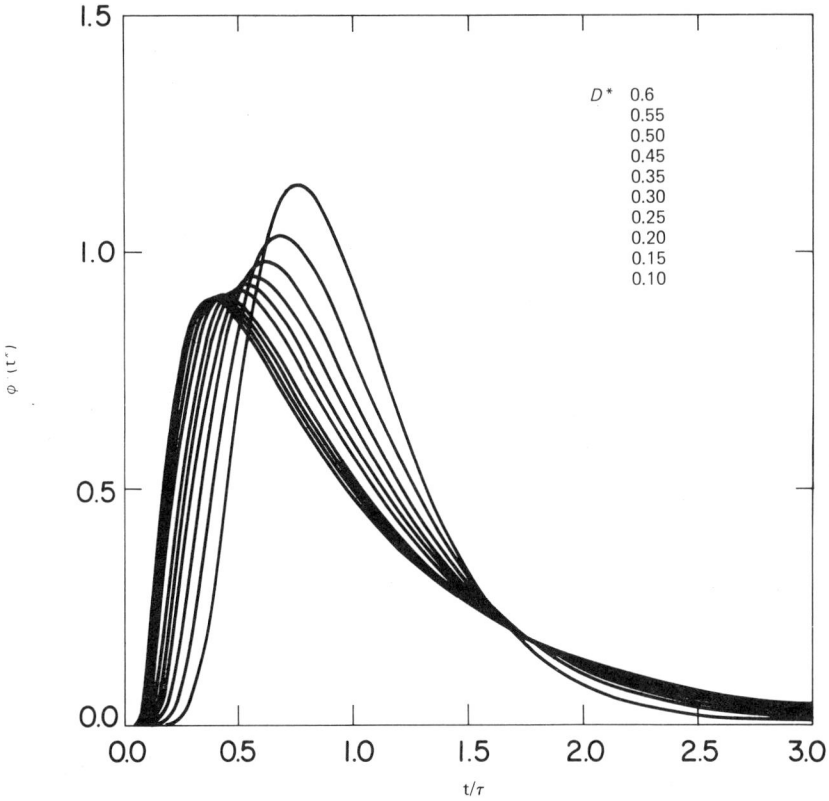

Fig. 3b. Dimensionless solution of diffusion model: complete boundary conditions, expanded scale for $D^* = 0.6$ to 0.1.

cross-section plane in the powder at $l = +0$ gives

$$-D\frac{\partial c}{\partial l}\bigg|_{l=0} + uc(0, t) = 0, \qquad t > 0 \tag{10a}$$

and through the cross-section plane in the powder at $l = L$,

$$\frac{\partial c}{\partial l}\bigg|_{l=L} = 0, \qquad t \geq 0 \tag{10b}$$

The initial condition is $c = 0$ for $0 < l \leq L$ and $c = c(0,0)$ at $l = 0$, and the proportional quantity of tracer is $c(0,0)\,\Delta t$ where Δt is the pulse duration time.

As usual, it is convenient to put the equations in dimensionless form, to reduce the computational effort. To avoid using a completely new set of symbols, reduced variables are defined by: $t^* = t/\tau$; $l^* = l/L$; $M_o^* = \int_0^\infty c(t)d(t/\tau) = M_o/\tau$; $\phi^* = c/M_o^* = \tau\phi$; $\Delta t^* = \Delta t/\tau$. The equations then become

$$\frac{\partial \phi^*}{\partial t^*} = D^* \frac{\partial^2 \phi^*}{\partial l^{*2}} - \frac{\partial \phi^*}{\partial l^*} \tag{9b}$$

$$\left. \begin{array}{ll} D^* \dfrac{\partial \phi^*}{\partial l^*} = \phi^*(0, t), & l^* = 0, t^* > 0 \\[2mm] \dfrac{\partial \phi^*}{\partial l^*} = 0, & l^* = 1, t^* \geq 0 \\[2mm] \phi^*(l^*, 0) = 0, & 0 < l^* \leq 1 \\[2mm] \phi^*(0, 0) = 1/\Delta t^* & \end{array} \right\} \tag{10c}$$

where D^* is the reduced diffusion coefficient D/uL (the Peclet number is uL/D). Thus the reduced equation is the expression for a reactor of unit length, with time in units of τ, at unit velocity, with diffusion coefficient D^*. The RTD is given by the values of c or ϕ^* at the mill exit, $l = L$ or $l^* = 1$.

A number of solutions of Eq. 9 for various other boundary conditions have been applied to tumbling ball mills, but the boundary conditions given above are the most correct expressions. For example, if the flow conditions in the mill corresponded to rapid plug flow with only a very slow diffusive process, the pulse of tracer could be carried down the mill faster than it could diffuse back. Then the spread of tracer along the mill before it reaches the other end would occur as if there were no end wall at the feed end. If it is also assumed that the wall at the discharge end does not disturb the mixing, the *fully infinite* solution of Eq. 9 can be used, giving

or

$$\left. \begin{array}{l} c(l, t) = \left(uM_o/2\sqrt{D\pi t}\right) \exp\left[-(l - ut)^2/(4Dt)\right] \\[2mm] \phi^*(l^*, t^*) = \left(1/2\sqrt{\pi D^* t^*}\right) \exp\left[-(l^* - t^*)^2/(4D^* t^*)\right] \end{array} \right\} \tag{11}$$

However, Austin, Luckie, and Ateya (1971) have shown that calculating the RTD by substituting $l = L$ in this equation is a poor approximation to the correct solution, for the range of u, D, and L for real milling conditions. The correct solution was obtained by a finite-difference numerical integration of the Fokker-Planck equation (note that the boundary conditions were given incorrectly in their paper, although the correct conditions were actually used in the computation).

Mori, Jimbo, and Yamazaki (1964, 1967) solved Eq. 9 using the *semi-infinite* boundary conditions, which assumes that the wall at the discharge end has no effect on the dispersion process. Thus their solution does not use the correct boundary condition of Eq. 10b, and leads to

$$\left. \begin{array}{l} \phi(t) = \dfrac{c(l, t)}{M_o} = \left(1/2\sqrt{\pi D t^3}\right) \exp\left[-(l - ut)^2/(4Dt)\right] \\[3mm] \phi^*(l^*, t^*) = \left(1/2\sqrt{\pi D^* t^{*3}}\right) \exp\left[-(l^* - t^*)^2/(4D^* t^*)\right] \end{array} \right\} \tag{12}$$

The method of solution using the correct boundary conditions is given in Appendix 5 (Trimarchi, 1979). Fig. 3 shows the dimensionless differential RTDs from this solution, expressed as functions only of the characteristic dimensionless parameter D/uL. For comparison, Figs. 4 and 5 show the results for the same conditions for the fully infinite and semi-infinite solutions. Comparison of the curves show that there are significant differences for the fully infinite solution at $D^* > 0.01$, but the semi-infinite solution is reasonably close to the correct solution for values of $D^* < 0.1$. The Mori semi-infinite solution, useful as an approximate simple solution, is used in several places in this book.

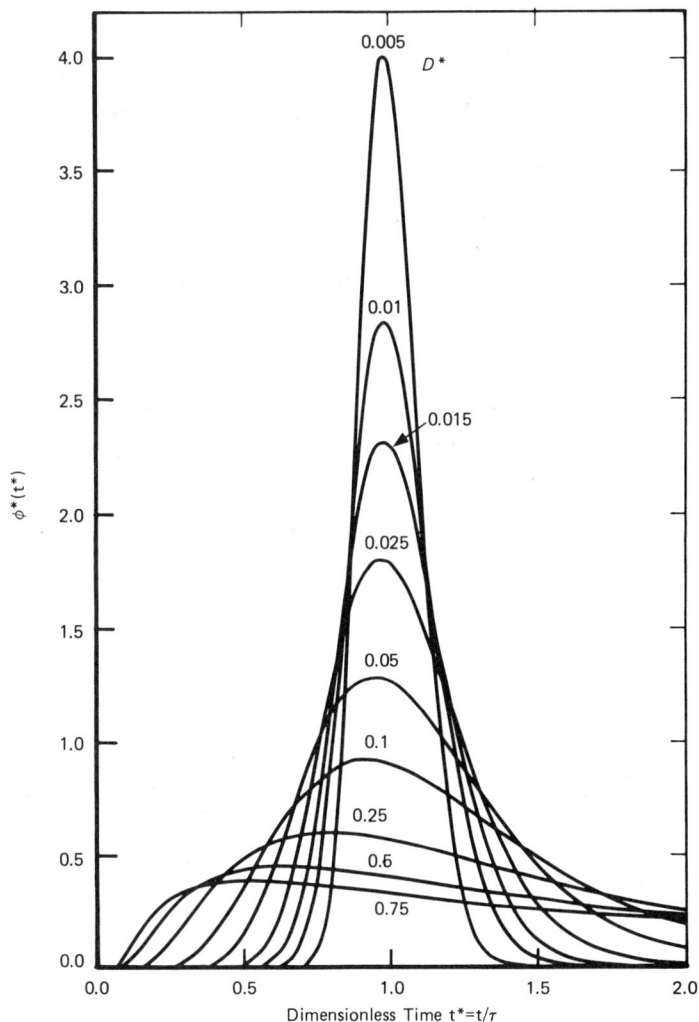

Fig. 4. Dimensionless solution of diffusion model: fully infinite boundary conditions.

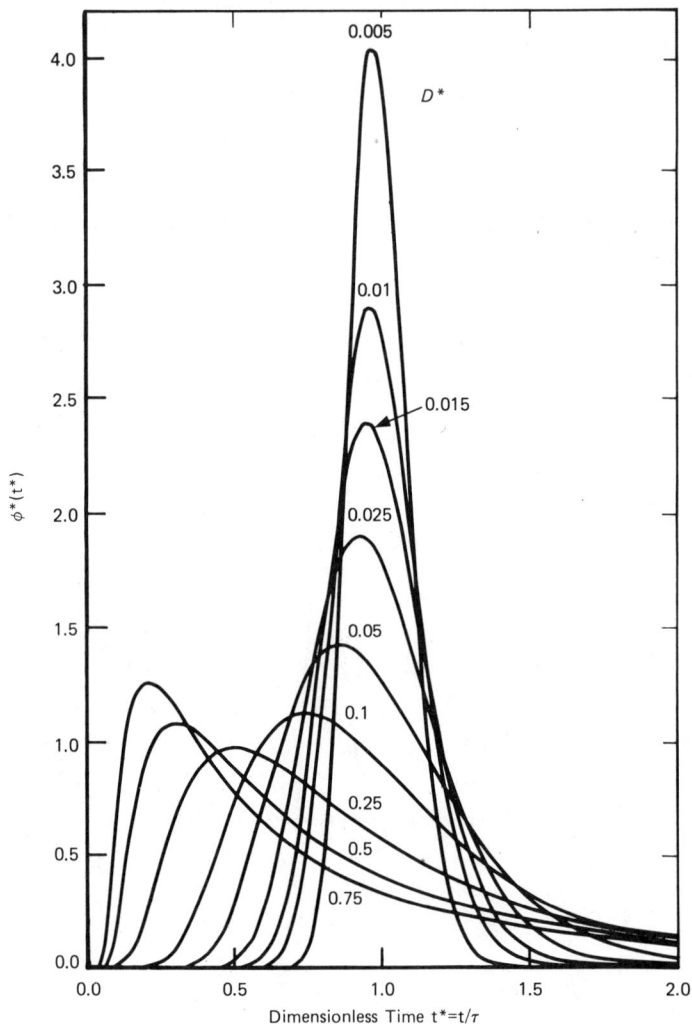

Fig. 5. Dimensionless solution of diffusion model: semi-infinite boundary conditions.

Fig. 6 gives the results of the computations in forms suitable for interpolation, or for measurement of D^* without requiring integration of experimental data to get areas under the $c(t)$ vs. t or $tc(t)$ vs. t curves. Experimental RTD results can be characterized in dimensionless form by defining a dimensionless parameter C

$$C = (t_{+1/2} - t_{-1/2})/t_p \qquad (13)$$

where t_p is the time at which the peak concentration occurs, $t_{+1/2}$ the time

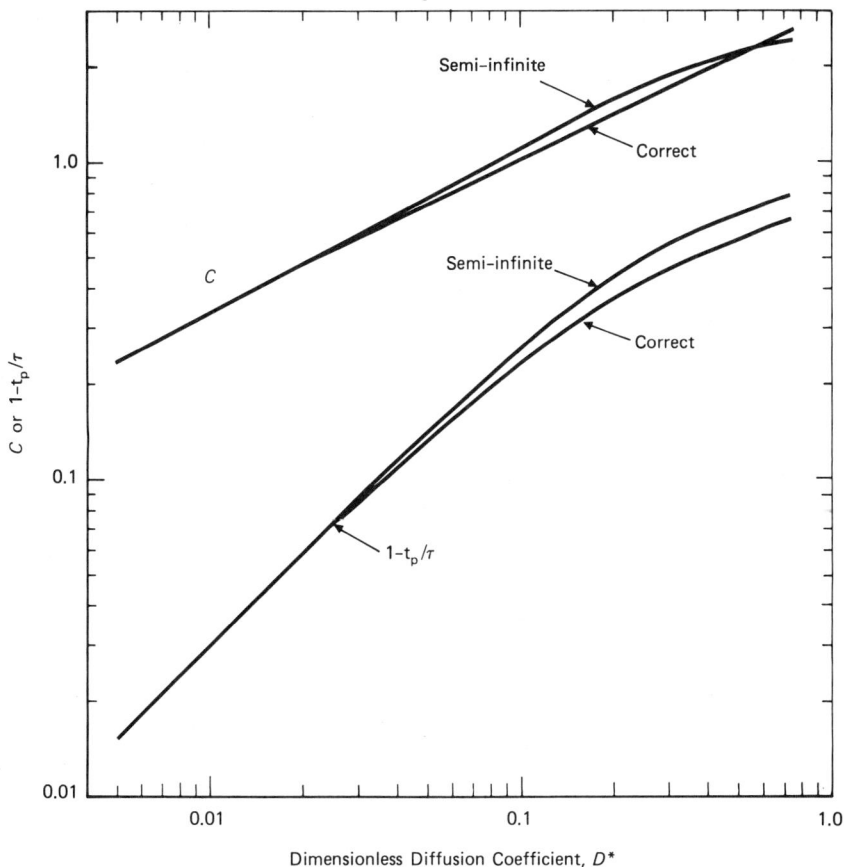

Fig. 6. The relation between $C = (t_{+1/2} - t_{-1/2}) / t_p$ and D^*, and $1 - t_p/\tau$ and D^*, for the fully correct and the semi-infinite solutions of the diffusion equation.

at which the concentration is half the peak after t_p, and $t_{-1/2}$ the time at which the concentration is half the peak before t_p. The value of C is uniquely related to D^*, which can be picked off the curve. In turn, D^* is uniquely related to $1 - t_p/\tau$ and since t_p is known, τ follows. Note that if a transport delay time occurs between the mill exit and the point where the tracer concentrations are measured, then this time should be subtracted from the test time before C is calculated.

Reactors-in-Series Models

The other method of establishing an empirical description of a residence time distribution is to consider the RTD as originating from a number of fully mixed reactors or tanks in series. The concept of tanks-in-series or stages-of-reaction, well known in chemical engineering, is stated as follows: the experimental RTD is matched against theoretical expressions derived for a number of fully mixed reactors in series and the number of proportional sizes giving the best match are chosen. Then the mill is considered to behave *as if* it were a corresponding number of mills in series (see Chapters 6, 7, and 17).

Consider m reactors in series, each with a differential RTD of $\phi_1(t)$, $\phi_2(t)$, $\phi_3(t)...\phi_m(t)$. What is the RTD of the total series? Considering the first reactor, the fraction of a pulse of tracer put in at time $t = 0$ which leaves in time θ_1 to $\theta_1 + d\theta_1$ is

$$\phi_1(\theta_1)\, d\theta_1$$

This fraction has stayed in the reactor for time θ_1 to $\theta_1 + d\theta_1$. It enters the second reactor and the fraction of this fraction which stays in the second reactor for an additional time θ_2 to $\theta_2 + d\theta_2$ is

$$\phi_2(\theta_2)\, d\theta_2$$

Consider time $t = \theta_1 + \theta_2$. The fraction of tracer which spends a sum time t to $t + dt$ in the two reactors is clearly

$$= \left[\int_{\theta_1=0}^{t} \phi_1(\theta)\phi_2(t - \theta)\, d\theta_1 \right] dt$$

where t is *constant* in the integration. In general, then,

$$\phi_m(t) = \int_{\theta_{m-1}=0}^{t} \int_{\theta_{m-2}=0}^{\theta_{m-1}} \cdots \int_{\theta_2=0}^{\theta_3} \int_{\theta_1=0}^{\theta_2} \phi_1(\theta_1)\phi_2(\theta_2 - \theta_1)...$$

$$\times ...\phi_m(t - \theta_{m-1})\, d\theta_1\, d\theta_2...d\theta_{m-1} \qquad (14)$$

where the first integration is performed with θ_1 as the variable, holding θ_2 constant, and so on.

For fully mixed reactors, $\phi_1(\theta) = (1/\tau_1)\exp(-\theta_1/\tau_1)$, etc., and

$$\phi_m(t) = \sum_{i=1}^{m} \frac{\tau_i^{(m-2)}\exp(-t/\tau_i)}{(\tau_i - \tau_1)(\tau_i - \tau_2)...(\tau_i - \tau_k)...(\tau_i - \tau_m)}, \quad i \neq k, \tau_i \neq \tau_k$$

$$(15)$$

Eq. 15 cannot be used for m *equal* fully mixed reactors in series, i.e., $\tau_1 = \tau_2 = \tau_3 = \cdots = \bar{\tau} = \tau/m$. Returning to Eq. 14 and using $\phi(t) = (1/\tau)\exp(-t/\tau)$ gives

$$\phi_m(t) = \frac{1}{\bar{\tau}^m} \frac{t^{m-1}}{(m-1)!} e^{-t/\bar{\tau}} \qquad (16)$$

The results for m equal fully mixed reactors in series are shown in Fig. 7.

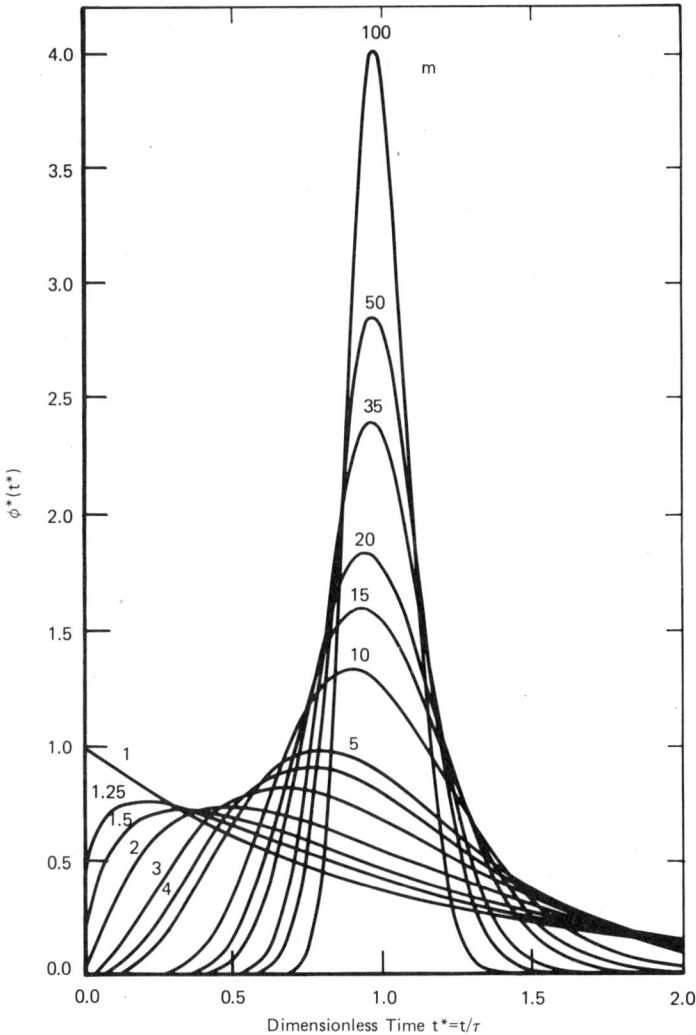

Fig. 7. Theoretical dimensionless residence time distributions for m equal fully mixed reactors in series.

Note that τ and t_p are related by

$$\bar{\tau} = t_p/(m-1)$$

or

$$\tau = t_p\left(\frac{m}{m-1}\right) \tag{17}$$

where τ is the total mean residence time.

Comparing Figs. 5 and 7 it is clear that the diffusion model and the "*m* equal fully mixed reactors in series" model do not give matching results except at small values of D^* and large values of m. The use of *unequal* fully mixed stages, as given by Eq. 15, gives greater flexibility in the shape of the RTDs. One form which has been used (Weller, 1980; Marchand, Hodouin, and Everell, 1980) is that resulting from one major fully mixed reactor of size τ_1 followed by two smaller equal fully mixed reactors of size τ_2, $\tau_1 \geq \tau_2$ (one-large/two-small model). This gives

$$\phi(t) = \frac{\tau_1}{(\tau_1 - \tau_2)^2}\left[\exp(-t/\tau_1) - \exp(-t/\tau_2)\right] - \frac{t}{(\tau_1 - \tau_2)\tau_2}\exp(-t/\tau_2)$$

$$(18)$$

Since $\tau_2 = (1 - \tau_1)/2$ in dimensionless form, the shape of $\phi^*(t)$ is only a

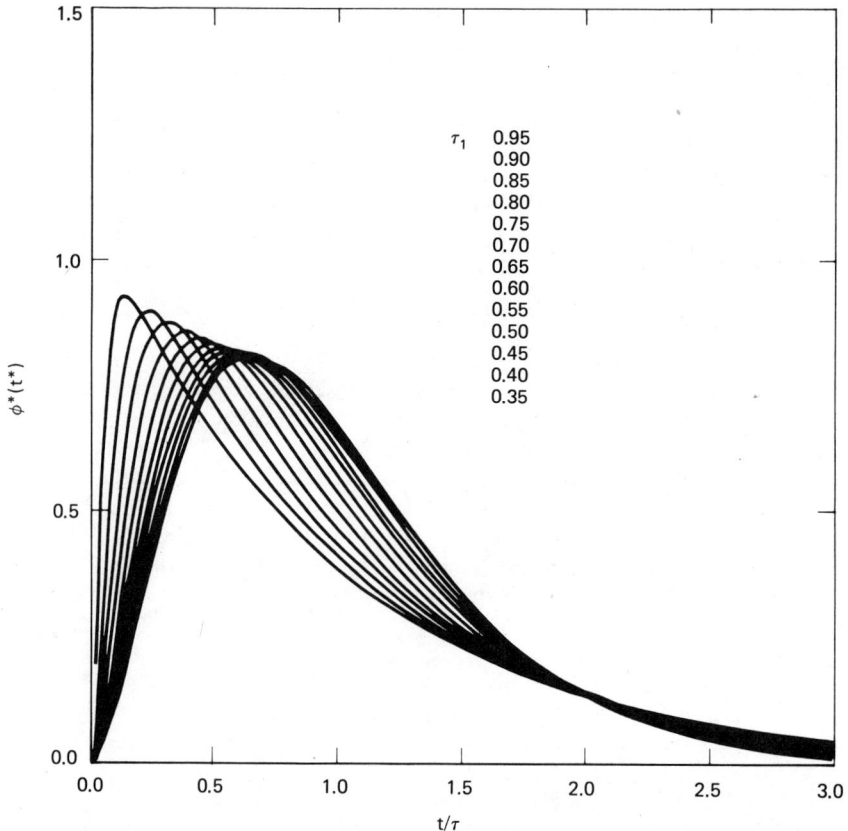

Fig. 8. Theoretical dimensionless residence time distributions for one-large / two-small fully mixed reactors in series model.

function of τ_1. The result of this computation is given in Fig. 8. It is frequently necessary to make this function more flexible by incorporating a plug flow section of τ_0, thus $\tau = \tau_0 + \tau_1 + 2\tau_2$ and t in Eq. 18 is replaced by $\tau - \tau_0$ on the right-hand side. The shape of the RTD is unchanged for a given τ_1, but the curve is shifted τ_0 to the right.

Rogers and Gardner (1979) have used a model derived from fully mixed reactors in series each connected in parallel to another fully mixed reactor. The rate of interchange between the reactors in parallel is an additional parameter. The equations for calculating $\phi(t)$ for this series are given in Appendix 6 for m stages: in practice they use $m = 4$. This model is more flexible than the two-parameter (e.g., τ, D^*) models given previously because it has one more adjustable parameter.

It is obvious that all of the tanks-in-series models can be modified by the incorporation of a plug flow reactor of time τ_p. Then the value of τ to be used in the above equations is $\tau - \tau_p$, and the RTD curves are shifted τ_p to the right.

Experimental Results for Tumbling Ball Mills

Before discussing the rather sparse data on mill RTDs in the literature, some points of philosophy should be made. First, if the value of D^* is changing along the mill, the resultant RTD at the mill exit can still be analyzed *as if* D^* were constant. If the shape of the exit RTD fits the solution of the diffusion equation reasonably well, then this is all that is necessary for solving the mill model equations. The fit of RTD (measured at the mill exit) to a diffusion model gives an effective *mean* value of D^* in the mill. Second, if a solution of the diffusion equation for false boundary conditions gives a reasonable match to the shape of the experimental RTD, then again the false expression is all that is necessary. The procedure is equivalent to fitting an empirical function to the experimental results. Third, in order to fit RTD data to a model it is necessary to estimate τ and D^*, or τ and m or τ and τ_1. However, there is no proof that the model is real for mixing in a mill, so the value of τ required to fit the model to the experimental data is not necessarily the same as the τ deduced by the integration of Eq. 5 or the same as the τ defined as W/F.

Fig. 9 shows the results of Shoji, Hogg, and Austin (1973) for diffusion in a small dry batch mill, with vinyl balls which caused negligible breakage $\left[D = 9.5 \text{ cm} \left(3\tfrac{3}{4} \text{ in.}\right), d \approx 6 \text{ mm} \left(\tfrac{1}{4} \text{ in.}\right) \right]$. The tracer was garnet powder, which could be separated from the charge of silicon carbide powder by a magnetic separator. A thin slice of garnet powder was placed in the center of the charge and when the mill was rotated the tracer dispersed in each direction along the axis. For a short time of mixing, the well-known source solution $c = (M_0/2\sqrt{\pi Dt}\,)\exp(- X^2/4Dt)$ was found to be applicable, where X is the distance along the axis away from the original tracer position. The experimental conditions are very unlike real mills, and there is no proof that diffusion in the absence of net flow along the axis is the same as diffusion in a flowing system. However, the general trends may be comparable to larger-scale mills. When the level of powder in the bed is lower than the

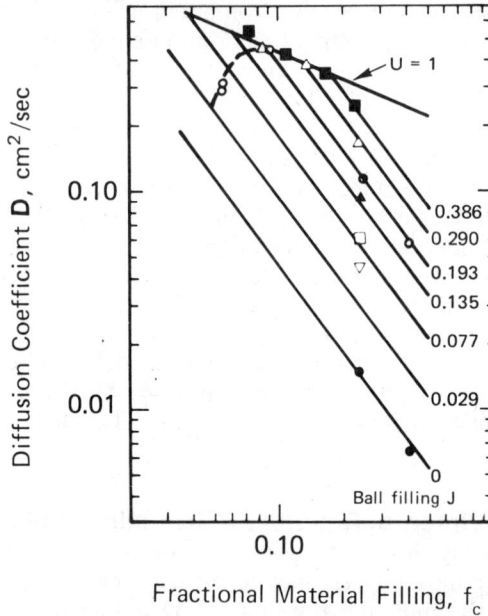

Fig. 9. Variation in the diffusion coefficient D with material filling f_c as a function of ball filling J.

balls ($U < 1$), the diffusion coefficient is not dependent on ball load J. The diffusion coefficient decreases slightly as powder filling is increased, but then drops sharply as the mill is overfilled ($U > 1$). For an overfilled mill, increasing the ball load tends to restore the higher diffusion coefficient.

Mori, Jimbo, and Yamazaki (1964) investigated the dry continuous grate discharge mills described in Table 1, using copper sulfate as a tracer with limestone powder. Unfortunately, they did not report the measured RTDs, but only the characteristic parameters (τ, D^*, D) of the data fitted to the semi-infinite model, although they did measure mean mill hold-up for each test. The results are shown in Table 2. Comparing the τ defined by W/F

Table 1. Mill Conditions Tested by Mori, Jimbo, and Yamazaki (1964)

	Test series	D, mm	L, mm	Grate area slit ratio, %	Ball size, mm	Feed size, mm	Mill speed, rpm	Critical speed, rpm
Pilot Mill	1	545	1980	2.8	30	0 by 5	30.6	57.3
	2	545	1980	11.3	30	0 by 5	30.6	57.3
	3	545	1980	11.3	40	0 by 5	30.6	57.3
Laboratory Mill	4	254	495	18.8	16	2.5 by 1.2	25 ~ 58	83.9

Table 2. RTD Results of Mori, Jimbo, and Yamazaki (1967)

	F kg/min	J %	W kg	f_c %	$\tau = \dfrac{W}{F}$ min	D^*	Model τ min	Model D cm²/min	u cm/min
SERIES 1									
d = 30 mm	1.17	10	18.0	2.3	15.4	0.076	16.4	181	12.1
2.8% grate	3.69	10	36.6	4.7	9.9	0.034	8.8	140	22.5
area	3.84	20	53.5	6.9	13.9	0.066	12.0	217	16.5
	8.15	20	79.3	10.2	9.7	0.033	8.5	153	23.4
	3.86	30	69.0	8.9	17.9	0.079	17.2	181	11.5
	7.94	30	86.2	11.1	10.9	0.039	11.0	139	18.0
	4.07	40	93.1	12.0	22.9	0.054	19.8	107	10.0
	5.62	40	98.8	12.7	17.5	0.038	16.0	94	12.4
	8.47	40	107.0	13.8	12.6	0.030	12.2	95	16.2
SERIES 2									
d = 30 mm	1.08	10	17.7	2.3	16.4	0.043	13.5	124	14.6
11.3% grate	2.31	10	23.0	3.0	10.0	0.024	7.8	121	25.1
area	4.11	10	31.6	4.1	7.7	0.021	6.7	121	29.4
	5.66	10	41.5	5.3	7.3	0.016	6.3	99	31.4
	7.54	10	57.1	7.4	7.6	0.026	7.1	147	28.0
	4.33	20	39.5	5.1	9.1	0.025	7.8	125	25.4
	6.40	20	50.6	6.5	7.9	0.026	6.7	151	29.7
	7.72	20	56.6	7.3	7.3	0.022	6.5	134	30.3
	4.08	30	56.0	7.2	13.7	0.039	11.4	134	17.3
	6.40	30	65.7	8.5	10.3	0.021	8.4	99	23.6
	7.63	30	73.3	9.4	9.6	0.022	8.2	107	24.2
	4.14	40	83.8	10.8	20.2	0.073	17.6	162	11.2
	6.02	40	95.7	12.3	15.9	0.062	13.0	186	15.2
	8.15	40	110.1	14.2	13.5	0.061	12.7	199	15.6

Table 2. RTD Results of Mori, Jimbo, and Yamazaki (1967)

						Model				
F kg/min	J %	W kg	f_c %	$\tau = \dfrac{W}{F}$ min	D^*	τ min	D cm^2/min	u cm/min		ϕ_c %
SERIES 3										
d = 40 mm										
11.3% grate										
area										
0.83	10	19.4	2.5	23.3	0.037	14.1	382	14.1		
1.83	10	22.2	2.9	12.1	0.060	9.0	263	22.0		
3.14	10	29.4	3.8	9.4	0.050	7.5	258	26.3		
3.88	10	31.7	4.1	8.2	0.041	6.7	243	29.7		
5.80	10	42.9	5.5	7.4	0.032	6.4	197	20.9		
8.05	10	59.5	7.7	7.4	0.046	6.2	292	32.1		
4.42	20	39.0	5.0	8.8	0.058	7.6	297	26.1		
5.70	20	47.3	6.2	8.4	0.068	7.0	381	28.2		
8.01	20	57.3	7.4	7.1	0.055	6.5	333	30.3		
4.23	30	53.9	6.9	12.7	0.078	11.0	276	17.9		
6.00	30	61.8	8.0	10.3	0.064	8.8	287	22.5		
8.08	30	69.4	8.9	9.8	0.059	7.9	291	25.0		
3.63	40	82.9	10.7	22.8	0.12	19.6	245	10.1		
6.41	40	92.2	11.9	14.4	0.10	13.3	305	14.9		
8.13	40	100.4	12.9	12.3	0.088	11.8	273	16.7		
SERIES 4										
(small										
mill)										
0.12	20	2.09		17.4	0.19	12.1	38.2	4.1		30
0.14	20	2.98		21.2	0.22	12.9	42.4	3.8		40
0.13	20	3.02		23.2	0.31	13.1	57.3	3.8		50
0.12	20	2.97		24.8	0.34	16.1	52.2	3.1		60
0.12	20	3.06		25.5	0.36	16.3	53.9	3.0		70

with the model τ values it is seen that the model frequently underestimated τ, thus showing that the model fit is only approximate. Calculations indicate that most of the tests were for a mean powder filling of $U < 1$. In agreement with the previously quoted results of Shoji, Hogg, and Austin, there appeared to be no strong effect of ball loading on the value of D. Comparing series 1 and 2, there was little effect of increased grate area. The most striking effect on D was that of ball diameter, indicating that balls of 40 mm (1.6 in.) diam almost doubled the diffusion coefficient for 30 mm (1.2 in.) balls, with the mean hold-up consistently somewhat less for the larger balls under comparable conditions. For the small laboratory mill, the values of W/F were about 50% higher than the model τ, indicating a very poor fit of the model to the results.

Kelsall, Reid, and Restarick (1969–70) measured values in a laboratory wet overflow ball mill of 305 mm diam by 305 mm long (12 by 12 in.), at a τ of 3.6 min, using a tracer of quartz in a calcite feed. The shape of their RTD is consistent with a D^* of about 0.4, that is, closer to the fully mixed condition, with $D = 0.012$ m^2/min (0.13 sq ft/min).

Austin, Luckie, and Ateya (1971) analyzed fluorescein tracer RTD data from a two-compartment cement mill of 4 m (13 ft) diam by 10 m (34 ft) long, with a flow rate through the mill of 195 t/h. Unfortunately, the data was not complete enough to allow for proper correction for recycled tracer, but an approximate correction gave a primary RTD which could be fitted with reasonable accuracy by the diffusion model with $D^* = 0.07$ to 0.10, $\tau = 5.6$ to 5.9 min, and D about 1.5 m^2/min (14 sq ft/min). As in most large-scale tests, it was not possible to compare τ with W/F because the mill hold-up was not known.

The most accurate available test data on large-scale mill RTD are those of Rogers and Gardner (1979), performed using radiotracing of solid particles. Three wet overflow mills (see Table 3) were tested in open circuit, requiring only a correction for the sump RTD. They concluded that the

Table 3. RTD Results from Three Wet Overflow Ball Mills*

D, m	L, m	$\dfrac{L}{D}$	F, kg / s	τ min	W, t	$t_p,$ min	D^*	D m^2 / min
1.83	2.74	1.5	7.6	8.9	4.06	4.4(5)	0.25	
			8.6	8.5	4.39	4.1(5)	0.25(5)	0.23
			9.6	7.9	4.55	3.8(5)	0.26(5)	
			10.6	7.2	4.58	3.4	0.27(5)	
3.81	5.18	1.36	17.7	20.9	22.2	11.0	0.25	
			21.4	15.3	19.6	9.0	0.19	0.33
			25.2	13.7	20.7	8.0	0.18	
4.57	9.20	2.0	40.4	22.0	53.3	12.5	0.22(5)	
			44.1	19.8	52.4	12.0	0.19	0.85
			51.7	17.3	53.6	11.0	0.17	

*Rogers and Gardner (1979).

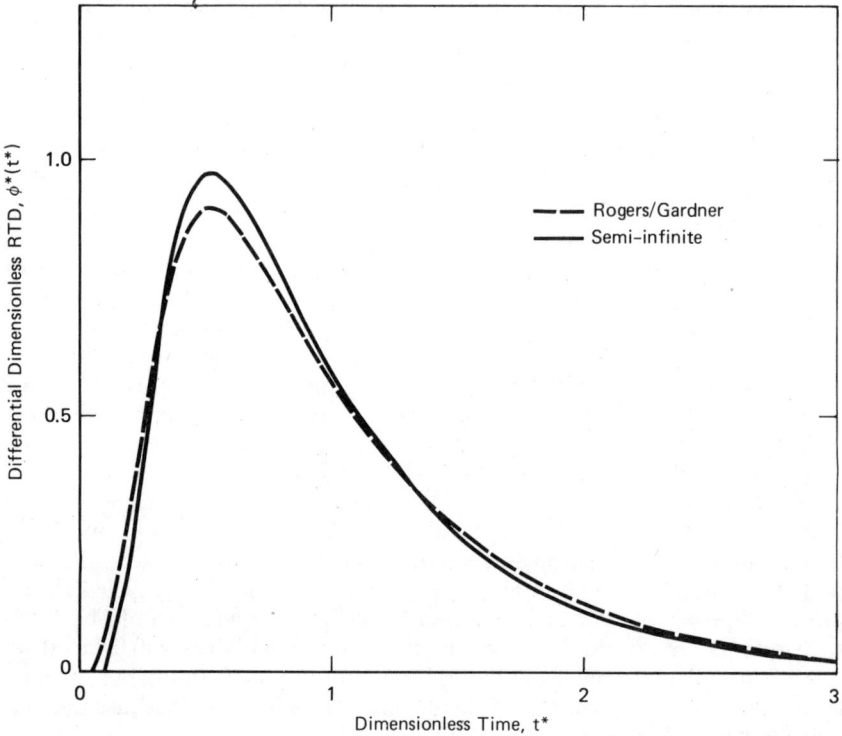

Fig. 10. Comparison of RTDs for Rogers / Gardner model ($m = 4$, $F_a = 0.677$, $F_b = 0.304$, $F_c = 0.019$, $k = 0.217$) and semi-infinite diffusion model $D^* = 0.23$.

Table 4. One-Large/Two-Small Model RTD Results for Wet Overflow Ball Mills*

D, m	L, m	$\dfrac{L}{D}$	F, t/h	τ min	τ_1 min	τ_2 min	$\tau_{\text{plug flow}}$ min	Dimensionless		
								τ_1	τ_2	τ_p
2.93	2.32	0.79	375	1.76	0.95	0.38	0.05	0.54	0.22	0.03
1.83	3.66	2.0	114	3.47	1.98	0.53	0.43	0.57	0.15	0.12
2.74	3.66	1.34								
		open	98	7.59	5.05	0.84	0.86	0.67	0.11	0.11
		closed[†]	296	4.38	1.65	1.25	0.23	0.38	0.29	0.05
		closed	105	7.23	3.23	1.97	0.06	0.45	0.27	0.01

*Weller, 1980; Marchand, Hodouin, and Everell, 1980. See also Table 3 in Chapter 5.
[†] High hold-up, overfilled.

**Table 5. Comparison of Simulated Open-Circuit Size Distribution for
Two Different RTDs for Mill Product 74% < 212 μm (70 mesh)**

Interval No.	Size Interval Mesh	% < size Feed	% < size Product	
			$D^* = 0.23$	Rogers/Gardner
1	16 × 20	100	100	100
2	20 × 30	92	99.1	99.0
3	30 × 40	80	96.3	96.1
4	40 × 50	68	91.2	91.0
5	50 × 70	55	83.4	83.3
6	70 × 100	44	74.0	74.0
7	100 × 140	33	63.0	63.0
8	140 × 200	24	52.0	52.1
9	< 200	18.5	43.0	43.2
			τ, min. = 3.26	3.37

Note: Breakage parameters $\alpha = 0.9$, $\beta = 2.9$, $\gamma = 0.55$, $\Phi = 0.66$, $\delta = 0$, $S_1 = 1$ min^{-1}.

RTDs of a mill were normalized with respect to mean residence time, and that the same normalized RTD applied for all the mills, within the accuracy of the determination. The RTD they use is shown in Fig. 10. However, analysis of their data indicated that the test results from the 1.83 m (6 ft) diam mill gave longer tails to the differential RTDs than would be predicted by Fig. 10 and the RTDs from the 4.57 m (15 ft) diam mill had shorter tails. Mean residence time for water was about 15% less than for solid, indicating a higher overall pulp density in the mill than that of the feed and exit streams.

Weller (1980) and Marchand, Hodouin, and Everell (1980) used water tracing with tritiated water and concluded that the RTDs of wet overflow mills (in closed circuit) could be fitted by the one-large/two-small model, with a small plug flow addition. The results are shown in Table 4 and Table 3 in Chapter 5.

Results have been obtained from a 0.3 m (1 ft) i.d. by 0.6 m (2 ft) long wet overflow mill at open circuit using radiotraced quartz at about 52 weight percent solids (Austin, et al., 1983). Fig. 11 gives the RTDs measured by detectors placed (outside of the mill case) at $L/3$, $2L/3$, and at the mill exit. Also shown are the RTDs predicted from the convective-mixing model using $D^* = 0.4, 0.5, 0.6$. To a reasonable approximation the value of $D^* = 0.5$ gives agreement with the experimental data. This suggests that the effective value of D^* does not vary much along the mill. Tests on irradiated size fractions of 16 × 20 mesh (1.18 mm × 850 μm), 40 × 50 mesh (425 × 300 μm), and 100 × 140 mesh (150 × 106 μm) showed no detectable influence of traced particle size on RTD, even though there was relatively little size reduction under the test conditions. This confirms that these sizes move through the mill with the same RTD, at least as a first approximation. Similar results were obtained on a pilot-scale mill, and the ratio of τ for

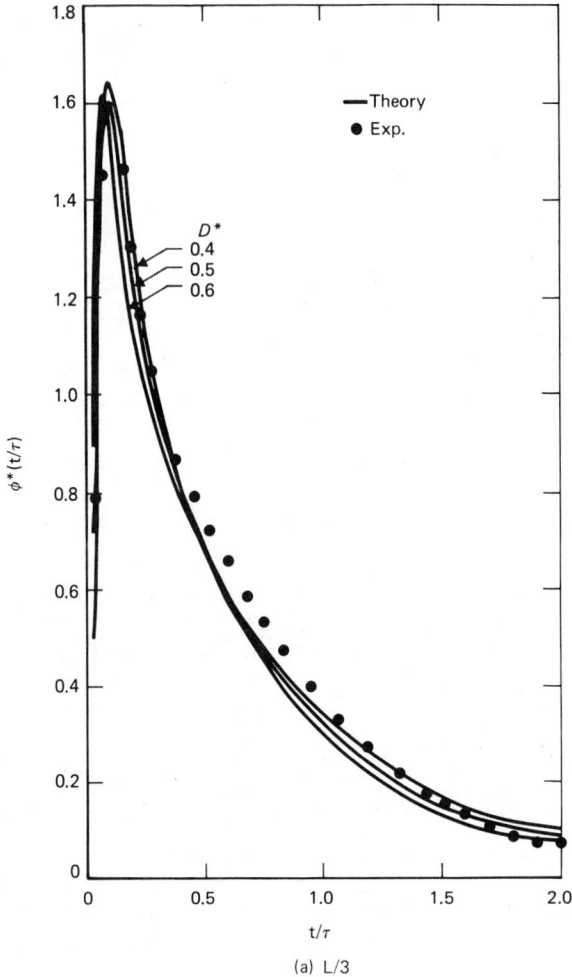

Fig. 11. Apparent residence time distributions (internal age distributions) at various positions along a 0.3 m i.d. by 0.6 m long (1 by 2 ft) wet continuous overflow mill: $J = 0.34$, $d = 9.6$ mm (3 / 8 in.), $F = 39.5$ kg / hr (87 lb / hr), $\phi_c = 0.7$; quartz, 52 weight percent solids in feed.

water to τ for solid was 0.84 to 0.91, that is, the values for water were about 10 to 15% less than for solid.

It is clear that there is no agreement on a single normalized RTD to describe all mills, although the results for wet overflow ball mills have D^* in the range 0.15 to 0.50. There appears to be no clear pattern in the change of RTD with mill dimensions and flow rate. It is difficult to get controlled

(b) 2L/3

(c) L

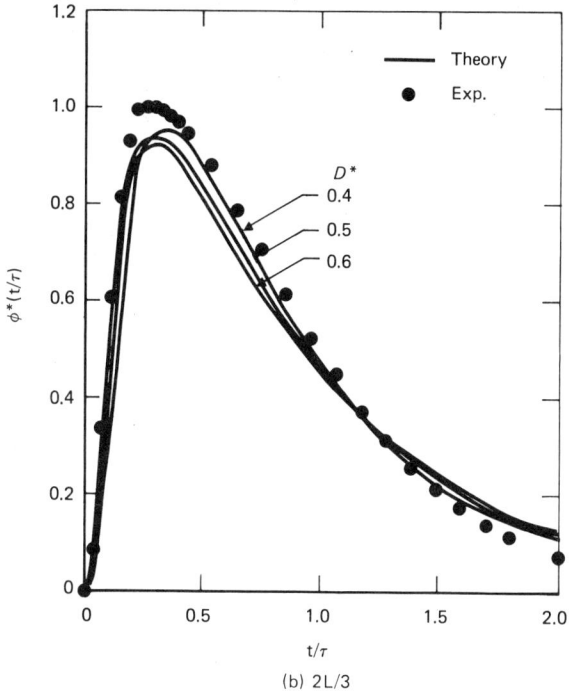

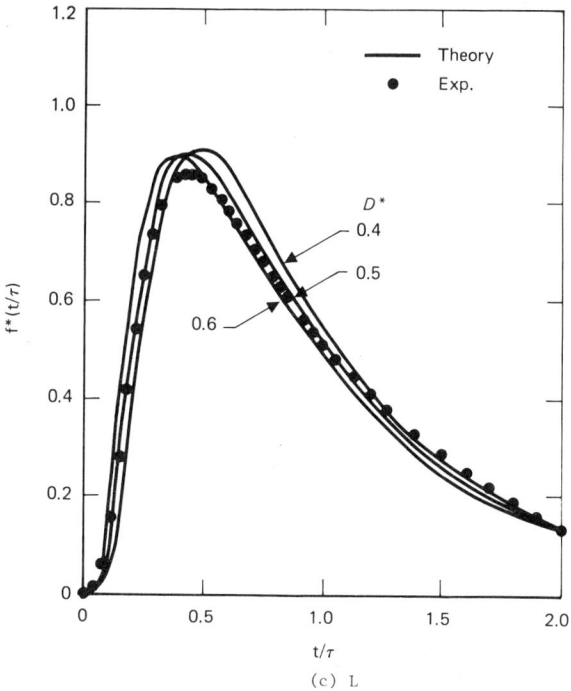

Fig. 11. Continued.

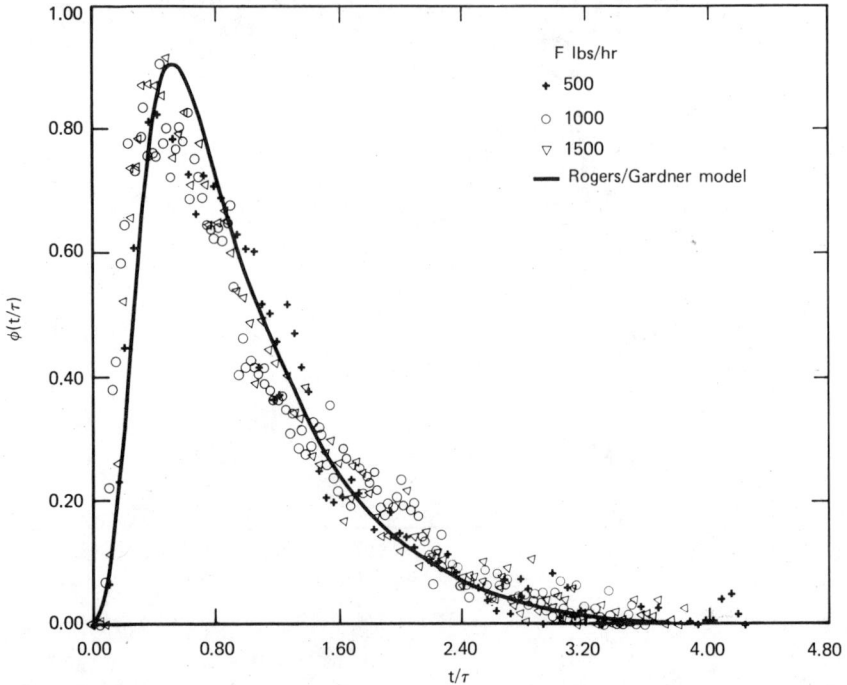

Fig. 12. Dimensionless residence time distributions for 0.9 m i.d. by 1.5 m long (3 by 5 ft) wet ball mill at three different feed rates.

variations of flow rate over a wide range in industrial-scale mills, so experiments were performed in a pilot-scale wet overflow mill of 0.8 m (2.6 ft) i.d. by 1.5 m (5 ft) length. The RTDs were measured at open circuit with an NaI detector at the mill exit, using irradiated quartz tracer, at flow rates of quartz in the ratio $1:2:3$, with 65% weight percent solids in slurry. The results are shown in Fig. 12. Any variations of the shape of the dimensionless RTD with flow rate over the $3:1$ change of flow rate is too small to be statistically significant. The results can be fitted quite well with a single value of D^* of 0.5 (see Fig. 3). Tests also showed that the value of τ for water was about 15% smaller than for solid, but with the same dimensionless RTD.

One factor of obvious importance to the mass transfer in a wet mill is the slurry density or, more precisely, the slurry rheology. Rogers, Bell, and Hukki (1982) have determined RTDs for grinding of a feed of minus 4 mesh (− 4.75 mm) limestone in a 0.9 m (3 ft) i.d. by 1.5 m (5 ft) long mill, giving the results shown in Fig. 13. It appears from this data that slurry density affects the dimensionless RTD far more than flow rate or hold-up. The lower the slurry density, the closer the RTD is to fully mixed; however, it

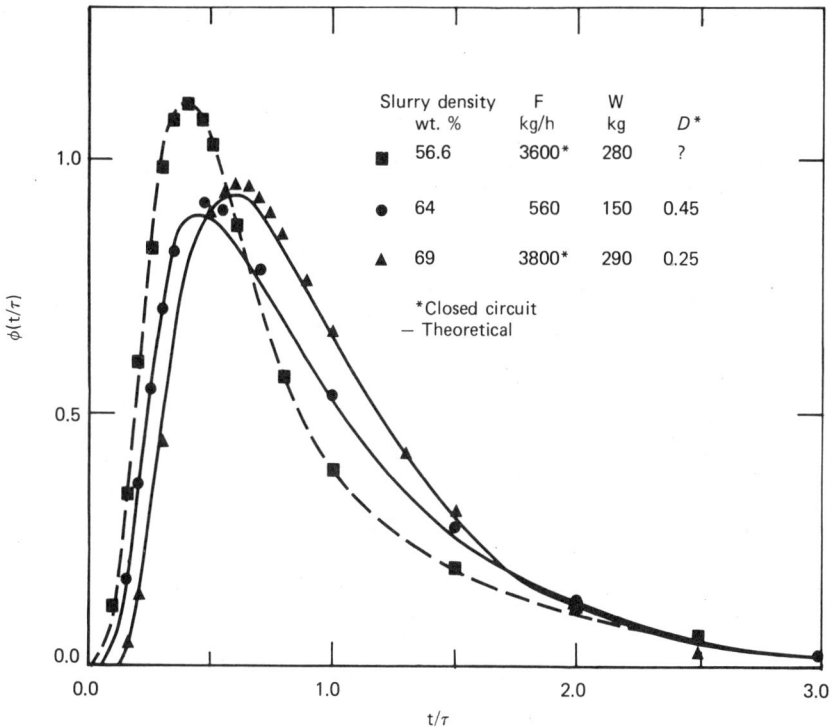

Fig. 13. Dimensionless residence times measured in 0.9 m (3 ft) i.d. mill at different slurry densities [ϕ_c = 0.73; limestone, ball charge 50.8 mm (2 in.) top size (Rogers, Bell, and Hukki, 1982)].

was not possible to fit the result at the lowest slurry density to the mixing model.

It is advantageous to have some formal method of comparing RTDs from the different models. The variance (σ^2) of a distribution [defined as $\int_0^\infty (t - \tau)^2 \phi(t) \, dt$] is a measure of the spread of the distribution about the mean. The equations for variance are, in dimensionless form,

Diffusion Model (Correct Solution)

$$\sigma^2 = 2D^*\left[D^*\left(\exp(-1/D^*) - 1\right) + 1\right] \qquad (19)$$

Diffusion Model (Semi-infinite)

$$\sigma^2 = 2D^* \qquad (20)$$

m Equal Fully Mixed Reactors

$$\sigma^2 = 1/m \qquad (21)$$

One-Large/Two-Small Reactors

$$\sigma^2 = \left(\tau_1^2 + 2\tau_2^2\right) = \tau_1^2 + \left(1 - \tau_1\right)^2/2 \qquad (22)$$

Rogers/Gardner Model (m = 4)

$$\sigma^2 = \frac{1}{m}\left[\left(F_A + F_B\right)^2 + \left(2F_B^2/k\right)\right] \qquad (23)$$

Fig. 14 shows a comparison for one-parameter (dimensionless) models. For example, a variance of 0.4 corresponds to $D^* = 0.28$; or $\tau_1 = 0.55$, $\tau_2 = 0.225$; or $m = 2.5$. Unfortunately, the shapes of the RTD curves are different for each model (except for $n > 6$ and $D^* < 0.1$) so models with the same σ^2 do not generally give the same RTD or the same mill simulation.

It has already been demonstrated in Chapter 7 that a mill simulation is sensitive to the value of RTD in open circuit. However, the sensitivity of overall mill capacity to the RTD decreases considerably for closed-circuit operation (see Fig. 8 in Chapter 7). The sensitivity of open-circuit mill simulation to the shape of the RTD is shown in Table 5, in which the effect of using the Rogers/Gardner RTD is compared to using a semi-infinite D^* of 0.23 (see Fig. 10). Although the shapes of the RTDs are somewhat different, the results are within 3% of each other in terms of mill capacity, at almost the same size distribution (slightly steeper for the diffusion model

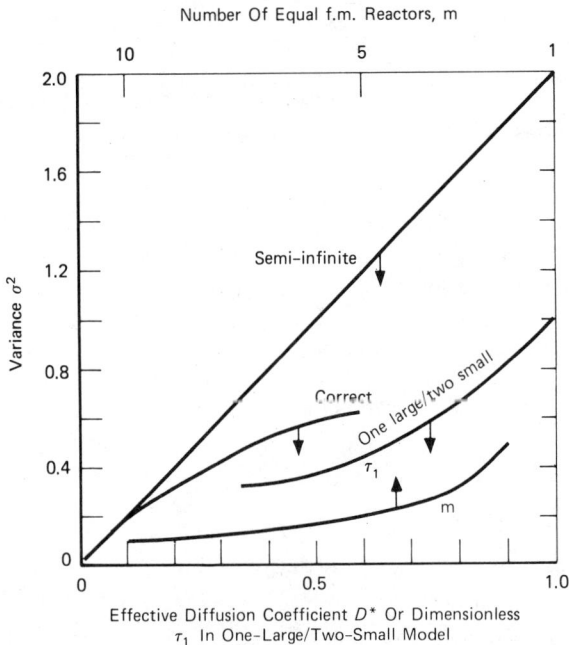

Fig. 14. Variance σ^2 for various one-parameter RTD models.

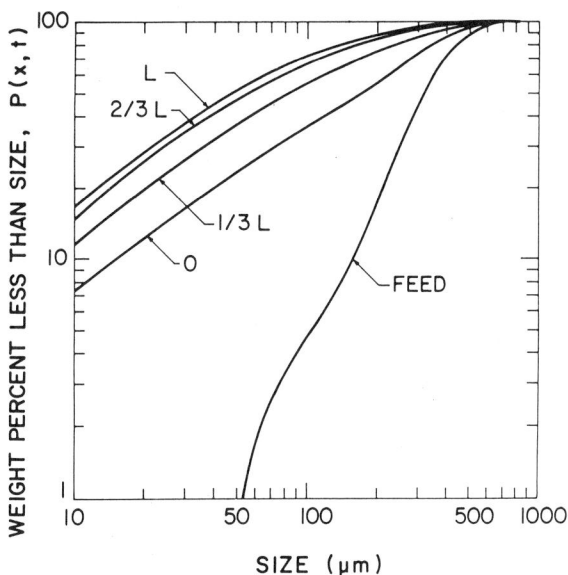

Fig. 15. Model predictions of size distribution along a mill.

RTD). The difference in capacity would be much less for a closed-circuit simulation.

One interesting conclusion from the RTD data is that the equivalent value of D^* enables the size distributions along the mill axis to be calculated (Rogovin, 1983), since results seem to be consistent with an axial convective-mixing model with constant D^*. Fig. 15 shows a typical result. It is seen that the size distributions are much closer to the product size distribution than to the feed. This is due to the boundary conditions of walls at each end, which give substantial back-mixing into the mill. These size distributions define the slurry rheology in the mill (see "Effect of Environment in the Mill" in Chapter 5).

Mass Transfer and Mixing in Ball Mills

It seems reasonable that one rotation of a dry ball mill will throw material at position l (to $l + dl$) in the mill backward and forward along the axis, according to a symmetric distribution, as illustrated in Fig. 16a. A mixing process such as this can be formally analyzed as comparable to a conventional diffusion process, as in Fig. 16b. All sections close enough to X will throw some material to the plane, from either side. This material can be treated as material originating from $X \pm \lambda$, where λ is an effective mean free path; thus,

$$m_1 - m_2 = \rho \bar{v}_1 f_{c1} - \rho \bar{v}_2 f_{c2}$$

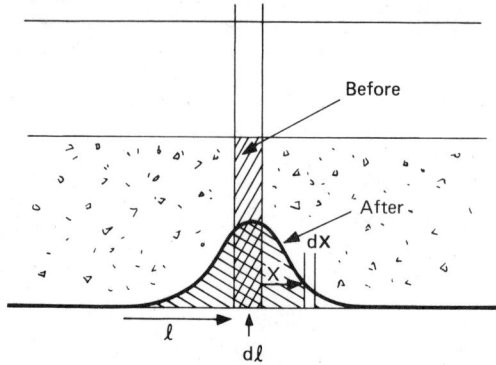

Fig. 16a. Illustration of some fraction of hold-up at l to $l + dl$ dispersed by one mill revolution.

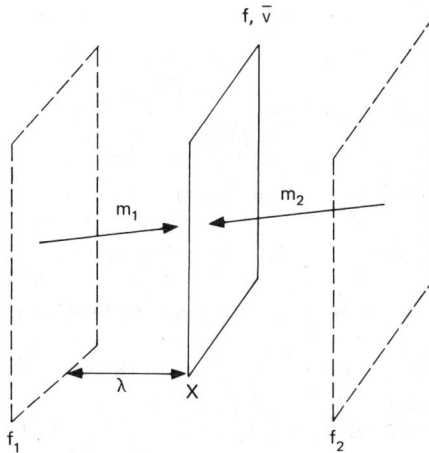

Fig. 16b. Illustration of mass reaching unit area at X, from either side.

or

$$\text{net rate of transfer per unit area} = \left(\bar{v} - \frac{d\bar{v}}{dl}\lambda \right)\left(f_c - \frac{df_c}{dl}\lambda \right)$$

$$- \left(\bar{v} + \frac{d\bar{v}}{dl}\lambda \right)\left(f_c + \frac{df_c}{dl}\lambda \right)$$

$$= -2\lambda\rho f_c \frac{d\bar{v}}{dl} - 2\lambda\rho\bar{v}\frac{df_c}{dl} \qquad (24)$$

where $\bar{v}$ is the mean velocity of material thrown in the l direction and ρ is the bulk density of the powder.

Under steady-state batch conditions $d\bar{v}/dl = 0$ and $df_c/dl = 0$ if there is a range of constant filling-mixing conditions in the mill, and there is no net mass transfer. However, if part of the powder is traced then the movement of traced material can be followed under these conditions even though $d\bar{v}/dl = 0$. In the mass balance on tracer, the concentration of tracer c replaces ρf_c and

$$\text{net flux of tracer} = -D\, dc/dl \qquad (25)$$

where $D = 2\lambda\bar{v}$ and is expected to be a function of f_c, ball charge, ϕ_c, ball size, etc. The spread of tracer (located as in Fig. 16a) in a region of uniform f_c and zero net mass transfer can be used to calculate D from

$$\text{concentration at } X, t \propto \left(1/2\sqrt{\pi Dt}\right)\exp\left(-X^2/4Dt\right)$$

In this way D can be determined as a function of f_c, $D = D(f_c)$. Substituting $\bar{v} = D/2\lambda$, Eq. 24 can now be put as:

$$\text{net mass flux at steady state} = -c\frac{dD}{dl} - D\frac{dc}{dl} \qquad (24a)$$

where D is expected to be a function of f_c (and J, ball size, etc.) and c is the total concentration of powder at position l.

If there are end effects in the mill, then D can be a function of l as well as f_c. If there are no end effects, the solution to Eq. 24a for a batch mill

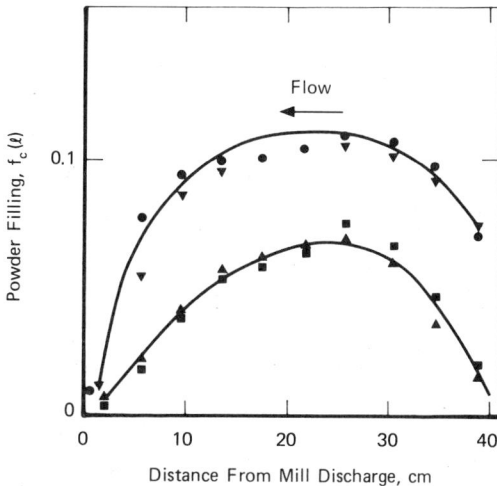

Fig. 17. Powder levels at steady state along a laboratory dry ball mill (Pour-Madani, 1982). $D = 152$ mm (6 in.); ■, batch, mean $\bar{f}_c = 0.044$; ▲, continuous, 0.8 g / s (0.03 oz./ s), $\bar{f}_c = 0.044$; ●, batch, $\bar{f}_c = 0.085$; ▼, continuous, $\bar{f}_c = 0.082$.

(flux = 0) is obviously c = constant, D = constant along the mill. However, Pour-Madani (1982) experimentally determined the filling level (dry) to be as shown in Fig. 17, demonstrating that end effects exist. It is clear that the throwing action is higher at the mill ends and D is greatest at $l = 0$ and decreases toward the middle of the mill, so that dD/dl is negative for $0 < l < L/2$. Eq. 24a for the batch condition is

$$D\frac{dc}{dl} = -c\frac{dD}{dl}$$

so that dc/dl is positive for $0 < l < L/2$: this explains the underfilling at the mill ends. In the continuous measurements at low flow rate shown in Fig. 17, the flow at the feed end of the mill moves *up* the filling level due to this effect.

Fig. 18 shows results at higher flow rates, up to flow rates so high that overfilling at the feed end expands the ball bed. When the bed expands, the balls also move down the mill, to give lower ball filling at the end where there is most powder and higher ball filling at the discharge end. In an industrial mill the discharge grate is designed to maintain the filling level at the discharge end at a reasonable value, so the filling levels do not change as much from discharge to feed as in Fig. 18. However, the decrease of filling at the feed end has been observed in cement mills (von Seebach, 1978).

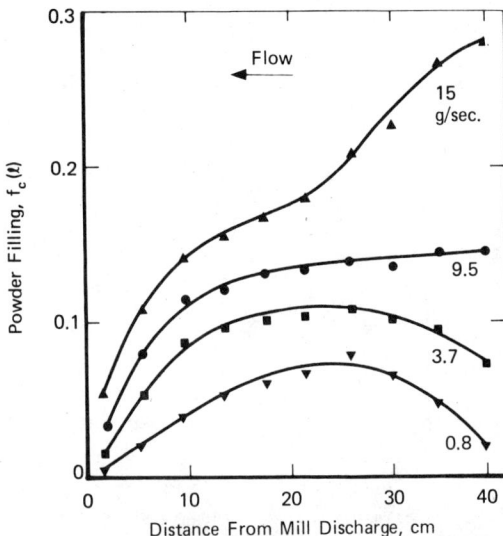

Fig. 18. Fractional powder filling along a ball mill at various feed rates of dry powdered quartz; D = 152 mm (6 in.); J = 0.2, ϕ_c = 0.7, d = 10 mm (3/8 in.), open screen discharge (Pour-Madani, 1982).

Marstiller (1979) has reported the accumulation of larger balls and the depletion of smaller balls at the mill ends for the second compartment of a cement mill of 3.8 m (12.5 ft) diam by 7.4 m (24.3 ft) long, with smaller balls accumulating in the middle. Presumably this is due to a different degree of end effect on the different ball diameters.

Hogg and Rogovin (1982) have proposed a model of transport and mixing in wet grinding mills, as illustrated in Fig. 19. They assume that the major mechanism of mass transfer for relatively dilute slurries is via hydraulic flow in the discrete pool of slurry, with hold-up controlled by a weir action at the overflow. Even if there were no axial mixing at all in the tumbling ball charge, the movement of particles in and out of the ball charge from the pool would give rise to dispersion of a pulse of tracer, very much like the retention and dispersion of a pulse of adsorbable material admitted to a chromatographic column. The more powder contained within the ball charge compared to that contained within the pool, the greater would be the dispersion at a given overall axial velocity of solid through the mill. Qualitatively, it is to be expected that a denser slurry with a higher viscosity would give more material in the charge and less in the pool, thus leading to larger apparent values of D^*. A combination of this type of effect with superimposed axial mixing caused by the tumbling action might well explain the results of Fig. 13.

Fig. 20 from work by Horst and Freeh (1972) shows the variation of mean solid filling level (hold-up) with slurry density at a fixed feed rate, for a grate discharge mill of about 0.4 m (1.3 ft) i.d. by 0.4 m (1.3 ft) long [$J = 0.4$, ball mix of size 20 to 64 mm (0.8 to 2.5 in.) $\phi_c = 78\%$]. The large

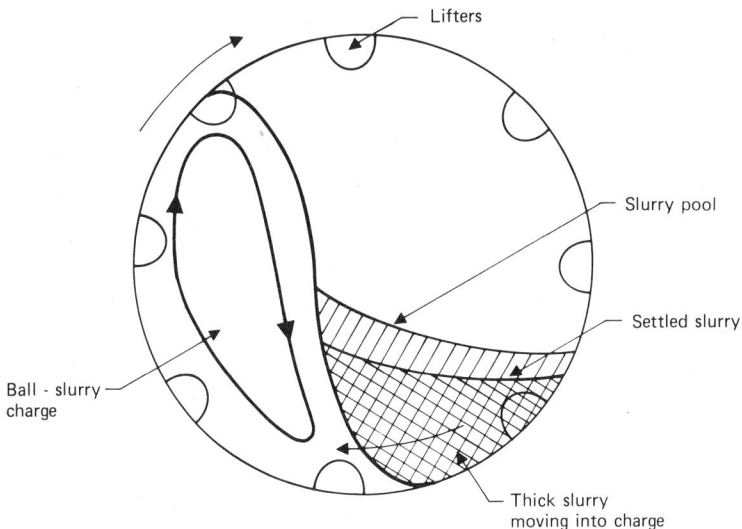

Fig. 19. Illustration of Hogg / Rogovin model for slurry transport in a tumbling ball mill.

increase in hold-up at higher slurry density might be explainable as due solely to the increased viscosity of the slurry. However, Hogg and Rogovin suggest that the discrete pool of relatively dilute slurry (see Fig. 19) is reduced by high slurry density (less water per unit of solid hold-up), so that solid hold-up must increase proportionately to give a desired flow rate.

The complexity of the mass transfer and mixing processes in dry and wet ball milling is evident. However, it seems only reasonable that low flow rates in small L/D (short) mills will give closer to a fully mixed condition, whereas high flow rates in a large L/D (long) mill will give closer to plug flow. A possible intuitive treatment based on mixing concepts follows. Assume that the flow and mixing processes in ball mills are geometrically similar between mills of different diameters, per revolution, providing ϕ_c, J, d, f_c, etc. are identical in the comparison. In particular, imagine the axial spread of a thin line of tracer at the mill center, as the mill rotates in a batch (no-flow) condition. For a long mill with no influence of back-mixing from the end walls, geometric similarity would mean that the same fractional concentration of tracer would occur at the same proportional length l/D from the center after the same number of revolutions. If a diffusion coefficient (D', say) is defined in units of m^2 per revolution, the relation between the degree of spreading and number of revolutions, r, is

$$\overline{x^2} = 2D'r \tag{26}$$

where $\overline{x^2}$ is the mean squared displacement from the origin.

Since $\overline{x^2}$ at a given r is proportional to D^2 by the argument of geometrical similarity, $D^2 \propto D'$; since $D = D'x(\text{rps})$ and rps $\propto 1/\sqrt{D}$ at

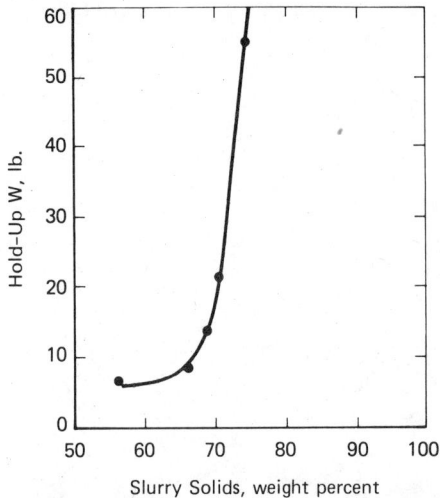

Fig. 20. Relation between hold-up and slurry density at constant feed rate (Horst and Freeh, 1972).

the same fraction of critical speed,

$$D \propto D^{1.5} \qquad (27)$$

If it is now assumed that for a given mill D is *not* dependent on flow rate u,

$$D^* = D/uL \propto D^{1.5}/uL \qquad (28)$$

Thus mills would tend more to plug flow (small D^*) at higher flow velocity, smaller D, and longer L. However, Fig. 13 shows that even over a wide range of feed rate the value of D^* did not change. This implies immediately that D is proportional to u for a given mill, which in turn implies that the physical process of bulk mass transfer along the mill axis is directly related to the tumbling action of the balls and charge, which also causes mixing. Using geometric similarity again, comparable conditions in two mills of different diameter would give $u \propto L/\text{rps}$ or $u \propto L/\sqrt{D}$. Thus Eq. 27 becomes

$$D^* \propto D^2/L^2 \qquad (29)$$

In this case, the only tendency to give closer to fully mixed or closer to plug flow would be the *ratio L/D*; short mills (small L/D) would be closer to fully mixed.

In the absence of a definitive theory of mixing and mass transfer in tumbling ball mills, either dry or wet, it is necessary to resort to empirical correlations. Chapter 5 has given the empirical correlation between overall mean hold-up W and flow rate deduced by Marchand, Hodouin, and Everell (1980), in the form (Eq. 22 in Chapter 5):

$$U = \left[(F/W_1)/(F_1/W_1) \right]^{0.5}, \qquad U \geq 1.0$$

where U is overall mean interstitial ball filling by powder, W_1 is the hold-up at $U = 1$ and F_1 is the solid flow rate through the mill which gives $U = 1$ and W_1. Chapter 7 has shown that this appears to be compatible with the Bond method of sizing mills. In order to achieve this compatibility for different mill diameters, it is necessary that F_1 varies with mill diameter according to

$$F_1 \propto \rho_s \phi_c D^{3.5}, \qquad D < 3.8 \text{ m (12.5 ft)}$$

The results in Fig. 3 in Chapter 12 for wet overflow ball mills suggest that

$$F_1 = 1.7 \phi_c D^{3.5} (\rho_s/2.65)(L/D), \text{ t/h}$$

where ρ_s is the solid density in t/m^3 and D is in meters.

Rogers (1983) has pointed out that this equation implicitly assumes that the ball loading in the mill corresponds to the overflow trunnion, so that overflow corresponds to $U = 1.0$. If the ball load is lower, overflow obviously corresponds to $U > 1.0$. He corrected for this factor by defining a corrected U value, U_c, by $U_c = UJ/J_o$, where J_o is the ball load required to reach the trunnion level; U_c is the filling level if $J = J_o$. Thus the empirical mass transfer equation becomes

$$U_c = (F/F_1)^{0.5}$$

that is,

$$U = (J_o/J)\left[F/(0.63\rho_s\phi_c D^{3.5}L/D)\right]^{0.5}, \quad F \geq F_1$$

$$= J_o/J, \quad F \leq F_1$$

where

$$F_1 = 0.63\rho_s\phi_c D^{3.5}L/D \qquad (30)$$

Rogers and Austin (1983) have used this correction on data from a number of mills, as shown in Fig. 21. Their data is consistent with $F_1 = 0.5\rho_s\phi_c D^{3.5}L/D$. However, it is certain that the constant in this equation must vary with factors such as slurry rheology, ball mix, lifter design, and it seems likely that it will change for large D and large L/D.

The estimates of powder filling f_c given in Table 3 in Chapter 5 imply that mills can be operated with powder filling perhaps as high as 0.45 volume fraction. For a ball load of $J = 0.4$, the actual volume of balls would be $0.6J$, that is, 0.24. Thus the total volume of balls and powder might be as

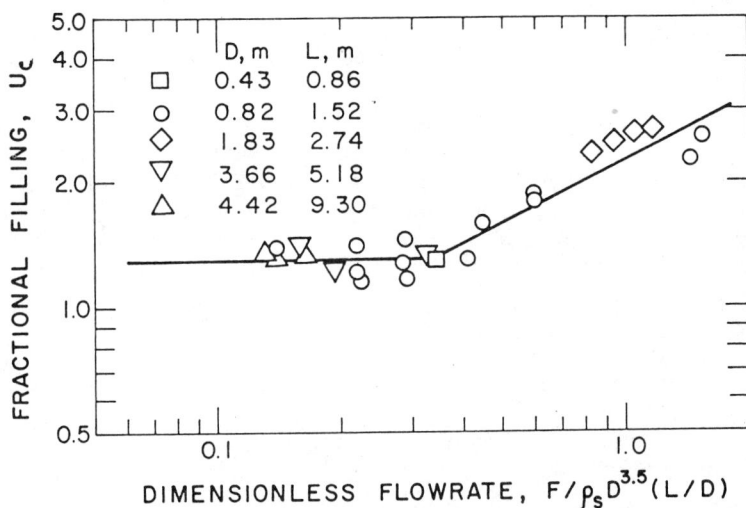

D, m	L, m
0.43	0.86
0.82	1.52
1.83	2.74
3.66	5.18
4.42	9.30

Fig. 21. Corrected interstitial ball filling U_c as a function of solid flow rate for mills of different diameter.

high as 0.7, corresponding to $U = f_c/0.4J = 2.8$. Assuming that $U > 2.8$ corresponding to a choked mill, Eq. 30 predicts a maximum rate of throughput of

$$F_{max} \approx (4 \text{ to } 5)\rho_s\phi_c D^{3.5}L/D, \qquad t/h \qquad (31)$$

For example, for $\rho_s = 2.7$, $\phi_c = 0.7$, $D = 3$ m, $L/D = 1.5$, the maximum predicted throughput is about 520 to 650 t/h.

An even more complex problem is the description of unsteady-state mass transfer. For example, the effect of a 10% step change in makeup feed rate to a cement clinker mill of 2.0 m (6.6 ft) diam by 11.5 m (37.8 ft) length has been investigated by von Seebach (1974). This mill was instrumented as a test facility, with mass flow measurement of makeup feed and recycle. The change in feed rate to the mill was detected as the start of a change in recycle rate only after about 10 min. Impulse residence time tracer tests on the mill gave the peak tracer concentration also at about 10 min. This suggests that the step of increased mass flow rate into the mill caused a step of increased filling level to move down the mill until it reached the mill exit, much like a pulse of traced material would move through the mill charge under steady-state flow conditions. This type of behavior is quite unlike that expected from any hydraulic analogy, where increased input flow rate would cause an immediate response in the exit flow rate.

In summary, it must be concluded that the laws of mass transfer in dry and wet ball milling are essentially unknown. This lack of knowledge leads to the following problems in process engineering design:

1. It is not possible to predict the effect of operating variables on the filling levels along a mill.

2. It is not possible to predict the maximum flow rate which any mill can handle, either for dry or wet grinding.

3. There are no equations available in the open literature which enable the calculation of the correct amount of open area for a grate or peripheral discharge mill for a given flow rate and optimum filling level.

4. It is not possible to predict the effect of slurry density and viscosity or the effect of cohesiveness of fine dry powders on the RTDs of mills.

5. Mill control systems must allow empirically for dead time in the response to a change in flow rate.

As an example of the difficulty of performing accurate process design calculations for certain cases, consider the following specific problem. Product and process specifications required the mill to perform a relatively small size reduction on an easily broken material, with a high circulating load to eliminate excess fines, by dry grinding. Simulation showed the required mean residence time to be small, so that very high flow rates through the mill were required. A central peripheral discharge mill was employed, since for the required volume of breaking media the path of feed through such a mill is half that of a normal mill of the same diameter and volume (and, hence, size reduction capability). Even so, the mill could not be forced to operate at a sufficiently high flow rate to avoid excess fines. In the absence of mass transfer laws, it could not be determined whether this was due to flow restrictions of the discharge ports or restrictions to flow through the ball charge.

Again, Rowland and Kjos (1980) have reported that the power drawn by a low level wet grate discharge mill is 1.16 times that of an equivalent wet overflow mill, with a corresponding increase of capacity. They ascribe this to a higher slurry density in the mill as the grate will allow relatively more water to leave the charge during the time the mill takes to come to steady state. There appears to be no way of predicting the true slurry density within a mill, yet breakage is certainly dependent on slurry density, especially for higher slurry density.

It must be emphasized that the analysis of large-scale mill test data is inevitably restricted by the lack of: (1) residence time distribution data; (2) solid hold-up deduced from mean residence time τ_{solid}; and (3) slurry density in the mill deduced from τ_{solid} and τ_{water}.

Summary

Simulations show that the circuit product size distribution and mill capacity for closed-circuit grinding are not very sensitive to RTD, although the value of circulating load and the size distributions of mill feed, tailings, and mill product are more sensitive to the RTD. Thus an accurate mill simulation requires an accurate RTD, but an overall design can be performed with an estimate of RTD.

The measurement of RTD on a mill using a tracer added to the feed requires correction for recycled tracer when the mill is in a closed circuit. This means that the concentration of tracer in the recycle stream must be measured in addition to the tracer in the mill exit. However, this is often difficult to do in practice. This problem is avoided if the traced material is made sufficiently small so that tracer at the mill exit is recycled solely via the apparent bypass fraction of the classifier, because the fraction of recycled material is then a known fraction of the exit material. The method of extracting the primary RTD from data with recycled tracer is outlined. In principle, recycled tracer leads to a series of peaks in tracer concentration at the mill exit, but in practice they are not usually detectable.

For open-circuit tests the cumulative RTD is calculated from

$$\Phi(t) = \int_0^t c(\theta)\, d\theta \Big/ \int_0^\infty c(\theta)\, d\theta \tag{S1}$$

where $c(\theta)$ is the proportional concentration of tracer at the mill exit at time θ after admission of a pulse of tracer. The differential residence time $\phi(t) = d\Phi/dt = c(t)/\int_0^\infty c(\theta)\, d\theta$. The value of mean residence time is obtained from

$$\tau = \int_0^\infty t\phi(t)\, dt \tag{S2}$$

The integrations are performed numerically, or analytically by fitting a function to the values of $c(t)$ or $\phi(t)$.

In wet systems, the water can be traced using chemical tracers, but it is better to trace the solid using radiotracing of the solid. This method is also preferred for dry grinding, although microporous solids such as cement clinker can be traced with fluorescein dye solution absorbed on the solid. Solid is irradiated by thermal neutrons in a nuclear reactor and the count rate measured by NaI-photomultiplier detectors. A small sample of traced solid is added to the mill feed and the proportional concentration of tracer leaving the mill is detected directly and continuously, or representative samples are collected at the mill exit at various times after admission for analysis of tracer content by counting a standard volume.

A simple treatment of flow and mixing of tracer along the mill axis leads to the convective-mixing equation

$$\frac{\partial c}{\partial t} = D\frac{\partial^2 c}{\partial l^2} - u\frac{\partial c}{\partial l} \tag{S3}$$

where c is tracer concentration at time t and position l along the mill, and D is a mixing factor equivalent to a diffusion coefficient. This equation in dimensionless form was solved for a pulse of tracer, for suitable boundary conditions, and the variation of $\phi^*(t)$ vs. t^* is shown in Figs. 3a and 3b, for various values of D^*: $\phi^*(t^*)\,dt^*$ is the fraction of feed which emerges from the mill at time t^* to $t^* + dt^*$ after admission; t^* is t/τ; $D^* = D/uL = D\tau/L^2$, L being mill length. A convenient approximate solution is the Mori semi-infinite solution

$$\phi^*(t^*) = \left(1/2\sqrt{\pi D^* t^{*3}}\right)\exp\left[-(1 - t^*)^2/4D^* t^*\right] \tag{S4}$$

Another method of treating RTD is to consider the RTD as equivalent to that from a number of reactors or tanks in series. For m unequal fully mixed reactors

$$\phi(t) = \sum_{i=1}^{m} \frac{\tau_i^{m-2}\exp(-t/\tau)}{(\tau_i - \tau_1)(\tau_i - \tau_2)\ldots(\tau_i - \tau_m)}, \qquad i \neq k,\, \tau_i \neq \tau_k \tag{S5}$$

where τ_i is the mean residence time in reactor i, $\tau = \Sigma\tau_i$. For equal reactors of mean residence time $\bar{\tau}$ each

$$\phi(t) = \frac{1}{\bar{\tau}^m}\frac{t^{m-1}}{(m-1)!}\exp(-t/\bar{\tau}) \tag{S6}$$

For one major fully mixed reactor, τ_1, followed by two smaller equal fully mixed reactors, τ_2, (one-large/two-small model, $\tau = \tau_1 + 2\tau_2$)

$$\phi(t) = \frac{\tau_1}{(\tau_1 - \tau_2)^2}\left[\exp(-t/\tau_1) - \exp(-t/\tau_2)\right] - \frac{t}{(\tau_1 - \tau_2)\tau_2}\exp(-t/\tau_2)$$

$$\tag{S7}$$

The Rogers/Gardner model is of flow through four equal stages in series, each stage consisting of a fully mixed reactor connected to a dead-space reactor in parallel, followed by a plug flow reactor. This model has two parameters in addition to τ, so it is more flexible than the above models which have only one parameter in addition to τ. It is, however, not suitable for comparison of the dimensionless RTD with mill conditions whereas the one-parameter model can be used to compare D^*, m or τ_1 with, for example, slurry density.

It must be recognized that these models are only convenient expressions for fitting experimental RTD information. If they fit, it does not mean that the models have any physical reality. The value of τ required to give the best fit of data to a model is not necessarily the same as W/F.

Small-scale results suggest that D does not vary much with ball filling for $U < 1$, but it decreases rapidly for $U > 1$; it decreased with increase of powder filling at a given ball charge. Tests by Mori on dry ball milling in a pilot-scale mill showed little effect of grate discharge area, but greater mixing for larger ball diameter. Tests on wet overflow ball mills gave similar normalized RTDs for a range of mill sizes. The normalized RTD did not appear to change with flow rate. There appeared to be little effect of particle size, but water RTDs consistently gave about 10 to 15% lower mean residence time. Counting of a radioactive tracer through the mill case gave dimensionless results consistent with a constant value of the mixing factor D^* along the mill, for a 0.3 m (1 ft) i.d. by 0.6 m (2 ft) long laboratory wet overflow mill.

A greatly oversimplified analysis using geometrical similarity of mixing per mill revolution suggests that the mixing factor D is proportional to $D^{1.5}$. However, dimensionless RTDs do not vary much with flow rate for a given mill and this suggests that the mass transfer mechanism and the mixing process are linked, so that D is also proportional to the flow velocity u. Then $D^* \propto (D/L)^2$, and it is expected that short L/D mills would be closer to fully mixed and long L/D mills closer to plug flow. However, other factors such as the effect of slurry density, ball size, appear to obscure this result.

There appears to be no satisfactory theory of mass transfer in ball mills, either for dry or wet grinding. Tests in small dry mills show that both ends of the mill can be underfilled, presumably because end effects throw material toward the middle of the mill. High flow rates require a gradient of filling level along the mill.

There is no doubt that overall hold-up increases with flow rate in wet milling, and that higher slurry densities require higher solid hold-up to give the same solid flow rate. In order for simulation models to be compatible with the Bond method of sizing wet overflow ball mills, the overall mean hold-up is related to solid mass flow by the empirical equation

$$U = \left(J_o/J\right)\left(F/0.63\rho_s\phi_c D^{3.5}L\right)^{0.5}, \quad D < 3.8 \text{ m (12.5 ft)}, \quad U \geq 1.0$$

$$(S8)$$

where J_o is the ball load to fill the trunnion level. The numerical constant in this equation must vary with slurry density, especially for high slurry density.

The dead time for response to change in feed rate appears to be the same magnitude as the mean residence time, for dry ball milling.

There are no published equations which can be used to calculate maximum flow rates, filling levels along a mill, or the correct grate discharge areas for various mill conditions. It is recommended that mill trials should routinely incorporate the measurement of RTDs, until sufficient information becomes available to enable the calculation of RTD hold-up and internal slurry density for any mill conditions.

References

Austin, L. G., Luckie, P. T., and Ateya, B. G., 1971, *Residence Time Distribution in Mills*, *Cement and Concrete Research*, Vol. 1, Pergammon Press, pp. 241–256.

Austin, L. G., et al., 1983, "The Axial Mixing Model Applied to Ball Mills," *Powder Technology*, Vol. 36, pp. 119–126.

Gardner, R. P., Rogers, R. S. C., and Verghese, K., 1977, "Short-Lived Radioactive Tracer Methods for the Dynamic Analysis and Control of Continuous Comminution Processes by the Mechanistic Approach," *International Journal of Applied Radiation and Isotopes*, Vol. 28, pp. 861–871.

Gardner, R. P., Verghese, K., and Rogers, R. S. C., 1980, "On-Stream Determination of Large Scale Ball Mill Residence Time Distributions with Short-Lived Radiotracers," *Mining Engineering*, Vol. 32, pp. 422–431.

Hogg, R. and Rogovin, Z., 1982, "Mass Transport in Wet Overflow Ball Mills," Preprint, 14th International Mineral Processing Congress, Canadian Institute of Mining and Metallurgy, pp. I-7.1–7.19.

Horst, W. E. and Freeh, E. J., 1972, "Mathematical Modeling of a Continuous Comminution Process," *Trans. SME-AIME*, Vol. 252, pp. 160–167.

Kelsall, D. F., Reid, K. J., and Restarick, C. J., 1969–70, "Continuous Grinding in a Small Wet Ball Mill. Part III. A Study of the Distribution of Residence Time," *Powder Technology*, Vol. 3, pp. 170–178.

Marchand, J. C., Hodouin, D., and Everell, M. D., 1980, "Residence Time Distribution and Mass Transport Characteristics of Large Industrial Grinding Mills," *Proceedings*, Third IFAC Symposium, J. O'Shea and M. Polis, eds., Pergammon Press, pp. 295–302.

Mardulier, F. J. and Wightman, D. L., 1971, "Efficient Determination of Mill Retention Time. Parts I, II, and III," *Rock Products*, Vol. 74, No. 6, pp. 74–75, 90–91; No. 7, pp. 78–79, 108–110; No. 8, pp. 60–61, 86–88.

Marstiller, S., 1979, "Cost and Up-time Considerations for Selection of Mill Liners and Grinding Balls," American Magotteaux Corp., Sept.

Mori, Y., Jimbo, G., and Yamazaki, M., 1964, "On the Residence Time Distribution and Mixing Characteristics of Powders in Open-Circuit Ball Mill," *Kagaku Kogaku*, Vol. 28, pp. 204–213.

Mori, Y., Jimbo, G., and Yamazaki, M., 1967, "Flow Characteristics of Continuous Ball and Vibration Mills," *Proceedings*, Second European Symposium Zerkleinein, H. Rumpf and W. Pietsch, eds., Dechema Monographien 57, Nr. 605–632, Verlag Chemie, Weinheim, pp. 605–632.

Pour-Madani, M., 1982, "Transport of Powder in Dry Ball Milling," Ph.D. Thesis, The Pennsylvania State University, University Part, PA, Aug.

Rogers, R. S. C., 1983, Kennedy Van Saun Corp., Danville, PA, private communication.

Rogers, R. S. C. and Gardner, R. P., 1979, "Use of a Finite-Stage Transport Concept for Analyzing Residence Time Distributions of Continuous Processes," *Journal of the American Institute of Chemical Engineers*, Vol. 24, pp. 229–240.

Rogers, R. S. C. and Austin, L. G., 1983, "Residence Time Distributions in Ball Mills," submitted to *Journal of Fine Particle Society.*

Rogers, R. S. C., Bell, D. G., and Hukki, A. M., 1982, "A Short-Lived Radioactive Tracer Method for Measuring Residence Time Distributions of Closed Circuit Ball Mills," *Powder Technology*, Vol. 32, pp. 245–252.

Rogovin, Z., 1983, "The Transport and Grinding of Material in a Continuous Wet, Overflow Ball Mill," Ph.D. Thesis, The Pennsylvania State University, University Park, PA, Mar.

Rowland, C. A., Jr. and Kjos, D. M., 1980, "Rod and Ball Mills," *Mineral Processing Plant Design*, A. L. Mular and R. B. Bhappu, eds., 2nd ed., AIME, New York, pp. 239–278.

Shoji, K., Hogg, R., and Austin, L. G., 1973, "Axial Mixing of Particles in Batch Ball Mills," *Powder Technology*, Vol. 7, pp. 331–336.

Trimarchi, T., 1979, "Mathematical Treatments of Some Problems in Fuel Science," Ph.D. Thesis, The Pennsylvania State University, University Park, PA.

von Seebach, H. M., 1974, Dyckerhoff Zementwurke, Neubeckum, West Germany, private communication.

von Seebach, H. M., 1978, Polysius Corp., Atlanta, GA, private communication.

Weller, K. R., 1980, "Hold-up and Residence Time Characteristics of Full Scale Grinding Circuits," *Proceedings*, Third IFAC Symposium, J. O'Shea and M. Polis, eds., Pergammon Press, pp. 303–309.

15

Grinding Aids

Introduction

Laboratory and industrial grinding tests have shown that the process of size reduction can be significantly influenced by chemicals added to the powder or slurry being ground. The terms *grinding aid* or *grinding additive* refer to a substance which when mixed into the mill contents causes an increase in the rate of size reduction. The increased rate can be used to grind a higher feed rate to the desired product size or to produce a finer product size at a fixed feed rate. Whether the use of a grinding aid is justified in any given situation depends on the cost of the substance vs. the improvement of output or product quality obtained with its use. Obviously, an expensive chemical must be effective in very small concentrations if it is to be economically justifiable. The cost criterion is calculated on the basis of the cost of the grinding additive per ton of material ground.

Although there is direct experimental verification of the advantageous effect of grinding additives, no sound scientific explanation has yet been offered to explain or predict the general behavior of additives. Rose and Sullivan (1958) have summarized most of the work reported prior to 1950; Hartley, Prisbrey, and Wick (1978), and Anon. (1981) have recently prepared an updated listing of the grinding additive literature. Unfortunately, none of the above summaries offers a coherent picture that the mineral processing engineer can use to choose and apply grinding aids on a consistent basis. Many of the studies reported consisted of subjecting materials with simple geometric shapes to some type of hardness or controlled single fracture test in the presence of different chemical environments. On the other hand, a number of the studies were tests on operating industrial-scale mills in which there was neither careful control of grinding conditions nor precise monitoring of the effect of the additives.

Out of this work have come several hypotheses to explain the action of grinding aids. A mechanism which is often quoted is the Rehbinder (1932) effect: Rehbinder suggested that the adsorption of additive on the surface of a solid lowers the cohesive force which bonds the molecules of the solid together (see Fig. 3 in Chapter 2). In particular, adsorption on the surfaces of a flaw in the surface of a solid could affect the bonding forces and surface energy at the point where fracture initiates (Somasundaran and Lin, 1972; Westbrook, 1966), as illustrated in Fig. 1. Westwood and Stoloff (1966) and Westwood (1974) have demonstrated the effect of adsorbed molecules on

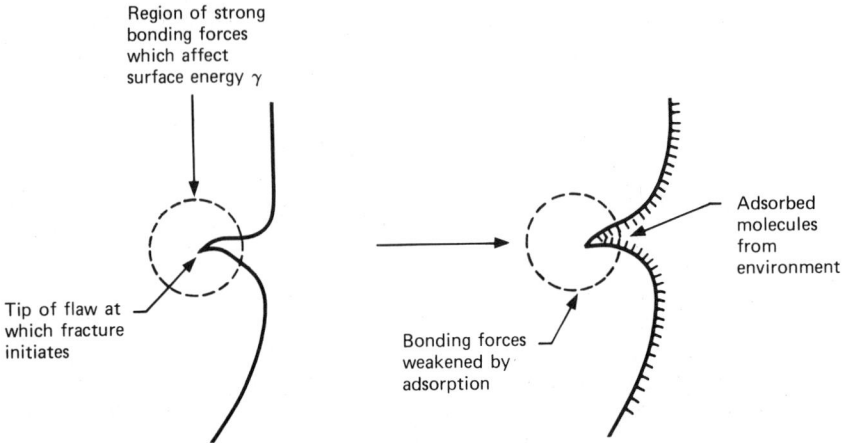

Fig. 1. Illustration of the explanation of the Rehbinder effect by adsorption in a surface flaw.

various surface mechanical properties and they refer to the phenomena in general as *chemomechanical effects*. They suggest that the adsorbed molecules may *pin* dislocations near the surface thus preventing easy movement of dislocations under stress gradients. Since plasticity is due to movement of the dislocations, the region near the surface of the solid is thus rendered more brittle.

The surrounding molecular environment can certainly affect the critical stress-strain required to produce fracture under conditions where the fracture initiates from a flaw in the surface. Experimental examples are slow compressive, tensile or bending tests on some materials, e.g., the tensile strength of glass fibers is strongly dependent on damage to the surface and is affected by immersion in different chemical environments. However, the normal action in a grinding mill is more comparable to striking a small piece of solid with a large hammer, as fracture will occur whether the surface is flawed or not. It is likely that the stress waves produced by massive high-speed impact will activate flaws throughout the solid; then a maze of fractures will propagate at high speed (near the speed of sound in the material). Simple calculation shows that molecules of an additive cannot diffuse down the cracks at rates comparable to the speed of sound, so they cannot affect the crack energy at the propagation tip of the crack.

Thus we will argue that it is highly improbable that the Rehbinder chemomechanical effect could explain the mode of operation of grinding aids in ball mills, roller mills, and others. Similarly, chemomechanical effects due to the movement of dislocations can only occur in the time scale of such movements, which are far slower than the speed of brittle crack propagation, and chemomechanical effects have only been demonstrated in processes where local plastic flow is important. Locher and von Seebach (1972) have given strong experimental evidence against chemomechanical effects in dry grinding of cement clinker using grinding aids (see "Chemical Additives for

Dry Grinding of Powders" in this chaper). Thus one must look for explanations of the effect of grinding aids other than those involving effects on the fracture energy or degree of surface plasticity.

There is substantial evidence that additives operate by changing the fluidity of slurries and the flowability of dry powders. Additives which affect fluidity obviously alter the mass transfer through a continuous mill and thus affect the grinding process. In addition, a change in fluidity may alter the manner in which particles distribute in the mill and get nipped by tumbling balls; this effect is detected in batch grinding tests. There is also evidence that very fine grinding can cause agglomeration of some materials to, in effect, regrow particles. Additives may prevent the formation of these strong agglomerates by keeping the particles well dispersed and/or preventing fusion of particle surfaces together.

The following sections present recent experimental information and the next-to-last section returns to the mechanism by which additives operate.

Chemical Additives for Dry Grinding of Powders

In the dry ball milling of fine powders, a common phenomenon is the slowing-down of the rates of breakage as the powder in the bed becomes finer. This may be due to the regrowth or rebuilding of particles from smaller particles by either agglomeration involving van der Waal's forces or by direct briquetting, or coating of the balls to give soft surfaces. It may be the effect discussed in "Effect of Environment in the Mill" in Chapter 5, where there was no evidence of particle growth or ball coating. For example, the fine grinding of cement, limestone, coal, chemical pigments, and plastics in ball mills gives size limits beyond which it is impractical to grind finer in the dry state, even though the chemical reactivity and/or the end use of the materials require dry grinding. With some materials complete drying or reduced temperature can aid dry grinding of fines. In some cases it may be necessary to use different breakage machines to obtain a sufficiently fine grind; these other machines are usually more expensive, require greater specific grinding energy, have greater maintenance, and may not have high capacity for a single unit.

An alternative that is economical for industrial application is the use of chemical additives. Most of the dry grinding additives are simple organic compounds which undergo chemisorption onto the particle surfaces in sufficient quantities to prevent adhesion forces from developing between particles. Compounds such as ethylene glycol, propylene glycol, triethanolamine, oleic acid, and aminoacetates have been used in fine cement and limestone grinding for a number of years. Batch grinding tests in a laboratory mill can be used to screen additives to get some idea of optimum concentrations, providing they are performed at material temperatures and dryness and levels of size reduction which are realistic for full-scale use. Finally, however, plant trials normally are used for the selection and determination of effective quantities, which are usually in the range of 100 to 500 ppm by weight.

A definitive investigation into the use of a variety of organics for dry cement grinding has been carried out by Locher and von Seebach (1972), giving conclusions which are consistent with industrial observations. They conclude that the use of chemical additives does not affect the breakage of coarse material, but only becomes a factor when fine material builds up in the mill (that is, they are effective for cements with greater than about 3200 Blaine, cm^2/g). The chemicals must be added under conditions leading to chemisorption on particle surfaces, which for some organics may require elevated temperatures, e.g., 120°C for ethylene glycol and 70°C for butylamine on cement clinker. Sufficient additive must be used to form the adsorbed layer on all the area of fine particles produced in the grinding. The effect of the additives is to decrease the van der Waal's adhesion force between fine particles, thus allowing more efficient breakage interactions between grinding media and particles by a mechanism which is not completely understood, and reducing agglomerate formation where this is a problem. It seems likely that the additive reduces the slowing-down effect discussed in "Effect of Environment in the Mill" in Chapter 5. Note the data for quartz in that section showed slowing-down with no indication of ball coating, suggesting that coating is not the basic cause of reduced breakage rates.

The adsorbed additive also gives increased flowability of the powder. A reduction in mean residence time in a grate discharge mill at a given feed rate is observed, which means that hold-up decreases. Recently reported tests on the use of additives for high area cement production state that the following phenomena are observed in the first 15 to 20 min after the start of chemical addition (Decasper, 1982). The noise level from the first compartment increases, while that from the second compartment decreases. This probably indicates more rapid removal of fines from the first compartment, and a lessening of cushioning effects by deagglomeration of powders in the second compartment. The circulating load at first increases due to the mill emptying out, then settles down to a lower value than for steady state without aid. This indicates a finer product and less recycle of agglomerated fines. It should be noted that the mill circuit, therefore, should be readjusted to bring the residence time and level of powder in the mill back to the proper values. It is also recommended that the mixture of balls in the second compartment be reoptimized (Decasper, 1982). The better flow characteristic of dry powders after grinding with additive sometimes has additional advantages, e.g., easier filling of bags of cement and less packing in storage. Milling systems which are dry grinding to fine powder frequently employ high levels of gas sweeping to blow fines out of the mill, to prevent accumulation of fines in the mill (see Chapter 17). Even if airsweeping is used, it may still be advantageous to use dry grinding additives.

Wet Grinding of Ores

The most important *grinding additive* in tumbling ball mills is, of course, water. It is hardly necessary to look for any chemomechanical action here, however, since the effect must surely be one of bringing and keeping the particles in advantageous positions to receive a breakage action. Meloy and Crabtree (1967) examined the effect of a variety of liquids on ball milling and concluded that the action of a liquid was to coat colliding surfaces with particles, thus reducing steel-to-steel contact which converts the energy of the tumbling balls to heat without causing breakage.

"Effect of Environment in the Mill" in Chapter 5 discussed the various regimes which occur in wet grinding. To summarize: larger sizes can break in a batch test with accelerating rate, decelerating rate, or first-order rate depending on the loading conditions, slurry density, and feed size distribution. Smaller sizes break in a first-order manner, with a constant value of α in $a(x_i/x_o)^{\alpha}$; the value of aW can be used as an index of the absolute breakage rates in the mill. The value of aW is almost constant for the range of normal slurry densities where the slurry is still fluid. A small maximum in breakage rate occurs at optimum filling at a higher slurry density as the slurry develops near-Newtonian pseudoplastic rheology (Region B). The optimum filling condition is higher for higher slurry density. At still higher slurry density, the rheology becomes pseudoplastic with yield stress (Region C), and breakage rates decrease. In all cases, the development of a fine size distribution as grinding proceeds can lead to a breakage region where the breakage rates slow down.

Klimpel et al. investigated in detail the region of slurry density between constant breakage rates and decreased breakage rates (Klimpel, 1980, 1982,

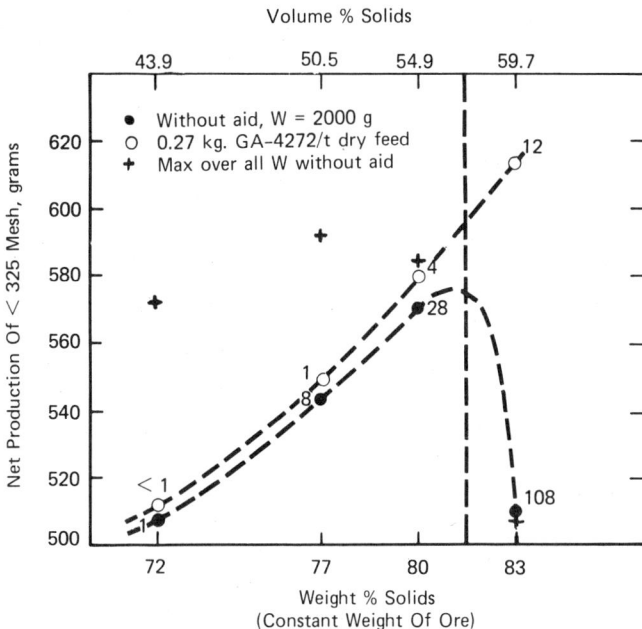

Fig. 2. Net production of $-45~\mu$m (-325 mesh) taconite in 30 min grind, at various weight percent solids (constant weight of ore $U = 0.7$); viscosities in Pascal seconds (Pa · s). The feed size was 12.0% $< 45~\mu$m (325 mesh), hold-up $W = 200$ g, mill i.d. = 200 mm (8 in.), ball diam = 25 mm (1 in.).

1982a, 1983; Klimpel and Manfroy, 1977, 1977a, 1978; Manfroy and Klimpel; Klimpel and Samuels, 1979; and Katzer, Klimpel, and Sewell, 1981). They showed that the highest breakage rates occurred for fairly dense slurries under conditions which gave pseudoplastic rheology without associated high yield stress. This cannot be accomplished by water addition since more water reduces viscosity and slurry density, changing the rheology toward dilatant. On the basis of this observation, Klimpel et al. investigated the use of chemical additives to reduce slurry viscosity while still maintaining a high fractional volume of solids. Additives were chosen which would act as fluidity modifiers under the high shear conditions in a ball mill. Rheological studies were used to screen the additives. Fig. 2 shows a typical result. As slurry density was increased, the production of $-45 \mu m$ (-325 mesh) material also increased, but between 80 and 83 weight percent solids, it fell sharply. Comparative slurry viscosities were measured with an RVT Brookfield viscometer under standard conditions (size B T-spindle, 20 rpm), and the values for the slurries at the end of the grinding time are shown as

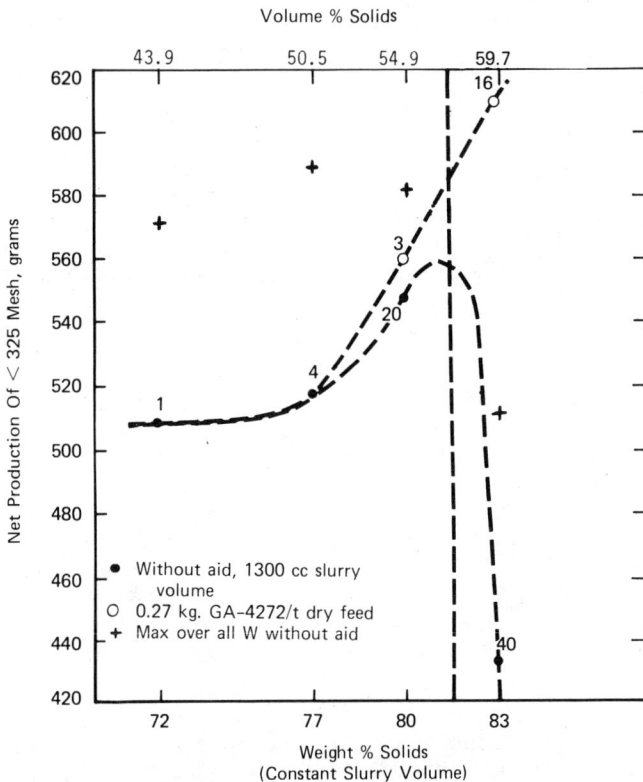

Fig. 3. Net production of $-45 \mu m$ (-325 mesh) taconite at various weight percent solids (constant slurry volume), viscosities in Pascal seconds, Pa · s (see Fig. 2).

the numbers on the figure. The change in comparative slurry viscosity between 80 and 83 weight percent solids was from 28 Pa · s to 108 Pa · s. They concluded that it was desirable to operate at as high a slurry density as possible, providing the development of viscosity as fines accumulated did not make the pulp too viscous.

Figs. 2 and 3 shows a result with Dow Chemical GA-4272, which is a low molecular weight (< 20,000) water-soluble anionic polymer. It is clear that the additive had only a small effect up to 80 weight percent solids, but enables an 83 weight percent solids slurry to be used without the development of high viscosity, giving the highest grinding rate. Note that results of

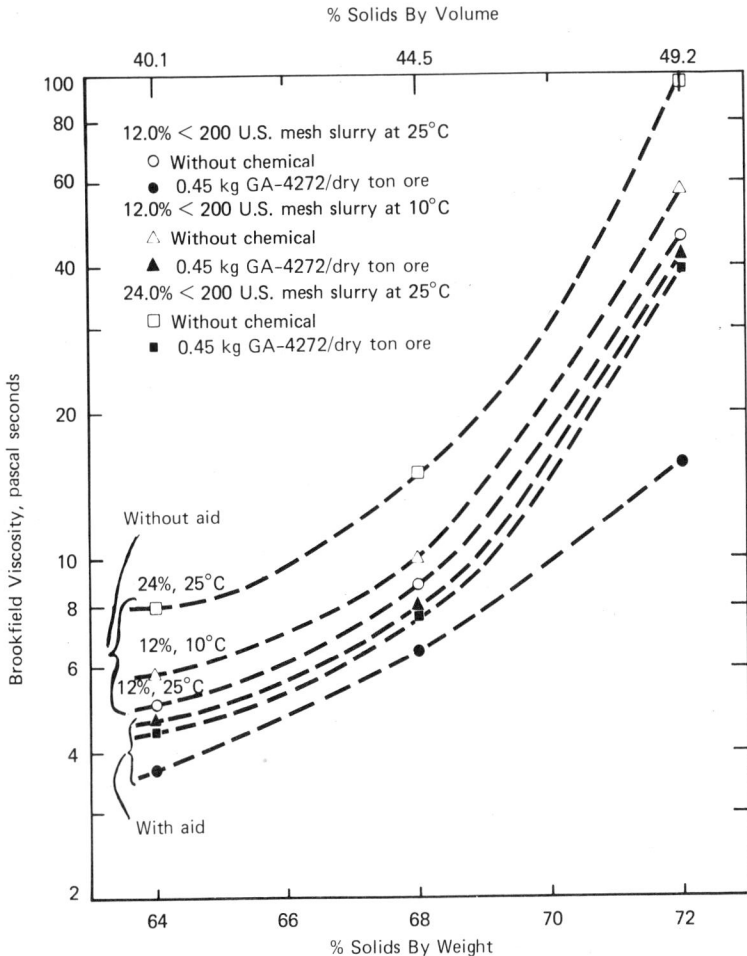

Fig. 4. Comparative viscosity of a slurry of copper ore as a function of percent solids, particle size, slurry temperature, and dosage of GA-4272 (Schuhmann slope ≈ 1).

this kind of test overemphasize the maximum in breakage rates in comparison with tests where optimum filling conditions are used (see Fig. 14 in Chapter 5). However, the increased rate with chemical is significantly higher than the maximum without chemical. The increased grinding rate with chemical addition was accompanied by an increase in net power drawn by the mill (Katzer, Klimpel, and Sewell, 1981), suggesting that the high slurry density with reduced viscosity causes an increase in tumbling action of the balls. Similar results were found with other additives (Manfroy and Klimpel).

Fig. 4 shows typical viscosity data for a slurry of copper ore, as a function of percent solids, slurry temperature, particle fineness, and grinding aid. It is apparent that increased slurry viscosity follows when percent solids is increased, the temperature is lowered, the slurry fineness is increased, or the chemical additive is not used. The influence of the temperature of the slurry being ground on net production is illustrated in Fig. 5. A temperature drop from 70° to 50°F lowered the maximum net production of -45 μm (-325 mesh) in 45 min of grinding by almost 10%. Sodium silicate was not effective as a grinding aid.

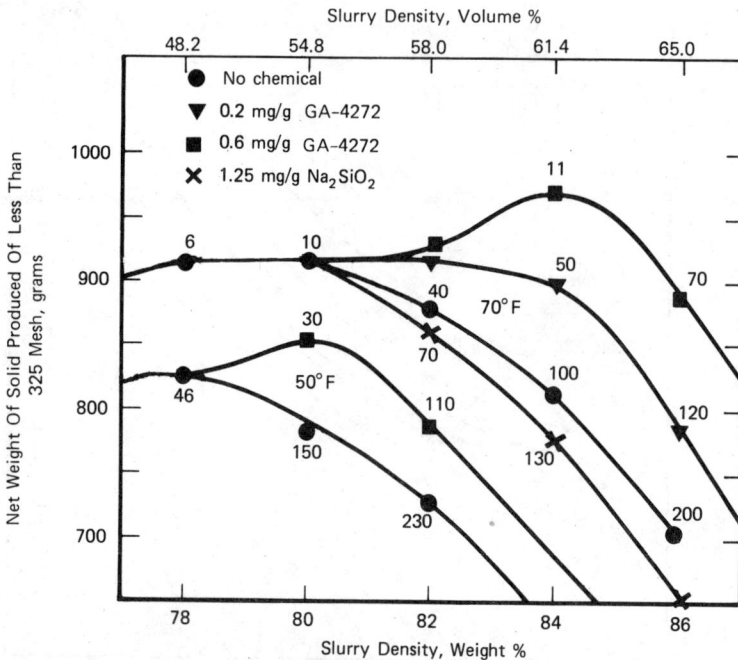

Fig. 5. Net production of -45 μm (-325 mesh) taconite in 45 min batch grind, as a function of slurry density and grind temperature (constant slurry volume); viscosities in Pascal seconds, Pa · s (see Fig. 2).

Fig. 6 shows further comparison of the effect on slurry viscosity of a simple dispersion agent, sodium silicate, compared to the fluidity modifier (Klimpel, 1982a, 1983). Sodium silicate reduced viscosity only at concentrations greater than 0.4 kg/t (0.9 lb/ 1.1 st), and actually increased viscosity in the presence of calcium ion unless greater than 0.5 kg/t (1.1 lb/ 1.1 st) was used. Calcium also decreased the effectiveness of GA-4272, but the aid still reduced viscosity at all concentrations. Obviously, higher concentrations of grinding aid are expected to be more effective and this was confirmed by rate studies.

A more complete rate analysis has been performed under the conditions shown in Table 1. Estimates were made of the B values and the α value in the first-order breakage region, and the values are given in Table 2. The major parameter varying as a result of slurry density change with or without chemical was the value of S. The relative values of S as a function of size did not change (α was constant) so no changes in preferential grinding with respect to particle size were observed, only changes in absolute production. These values of α and $B_{i,j}$ were used on data from batch grinding of a similar ore to back-calculate S values. In this case, the starting feed consisted of a size distribution, 100% −2 mm (−10 mesh) and 6.3% −45 μm (−325 mesh). After 15 and 30 min of grinding, the first-order breakage hypothesis predicted the results to a reasonable approximation for all conditions (see Fig. 7). After 60 min, however, the experimental values were much lower than the predicted values, for 82 and 84% slurry density when no aid was present, showing the slowing-down of breakage under these

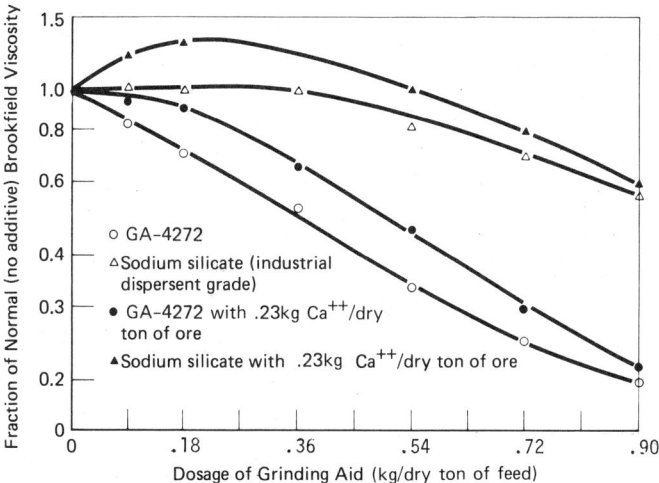

Fig. 6. Viscosity reduction of a slurry of copper ore (see Fig. 4) at 72% solids by weight, as a function of chemical type, dosage, and calcium ion concentration; 12.0% < 75 μm (200 mesh) slurry at 25°C.

Table 1. Experimental Parameters for Taconite Grinding Study

True specific gravity	3.308
Ball charge	
Bulk vol, cm^3	2000
True vol, cm^3	1130
Void vol, cm^3	870
Ball diam, mm	25.4
Percentage voids	43.5
Slurry volume, cm^3	1044
Ball mill	
Vol, cm^3	5790
Diam, mm	200
rpm	60
Slurry concentrations	
80% solids by wt	
Grams of solids, g	1886
Vol of solids, cm^3	572
Vol of water, cm^3	472
Vol % solids	54.8
82% solids by wt	
Grams of solids, g	1997
Vol of solids, cm^3	605
Vol of water, cm^3	439
Vol % solids	58.0
84% solids by wt	
Grams of solids, g	2115
Vol of solids, cm^3	641
Vol of water, cm^3	403
Vol % solids	61.4
Grinding times, 15, 30, 60 min	

Table 2. Breakage Parameters for Taconite, $S_i = ax_i^\alpha$

	No chemical			With chemical*		
Wt % solids =	80	82	84	80	82	84
Vol % solids =	54.8	58.0	61.4	54.8	58.0	61.4
$S_{10\times14}$, min^{-1} =	0.70	0.69	0.60	0.76	0.73	0.65
$WS_{10\times14}$, kg / min =	1.32	1.38	1.27	1.43	1.46	1.38
α	0.95	0.95	0.95	0.95	0.95	0.95
γ	0.65	0.65	0.61	0.65	0.65	0.65
β	3.1	3.1	3.1	3.1	3.1	3.1
Φ	0.31	0.31	0.34	0.31	0.31	0.31
δ	0	0	0	0	0	0

*0.45 kg GA-4272 / t dry feed.

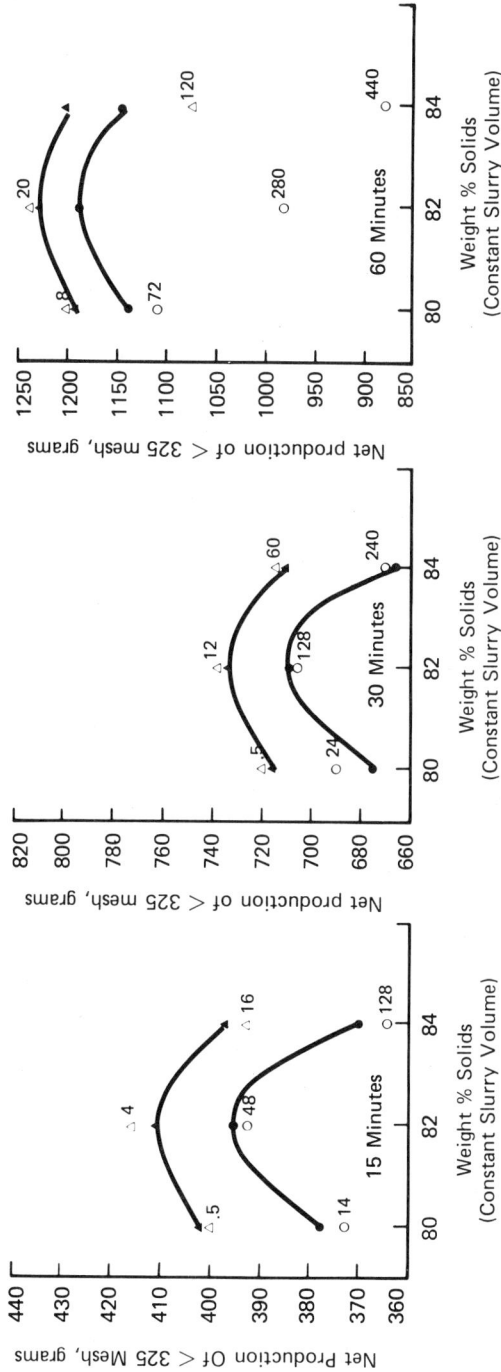

Fig. 7. Net production of −45 μm (− 325 mesh) taconite at various times of grind (see Table 1) in 200 mm (8 in.) mill at various weight percent solids [constant slurry volume, feed 6.3% < 45 μm (325 mesh)]; viscosities in units of Pascal seconds, Pa · s.

conditions. The slowing-down effect was evident even in the presence of aid for the 84% slurry density.

The rheological character as a function of the presence or absence of chemical is shown in Fig. 8, for slurries of a single $\sqrt{2}$ size fraction or of a broad distribution with a Schuhmann coefficient (α_s) of one (Klimpel, 1982, 1982a, and 1983). It is clear that the presence of chemical reduces the viscosity at any shear rate and also reduces the yield stress. The fractional effect of the chemical is greater for: (1) higher slurry density; (2) a broad distribution compared to a narrow distribution; and (3) a finer distribution. Klimpel and co-workers found that similar patterns of results existed for the other ores tested (copper, molybdenum, gold, and tin) although the optimum slurry density for maximum breakage rates occurred at different slurry densities. They also concluded (Klimpel, 1982a, 1983) that a grinding aid would be effective if: (1) the mill is operated at high slurry density; (2) the solids have sufficient adsorption sites to adsorb enough chemical to affect slurry dispersion; (3) the chemical acts consistently to lower or control viscosity over the encountered range of pH, impurity type and amount (e.g., Ca^{++} ion), and shear level; and (4) the chemical does not adversely affect downstream processing operations such as flotation, filtration, thickening, pelletization, and classification. Only a few natural or synthetic dispersants will meet all of these requirements for any given ore and grinding conditions.

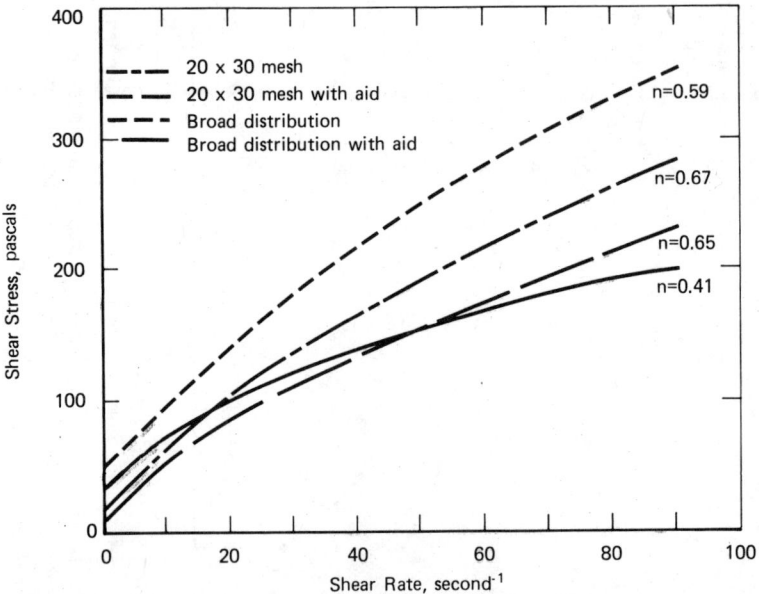

Fig. 8. Effect of fluidity modifier (0.5 kg / t GA-4272) on shear stress vs. shear rate for coal-water slurries: 850 μm (20 mesh) largest particle size: 72.3% solids by weight (66% by volume).

One further factor influencing the effectiveness of chemical additives in wet grinding is the nature of the way in which the particular material being ground breaks as described by the B function (the γ value) (Klimpel, 1982a, 1984). Fig. 9 quantitatively demonstrates this relationship between net production achievable vs. the relative amount of fines produced by breakage of larger particles. The smaller the γ value, the higher is the relative amount

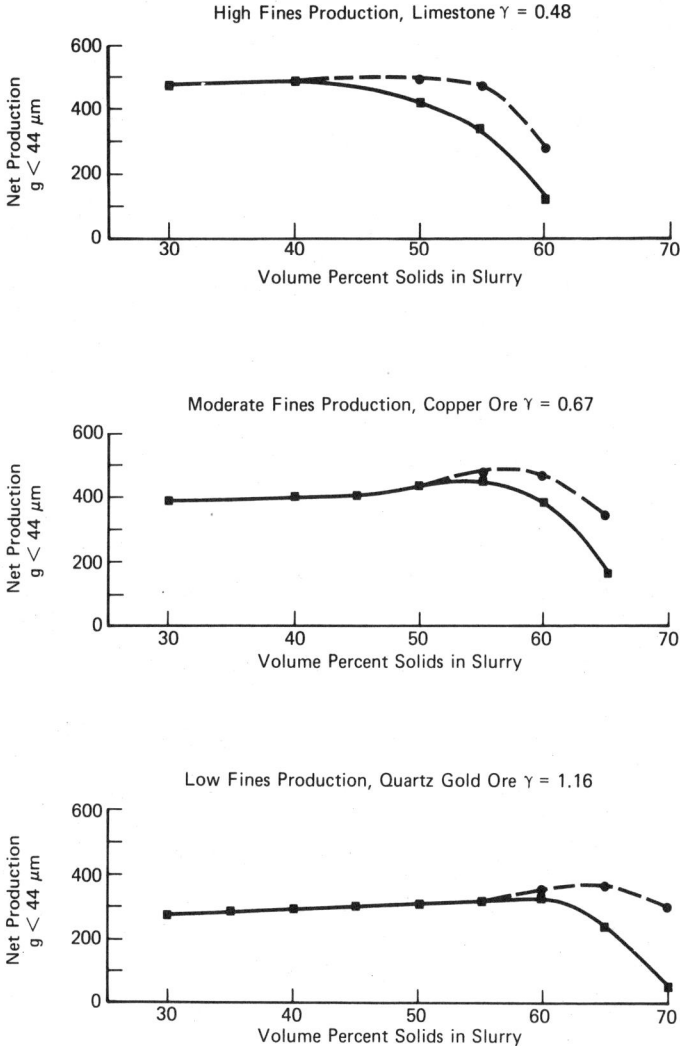

Fig. 9. The influence of the primary breakage distribution (indexed by γ) on the effectiveness of grinding additive: net production from a feed of Schuhmann slope = 1.0 ground for 20 min; D = 200 mm (8 in.), d = 25 mm (1 in.).——— no additive; ----- with GA-4272 at 1 kg / t.

of fines being produced from breakage. With materials which give high fines production, $\gamma < 0.5$, the change of slurry rheology as percent solids is increased is from dilatant directly to a yield value slurry, with essentially no clearly defined region of pseudoplasticity—without yield. The use of additives in this case merely extends the flat portion of the net production curve vs. solids content to higher percent solids, without necessarily increasing the net production achievable. With materials exhibiting more moderate fines production, $0.5 < \gamma < 0.9$, the rheological transformations described in Chapter 5 are followed, and the use of additives generally gives significant effects over relatively small changes in percent solids (e.g., 3 to 8% by volume). In the case of materials having low fines production, $\gamma > 0.9$, the complete rheological transformation again occurs but the region of change associated with pseudoplastic—without yield—slurry behavior can be very exaggerated (e.g., 8 to 20% by volume). This leads to a net production vs. percent solids curve having a very long region of gradually increasing production.

It is important to realize that conclusions from experiments on the effect of grinding additives will depend on this property of the material, which undoubtedly explains some of the confusion about additive effects present in the literature.

Continuous Industrial Scale Tests

Several industrial grinding circuit trials involving rod and ball mills with overflow discharges have been performed (Klimpel and Manfroy, 1978; Klimpel and Samuels, 1979; Klimpel, 1980, 1982a, 1983), and have shown 4 to 15% increased rates of production due to the use of grinding aid, in both open and closed-circuit operations. Increased rates were obtained while maintaining a similar product size distribution. Unlike the laboratory batch tests, the mill power did not change with the use of additive, so that specific grinding energy was reduced.

Fig. 10 shows a typical result from one of these trials on a ball mill grinding copper ore, in closed circuit with a classifier. The rate of makeup feed and the product size distribution were monitored continuously. All other factors such as percent solids, classifier control, were kept constant over the duration of the test. With approximately constant feed rate, the period from 2 to 3, with aid, produced an average size distribution which was 4% finer than the desired control size as compared to the period 1 to 2. When the additive was shut off, the percent greater than the control size went back to its original level. At time 4, the additive was again used, with an increase in fineness of grind. After time 5, the feed rate was slowly increased. After 3 hr with the aid, the feed rate was 10% higher than earlier, but with the product size being the same as the base case without additive. When the additive was shut off, and the feed rate reduced, the mill output returned to approximately its earlier operating condition. During the period of increased feed rate with the grinding additive, the mill power draw was identical to the base case without chemical.

Fig. 11 shows the net production rate of $-500\ \mu\text{m}$ (-35 mesh) material in an open-circuit continuous rod mill operating on copper ore (Klimpel

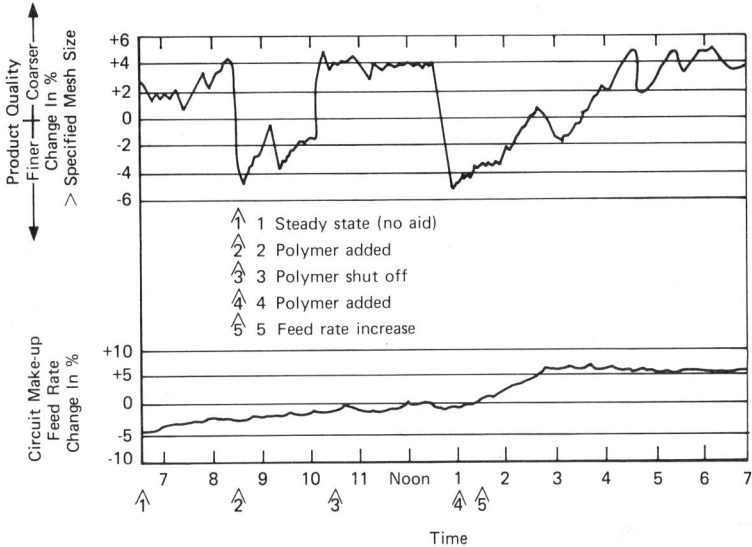

Fig. 10. Product response curve for industrial-scale mill, normal circuit capacity 125 t / h, with and without polymeric grinding aid (GA-4272).

and Samuels, 1979). The water rate is held constant so that increased tonnage means increased percent solids in the slurry, giving the continuous counterpart of the batch data of Figs. 2 and 3. Other tests run on this same mill demonstrated that the optimum operating condition without chemical was between 95 and 100 t/h at the water level chosen and this increased to about 110 t/h with the additive. Fig. 12 shows the use of GA-4272 in a circuit involving a rod mill followed by a ball mill in closed circuit. The water rate to the total circuit was held constant, but the ore hardness varied widely during the test. The chemical was evenly split between the rod and ball mills which was later shown to be not an optimal use of chemical. However, the increase in output is clearly demonstrated. Fig. 13 shows the increased fineness of grinding that was achieved with additive in an operation grinding copper sulfide ore at two different feed rates (Klimpel, 1980).

In industrial practice, a series of tests are required, with complete size distribution sampling around the circuit, to identify the optimum additions of water and chemical. The correct operating conditions are indicated by the slurry having Region B characteristics. Best results are obtained for well-controlled plants where it is possible to maintain the slurry conditions at the end of Region A or in Region B. The effective amounts of grinding aid used in the batch tests described earlier are higher than the effective amounts in the continuous industrial tests because the final size distribution of the batch tests is considerably finer (has more adsorptive surface area) than the product of the continuous tests. There are several processing plants which have run at least one million tons of ore through the grinding circuits with

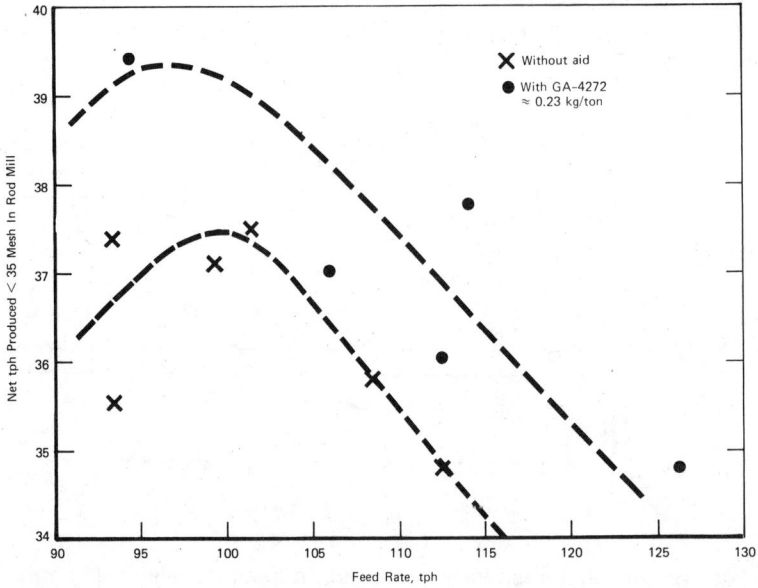

Fig. 11. Net production rate in t / h of −500 μm (− 35 mesh) material in an open-circuit continuous rod mill with copper ore, holding water flow rate constant.

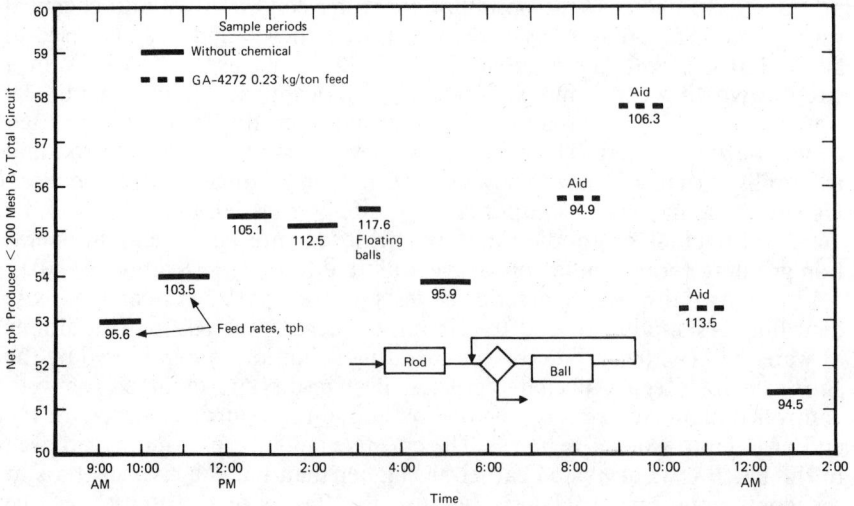

Fig. 12. Net production rate in t / h of − 75 μm (− 200 mesh) material produced in a total circuit with copper ore, holding water flow constant.

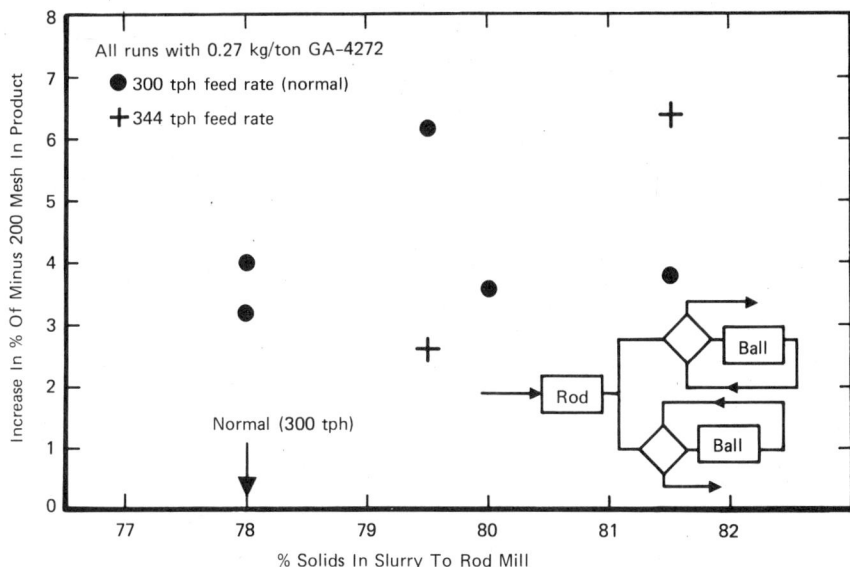

Fig. 13. Increase in percentage of −75 μm (− 200 mesh) in the product to flotation from a circuit grinding a copper sulfide ore with constant feed rate using grinding additive.

an average increase of 8% of throughput at dosage levels of 0.2 to 0.3 kg (0.4 to 0.7 lb) of GA-4272 per ton of ore.

The choice of whether to use grinding aids to achieve increased throughput at a constant product size or a finer product at a constant throughput, or some combination of increased throughput and fineness, depends on the economics of the particular grinding operation (Klimpel, 1980). The choice depends to a large measure on the liberation achieved at the normal grinding conditions, on what finer grind could be achieved economically with additive, and on what changes in liberation accompany this finer grind. As an example, Fig. 14 shows the increased liberation of copper as a function of fineness of grinding as indexed by the weight percent less than 75 μm (200 mesh), at two different mines. This data, plus the increased percent fineness produced by the additive and the financial value of the increased recovery of copper, enables a first estimate to be made of the breakeven cost which can be paid for the chemical at the two mines. If the chemical costs less than this, the use of additive will certainly be advantageous. If the cost of the chemical is somewhat higher than this breakeven value, the use of additive might still be justifiable by increased throughput, but the calculation of economic return is more complex because of cost allocation for extra mining, flotation, chemicals, tailings disposal, etc. When a grinding circuit is operating near full capacity, and more capacity is desired, the use of the grinding aid is normally economic in

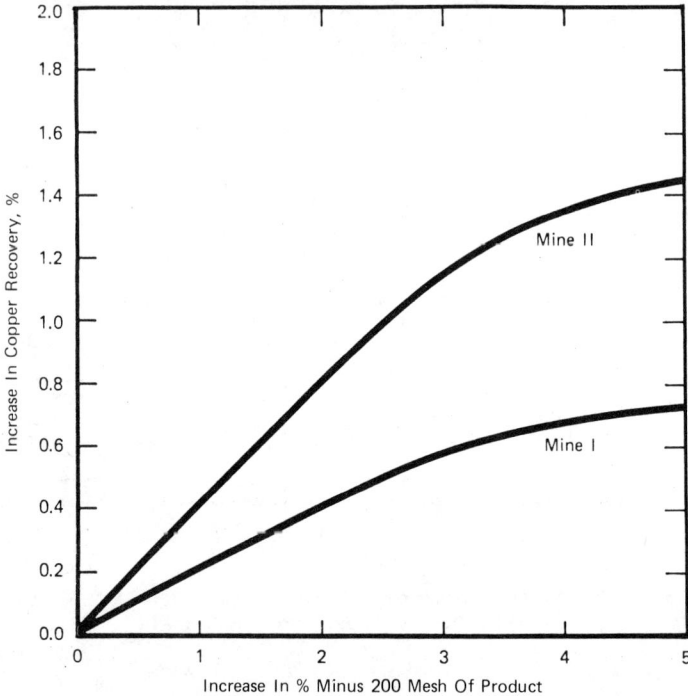

Fig. 14. Variation of copper liberation with increased fineness
of grind, at two mines.

comparison to the alternative of expansion of capacity by further capital
expenditure.

It has been estimated (Klimpel and Samuels, 1979) that the line of
fluidity modifiers developed by the Dow Chemical Co. may be economic in
many operations with on-site additive costs of 12¢ to 30¢ US (1980) per ton
of dry feed ore giving a 6 to 12% increase in throughput at constant grind or
a 2 to 4% finer grind [-75 μm (-200 mesh)] at constant throughput. For
example, an increase of output of 10% at a chemical cost per ton of ore of
25¢ represents a cost of $(1.1)(25\text{¢})/(0.1) = \$2.75$ per ton of increased
output. This can be compared with the costs of grinding quoted in Table 2
in Chapter 18, assuming an increase in throughput is possible with existing
equipment.

A method to predict the effect of chemical usage by scale-up from
laboratory tests to industrial scale has been developed by Klimpel and
Manfroy (1978). It consists of estimating an approximate S curve for the
industrial circuit using the analytical solution for the continuous plug flow
case based on the compensation approximation, Eq. 9 in Chapter 8. The
feed and output size distributions $P(x, 0)$ and $P(x, \tau)$ were used to obtain a
set of formal values of S_i using this equation. Then the proportional
influence of grinding additive measured in a laboratory test was used to

Table 3. Prediction of Rod Mill Performance in Terms of Increase in Throughput with Grinding Additive GA-4272

Slurry density wt % solids	Chemical addition rate, lb/t*	Product size % < 35 mesh	Feed rate t/h	% increase	Product size %35 mesh	Feed rate t/h	% increase
84	0	40.0	410	—	40.0	410	Base
	0.4	40.0	410	0	40.0	410	0
	0.6	40.0	410	0	40.0	410	0
	0.85	40.0	410	0	40.0	410	0
	1.7	40.0	410	0	40.0	410	0
86	0	37.2	410	0	40.0	404	−1.5
	0.4	37.2	428	4.4	40.0	422	2.9
	0.6	37.2	437	6.6	40.0	431	5.1
	0.85	37.2	448	9.3	40.0	442	7.8
	1.7	37.2	478	16.6	40.0	471	14.9
88	0	34.8	410	0	40.0	392	−4.4
	0.4	34.8	450	9.8	40.0	430	4.9
	0.6	34.8	470	14.6	40.0	449	9.5
	0.85	34.8	494	20.5	40.0	472	15.1
	1.7	34.8	526	28.3	40.0	503	22.7

*Metric equivalents: lb × 0.453 = kg; 35 mesh = 500 μm.

scale the S values, enabling an estimate to be made of the increase in feed rate (reduction in τ) at a constant production specification, or of the increase in fineness at the same τ. Table 3 gives the results of this technique for an open-circuit rod mill grinding taconite ore. For a given slurry density and chemical addition rate, if the feed-rate is less than the figure stated, the product will be finer than the base case and vice versa. (Note that the slurry density used in rod milling tends to be 2 to 6 weight percentage points higher than in ball milling the same ore.) Comparison with actual test data on the plant showed the predictions were qualitatively correct up to circuit feed rates of 500 t/h.

It is possible, of course, to apply more sophisticated mill models (see Chapter 19) and to allow for the effects of the additive on classifier performance (see Chapter 13).

Mechanism of Wet Grinding Additives

The experimental data indicate that there are two factors which lead to optimum breakage rates. The first is higher slurry density, which tends to increase breakage rates, possibly by increasing the lift of balls, as suggested by an increase in power draw. The second is slurry viscosity, with higher viscosity tending to decrease breakage rates, presumably by increased drag and reduced impact as the balls come together with slurry between them. These two factors acting against one another lead to an optimum slurry density. The function of the grinding aid is to allow greater advantage to be taken of higher slurry density by reducing the associated effect of higher viscosity.

It follows that the development of higher and higher viscosity as the size distribution becomes finer in a slurry of constant solid content (see Fig. 17 in Chapter 5) will eventually lead to a slowing-down of breakage rates and, hence, the appearance of the non-first-order, slowing-down effect. The grinding aid will delay the development of the slowing-down effect.

In laboratory batch tests, the rheology for optimum breakage is associated with a higher net mill power. In continuous industrial tests, no change in total power draw was seen with or without additives. It is possible, therefore, that different mechanisms are involved. In continuous mill operation, the additive might act to reduce overfilling of the mill by improving mass transfer through the mill (see Chapter 14).

In the case of dry grinding additives, it can be hypothesized that the effect of the adsorbed additive is to change the surface chemistry and cohesive forces of the particles, thus reducing or eliminating the slowing-down effect observed in fine dry grinding (Locher and von Seebach, 1972; Decasper, 1982). See "Effect of Environment in the Mill" in Chapter 5. The exact mechanism of this slowing-down effect has not been deduced to date. However, the most noticeable change in the physical characteristics of the powder with additive is the increase in flowability. Cottaar and Rietmena (1983) have shown that breakage rates decrease when a mill is operated at a high vacuum and that the powder tends not to flow because the effect of gas in providing fluidity of the powder is removed in a vacuum. It can be hypothesized that the nonflowing powder does not tumble as well in the free-flowing surface of the ball charge where ball-ball collisions are most frequent.

Summary

It has been shown in both laboratory and industrial tests that significant increases in the rate of size reduction can be achieved by the addition of chemical grinding aids. This increased rate can be used to grind a higher feed rate to the desired product size or it can be used to produce a finer product size at a fixed feed rate.

In the case of dry milling of fine powders, the action of effective chemicals involves adsorption on solid surface, resulting in the lowering of van der Waal's adhesion forces between fine particles. This allows more efficient breakage interactions between media and particles by a mechanism not fully understood, but probably involving reduction of the slowing-down effect. Typical compounds used are ethylene glycol, propylene glycol, triethanolamine, oleic acid, and aminoacetates; they must be used at temperatures sufficient to give adsorption. The adsorbed additives also give increased flowability of the powder and a reduction in mean residence time at a given feed rate is observed in grate discharge mills. The mill hold-up must then be brought back to the correct filling level (steel-to-clinker ratio). It can be hypothesized that decreased flowability due to cohesive forces prevents fine powder from moving round into the region of cascading balls where ball-ball collision is most effective.

The most important grinding additive in tumbling media mills is water. The rates of grinding in water can be from 1.2 to 2.0 times as great as that of dry grinding. The mechanism of water addition is one of bringing and keeping particles in advantageous positions to receive a breakage action. The various viscosity regimes which occur in wet grinding were discussed in "Effect of Environment in the Mill" in Chapter 5. Normal grinding in a highly fluid environment (low viscosity dilatant slurries, denoted as Regime A) gives first-order breakage of small sizes. As slurry density is increased, a higher viscosity pseudoplastic slurry (Regime B) is obtained, which shows a small increase in breakage rate over that achievable with Regime A. Finally, at even higher slurry density, the slurry is pseudoplastic with a yield value (Regime C), and the breakage rate decreases below that of A and B. In Regime C, grinding is initially first-order until sufficient fine material is produced to give rise to the slowing-down effect.

Klimpel (1982a, 1983) has shown that chemicals can be selected which effectively control slurry rheology and give higher grinding rates than possible without chemical addition. The mechanism involves the two factors which lead to optimum breakage rates. The first factor is the tendency of higher slurry densities to give somewhat higher breakage rates. The second factor is the decrease of breakage rates caused by too high slurry viscosity. The function of a grinding additive is to allow the use of higher slurry density with reduction of the associated effect of higher viscosity, that is, maintaining a pseudoplastic rheological character without excessive yield.

It was found that wet grinding aids would be effective in wet grinding if: (1) the mill is operated at high slurry density; (2) the solids have sufficient adsorption sites to adsorb enough chemical to affect slurry dispersion; (3) the chemical acts consistently to lower viscosity over the encountered range of pH, impurity type and amount (e.g., Ca^{++} ion), and shear level; (4) the chemical does not adversely affect downstream processing operations such as flotation, filtration, thickening, pelletization, and classification. The grinding aids which fulfill these conditions are low molecular weight, water-soluble anionic polymers (Manfroy and Klimpel).

For the successful implementation of grinding aids on an industrial scale, experience has shown that it is necessary to have reasonable control of ore feed rate and quality (both mineralogical and feed particle size distribution) and water feed rate and quality. If it is not possible to routinely maintain an operating regime near the end of Regime A or in Regime B in the normal operation without chemical, then it is likely that the use of grinding aid chemicals will not be economically advantageous.

In batch grinding tests, the use of additives which affect the rheology increases the net power drawn by the mill, leading to a constant specific grinding energy, while in continuous large-scale tests, the total mill power draw was essentially constant with or without chemical. Thus in continuous tests with increased throughput obtained with chemical, the specific grinding energy was reduced. The effect might be reduction in the extent of overfilling of the mill, since lower viscosity slurry will flow more easily through the mill.

References

Anon., 1981, "Comminution and Energy Consumption," Report No. NMAB-364, Committee on Comminution and Energy Consumption, US National Academy of Sciences.

Cottaar, W. and Rietmena, K., 1983, "Effect of Interstitial Gas on Milling," submitted to *Powder Technology*.

Decasper, J., 1982, "Optimization of Cement Grinding Plants with Grinding Aids and Higher Air Flow Rates Through the Mill," *Zement-Kalk-Gips*, No. 11, pp. 565–570.

Hartly, J., Prisbrey, K., and Wick, O., 1978, "Chemical Additives for Ore Grinding: How Effective Are They?" *Engineering and Mining Journal*, Oct., pp. 105–111.

Katzer, M., Klimpel, R. R., and Sewell, J., 1981, "An Example of the Laboratory Characterization of Grinding Aids in the Wet Grinding of Ores," *Mining Engineering*, Vol. 33, No. 10, pp. 1471–1476.

Klimpel, R. R., 1980, "The Engineering Analysis of Dispersion Effects in Selected Mineral Processing Operations," *Fine Particles Processing*, Proceedings of the International Symposium on Fine Particles Processing, P. Somasundaran, ed., Vol. 2, AIME, New York, pp. 1129–1152.

Klimpel, R. R., 1982, "Laboratory Studies of the Grinding and Rheology of Coal-Water Slurries," *Powder Technology*, Vol. 32, pp. 267–277.

Klimpel, R. R., 1982a, "Slurry Rheology Influence on the Performance of Mineral/Coal Grinding Circuits," Part 1, *Mining Engineering*, Vol. 34, No. 12, pp. 1665–1668.

Klimpel, R. R., 1983, "Slurry Rheology Influence on the Performance of Mineral/Coal Grinding Circuits," Part 2, *Mining Engineering*, Vol. 35, No. 1, pp. 21–26.

Klimpel, R. R., 1984, "The Influence of Material Breakage Properties on the Slurry Rheology Transformations Occurring in Wet Grinding," *Particulate Science and Technology*.

Klimpel, R. R. and Manfroy, W., 1977, "Computer Analysis of Viscosity Effects on Selection for Breakage and Breakage Distribution Parameters in Wet Grinding of Ores," *Proceedings, 14th APCOM Symposium*, R. V. Ramani, ed., AIME, New York, pp. 197–206.

Klimpel, R. R. and Manfroy, W., 1977a, "Development of Chemical Grinding Aids and Their Effect on Selections-for-Breakage and Breakage Distribution Parameters in the Wet Grinding of Ores," *Proceedings*, 12th International Mineral Processing Congress, Vol. 1, Sao Paulo, Brazil, pp. 65–91.

Klimpel, R. R. and Manfroy, W., 1978, "Chemical Grinding Aids for Increasing Throughput in the Wet Grinding of Ores," *Industrial and Engineering Chemistry, Process Design and Development*, Vol. 17, pp. 518–523.

Klimpel, R. R. and Samuels, R., 1979, "Examples of the Use of Grinding Aids in Industrial Operations," *Proceedings*, 11th Annual Meeting of Canadian Mineral Processors, Ottawa.

Locher, F. and von Seebach, H. M., 1972, "Influence of Adsorption on Industrial Grinding," *Industrial and Engineering Chemistry, Process Design and Development*, Vol. 11, pp. 190–197.

Manfroy, W. and Klimpel, R. R., US Patents 4,126,276; 4,126,277; 4,126,278; 4,136,830; 4,162,044; 4,162,045; 4,274,599.

Meloy, T. and Crabtree, D., 1967, "Surface Tension and Viscosity in Wet Grinding," Second European Symposium Zerkleinein, H. Rumpf and W. Pietsch, eds., Dechema Monographien 57, Nr. 993–1026, Verlag Chemie, Weinheim, pp. 405–426.

Rehbinder, P. and Kalinkovskay, N., 1932, *Journal of Technology and Physics*, Vol. 2, USSR, pp. 726–755.

Rose, H. and Sullivan, R. M., 1958, *Ball, Tube, and Rod Mills*, Chemical Publishing Co., New York, pp. 236–251.

Somasundaran, P. and Lin, I., 1972, "Effect of the Nature of Environment on Comminution Processes," *Industrial and Engineering Chemistry, Process Design and Development*, Vol. 11, pp. 321–331.

Westbrook, A., 1966, *Environment Sensitive Mechanical Behavior*, Gordon and Breach Publishing, New York, pp. 247–266.

Westwood, A., 1974, "Tewksbury Lecture: Control and Application of Environment-Sensitive Fracture Processes," *Journal of Materials Science*, Vol. 9, pp. 1871–1895.

Westwood, A. and Stoloff, N., 1966, *Environment Sensitive Mechanical Behavior*, Gordon and Breach Publishing, New York, pp. 1–65.

16

Ball Wear and Ball Size Selection

Introduction

The discussion of the economics of grinding processes given in Chapter 18 (see also "Optimization of Mill Power and Ball Loading for Tumbling Ball Mills" in Chapter 11) shows that steel loss of media and liners during grinding is a substantial fraction of the total cost of grinding. In order for a mill to produce at a steady optimal rate it is desirable to start a new mill charge with a ball size distribution close to the equilibrium mixture of balls produced by natural wear, with addition of makeup balls to give a correct equilibrium ball mix. In addition, the determination of $\bar{S}_i$ and $\bar{B}_{i,j}$ values to be used in mill simulations requires that an estimate be made of the distribution of ball sizes in the large-scale mill. This chapter addresses these particular problems.

The abrasiveness of a particular material is usually determined by some form of an empirical abrasion test (Stern, 1962; Marshall, 1975). Different manufacturers have developed their own tests; some users also have developed tests specific to their particular needs. A discussion and comparison of such tests is beyond the scope of this book, especially since there is little information on tests on the same material in different abrasion testers. A typical test is that developed by Bond (1963) from an original test reported by Pennsylvania Crusher. To quote:

> A flat paddle 3 in. $\times$ 1 in. $\times$ 1/4 in., of SAE 4325 chrome-nickel-molybdenum steel hardened to 500 Brinell, is inserted for one inch into a rotor 4.5 in. in diameter, which rotates on a horizontal shaft at 632 through falling ore particles. Two square inches of paddle surface are exposed to abrasion, and the paddle tip, with a radius of 4.25 in., has a linear speed of 1410 feet per minute; sufficient for a good impact blow.
>
> The rotor is enclosed by a concentric drum 12 in. in diameter and 4.5 in. deep, which rotates at 70 rpm, or 90% of critical speed, in the same direction as the paddle. The inner circumference of the drum is lined with perforated steel plate to furnish a rough surface for continuously elevating the ore particles and showering them through the path of the rotating paddle.
>
> In operation, screened particles passing 3/4 in. square and retained on 1/2 in. square are used as feed. Four hundred grams of 3/4 in. $\times$ 1/2 in. feed are placed in the drum, the end cover is attached, and abrasion is

continued for 15 min; then the drum is emptied, another 400 grams are added, and the abrasion continued. In each complete test four 400 gram samples are each abraded for 15 min. Thus the paddle is abraded for a total of one hour, after which it is weighed to the tenth of a milligram. The loss of weight in grams is the abrasion index A_i of the material.

Based on averages of large numbers of tests compared to collected plant experience, Bond gave the following average wear loss formulae:

Wet ball mills

   Balls:

$$kg/kWh = 0.16(A_i - 0.015)^{1/3} \tag{1}$$

   Liners:

$$kg/kWh = 0.012(A_i - 0.015)^{0.3} \tag{2}$$

Dry ball mills (grate discharge)

   Balls:

$$kg/kWh = 0.023A_i^{0.5} \tag{3}$$

   Liners:

$$kg/kWh = 0.0023A_i^{0.5} \tag{4}$$

Table 1. Abrasion Index Averages

Material	Specific gravity	A_i
Dolomite	2.7	0.016
Shale	2.62	0.021
L.S. for cement	2.7	0.024
Limestone	2.7	0.032
Cement clinker	3.15	0.071
Magnesite	3.0	0.078
Heavy sulfides	3.56	0.128
Copper ore	2.95	0.147
Hematite	4.17	0.165
Magnetite	3.7	0.222
Gravel	2.68	0.288
Trap rock	2.80	0.364
Granite	2.72	0.388
Taconite	3.37	0.624
Quartzite	2.7	0.775
Alumina	3.9	0.891

Source: Marshall, 1975.

However, no standard deviation was reported for the accuracy of these formulae. Table 1 gives average abrasion indices for a number of materials (Marshall, 1975).

The typical appearance of balls from a dry grinding ball mill shows surface scratches, indicating wear by abrasion (Bond, 1963). Balls from wet grinding operations are smoother but pitted, indicating the role of corrosion in metal loss. There is little doubt that microsurfaces formed by abrasion under mechanical stress are highly reactive until the chemical bonds at the surface have been stabilized by reaction with the grinding fluid (Lin and Nadir, 1979). It is expected, therefore, that metal wear rates in wet grinding would be highly variable as compared to a dry abrasion test, depending on the corrosive (electrochemical) properties of the system (Natarajan, Riemer, and Iwasaki, 1983).

Ball Wear and Ball Size Distributions

In order to solve the problem of choosing the best mixture of makeup balls to add to a mill, it is necessary to consider the process of wear and the establishment of a pseudo steady-state (equilibrium) mix of ball diameters in a mill. The treatment by Bond (1960) makes two major assumptions: (1) that the wear rate of a ball is proportional to its surface area, and (2) that ball makeup consists only of a single large size of ball. The formulation given below (Austin and Klimpel, 1984) extends this treatment to allow for other cases of wear laws and ball additions.

In ball wear, there is no problem in distinguishing between balls and the wear powder, so the wear products can be considered simply as mass lost from the ball charge. Consider unit mass of balls in the mill, containing a total number of balls of N_T, with a cumulative fractional number-size distribution of $N(r)$, r being ball radius. Let n_T be the number rate of addition of fresh balls per unit time, with a cumulative fractional number size distribution of $n(r)$. Consider balls of size r to $r + dr$ in the steady-state charge. A *number* balance on this size interval is "number rate of balls entering by wear of size $r + dr$ balls equals number rate of balls wearing out of the size interval plus rate of addition of balls of this size range from makeup."

The fraction of balls greater than size r in the makeup is $1 - n(r)$ and at steady-state the number of balls added per unit time must equal the number per unit time which wear to below size r (see Fig. 1). Thus the verbal balance in symbolic form is

$$n_T[1 - n(r + dr)] = n_T[1 - n(r)] + n_T \frac{dn(r)}{dr} dr \tag{5}$$

It is assumed that the fresh balls entering the size element are indistinguishable from those wearing in. The difference in mass between balls entering the element and leaving the element is

$$\text{rate of loss of mass} = n_T[1 - n(r)]\rho_b 4\pi r^2 \, dr \tag{6}$$

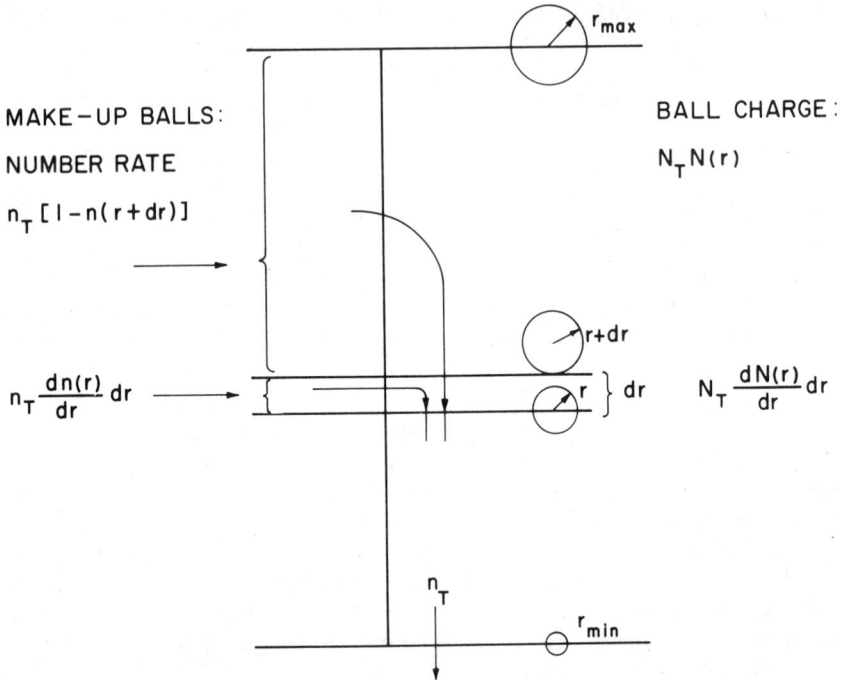

Fig. 1. Illustration of number balance of equilibrium ball wear system.

which must equal the ball wear rate for this size. Let rate of wear (mass per unit time) be a function of ball size, $f(r)$

$$\text{rate of wear of each ball} = f(r) \qquad (7a)$$

then

$$\text{rate of loss of mass} = N_T f(r) \frac{dN(r)}{dr} dr \qquad (7b)$$

Equating to Eq. 6 gives

$$\frac{dN(r)}{dr} = \left(\frac{n_T \rho_b 4\pi}{N_T} \right) \left(\frac{[1 - n(r)] r^2}{f(r)} \right) \qquad (8)$$

This is the basic differential equation defining the distribution of ball sizes $N(r)$, with boundary conditions of $N(r_{min}) = 0$ and $N(r_{max}) = 1$, where r_{min} is the minimum size of ball which can exist in the mill and r_{max} is the maximum size of ball added. The equation implicitly defines the relation between ball addition rate n_T and wear rate. Experimental measurements of

$N(r)$, N_T, and n_T for a known addition of balls $n(r)$ enables $f(r)$ to be calculated.

The relation between cumulative mass fraction $M(r)$ in the charge and number fraction $N(r)$ is

$$dM(r) = \tfrac{4}{3}\pi r^3 \rho_b N_T \, dN(r)$$

Also, mass and number fraction in the makeup are related by

$$m_T dm(r) = \tfrac{4}{3}\pi r^3 \rho_b n_T \, dn(r)$$

where $m(r)$ is the cumulative mass fraction of balls less than size r in the makeup, m_T is the mass rate of makeup (per unit mass of balls). These convert Eq. 8 to

$$\frac{dM(r)}{dr} = m_T K 4\pi \rho_b \frac{[1 - n(r)] r^5}{f(r)} \tag{8a}$$

where, since $\tfrac{4}{3} n_T \pi \rho_b = m_T \int (1/r^3) dm(r)$,

$$K = \int_{r_{\min}}^{r_{\max}} \left(\frac{1}{r^3} \right) dm(r) \tag{9}$$

A convenient method of analysis is to assume that the variation of $f(r)$ with r can be approximated by a power function $r^{2+\Delta}$

$$\text{wear rate} = f(r) = \kappa \rho_b 4\pi r^{2+\Delta}$$

or

$$f(r) = \left(\rho_b 4\pi r^2 \right) \left(\kappa r^\Delta \right) \tag{10}$$

where Δ can be positive or negative. Since wear rate is $-d(4\pi r^3 \rho_b/3)/dt$, $\kappa \rho_b 4\pi r^{2+\Delta} = (-4\pi \rho_b/3) dr^3/dt = (-4\pi \rho_b r^2) dr/dt$ and the *wear distance per unit time* $(-dr/dt)$ is κr^Δ (for $\Delta = 0$ the wear distance per unit time $= \kappa$). Eq. 8a becomes

$$M(r) = (m_T K/\kappa) \int_{r_{\min}}^{r} [1 - n(r)] r^{3-\Delta} \, dr \tag{11}$$

If the makeup is in definite sizes of balls of $r_1, r_2, \cdots, r_k, \cdots r_m$, and if the sizes are ordered $r_{\max} = r_1 > r_2 \cdots > r_k > \cdots > r_m \geq r_{\min}$, and m_k is the weight fraction of makeup of size r_k, Eq. 9 becomes

$$K = \sum_k \left(m_k/r_k^3 \right) \tag{9a}$$

Similarly, n_k is the number fraction of balls of size r_k, where $n_k = (m_k/r_k^3)/K$. Then

$$n(r) = \begin{cases} n_m + n_{m-1} + \cdots + n_k, & r_k < r \le r_{k-1} \\ 0, & r \le r_m \end{cases}$$

Eq. 11 is then readily integrated to

$$M(r) = [m_T K/(4 - \Delta)\kappa] [r_m^{4-\Delta} - r_{\min}^{4-\Delta}) + (r_{m-1}^{4-\Delta} - r_m^{4-\Delta})(1 - n_m)$$

$$+ \cdots (r^{4-\Delta} - r_k^{4-\Delta})(1 - \overline{n_m + \cdots + n_{k+1}})],$$

$$r_{k+1} \le r < r_k, m \ge k \ge 1 \tag{12}$$

and m_T follows from Eqs. 9a and 12 using $M(r_1) = 1$. In general, the choice of makeup ball size distribution $n(r)$ to give the approach to a desired $M(r)$ is a trial-and-error calculation which requires a knowledge of κ and Δ.

For a *single size of makeup ball* $n(r) = 0$ for $r < r_{\max}$ and

$$M(r) = [m_T K/(4 - \Delta)\kappa] (r^{4-\Delta} - r_{\min}^{4-\Delta})$$

Since $M(r_{\max}) = 1$, $K = 1/r_{\max}^3$,

$$m_T = \frac{(4 - \Delta)\kappa r_{\max}^3}{r_{\max}^{4-\Delta} - r_{\min}^{4-\Delta}} \tag{13}$$

and

$$M(d) = \frac{d^{4-\Delta} - d_{\min}^{4-\Delta}}{d_{\max}^{4-\Delta} - d_{\min}^{4-\Delta}} \tag{14}$$

where $d = 2r$. For the Bond assumption that wear is proportional to the surface area, $\Delta = 0$. For some reason, Bond approximated Eq. 14 by $M(d) = (d/d_{\max})^{3.8}$. Note that m_T is the *fraction* of ball charge replaced per unit time.

Davis (1919) gave data from a Hardinge conical wet mill and a 1.8 m (6 ft) diam × 2.4 m (8 ft) dry ball mill which indicated $\Delta = 1$, corresponding to a wear law of "wear rate ∝ ball weight, d^3." Lorenzetti (1980) has found that Eq. 10 with $\Delta = 0$ is a good assumption for wet milling and he gives κ values ranging from 3.8 μm/hr for a relatively soft iron ore to 15.4 μm/hr for an extremely hard copper ore, for Armco Moly-Cop balls.

Table 2. Ball Wear Data on 4.3 m Diam by 5 m Long Wet Overflow Mill Grinding Abrasive Inorganic Material

Type of balls	= steel, specific gravity 8.5, 600 Brinell Hardness	
Size of makeup ball, mm	= 100	75
Media weight, t	= 110	110
Daily addition, t	= 2.0	2.7
Daily throughput, t	= 1600	1480
Steel per ton throughput, kg / t	= 1.25	1.82
Steel per kWh, kg / kWh	= 0.0676	0.0984

On the other hand, Austin and Klimpel (1984) have analyzed data from a wet industrial ball mill where makeup was of a single size of ball. Table 2 gives the milling conditions. Fig. 2 shows the cumulative weight fraction of balls vs. ball diameter, as measured by emptying the mill contents and counting and sizing balls using calipers. Two different sizes of makeup balls were tested. It appears that $\Delta = 2$ in this case, since the data agreed reasonably well with the form of Eq. 14 with $\Delta = 2$, that is, wear rate in mm/hr $\propto r^4$. The values of κ obtained from Eq. 13 using this value of Δ were $(26)(10^{-6})$mm^{-1}h^{-1} for the ball charge originating for 100 mm (4 in.) makeup balls and $(43)(10^{-6})$mm^{-1}h^{-1} for that originating from 75 mm (3 in.) makeup balls. It should be noted that if the smaller balls formed by wear were softer because of the loss of a hardened outer layer, then they would wear faster, giving $\Delta < 0$.

In this test it appeared that larger balls wear much faster with respect to smaller balls than predicted by the Bond expression, leading to a much flatter ball size distribution. It also appeared that wear rates were faster in the mix of smaller balls. For example, the linear wear rate of 75 mm (3 in.) balls was 37 μm/hr for the larger ball mix and 60 for the smaller, corresponding to wear rates of 5.5 and 9.1 g per ball per hour. This is perhaps due to the greater number of ball-ball collisions per unit time for the smaller ball mix. The total number of balls per unit mass of charge, N_T, is given by

$$(1/6)\rho_b \pi N_T = \left(\frac{4 - \Delta}{1 - \Delta} \right) \frac{d_{max}^{1-\Delta} - d_{min}^{1-\Delta}}{d_{max}^{4-\Delta} - d_{min}^{4-\Delta}} \qquad (15)$$

For $\Delta = 2$, the ratio of N_T for $d_{max} = 75$ mm (3 in.) to N_T for $d_{max} = 100$ mm (4 in.), assuming $d_{min} \approx 12$ mm (0.5 in.), is 1.7: the wear ratio of 9.1 to 5.5 is ≈ 1.7 also.

As an illustration of the method of following ball wear with time (Lorenzetti, 1980), Austin and Klimpel (1984) also report ball wear analyses

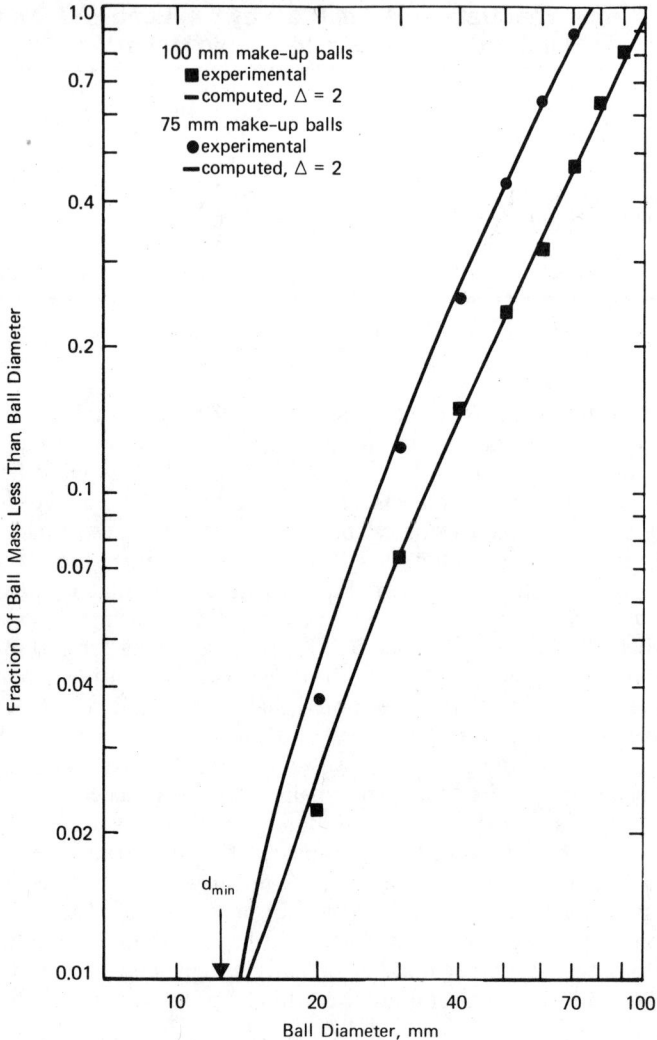

Fig. 2. Cumulative ball size distribution at steady state (see Table 2).

performed by tagging balls and measuring the mass and diameter of the balls after various periods of grinding. Using Eq. 10, integration of $-d(4\pi r^3 \rho_b/3)/dt = \kappa \rho_b \pi r^{2+\Delta}$ gives

$$1 - (r/r_o)^{1-\Delta} = (1 - \Delta)\kappa t/r_o^{1-\Delta} \qquad (16)$$

where r_o is the initial size of the ball at time $t = 0$. For a continuous mill of 4 m (13 ft) i.d. by 4.8 m (15.7 ft) long wet grinding a hard gold ore at over 1000 tons per day, the variation of the radius of balls as a function of

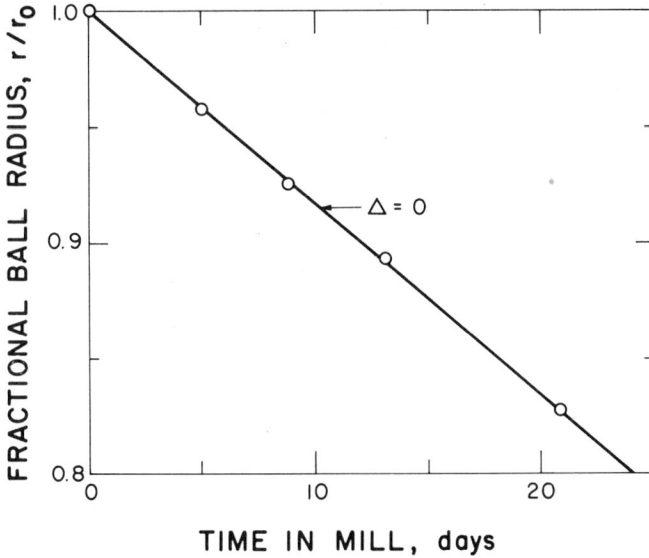

Fig. 3. Change of ball dimension with time in mill, $r_o = 100$ mm (4 in.).

grinding time is shown in Fig. 3. It is clear that this result is consistent with $\Delta = 0$: the loss of 0.8 of ball radius represents a loss of half of the ball weight, and gives a wear rate κ of about 35 μm/hr.

Eq. 14 can be used to calculate $\bar{S}_i$ values in the *normal breakage region* if $S_i \propto 1/d^{N_o}$, for this type of mix of balls. Then,

$$\bar{S}_i = S_i(d_s)d_s^{N_o}\Sigma\left(m_k/d^{N_o}\right)$$

$$= S_i(d_s)d_s^{N_o}\int_{d_{min}}^{d_{max}}\left(\frac{1}{d^{N_o}}\right)dM$$

$$= \left(\frac{4-\Delta}{(4-N_o)-\Delta}\right)S_i(d_s)d_s^{N_o}\frac{d_{max}^{(4-N_o)-\Delta}-d_{min}^{(4-N_o)-\Delta}}{d_{max}^{4-\Delta}-d_{min}^{4-\Delta}} \qquad (17)$$

where $S_i(d_s)$ is the specific rate of breakage measured for standard balls of d_s diameter. This is conveniently put as

$$\bar{S}_i/S_i(d_s) = \left(\frac{4-\Delta}{(4-N_o)-\Delta}\right)\left(\frac{d_s}{d_{max}}\right)^{N_o}\left(\frac{1-\bar{d}^{(4-N_o)-\Delta}}{1-\bar{d}^{4-\Delta}}\right) \qquad (17a)$$

where $\bar{d} = d_{min}/d_{max}$. Table 3 gives values of the ratio $\bar{S}_i/S_i(d_s)$ for various values of makeup size (a single makeup size) and $\Delta = 0$, 1, and 2; $N_o = 1.0$ and $d_{min} = 12.7$ mm (1/2 in.).

Table 3. Ratio of Mean $\bar{S}_i$ Values to Standard 25 mm (1 in.) Ball Diam S_i Values in the *Normal* Breakage Region for Various Ball Mixture Conditions*

Makeup ball diam			Ratio $\bar{S}_i/S_i(d_s)$ for		
in.	mm	$\dfrac{d_{max}}{d_s}$	$\Delta = 0$	$\Delta = 1$	$\Delta = 2$
3	76.2	3	0.44	0.49	0.57
2 1/2	63.5	2.5	0.53	0.58	0.67
2	50.8	2	0.66	0.71	0.80
1 3/4	44.5	1.75	0.75	0.81	0.89
1 1/2	38.1	1.5	0.87	0.92	1.00

*Assuming $N_o = 1.0$; $d_s = 25$ mm (1 in.), $d_{min} = 12.7$ mm (1 / 2 in.).

For makeup of balls consisting of two ball sizes d_1 and d_2 ($d_1 = d_{max}$), of mass fraction m_1 and m_2, Eq. 14 becomes

$$M(d) = \begin{cases} \dfrac{d^{4-\Delta} - d_{min}^{4-\Delta}}{K_1 d_1^{4-\Delta} + (1 - K_1) d_2^{4-\Delta} - d_{min}^{4-\Delta}}, & d_{min} \leq d < d_2 \\[2ex] \dfrac{K_1 d^{4-\Delta} + (1 - K_1) d_2^{4-\Delta} - d_{min}^{4-\Delta}}{K_1 d_1^{4-\Delta} + (1 - K_1) d_2^{4-\Delta} - d_{min}^{4-\Delta}}, & d_2 \leq d \leq d_1 \end{cases} \quad (14a)$$

where

$$K_1 = 1 \Big/ \left(1 + \left(\frac{m_2}{m_1}\right)\left(\frac{d_1}{d_2}\right)^3\right)$$

and thus lies between 0 and 1.0. Eq. 17 becomes

$$\bar{S}_i = S_i(d_s)\left(\frac{4 - \Delta}{(4 \quad N_o) \quad \Delta}\right)\left(\frac{d_s}{d_1}\right)^{N_o}$$

$$\times \left(\frac{K_1 + (1 - K_1)\left(\dfrac{d_2}{d_1}\right)^{(4-N_o)-\Delta} - \left(\dfrac{d_{min}}{d_1}\right)^{(4-N_o)-\Delta}}{K_1 + (1 - K_1)\left(\dfrac{d_2}{d_1}\right)^{4-\Delta} - \left(\dfrac{d_{min}}{d_1}\right)^{4-\Delta}}\right) \quad (17b)$$

For example, for $d_s = 1$ in., $d_1 = 2$ in., $d_2 = 1$ in., $m_1 = 0.5$, and $m_2 = 0.5$, and $d_{min} = 1/2$ in., then $K_1 = 1/9$. Then $\bar{S}_i = 0.913 S_i(d_s)$ for $\Delta = 0$, or $\bar{S}_i = 1.13 S_i(d_s)$ for $\Delta = 2.0$.

Ball Size Effects on *B* Values

Normal Breakage

Chapter 5 has already given information on the variation of *B* values of crystalline quartz in the normal breakage region as a function of ball diameter. Fig. 4 shows the experimental values for dry grinding of this quartz in a batch mill of 0.6 m (2 ft) i.d. Fragmentary results from batch wet grinding in the same mill appear to follow the same trends, although the value of γ is somewhat larger. Thus the *B* values changed slightly but significantly with ball diameter, with larger ball diameter giving proportionally more fines. Although the trend seems to be quite real, the measurement of *B* values is not precise enough to enable quantitative relations to be developed with accuracy. However, the proportional variation of γ and Φ estimated from Fig. 4 can be used in simulations until more information becomes available. This effect gives somewhat flatter size distributions for mills containing larger balls than those predicted by using *standard B* values determined with 25.4 mm (1 in.) balls. It is possible that larger ball sizes

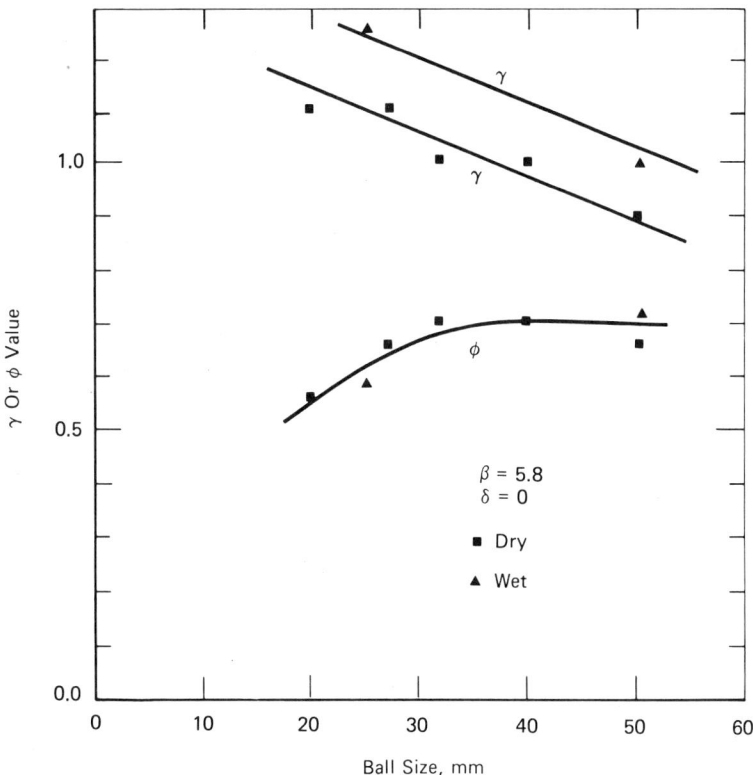

Fig. 4. Variation of *B* parameters with ball diameter: dry grinding of quartz [D = 0.6 m (2 ft), J = 0.2, U = 0.5, ϕ_c = 0.7].

[> 50 mm (2 in.)] will produce even more proportional quantities of fines than predicted by the standard B value set, because the higher impact produced by a larger ball can be expected to give a more catastrophic fracture, with a smaller fraction of particles in the larger fragments and more in the finer fragments.

If B values change with ball diameter, an accurate simulation model requires the use of appropriate mean B values. Consider a mixture of balls of m_1 fraction of size 1, m_2 of size 2, etc. The fraction of breakage of particle size j by size k balls, $m_{j,k}$, is

$$m_{j,k} = m_k S_{j,k}/\bar{S}_j \tag{18}$$

where

$$\bar{S}_j = \sum_k m_k S_{j,k}$$

Thus,

$$\bar{B}_{i,j} = \sum_k m_{j,k} B_{i,j,k}$$

$$= \sum_k \left(m_k S_{j,k} B_{i,j,k} \right) / \sum_k \left(m_k S_{j,k} \right) \tag{19}$$

Fig. 5 shows B values from batch grinding quartz with a mixture of ball sizes, as compared to values for grinding with 25 mm (1 in.) or 50 mm (2 in.) diameter balls. As expected the values lie between those for the extremes of ball sizes; the series of size distributions at various grinding times also showed Schuhmann slopes between those obtained with 25 mm (1 in.) and 50 mm (2 in.) diameter balls. This suggests that the mean $\bar{B}_{i,j}$ values might be fitted by the usual form of Eq. 5 in Chapter 5, with effective overall $\bar{\gamma}$, $\bar{\Phi}$, and $\bar{\beta}$ values.

This was tested by using the curves of Fig. 4 to estimate $\bar{B}_{i,j}$ values for various ball mixes, as follows. Eq. 19 becomes an integration for a continuum of equilibrium ball sizes in a mill:

$$\bar{B}_{i,j} = \int_{d_{\min}}^{d_{\max}} S_j(d) B_{i,j}(d) \, dM(d)/\bar{S}_j \tag{19a}$$

where $B_{i,j}(d)$ is the value of $B_{i,j}$ for ball size d. If $f(d, i, j)$ is defined by the ratio $B_{i,j}(d)/B_{i,j}(d_s)$, and if $S_j(d) \propto 1/d^{N_o}$ in the normal breakage region, Eq. 19a becomes

$$\bar{B}_{i,j} = B_{i,j}(d_s) S_j(d_s) d_s^{N_o} \int_{d_{\min}}^{d_{\max}} (1/d^{N_o}) f(d, i, j) \, dM(d)/\bar{S}_j$$

Replacing $M(d)$ with an equilibrium ball size distribution for a single size

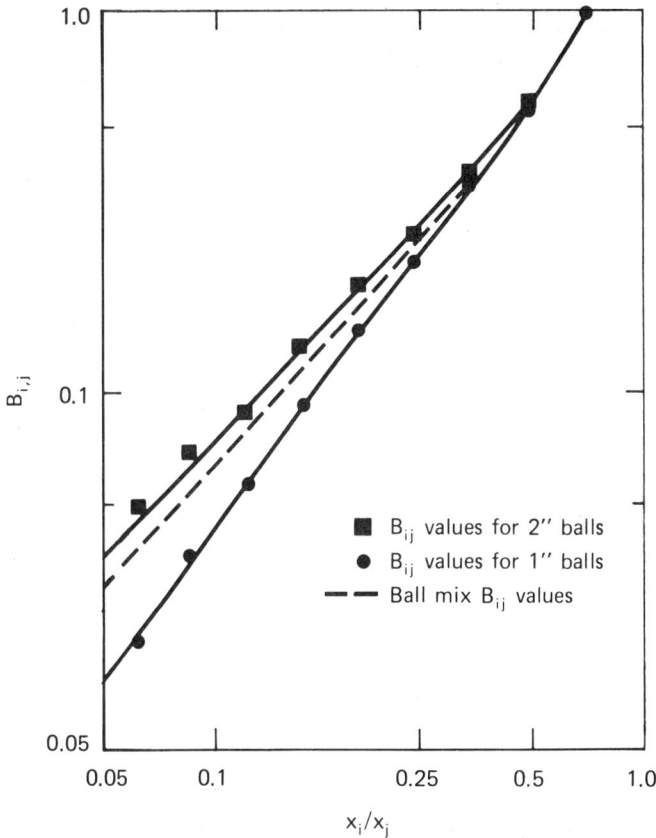

Fig. 5. Experimental B values for breakage of quartz in a mixture of balls: 50.8 mm (2 in.) = 0.4; 38.1 mm ($1\frac{1}{2}$ in.) = 0.45; 25 mm (1 in.) = 0.15 [40% solids by volume in water, $D = 0.6$ m (2 ft); $\phi_c = 0.7$].

of makeup ball, that is, Eq. 14, gives

$$\overline{B}_{i,j} = B_{i,j}(d_s)\left(\frac{S_j(d_s)}{\overline{S}_j}\right)\frac{4-\Delta}{(d_{max}/d_s)^{4-\Delta}-(d_{min}/d_s)^{4-\Delta}}$$

$$\times \int_{d_{min}}^{d_{max}} f(d,i,j)\left(\frac{d}{d_s}\right)^{3-\Delta-N_o} d(d/d_s) \tag{19b}$$

This equation was numerically integrated using the normalized B values of Eq. 5 in Chapter 5,

$$B_{i,j}(d_s) = \Phi\left(\frac{x_{i-1}}{x_j}\right)^{\gamma} + (1-\Phi)\left(\frac{x_{i-1}}{x_j}\right)^{\beta}$$

and calculating the $f(d, i - j)$ using the Φ and γ values of Fig. 4 for other ball diameters. The resulting values of $\bar{B}_{i,j}(= \bar{B}_{i-j})$ were then fitted to Eq. 5 in Chapter 5 to see if appropriate $\bar{\gamma}$, $\bar{\Phi}$, and $\bar{\beta}$ could describe the results.

It was found that Eq. 5 in Chapter 5 described the computed mean B values with considerable accuracy, over the 20 intervals tested. Table 4 gives the ratios of $\bar{\gamma}$, $\bar{\Phi}$, and $\bar{\beta}$ for the normalized mean values to the standard B parameters γ, Φ, and β. The value of β can be taken as unchanged since mill simulations are not very sensitive to β. However, as expected, the changes in $\bar{\gamma}$ and $\bar{\Phi}$ are significant: the ratio $\bar{\gamma}/\gamma$ is essentially the same for an ore with $\gamma = 0.61$ and quartz with $\gamma = 1.10$, so as a reasonable approximation the same set of correcting factors can be used for all materials, with values dependent on d_{max} and Δ.

Table 4. Variation of Parameters of Mean B Values with Ball Mixture in Mill*

	$\dfrac{d_{max}}{d_s}$	Δ	$\dfrac{\bar{\gamma}}{\gamma}$	$\dfrac{\bar{\Phi}}{\Phi}$	$\dfrac{\bar{\beta}}{\beta}$
Copper ore	3	0	0.72	1.08	0.96
		1	0.77	1.05	0.96
$\gamma = 0.61$		2	0.80	1.02	0.96
$\Phi = 0.63$	2.5	0	0.82	1.08	0.98
		1	0.85	1.05	0.98
$\beta = 2.9$		2	0.88	1.02	0.98
	2.0	0	0.90	1.06	1.0
		1	0.92	1.03	1.0
		2	0.93	1.00	1.0
	1.75	0	0.93	1.05	1.0
		1	0.95	1.02	1.0
		2	0.97	0.97	1.0
	1.5	0	0.97	1.02	1.0
		1	0.98	0.98	1.0
		2	0.98	0.94	1.0
Quartz	3	0	0.73	1.08	0.92
		1	0.75	1.05	0.92
$\gamma = 1.10$		2	0.81	1.00	0.92
$\Phi = 0.65$	2.5	0	0.83	1.08	0.97
		1	0.85	1.05	0.97
$\beta = 5.8$		2	0.88	1.02	0.97
	2.0	0	0.91	1.06	0.98
		1	0.92	1.03	0.98
		2	0.94	1.00	0.98
	1.75	0	0.94	1.05	1.0
		1	0.95	1.01	1.0
		2	0.96	0.97	1.0
	1.5	0	0.96	1.02	1.0
		1	0.97	0.98	1.0
		2	0.98	0.94	1.0

*See Table 3.

Large Particle Size Effects

Chapter 5 gave a discussion of the maximum in S_i values when plotted vs. x_i values (see Fig. 2 in Chapter 5), and the variation of the position of the maximum, x_m, with ball and mill diameter. It could be argued that industrial practice uses large enough balls and small enough feed sizes that breakage is always to the left of the maximum. For example, Bond (1958) gives the following empirical equation for calculating the proper size of makeup ball in inches

$$d_m = C(x_G)^{1/2}(\sigma W_i/100\phi_c)^{1/3}/D^{1/6} \qquad (20)$$

where x_G is the 80%-passing size of the makeup feed in micrometers, σ the true specific gravity of the material, W_i the Bond Work Index in kWh/st, ϕ_c the fraction of critical speed. C is 0.0535 for d in in. and D in ft and 1.114 for d in mm, D in m for wet grinding; C is 0.055 for d in in. and D in ft and 1.139 for d in mm and D in m for dry grinding. However, recent practice has been to eliminate the use of rod mills from mill circuits and use a crushed feed to the ball mill. Thus feed containing significant fractions of 5 to 50 mm (0.2 to 2 in.) material is not uncommon, and Eq. 20 will predict a makeup ball size of up to 160 mm (6.3 in.). It is not usual to use balls greater than about 90 mm (3.5 in.) due to damage to the liners caused by cataracting heavy balls.

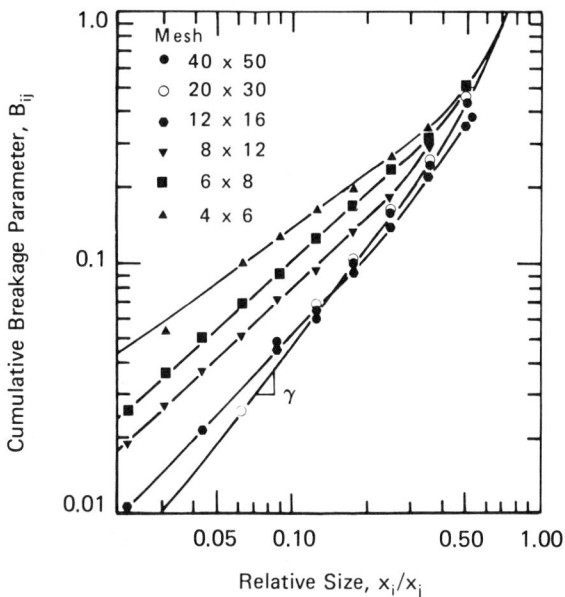

Fig. 6. Experimental B values for dry grinding of quartz in a laboratory mill, $D = 200$ mm (8 in.).

A mill model to allow for such large sizes must not only include S_i values passing through a maximum but also the primary progeny fragment distributions of these large sizes. Fig. 6 shows the B values for large sizes of quartz compared to the normal B values. The value of γ decreases for larger sizes. Not all materials show this effect and weak coals give the same B values for larger sizes as for the normal breakage region. However, it seems likely that strong materials such as ores and cement clinker always give B values for larger particles which contain a larger proportion of fines, due to a larger component of chipping and abrasion in the fracture of these particles, and hence, smaller γ.

Fig. 7 shows similar results from grinding quartz in a batch mill of 0.6 m (2 ft.) i.d. The pattern of results is quite clear: when the breaking particle becomes larger in relation to the ball size, the primary fragment distribution tends to a bimodal distribution, with most of the broken mass going into large fragments, less into medium-size fragments, and the fine sizes fitting a power relation. This is what would be expected if the balance between fracture and chipping-abrasion mechanisms changes toward a bigger component of chipping-abrasion.

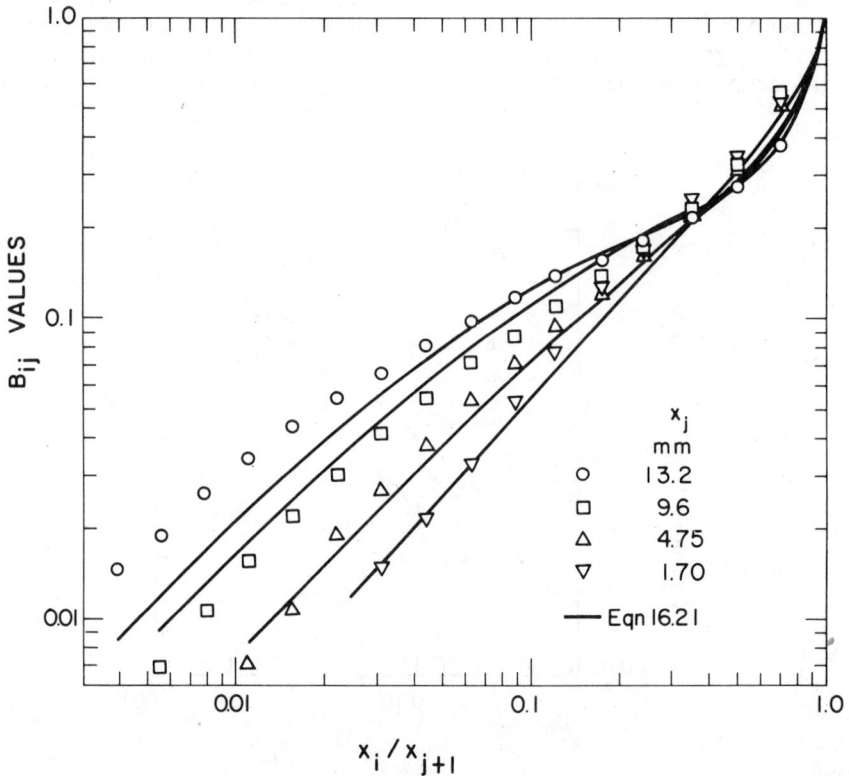

Fig. 7. Variation of B values for breakage to right of the maximum in S [quartz, $D = 0.6$ m (2 ft), $d = 25$ mm (1 in.), $\phi_c = 0.7$].

There is a clear need for investigation of primary breakage distributions of large sizes as a function of ball diameter and mill diameter. However, in the lack of more data, it can be assumed that these *abnormal B* values depend on the ratio of the breaking size, x_j, to the size for maximum breakage, x_{max}, so that the B_{i-j} values are the same for the same x_j/x_{max}. The values of Q_j in Eq. 2 in Chapter 5 also depend on x_j/x_{max} (since $x_{max} \propto \mu$)

$$Q_j = \frac{1}{1 + (x_j/\mu)^\Lambda}$$

A description of these B values requires a function which reduces to the normal function for $Q_j = 1$, which gives smaller values for Φ and γ as Q_j goes to zero, and which introduces a flattening of the middle of the size distribution. An empirical function (Miles, Shah, and Austin) which satisfies these conditions is

$$B_{i,j} = \Phi_j \left(\frac{x_{i-1}}{x_j}\right)^{\gamma(1-aP_j)} \frac{\left\{1 - \exp\left[-\left(\frac{x_{i-1}}{x_o}\right)^{a\gamma P_j}/k_j\right]\right\}}{\left\{1 - \exp\left[-\left(\frac{x_j}{x_o}\right)^{a\gamma P_j}/k_j\right]\right\}} + (1 - \Phi_j)\left(\frac{x_{i-1}}{x_j}\right)^\beta$$

$$(21)$$

where

$$\Phi_j = \Phi_o^{(1+bP_j)}$$

$$k_j = c^{1+dP_j}$$

$$P_j = 1 - Q_j$$

This involves the introduction of four extra positive parameters, a, b, c and d. Since P_j goes from 0 to 1 as the breaking size increases, this function gives the normal B values for $P_j = 0$. As the breaking size increases, it gives smaller effective γ values, a longer plateau region, and smaller Φ values.

The solid lines in Fig. 7 are for values of $a = 0.64$, $b = 1.45$, $c = 0.77$, and $d = 2.75$, with the normal values of $\gamma = 1.1$, $\Phi = 0.65$, $\beta = 5.8$, $\Lambda = 2.9$, and $\mu = 4.3$ mm (0.17 in.) for the test ball size of 25.4 mm (1 in.).

This relation automatically includes the effect of ball diameter on a given size x_j via the change in Q_j values as a function of ball diameter (and density). Then the mean overall value of $\bar{B}_{ij}$ for a mixture of ball sizes is obtained from Eq. 18. Fig. 7 shows that the higher proportional production of fines from breakage of larger sizes tends to compensate for the lower specific rates of breakage of larger sizes.

Ball Charge Optimization: Lifter Design

It is apparent that the grinding efficiency can be directly affected by the choice of mix of balls in the charge for dry grinding using slowly wearing hard alloy steel balls (e.g., cement clinker grinding), or the equilibrium mix of balls produced by wear and makeup in dry or wet grinding with softer steel. However, there can be no simple rule which is accurate for all conditions because the optimum mix depends on: (1) the maximum particle size in the feed; (2) the feed size distribution; (3) the desired product size distribution; (4) the classifier parameters and degree of recycle; (5) the breakage parameters of the material, including $B_{i,j,k}$ values, α, μ, and Λ; (6) the effects of slowing-down and acceleration processes; and (7) lifter design and rotational speed. There is only one way to optimize the ball mix and that is by constructing the most realistic simulation model possible and simulating the mill and mill circuit with different ball mixtures until the maximum output is obtained for the specific feed, breakage parameters, and desired product.

Using the laws of breakage developed in this book, a simulation can be constructed as a computer program which searches to obtain the condition of maximum capacity and hence, minimum specific energy, for a given ball

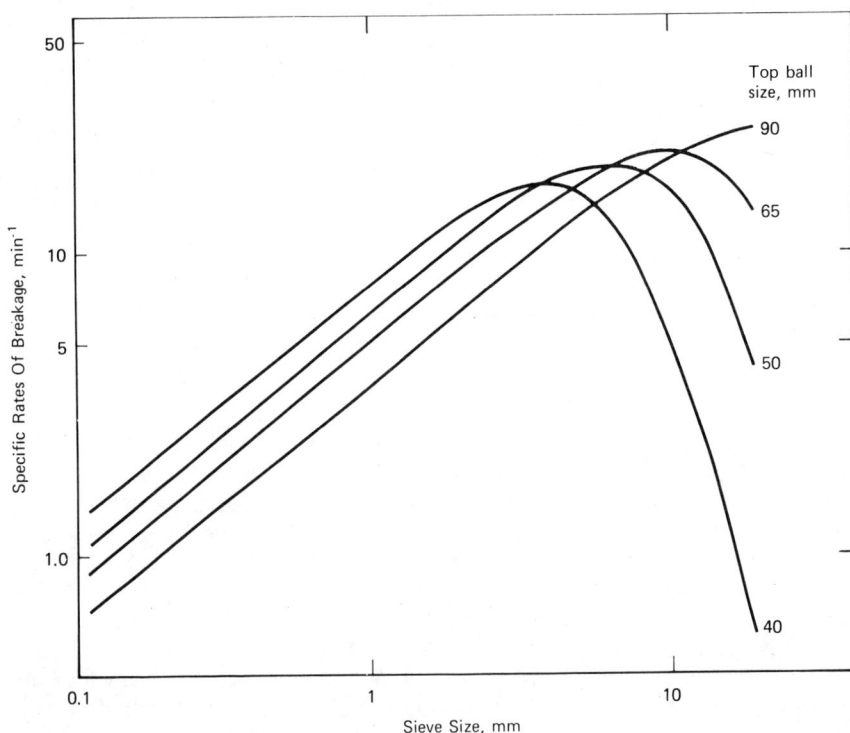

Fig. 8. Specific rates of breakage of quartz for Bond ball mixture for various makeup ball sizes.

loading. Figs. 8 and 9 give an example for conditions where no slowing-down, acceleration of rate, or overfilling effects are present, using the breakage rates and B values of quartz in a mill of 0.6 m (2 ft) i.d. The feed was minus 20 mm (-0.8 in.) and the rate was adjusted to give a control point of 50% -150 μm (-100 mesh). The ball mix of Eq. 14 with $\Delta = 0$ was assumed, with a single size of makeup ball, and the mill was operated in open circuit. The use of too large top ball size [65 mm (2.5 in.)] gave a size distribution without coarse material but at a capacity of 14% less than optimum. The use of too small top ball size [40 mm (1.6 in.)] gave only a 3% drop in capacity, but also gave about 15% of the product in size greater than 5 mm (0.2 in.). The optimum top ball size for capacity, 50 mm (2 in.), gave a size distribution without coarse material.

Clearly, the use of closed circuit would enable product specification to be met even though the mill product contained excessive coarse material.

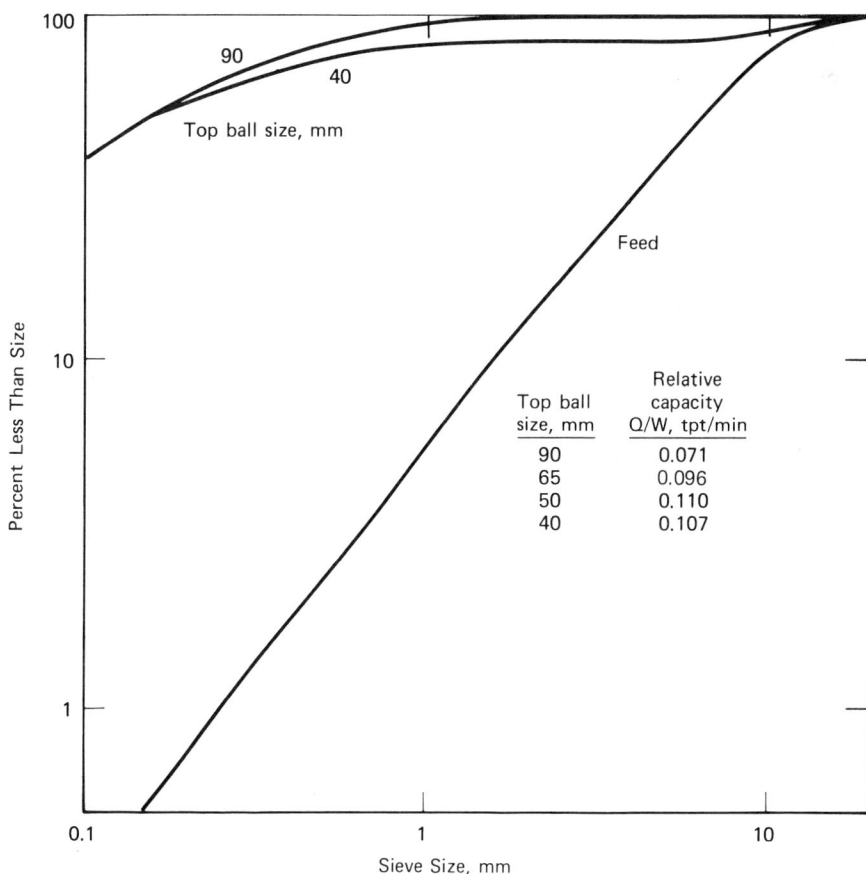

Top ball size, mm	Relative capacity Q/W, tpt/min
90	0.071
65	0.096
50	0.110
40	0.107

Fig. 9. Simulated results for grinding of quartz as a function of makeup ball size.

However, the accumulation of recycled too large material in the mill would still represent inefficient breakage.

It is clear that it is more efficient to break large particle sizes with large balls and smaller sizes with small balls. Cement clinker grinding mills are usually in two compartments in series, connected by a grate (partition or diaphragm); see Figs. 10 and 11. The first compartment receiving feed clinker up to 50 mm (2 in.) in diameter is filled with a mix of large balls and the second compartment is filled with a mix of smaller balls. Table 5 in Chapter 12 gives typical values for a mill of 4.2 m (13.7 ft) diam with a first compartment 5 m (16 ft) long and a second compartment 10.25 m (33.6 ft) long. When hard alloy steel balls are used, the optimum ball mix retains its size distribution for periods of operation of up to two years because of low wear rates.

Another parameter which cannot at present be incorporated quantitatively into the mill simulation is that of lifter design. Fig. 12 illustrates various lifter configurations (Marstiller, 1979), but there has been no systematic study of the quantitative effects of lifter configuration on specific rates of breakage and the position of $x_{\max}$ as a function of ball load and rotational speed. Qualitatively, the general rule is that larger and sharper-edged lifters (e.g., wedge bar, lorain, and block lifters) will tend to give a somewhat bigger proportion of cataracting, at a given fraction of critical speed (see "Discussion of Mill Power Expressions" in Chapter 11). They are used for larger balls and mill feeds containing large particle sizes. Lower lifters with more gradual profiles are used with smaller balls to give cascading, which gives the highest breakage rates for normal breakage of smaller sizes. Less cataracting can be obtained with the more rectangular lifters by operating at lower speeds, at the expense of less total tumbling action.

The use of classifying liners in dry grinding can also give an increase in direct efficiency (Slegten and Mattan, 1973). These liners distribute the balls, with larger balls concentrated at the feed end and smaller at the discharge end. Fig. 11 shows the configuration of such liners. These are viewed as they appear along the mill axis, not radially, so the shape of the liners functions to throw balls along the axis (the shape in the radial direction gives the normal lifting-keying action). Obviously, the effectiveness of a classified ball charge is highest for plug flow of powder down the mill and zero for perfect mixing. The long L/D of the second compartment of a two-compartment cement finish mill gives RTDs which are somewhat closer to the plug flow limit than short L/D. Unfortunately, there appears to be no published data on the variation of ball size distribution (as distinct from mean ball size) along mills with classifying liners, so it is not possible at the moment to use the simulation models to predict the improvement of performance produced by such liners.

Although this section is devoted to ball charge optimization, it must be remembered that it is implicitly assumed in the mill simulations that powder level (hold-up) is in the correct region. Too low powder level also leads to excessive ball wear. The three diaphragms shown in Fig. 10 illustrate the optimization of hold-up via diaphragm design (Cleeman), for dry milling. In

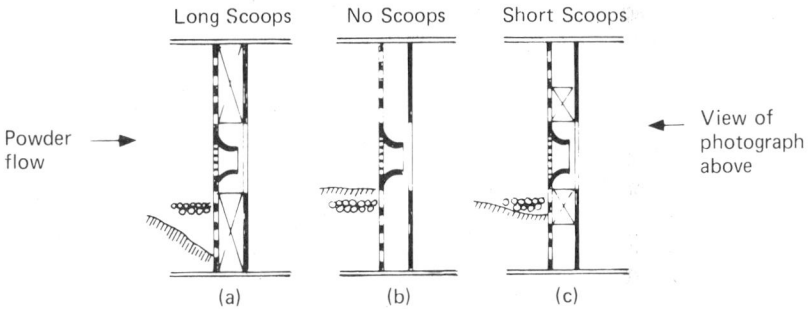

Fig. 10. Mill partitions used to connect compartments in cement tube mills.

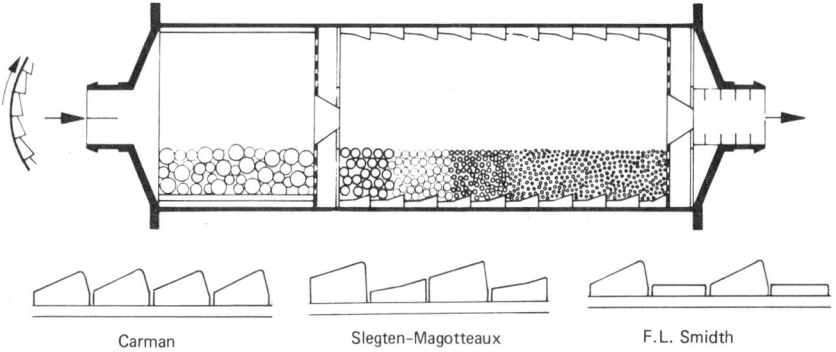

Fig. 11. Classifying liners used in cement tube mills to distribute larger balls to feed end and smaller ones to discharge end.

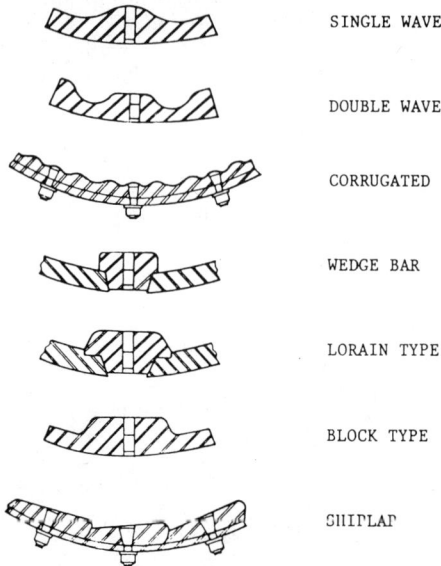

Fig. 12. Lifter configurations (Marstiller, 1979).

Part (a), the scoops moving powder from the first to the second compartment are too long; they remove powder too efficiently and starve the discharge end of the compartment (underfilling). In Part (b), the absence of scoops allows buildup of powder against the discharge opening, which reduces the flow of powder and leads to overfilling of the mill. Part (c) shows scoops adjusted in length to give a reasonable hold-up level at the compartment exit. Reducing the ball size to give increased resistance to flow through the mill is also used to maintain hold-up (or vice versa).

Summary

The empirical Bond wear laws for the kg of steel worn away per kWh of milling energy are:

Wet ball mills:

$$\text{balls, kg/kWh} = 0.16(A_i - 0.015)^{1/3} \tag{S1}$$

$$\text{liners, kg/kWh} = 0.012(A_i - 0.015)^{0.3} \tag{S2}$$

Dry ball mills:

$$\text{balls, kg/kWh} = 0.023A_i^{0.5} \tag{S3}$$

$$\text{liners, kg/kWh} = 0.0023A_i^{0.5} \tag{S4}$$

where A_i is an abrasion index determined by a standardized test in which a steel paddle is rotated in a tumbling dry powder charge. However, steel loss in wet grinding is strongly influenced by the corrosion conditions in the mill charge.

If it is assumed that the wear rate of a ball of size r is equal to $\kappa\rho_b 4\pi r^{2+\Delta}$, then the equilibrium distribution of ball sizes in the mill charge is given by

$$M(r) = (m_T K/\kappa) \int_{r_{min}}^{r_{max}} [1 - n(r)] r^{3-\Delta} \, dr \tag{S5}$$

with

$$K = \int_{r_{min}}^{r_{max}} (1/r^3) \, dm(r)$$

where $M(r)$ is the cumulative mass fraction less than ball radius r; κ is a wear factor; Δ is a constant; $n(r)$ is the cumulative number fraction and $m(r)$ is the cumulative mass fraction of balls less than size r in the makeup; and m_T is the mass rate of makeup per unit mass of balls in the mill. For the Bond wear law that wear is proportional to ball area, $\Delta = 0$ and $\Delta = 1$ for the Davis wear law that wear is proportional to ball volume. For a makeup of a single size of ball, the equation gives

$$M(d) = \frac{d^{4-\Delta} - d_{min}^{4-\Delta}}{d_{max}^{4-\Delta} - d_{min}^{4-\Delta}} \tag{S6}$$

Test data on different mills gave values of Δ of 0, 1, and 2. In a recent test where $\Delta = 2$ for wet ball milling, the value of κ was greater for a mixture of balls of smaller size in the same proportion as the number of balls. The weighted mean value of specific rate of breakage of particle size in the equivalent ball mix of Eq. S6 is

$$\bar{S}_i = \left(\frac{4 - \Delta}{4 - N_o - \Delta} \right) S_i(d_s) d_s^{N_o} \frac{d_{max}^{4-N_o-\Delta} - d_{min}^{4-N_o-\Delta}}{d_{max}^{4-\Delta} - d_{min}^{4-\Delta}} \tag{S7}$$

if S_i varies with ball diameter by $S_i \propto 1/d^{N_o}$, and $S_i(d_s)$ is the value for a standard ball diameter of d_s.

Since the primary progeny fragment distribution can change with ball and particle diameter, the mean $\bar{B}_{ij}$ value for a mixture of balls is to be calculated from

$$\bar{B}_{i,j} = \sum_k m_k S_k B_{i,j,k} \Big/ \sum_k m_k S_{j,k} \tag{S8}$$

where m_k is the mass fraction of ball charge of ball size k, that is,

$\bar{S}_j = \Sigma_k m_k S_{j,k}$. Numerical integration of this equation for an equilibrium continuum of ball size distribution from a single size of makeup ball showed that the overall $\bar{B}_{i,j}$ values in the normal breakage region also fitted the form

$$\bar{B}_{i,j} = \bar{\Phi}\left(\frac{x_{i-1}}{x_j}\right)^{\bar{\gamma}} + (1 - \bar{\Phi})\left(\frac{x_{i-1}}{x_j}\right)^{\bar{\beta}} \qquad (S9)$$

even when the γ and Φ values changed with ball diameter, for $\Delta = 0, 1,$ or 2. Under any given conditions, it is possible to estimate the mean parameters $\bar{\gamma}$ and $\bar{\Phi}$ knowing γ and Φ: $\bar{\beta} \approx \beta$.

Large particle sizes which have breakage rates to the right of the maximum in the plot of S_i vs. x_i for a given ball diameter give B values with a higher component of chipping-abrasion and less fracture for hard rock. The extra production of fines compensates somewhat for the lower rates of breakage of large sizes. An empirical equation to fit the data is given, but it requires four extra fitting parameters.

The empirical Bond equation for selection of the makeup ball diameter is

$$d_m = C(x^{1/2})(\sigma W_i/100\phi_c)^{1/3}/D^{1/6} \qquad (S10)$$

where σ is the true specific gravity of the ore and $C = 1.114$ for d_m in mm and D in m. However, the optimal ball mixture depends on the feed size distribution, the desired product size distribution, the degree of recycle, and lifter design and rotational speed. The optimum mixture of balls is best determined by an accurate mill simulation, which enables the effect of different ball mixtures to be tested. Then the makeup ball size(s) is chosen to give an equilibrium ball mix as close as possible to the predicted optimum. It is expected that liners with more lift will give more cataracting (as will higher rotational speed) which is advantageous for breaking large sizes in the mill, whereas cascading is better for breaking smaller sizes.

Tube mills for finish grinding of cement are designed as compartments in series separated by a diaphragm, with larger balls in the first compartment and smaller in the second. The scoops in a diaphragm must be designed to keep the powder in the front compartment at the proper filling level. Classifying liners are also used in the long second compartment to throw larger balls to the feed end and smaller to the discharge end.

References

Austin, L. G. and Klimpel, R. R., 1984, "Ball Wear and Ball Size Distribution," *Powder Technology*, in press.

Bond, F. C., 1958, "Grinding Ball Size Selection," *Mining Engineering*, Vol. 10, pp. 592–595.

Bond, F. C., 1960, "Crushing and Grinding Calculations," *British Chemical Engineering*, Vol. 6, pp. 378–391, 543–549.

Bond, F. C., 1963, "Metal Wear in Crushing and Grinding," AIChE Annual Meeting, Vol. 54.

Cleeman, J. O., "How to Improve Efficiency of Cement Grinding," literature provided by F. L. Smidth.

Davis, E. W., 1919, "Fine Crushing in Ball Mills," *Trans. AIME*, Vol. 61, pp. 250–296.

Lin, I. J. and Nadir, S., 1979, "Review of the Phase Transformations and Synthesis of Inorganic Solids by Mechanical Treatment," *Materials Science and Engineering*, Vol. 39, pp. 193–209.

Lorenzetti, J. J., 1980, "Ball Size Distribution—From Computer Simulation to Product," Third Symposium on Grinding, Armco Chile, S.A.M.I., Vina de Mar.

Marshall, V. C., ed., 1975, *Comminution*, Institute of Chemical Engineering, London.

Marstiller, S., 1979, "Cost and Up-time Considerations for Selection of Mill Liners and Grinding Balls," Cement Manufacturer Short Course, The Pennsylvania State University.

Miles, T. I., Shah, I., and Austin, L. G., "A Primary Breakage Function for Abnormal and Autogenous Breakage," submitted to *Powder Technology*.

Natarajan, K. A., Riemer, S. C., and Iwasaki, I., 1983, "Corrosive and Erosive Wear in Magnetic Taconite Grinding," SME-AIME Preprint No. 83-4, SME-AIME Annual Meeting, Atlanta, GA.

Slegten, P. and Mattan, J., 1973, "Cement Mill Liners Segregate Grinding Media," *Rock Products*, Vol. 76, pp. 68–71, 96–97.

Stern, A. L., 1962, "A Guide to Crushing and Grinding Practice," *Chemical Engineering*, Vol. 69, pp. 129–146.

17

Solution of the Equations for Grinding Circuits: More Complex Mill Models

Introduction

In previous chapters the basis has been laid for construction of models of grinding circuits, and some specific solutions and solution techniques have been outlined with the assumption that the models will be programmed for digital computation. In this chapter these techniques are explained in more detail, to the extent that a designer can construct and program his own models of circuits. There are several computer languages suitable for programming the models and the particular format of a model can be chosen to match the particular language to be used. However, to avoid the excessive length and confusion which would arise from a discussion of the various possibilities, we will consider only models and techniques especially compatible with FORTRAN language.

Some general comments should be made. First, matrix notation and matrix manipulations are avoided in the formulations because the use of explicit general terms produced by performing the manipulations algebraically has proved to give excellent results. Quite complex circuit simulations can be run for a few dollars of computer time, even on machines with relatively small capacity. Second, procedures which employ *modular* arrangements will be stressed. In effect, we recommend constructing a number of modular subroutines which can be linked in different ways by a simple main program. This provides extreme flexibility and economy of programming. Third, the techniques suggested have been particularly designed to give rapid computations with a minimum of storage requirements, so that they are adaptable to small machines.

Another general principle is linking modules together to construct a model which is as simple as possible for the job at hand. An excellent illustration of the varying levels of complexity possible has been given by Kelsall and Stewart (1971), who have created a synoptic table of the type and range of models of continuous grinding circuits, which is reproduced here as Fig. 1.

SINGLE COMPONENT SINGLE STAGE	MULTI–COMPONENT SINGLE STAGE	SINGLE COMPONENT MULTI–STAGE	MULTI–COMPONENT MULTI–STAGE
11 MILL & CLASSIFIER Feed → Product	12	13	14
21 MILL Classifier	22	23	24
31 Simple Matrix	32	33	34
41 Complex Matrix	42	43	44
51 Dynamic Flow	52	53	54
61 Different Flows	62	63	64

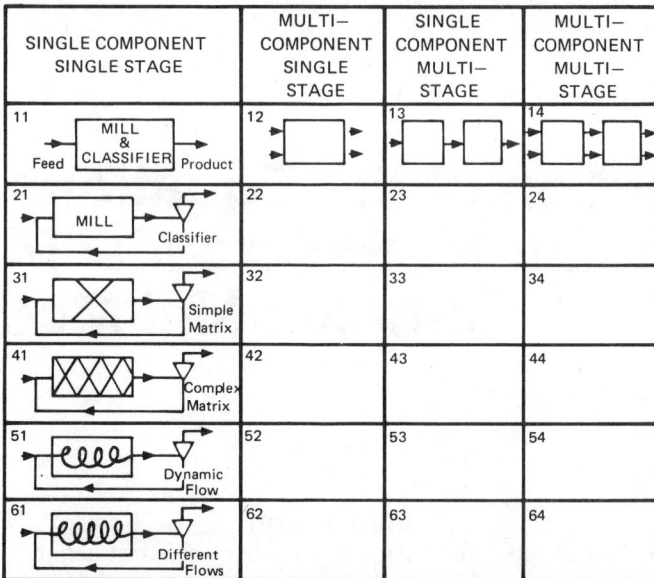

Fig. 1. Illustration of the increasing degrees of complexity of mill circuit modeling (Kelsall and Stewart, 1971).

In the figure, Row 1 treats a mill-classifier combination as an overall *black box* with calculations of product size from a given feed based, for example, on empirical relations for the 80%-passing sizes. Row 2 separates the mill and classifier action, with again the mill being treated by empirical correlations of products and feed. Row 3 treats the mill by applying an overall transfer matrix of feed to product, thus allowing for variations in feed size distribution. Row 4 treats the mill as a series of stages of breakage, implicitly assuming batch/plug flow. Row 5 introduces the concept of residence time distribution. Row 6 further refines the model by considering different sizes of particles to flow at different rates.

The models treated in this book are primarily in the 53 category. Model complexity increases with row number while system complexity increases with column number. Fig. 2 shows a general logic flowchart of a mill circuit simulator.

Simple Batch Grinding

As a general example of the philosophy of programming, examine the simple case of programming the solution of the batch grinding equation. We start with the Reid batch grinding solution (Eq. 7 in Chapter 6)

$$w_i(t) = \sum_{j=1}^{i} a_{ij} \exp(-S_j t), \quad n \ge i \ge j \ge 1$$

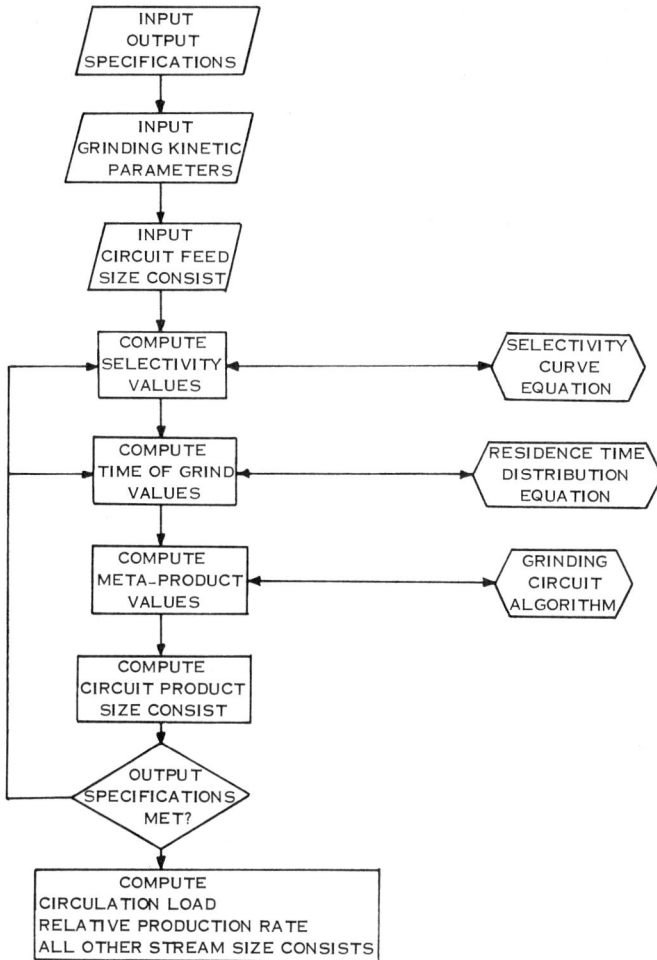

Fig. 2. General logic flowchart of the grinding circuit algorithm.

where the values of a_{ij} are as defined previously,

$$a_{ii} = w_i(0) - \sum_{\substack{k=1 \\ i>1}}^{i-1} a_{ik}, \quad i = j$$

$$a_{ij} = \frac{1}{S_i - S_j} \sum_{k=j}^{i-1} S_k b_{ik} a_{kj}, \quad i > j$$

The equations are computed sequentially starting with $i = 1$. The advantage of this particular form is that the time of grinding appears only in the exponentials.

A modified form which preserves this advantage is

$$w_i(t) = \sum_{j=1}^{i} c_{ij}a_{jj}\exp(-S_j t), \quad n \geq i \geq j \geq 1 \tag{1}$$

$$a_{jj} = w_j(0) - \sum_{\substack{k=1 \\ j>1}}^{j-1} c_{jk}a_{kk}$$

$$c_{ij} = \begin{cases} 1, & i = j \\ \dfrac{1}{S_i - S_j}\displaystyle\sum_{k=j}^{i-1} S_k b_{ik}c_{kj}, & i > j \end{cases}$$

In this case, the values of c_{ij} do not contain t or $w_i(0)$. The use of the equation in this form means that batch grinding results can be computed for various times for a given feed without recalculating c_{ij} and a_{jj} values. If the feed is changed, only a_{jj} values need to be recalculated. For these reasons it is called the *hybrid* solution (Luckie, 1976), since it has the advantages of both the Reid solution and the transfer function solution.

Considering the c_{ij} values:

$$c_{21} = S_1 b_{21}/(S_2 - S_1)$$

$$c_{31} = (S_1 b_{31} + S_2 b_{32} c_{21})/(S_3 - S_1)$$

$$\vdots$$

$$c_{n1} = (S_1 B_{n,1} + S_2 B_{n,2} c_{21} + \cdots + S_{n-1} B_{n,n-1} c_{n-1,1})/(S_n - S_1)$$

$$c_{32} = S_2 b_{32}/(S_3 - S_2)$$

$$\vdots$$

etc.

We now note that the values of b_{ij} can be destroyed as they are used during this calculation and the space used for the storage of the c_{ij} values. Hence, the storage requirements of the program can be greatly reduced since the same storage area can be used for b_{ij}, c_{ij} (and, in effect, a_{ij}) values.

The storage requirement can be further reduced, at the expense of a slight increase in execution time, by employing vector storage for b_{ij} (and, hence, c_{ij}) instead of a lower triangular matrix (see Fig. 6 in Chapter 4). This is achieved by locating the ij element of the matrix in the $(i(i-1)/2)$

$+j$ element of the vector. Then, the $1, 1$ matrix element becomes vector element 1, the $2, 1$ element becomes element 2, etc., down to the matrix n, n element which is the $n(n + 1)/2$ vector element; the dimension of the vector is clearly $n(n + 1)/2$. Thus, the B_{ij} values are defined and entered in a vector, $\underline{B}$, and the S_j values are also defined in a vector, $\underline{S}$; the cumulative feed size distribution is also defined in a vector, P.

The Reid solution and all other forms derived from it become unstable if any value of S equals any other value of S, since $S_i - S_j$ occurs in the denominator. In cases where this is a possibility, it is necessary to program a search on all $S_i - S_j$ values and if any are zero, replace one of the S values in the pair with $S(1 + \varepsilon)$ where ε is a small but finite number.

It is sometimes more convenient to solve the batch grinding equation using a *finite difference* solution, that is, a fully numeric solution. In addition, it may be necessary or convenient to use a fully numeric solution when S values are variable with grinding time. Luckie and Austin (1972) have described the method of solution as follows. The finite difference batch grinding equation over a short interval of grind time Δt is

$$\left[w_i(\Delta t) - w_i(0) \right]/\Delta t = \left(\sum_{j=1}^{i-1} S_j b_{ij} w_j \right) - S_i w_i \tag{2}$$

where Δt is chosen so that $w_i = [w_i(\Delta t) + w_i(0)]/2$ with reasonable accuracy. Our experience is that a suitable Δt value that always satisfies this condition is

$$\Delta t \le 0.05/S_{max} \tag{3}$$

where S_{max} is the maximum value of S for the set of data being considered.

Solving Eq. 2 for $w_i(\Delta t)$ gives

$$w_i(\Delta t) = \sum_{j=1}^{i} d_{ij}(\Delta t) w_j(0) \tag{4}$$

$$d_{ij}(\Delta t) = \begin{cases} \dfrac{(2 - S_j \Delta t)}{(2 + S_i \Delta t)}, & i = j \\ c_{ij}(\Delta t), & i > j \end{cases}$$

$$c_{ij}(\Delta t) = \begin{cases} 4/(2 + S_i \Delta t), & i = j \\ [1/(2 + S_i \Delta t)] \displaystyle\sum_{k=j}^{i-1} S_k \Delta t b_{ik} c_{kj}(\Delta t), & i > j \end{cases}$$

The calculation is then repeated using $w_i(\Delta t)$ as the new feed $w_i(0)$, and so on, for the desired number of steps to give the desired grind time. The computation is very rapid even for several hundred steps because the d_{ij} values are only calculated once and no evaluation of exponentials is required.

Mill Circuits Assuming First-Order and Simple RTD

Various Circuits

A grinding circuit is a collection of mills and classifiers; there are a variety of types in use in industry. For example, there are four single-stage circuits: (1) grinding circuits employing one mill in open circuit (Figs. 3a and d); (2) normal closed circuit (Fig. 3b); (3) reverse closed circuit (Fig. 3c); and (4) combined closed circuit (Fig. 4). The combined closed circuit is a sufficiently general single-stage configuration that the other three circuits are special variations of it, so it is only necessary to program this case. If the pre-classifier is set so that all the circuit feed material flows into the mill feed stream and the post-classifier set so that all the mill product flows into the circuit product stream, an open circuit is created. By setting the pre-classifier so that all the circuit feed material flows into the mill feed stream, a normal closed circuit is created. By making the pre-classifier and the post-classifier the same classifier, the reverse closed circuit is created. See Table 1.

The modeling of the combined circuit requires the mass balance equations, a mill model, and two classifier models. The nomenclature used in arriving at the mass balance equations is presented in Table 2. At steady state, $F = P$, $G = Q$, and $G' = Q'$. The pre-classifier selectivity values are denoted by s_{1i} while the post-classifier selectivity values are denoted by s_{2i}. The mass balance equations are:

the pre-classifier makeup

$$g_i = s_{1i} g_i' (G'/G)$$

where

$$(G/G') = \sum_j s_{1j} g_j'$$

the mill feed

$$f_i(1 + C) = g_i + s_{2i} p_i (1 + C)$$

the post-classifier

$$q_i = (1 - s_{2i}) p_i (1 + C)$$

the overall circuit product

$$q_i' = (1 - s_{1i}) g_i' + q_i (G/G')$$

Mill Simulators with Analytical RTDs

Gardner and Verghese (1975) and Gardner and Sukarijnajtee (1973) used the Reid model to develop a mill simulator convenient for use with analytical forms of the RTD. Substituting the Reid solution into $p_i = \int_0^\infty w_i(t)\phi(t)\,dt$, where $\phi(t)$ is the differential RTD gives $p_i = \int_0^\infty \Sigma_{j=1}^i a_{ij}e^{-S_j t}\phi(t)\,dt$ and

$$p_i = \sum_{j=1}^i a_{ij}e_j, \quad n \geq i \geq j \geq 1 \tag{5}$$

where

$$a_{ij} = \begin{cases} f_i - \displaystyle\sum_{k=1}^{i-1} a_{ik}, & i=j \\ \quad\quad i>1 & \\ \dfrac{1}{S_i - S_j}\displaystyle\sum_{k=j}^{i-1} S_k b_{ik}a_{kj}, & i>j \end{cases}$$

and

$$e_j = \int_0^\infty e^{-S_j t}\phi(t)\,dt \tag{6}$$

The same approach can be used with the transfer function model giving (Luckie and Austin, 1972)

$$p_i = \sum_{j=1}^i d_{ij}f_j, \quad n \geq i \geq j \geq 1 \tag{7}$$

where

$$d_{ij} = \begin{cases} e_j, & i=j \\ \displaystyle\sum_{k=j}^{i-1} c_{ik}c_{jk}(e_k - e_i), & i>j \end{cases}$$

and

$$c_{ij} = \begin{cases} -\displaystyle\sum_{k=i}^{j-1} c_{ik}c_{jk}, & i<j \\ 1, & i=j \\ \dfrac{1}{S_i - S_j}\displaystyle\sum_{k=j}^{i-1} S_k b_{ik}c_{kj}, & i>j \end{cases}$$

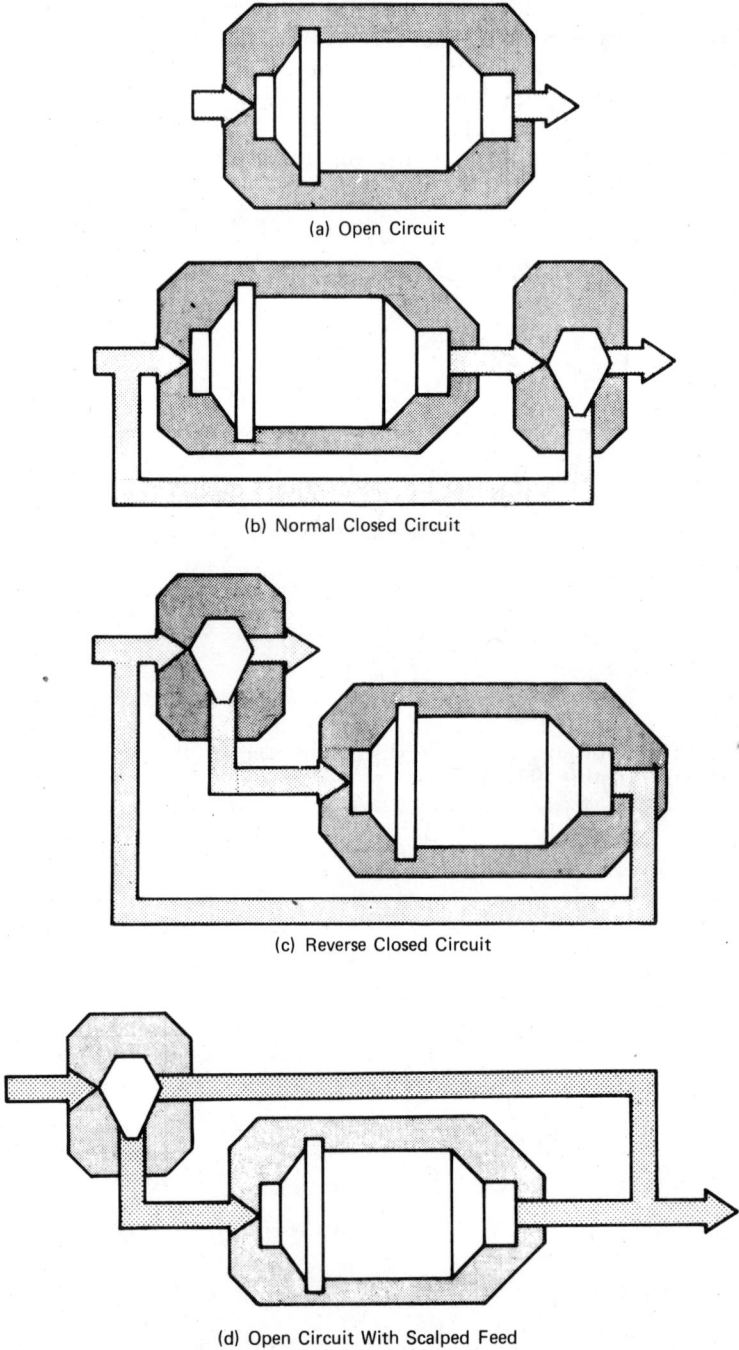

(a) Open Circuit

(b) Normal Closed Circuit

(c) Reverse Closed Circuit

(d) Open Circuit With Scalped Feed

Fig. 3. Single-stage grinding circuits.

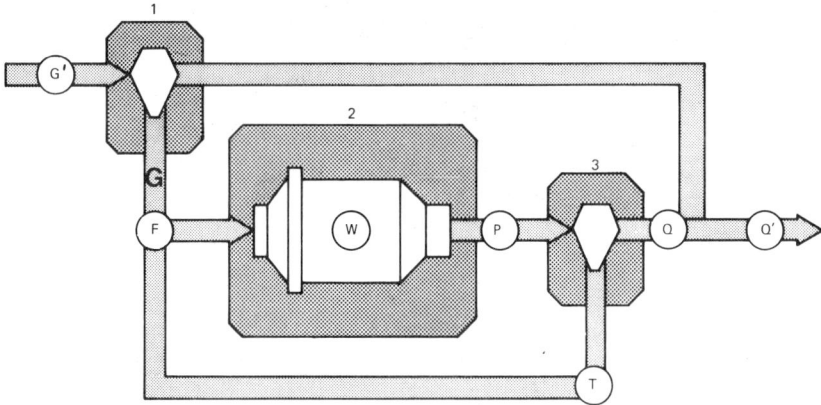

Fig. 4. General single-stage mill circuit with pre- and post-classifier: the combined closed circuit.

Note that solution of Eq. 7 requires a sequential computational algorithm of the form $((c_{ij}, i = 1, n), j = 1, n)$.

However, it may again be preferable to use the *hybrid* model, Eq. 1, because c_{ij} values are independent of time or feed size, giving

$$p_i = \sum_{j=1}^{i} c_{ij} a_{jj} e_j, \quad n \geq i \geq j \geq 1 \tag{8}$$

where

$$a_{ii} = f_i - \sum_{\substack{j=1 \\ i>1}}^{i-1} c_{ij} a_{jj}$$

Table 1. Reduction of Combined Closed Circuit to Other Circuits

	Selectivity Values	
Type of Grinding Circuit	s_{1i} **Pre-classifier**	s_{2i} **Post-classifier**
Open circuit	All equal to 1	All equal to 0
Scalped open circuit	All appropriately defined	All equal to 0
Normal closed circuit	All equal to 1	All appropriately defined
Reverse closed circuit	All equal to post-classifier values	All appropriately defined
Combined closed circuit	All appropriately defined	All appropriately defined

Table 2. Nomenclature for Mass Balance

Stream	Mass flow rate of material	Cumulative fraction of material less than size i	Fraction of material of size i
Circuit feed	G'	G_i'	g_i'
Circuit feed to mill	G	G_i	g_i
Mill feed	F	F_i	f_i
Mill product	P	P_i	p_i
Recycle	T	T_i	t_i
Circuit product for mill	Q	Q_i	q_i
Overall circuit product	Q'	Q_i'	q_i'

Note that solution of Eq. 8 requires a sequential computational algorithm of the form $(a_{ii}, p_i, i = 1, n)$.

The closed-circuit calculations are simplified by introducing the meta product, p_i^*, which is defined for this case by

$$p_i^* = p_i(1 + C)$$

Combining Eq. 8 with the post-classifier mass balance equation gives

$$q_i = (1 - s_{2i}) p_i^* \tag{9}$$

where

$$p_i^* = \frac{a_{ii}^* e_i + \sum_{\substack{j=1 \\ i>1}}^{i-1} c_{ij}\left(a_{jj}^* + s_{2j} p_j^*\right) e_j}{(1 - s_{2i} e_i)}, \quad n \geq i \geq j \geq 1 \tag{9a}$$

$$a_{ii}^* = g_i - \sum_{\substack{j=1 \\ i>1}}^{i-1} c_{ij}\left(a_{jj}^* + s_{2j} p_j^*\right)$$

$$g_i = s_{1i} g_i' \Big/ \sum_{j=1}^{n} s_{1j} g_j'$$

Note that solution of Eq. 9a requires a sequential computational algorithm of the form $(a_{ii}^*, p_i^*, i = 1, n)$. This is the preferred form of the three equations using the $\phi(t)$ function because of its flexibility and speed of computation.

The other stream size distributions can be calculated from the meta product values, from

$$f_i = \frac{g_i + s_{2i} p_i^*}{\sum p_j^*} \tag{10}$$

$$p_i = p_i^* / \sum p_j^* \tag{11}$$

$$t_i = \begin{cases} 0, & \sum p_j^* = 1 \\ \dfrac{s_{2i} p_i^*}{\left(\sum p_j^*\right) - 1}, & \sum p_j^* > 1 \end{cases} \tag{12}$$

and the circulating load calculated as (Eq. 14 in Chapter 6)

$$(1 + C) = \sum p_i^*$$

The overall circuit product size distribution is calculated from

$$q_i' = (1 - s_{1i}) g_i' + q_i \sum_{j=1}^{n} s_{1j} g_j'$$

If the plug flow RTD model is used in Eq. 6

$$e_j = \exp(-S_j \tau) \tag{13a}$$

If the fully mixed RTD is used

$$e_j = 1/(1 + S_j \tau) \tag{13b}$$

If the m equal fully mixed in series RTD model is used (see Eq. 16 in Chapter 14)

$$e_j = 1/(1 + S_j \tau/m)^m \tag{13c}$$

If the RTD model used is the Mori semi-infinite solution (Eq. 12 in Chapter 14)

$$e_j = \exp\left[-\left(\frac{\sqrt{1 + 4D^*\tau S_j} - 1}{2D^*}\right)\right] \tag{13d}$$

If the RTD is the one-large/two-small model (Eq. 18 in Chapter 14)

$$e_j = \frac{1}{(1 + S_j \tau_1)(1 + S_j \tau_2)^2} \tag{13e}$$

Of course, as Herbst, Grandy, and Mika (1971) have shown, e_j values can also be evaluated numerically for functional forms of $\phi(t)$ which do not result in an analytical solution.

It must be noted that the *sink* interval, n, is an exceptional interval because it is clearly not a normal geometric screen sequence interval. In dealing with this interval there are three factors to be considered: (1) $S_n = 0$, because there can be no breakage out of the sink interval; (2) $b_{n,j} = B_{n,j}$; and (3) the value of s_n in the closed-circuit computation of Eq. 13 in Chapter 6 must be the effective mean value for all sizes in the sink interval. This last feature is handled by mathematical extrapolation of S_i, $B_{i,j}$, and g_i' to give sufficient intervals so that the s_n for the sizes in the sink interval is constant (in the apparent bypass region, $s_n = a$) or that the nth interval has an insignificant effect on the amount of recycle.

Again, in all of these forms it is necessary that $S_i - S_j$ does not tend to zero, to avoid numerical instability. In some cases, setting $S_i - S_j$ equal to a small number ε will not avoid the problem because the Reid solution forms give multiples of terms in $1/(S_i - S_j)$. If there are too many terms where $S_i - S_j$ is small, the product of the reciprocal terms will again lead to numerical instabilities. This can occur for a large number of intervals and larger α, since S values at the smaller sizes become small. In these cases either the finite difference technique can be used or the m-stages-in-series technique described next can be used.

Mill Simulators for Stages-in-Series RTDs

It is often convenient to use the m-stages-in-series concept, which can be applied with different values of S and b in the various stages (Austin, Luckie, and von Seebach, 1976). In order to fit experimental RTD values it is generally necessary to allow each mill section to have a different fraction of the total mean residence time (see Chapter 14). The simple mill model describing grinding in a fully mixed mill is (Eq. 4 from Chapter 6)

$$p_i = \left(f_i + \tau \sum_{\substack{j=1 \\ i>1}}^{i-1} S_j b_{ij} p_j \right) \bigg/ (1 + \tau S_i)$$

The algorithm can be recursively applied to generate an algorithm describing grinding in m fully mixed stages in series as

$$p_i(m) = \left[p_i(m-1) + \tau(m) \sum_{j=1}^{i-1} S_j(m) b_{ij} p_j(m) \right] \bigg/ [1 + \tau(m) S_i(m)]$$

$$p_i(0) = f_i \tag{14}$$

where $p_i(m)$ is the weight fraction in the ith size interval of the product stream out of the mth stage, $S_i(m)$ is the specific rate of breakage value for the ith size interval in the mth stage, and $\tau(m)$ is the mean residence time for the mth stage, $\tau(m) = \theta(m)\tau$. Note that using this concept has automatically included the RTD information and eliminated the integrations involved in the use of $e_j = \int_0^\infty e^{-S_j t}\phi(t)\, dt$ in the previous section.

The recursive algorithm can be expressed as

$$p_i(m) = \left\{ f_i + \sum_{j=1}^{i-1} b_{ij} \sum_{k=1}^{m} \tau(k)S_j(k)p_j(k) \prod_{l=1}^{k-1} [1 + \tau(l)S_i(l)] \right\} \Big/$$

$$\prod_{k=1}^{m} [1 + \tau(k)S_i(k)]$$

Again, simultaneously solving with the post-classifier mass balance equation gives

$$q_i = (1 - s_{2i})p_i^*(m)$$

where

$$p_i^*(m) = \frac{g_i + \sum_{j=1}^{i-1} b_{ij} \sum_{k=1}^{m} \tau(k)S_j(k)p_j^*(k) \prod_{l=1}^{k-1} [1 + \tau(l)S_i(l)]}{\prod_{k=1}^{m} [1 + \tau(k)S_i(k)] - s_{2i}},$$

$$n \geq i \geq j \geq 1 \quad (15)$$

The intermediate mill meta products required are obtained from

$$p_i^*(m - 1) = [1 + \tau(m)S_i(m)]\, p_i^*(m) - \tau(m) \sum_{j=1}^{i-1} S_j(m)b_{ij}p_j^*(m),$$

$$n \geq i \geq j \geq 1 \quad (15a)$$

The computation starts at $i = 1$ and $p_1^*(m)$ is calculated from Eq. 15. Then $p_1^*(m - 1)$, etc., follow from Eq. 15a. These values are then used in the calculation of $p_2^*(m)$, etc. Although this algorithm may appear rather involved, it is computationally efficient since it starts at $i = 1$ and continues to build up information. As noted previously, the other stream size distributions can be readily calculated from the p_i^* values.

This algorithm has the advantage of accepting any S values without modification, even if all the S values are the same. Although the algorithm is limited to mill RTDs which can be represented by m fully mixed stages in series, this is not as restrictive as it might initially seem. Consider the case where the S values are the same in each stage and the mill RTD is the one-large/two-small model ($\tau = \tau_1 + 2\tau_2$). The solution is

$$p_i^*(3)$$

$$= \frac{g_i + \displaystyle\sum_{j=1}^{i-1} S_j b_{ij}\left[\tau_1 p_j^*(1) + \tau_2(1 + \tau_1 S_i) p_j^*(2) + \tau_2(1 + \tau_1 S_i)(1 + \tau_2 S_i) p_j^*(3)\right]}{(1 + \tau_1 S_i)(1 + \tau_2 S_i)^2 - s_{2i}}$$

where

$$p_i^*(2) = p_i^*(3)(1 + \tau_2 S_i) - \tau_2 \sum_{j=1}^{i-1} S_j b_{ij} p_j^*(3)$$

and

$$p_i^*(1) = p_i^*(2)(1 + \tau_2 S_i) - \tau_2 \sum_{j=1}^{i-1} S_j b_{ij} p_j^*(2)$$

(15b)

This particular solution can obviously be used for one, two, or three equal fully mixed stages by choosing τ_1/τ appropriately (1, 0, or 1/3).

A Mill Model with Internal (Exit) Classification

As discussed in Chapter 14, it is possible that wet overflow mills have the equivalent of an internal classification action at the mill exit. If coarser material cannot leave the mill as readily as fine material, it will build up in the discharge end of the mill. This cannot obviously be treated as a recycle to the feed end of the mill. However, the action can be approximated with sufficient accuracy by recycling to the feed to the third stage in the one-large/two-small reactors in series model (also see the section "Distributed Parameter Models," Eq. 23), as shown in Fig. 5. Defining the classification action by a set of classification numbers c_i, which represent the fraction of size i presented to the mill exit which is returned to the mill contents,

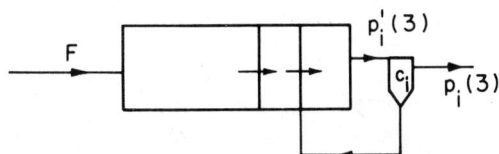

Fig. 5. Mill model of one-large / two-small equivalent RTD with internal classification treated as an exit classification.

gives

$$p_i^*(3)$$

$$= \frac{f_i + \sum_{j=1}^{i-1} S_j b_{ij} \left[\tau_1 p_j(1) + \tau_2 p_j(2)(1 + \tau_1 S_i) + \tau_3' p_j^*(3)(1 + \tau_1 S_i)(1 + \tau_2 S_i) \right]}{(1 + \tau_1 S_i)(1 + \tau_2 S_i)(1 + \tau_3' S_i - c_i)}$$

where $\tau_3' = \tau_3/(1 + C')$ and C' is the circulation ratio for the exit classification. This is determined from $1 + C' = \sum_i \bar{p}_i(3)$: the calculation is performed by setting $C' = 0$ and iterating until a stable C' is obtained.

$$p_i(2) = \left(1 + \tau_3' S_i - c_i\right) p_i^*(3) - \tau_3' \sum_{j=1}^{i-1} S_j b_{ij} p_j^*(3)$$

$$p_i(1) = \left(1 + \tau_2 S_i\right) p_i(2) - \tau_2 \sum_{j=1}^{i-1} S_j b_{ij} p_j(2) \qquad \left.\vphantom{\sum_{j=1}^{i-1}}\right\} \text{(15c)}$$

and

$$p_i(3) = \left(1 - c_i\right) p_i^*(3)$$

For closed circuit this becomes, as before,

$$p_i^*(3) =$$

$$\frac{s_{1i} \times g_i' + \sum_{j=1}^{i-1} S_j b_{ij} \left[\tau_1 p_j^*(1) + \tau_2 p_j^*(2)(1 + \tau_1 S_i) + \tau_3' p_j^*(3)(1 + \tau_1 S_i)(1 + \tau_2 S_i) \right]}{(1 + \tau_1 S_i)(1 + \tau_2 S_i)(1 + \tau_3' S_i - c_i) - (1 - c_i) s_{2i}}$$

where

$$p_i^*(2) = \left(1 + \tau_3' S_i - c_i\right) p_i^*(3) - \tau_3' \sum_{j=1}^{i-1} S_j b_{ij} p_j^*(3)$$

$$p_i^*(1) = \left(1 + \tau_2 S_i\right) p_i^*(2) - \tau_2 \sum_{j=1}^{i-1} S_j b_{ij} p_j^*(2) \qquad \left.\vphantom{\sum_{j=1}^{i-1}}\right\} \text{(15d)}$$

and

$$q_i' = \left(1 - s_{1i}\right) g_i' + \left(1 - s_{2i}\right)\left(1 - c_i\right) p_i^*(3)$$

One-Point Match and Two-Point Match Searching Procedures

As discussed in Chapter 7, the user of a circuit simulator will normally want to produce a suitable circuit output size distribution. A criterion which is frequently used to define a suitable output size distribution is to specify a fraction, ψ, less than a specified size x^*, e.g., 80% < 75 μm. In open-circuit grinding, there is only one τ value which will produce the value for all other parameters specified. In closed-circuit grinding, there is only one τ value for a given set of s_i values which will produce a size distribution passing through ψ, x^*, and there is a corresponding circulation ratio. However, by varying d_{50} the one-point match can also be obtained at a desired circulation ratio.

Because time occurs in algebraically awkward places in the grinding equations, it is not possible to solve directly for the τ value required to reach a specified product. Hence, a method is needed to search for the value which produces the desired specification. Fortunately, the cumulative fraction $Q'(x^*)$ less than the control size x^* varies monotonically-increasing with τ increasing, which means that too small a value of τ will give too small a value of $Q'(x^*)$ and vice versa. Consequently, very simple search methods, such as binary or Fibonacci search, can be used.

The first step of any of these methods is to bound the specifications, i.e.,

$$Q'\left(x^*, \underline{\tau}\right) < Q'\left(x^*, \tau\right) < Q'\left(x^*, \bar{\tau}\right)$$

where $\underline{\tau}$ and $\bar{\tau}$ are lower and upper bounds on τ, that is, any positive values which give

$$\left[Q'\left(x^*, \underline{\tau}\right) - \psi\right]\left[Q'\left(x^*, \bar{\tau}\right) - \psi\right] < 0 \tag{16}$$

Using binary search as an example, the next step is to estimate τ from

$$\tau = \left(\underline{\tau} + \bar{\tau}\right)/2 \tag{17}$$

Then the comparison $|Q'(x^*, \tau) - \psi| \leq \varepsilon$ is made, where ε is some small error allowed by the user. If the absolute difference is greater than the error and if $Q'(x^*, \tau) < \psi$, then τ must be made larger, so $\underline{\tau}$ is replaced with τ and a new estimate of τ is made using Eq. 17 again. If the absolute difference is greater than the error and if $Q'(x^*, \tau) > \psi$, then τ must be reduced, so $\bar{\tau}$ is replaced with τ, etc.

Specifying only one control condition on the circuit product size distribution is the *one-point match* criterion. The empirical mill sizing techniques discussed in Chapter 3 work reasonably well because specifying one point is not a difficult criterion to meet. However, there are times when a second control condition is specified in order to produce an advantageous size distribution. Specifying a suitable second control point, the *two-point match*, will virtually define the entire size distribution. To meet this second point criterion requires a second degree of freedom; this eliminates open-circuit grinding systems which have only flow rate as their single degree of freedom. Thus it is necessary to have a classifier in the circuit, but although necessary, it may not be sufficient, since it still may not be possible to fit both of the desired two points (see "The Behavior of Different Designs of Mill Circuits" in Chapter 7).

Because the second control point, $Q'(x_2^*) = \psi_2$, may not be simultaneously achievable with $Q'(x_i^*) = \psi_1$, several search strategies are possible. For example, simple search techniques can be nested, which means that the first control point, $Q'(x_1^*) = \psi_1$, is satisfied after each improvement in the variable controlling the second control point fit. This technique is useful when the user is not certain that the second control point is in the region of feasible solution and desires the closest fit for the second point while matching the first point exactly. If, for example, d_{50} is the second controllable variable, it is clear that a smaller d_{50} will give more recycle and a steeper size distribution, so that $Q'(x_2^*)$ is monotonically-increasing with d_{50} increasing. Thus the same type of equation as Eq. 17 can be applied to d_{50}. On the other hand, if the second point is known to lie in the feasible solution region or if the best fit does not require that either control point be met exactly, a least-squares criterion applied to both points gives a faster fitting routine.

For closed-circuit simulations there are other types of one-point and two-point fits of interest. For example, when comparing a normal vs. a reverse closed circuit with the same size mills, it would seem logical to set the same feed rate to the mills (i.e., the τ values would be the same) since this gives the same degree of overfilling and then achieve a one-point fit of the circuit product, $Q(x^*)$, by varying the d_{50} value of the classifier.

The same simple binary search procedure can be used when the desired response varies monotonically with changes in the controllable variable. In fact, the same algorithm can be used if the difference between the current response value and the desired response value is calculated so that the difference varies from negative to positive values as the value of the controllable variable increases.

Another response variable of interest for a fixed τ value is P/Q' or T/Q. Such a simulation would set the Q'/W or Q/W values, but not the circuit product size consist.

A two-point fit of interest, achievable by varying τ and d_{50} respectively in a nested search, is $Q'(x^*)$ and P/Q' or T/Q. Thus a user can examine the results of producing 80% less than 208 μm (65 mesh) while varying the circulating load. Another one of interest is $Q'(x^*)$ and Q'/W or Q/W.

Combining the circuit simulator with a search algorithm in this fashion reduces the number of runs that the user must perform to achieve a desired objective.

Unsteady-State First-Order Models

The general problem of unsteady-state mill simulation is complex, because solution requires a knowledge of mass transfer laws, plus methods of keeping track of feed material of different grindability when S_i of feed material is a function of time. However, a simplified model which is useful for some cases can be readily developed. Assume that hold-up in the mill does not vary with flow rate, or that the variation is so small that a formal value of W can be used (see "Solution of Grinding Equations" in Chapter 6).

Consider the case where only feed size distribution and grindability are functions of time, at a constant feed rate. Assume no non-first-order (slowing-down) effect along the mill. The quantity of feed from time t_1 to $t_1 + dt_1$ is $F dt_1$, and the quantity of this feed appearing in size i at a later time t to $t + dt$ is $F w_i(t - t_1, t_1)\, dt_1 \phi(t - t_1)\, dt$, with

$$w_i(t - t_1, t_1) = \sum_{j=1}^{i} d_{ij}(t - t_1) f_j(t_1) \tag{18}$$

where $d_{ij}(t - t_1)$ is the transfer function of size j to i for batch grind time $t - t_1$ and $f_i(t_1)$ is the feed size distribution at t_1. The total of material of size i appearing in product in time t to $t + dt$ is $F\int_{t_1=0}^{t} w_i(t - t_1, t_1)\phi(t - t_1)\, dt_1\, dt$ and dividing by the mass $F\, dt$,

$$p_i(t) = \int_{t_1=0}^{t} w_i(t - t_1, t_1)\phi(t - t_1)\, dt_1 \tag{19}$$

Since $\phi(t - t_1)$ tends to zero as $t - t_1$ becomes larger, $t_1 = 0$ can be replaced by t' where $t - t'$ equals, say, at least five mean residence times and the computation performed from $t - t' > 5\tau$. This avoids the problem of postulating an initial condition for size distributions along the mill at time 0. The integration is readily performed numerically.

Another simplified case is where feed rate changes relatively slowly compared to τ, so that there are only pseudo-steady-state changes of flow rate in the mill. In this case it can be assumed that the mean residence time τ at any instant of time in the mill is the same for all material in the mill. This means that the residence time distribution is a function of time, $\phi(t, t_1)$ and Eq. 19 becomes

$$p_i(t) = \int_{t_1=t'}^{t} w_i(t - t_1, t_1)\phi(t, t_1)\, dt_1$$

A model is required for $\phi(t, t_1)$ as a function of $F(t)$. Assuming a constant dimensionless residence time distribution of $\phi^*(t^*)$ the value of t^* is given by

$$t^* = \int_{\xi=t_1}^{t} \frac{1}{\tau(\xi)}\, d\xi$$

and

$$\phi(t, t_1)\, dt = \tau(t)\phi^*(t^*)\, dt$$

giving

$$p_i(t) = \int_{t_1 = t'}^{t} w_i(t - t_1, t_1)\tau(t)\phi^*\left(\int_{\xi = t_1}^{t} \frac{1}{\tau(\xi)}d\xi\right)dt_1 \qquad (20)$$

Distributed Parameter Models

Up to this point, there have been restrictive assumptions associated with the mill models presented, such as:

1. The primary breakage function B is the same throughout the grinding device.
2. The specific rates of breakage are constant along the grinding device or may vary only between each fully mixed reactor expressing the RTD.
3. The transport of each particle size through the grinding device is the same.

These restrictive assumptions make the previously presented models members of a class known as *lumped parameter* models. A more general equation of a comminution device can be devised by dividing the machine into short segments (Whiten and Lynch, 1967).

Then, for the lth segment, (element 1 is the feed end of the mill) the expression

$$\partial W_l w_{i_l}/\partial t = \left(\sum_{j=1}^{i-1} S_{j_l}b_{ij_l}w_{j_l}W_l\right) - S_{i_l}w_{i_l}W_l - \sum_m v_{i_{l,m}}w_{i_l} + \sum_k v_{i_{k,l}}w_{i_k} \qquad (21)$$

is readily derived for the ith size fraction. Here, $v_{i_{l,m}}$ represents the transfer rate of material of size i from the lth segment into any mth segment. By using differential segments this mill model allows variation of all the parameters along the continuous spatial axis of the device and defines the class of models known as *distributed parameter* models. For steady-state continuous flow $\partial W_l w_{i_l}/\partial t = 0$.

However, this expression is difficult to use practically because of the number of unknown factors involved. Hence, a number of simplifications have been used by various researchers. For example, it may be sufficient to assume that the mixing length is relatively short compared to the size of the device, so that mixing occurs only between adjacent elements. This means that

$$\sum_m v_{i_{l,m}}w_{i_l} \equiv v_{i_{l,l-1}}w_{i_l} + v_{i_{l,l+1}}w_{i_l}$$

$$\sum_k v_{i_{k,l}}w_{i_k} \equiv v_{i_{l-1,l}}w_{i_{l-1}} + v_{i_{l+1,l}}w_{i_{l+1}}$$

The transfer of material is a combination of flow and mixing: therefore it is possible that v can be expressed as

$$v_{i_{l,l+1}} = F_{i_l} + K_{i_l}W_l$$

$$v_{i_{l-1,l}} = F_{i_{l-1}} + K_{i_{l-1}}W_{l-1}$$

and since there is no flow backwards in the mill,

$$v_{i_{l,l-1}} = K_{i_{l-1}}W_l$$

$$v_{i_{l+1,l}} = K_{i_{l+1}}W_{l+1}$$

where K_{i_l} is a mixing factor for size i, defined as the fraction of mass W_l in the segment mixed into the following or preceding section per unit time, if all of W_l were of size i. Similarly, $F_{i_l}w_{i_l}$ is the mass flow rate of size i in section l. This gives

$$\partial W_l w_i / \partial t = F_{i_{l-1}}w_{i_{l-1}} - F_{i_l}w_{i_l} + K_{i_l}\left(W_{l+1}w_{i_{l+1}} - W_l w_{i_l}\right)$$

$$-K_{i_{l-1}}\left(W_l w_{i_l} - W_{l-1}w_{i_{l-1}}\right) + \sum_{j=1}^{i-1} S_{j_l}b_{ij_l}w_{j_l}W_l - S_{i_l}w_{i_l}W_l \tag{22}$$

Note that the third and fourth terms in the right side of the equation are the finite segment equivalent of the spatial continuous diffusion expression

$$\left[\partial\left(D_i\frac{\partial W_o w_i}{\partial x}\right)\Big/\partial x\right] dx$$

where W_o is the mass per unit length, while the first and second terms are the finite equivalent of the continuous convection expression

$$\left[-\partial\left(F_i w_i\right)/\partial x\right] dx$$

Other researchers have presented a conventional one-dimensional balance equation of a continuous mill using simple diffusion and convection terms. However, most of the solutions, obtained by various simplifying assumptions, either could have employed a lumped parameter model directly (Durando, Randolph, and Nuttal, 1970) or the results have been shown to be closely equivalent to those using a lumped parameter model (Herbst, Mika, and Rajamani, 1976).

There is one case where it is essential to use a distributed parameter model, that is, when the slowing-down effect discussed in Chapters 5, 12, and 15 is to be incorporated into the mill model. In this case, the values of S at any location in the mill vary depending on the amount of fines at that location; integration along the mill is necessary to determine the S values. Austin, Bell, and Rogers (1983) have used the convective-mixing equation for steady-state mill simulation (see Chapter 14).

$$0 = D\frac{d^2 w_i}{dl^2} - u\frac{dw_i}{dl} - S_i w_i + \left(\sum_{\substack{j=1 \\ i>1}}^{i-1} b_{i,j} S_j w_j \right), \quad n \geq i \geq j \geq 1 \quad (23)$$

where S_i can be a function of l via its relation to $w_n(l)$. Since material cannot diffuse through the mill end walls or back up the feed pipe, the boundary conditions are

$$f_i = w_i(0) - (D/u) \left. \frac{dw_i}{dl} \right|_{l=0} \quad (23a)$$

$$\left. \frac{dw_i}{dl} \right|_{l=L} = 0 \quad (23b)$$

The velocity of flow is given by $u = L/\tau = LF/W$. This equation set (in dimensionless form) was solved using a simple finite difference solution technique, to give $w_1(1), w_2(1), \cdots, w_n(1); w_1(2), \cdots, w_n(2)$, etc., with S_i values being dependent on the mean w_n over each finite element of mill length.

Classification at the mill exit can be readily incorporated into this model by variation of the discharge boundary conditions, which becomes

$$r(1 - c_i)w_i(L) = w_i(L) - (D/u) \left. \frac{dw_i}{dl} \right|_{l=L} \quad (23c)$$

where r is the rate of presentation of material to the discharge grate as a fraction of the flow rate, c_i is the fraction of size i in r which is returned to the mill, $1 - c_i$ the fraction of size i in r which passes through, and, clearly,

$$r = 1 / \sum_i (1 - c_i)w_i(L) \quad (24)$$

Since $0 \leq c_i \leq 1.0$ then $r \geq 1$. If there is no classification action $c_i = 0$, $r = 1$, and Eq. 23c reduces to Eq. 23b as logically required.

Two-Stage Grinding Circuits

Two-stage grinding circuits involve two mills in series, that is, a single-stage circuit feeding another single-stage circuit; see Fig. 6 for some examples. The two-stage grinding circuit can be readily simulated by using the appropriate single-stage simulators in series. However, as seen in the section "One-Point Match Searching Procedures," it is sometimes necessary to be able to search on τ to obtain a desired characteristic of the size distribution of the circuit output (one-point match). For two mills in series, there are two values of τ but only one final circuit product specification. The problem of searching is simplified by the following algebra. Let the relative size Ω of the two mills be defined by the ratio of the hold-up of the first

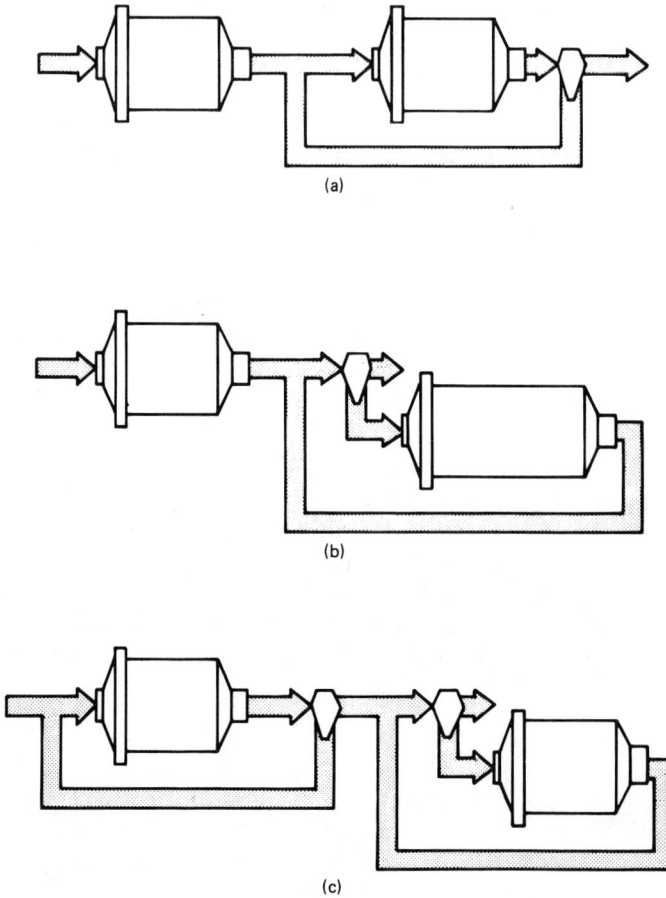

(a)

(b)

(c)

Fig. 6. Examples of two-stage grinding circuits (see also Fig. 12).

stage to the hold-up of the second stage, that is, $\Omega = W_1/W_2$. For a simple two-stage series at steady state the flow of mass, G', into the first stage equals the flow of mass into the second stage; then, since the actual flow rate to each mill is $F_1 = G_1(1 + C_1)$, $F_2 = G_2(1 + C_2)$ and $\tau_1 = W_1/F_1$, $\tau_2 = W_2/F_2$,

$$\Omega = \left(\frac{\tau_1}{\tau_2}\right) \frac{(1 + C_1)}{(1 + C_2)} \left(\frac{G_1}{G_2}\right)$$

or

$$\tau_2 = \left(\frac{\tau_1}{\Omega}\right) \frac{(1 + C_1)}{(1 + C_2)} \left(\frac{G_1}{G_2}\right) \tag{25}$$

This algebra shows that by setting the breakage, RTD, and classifier parameters in the system, *plus* Ω, a particular value of τ_1 fixes the entire system so that a search on τ_1 vs. $Q'(x^*)$ is now feasible. There is only one makeup feed rate and hence, only one circuit product rate which will lead to τ_1; C_1 is then also a single value and the feed to the second stage is a single value. There is only one value of τ_2 which will give grinding to produce a value of C_2 to satisfy Eq. 25. Unfortunately, the value of C_2 depends on the feed size distribution, S_2, B_2, and τ_2 values in a complicated manner and it is not possible to simultaneously solve Eq. 25 with the mill model to extract τ_2. Therefore it is necessary to search on values of τ_2 until the mill-classifier model generates C_2 to satisfy the equation.

Rearranging the equation gives

$$\tau_2(1 + C_2) - \Gamma = 0$$

where

$$\Gamma = \frac{\tau_1}{\Omega}(1 + C_1)\frac{G_1}{G_2} = \text{known}$$

since

$$\frac{G_1}{G_2} = \frac{\sum s_{1j_1} g'_j}{\sum s_{1j_2} q'_{j_1}}$$

A simple binary search can be conducted (starting with an initial lower value of τ_2 of zero and increasing τ_2 in stages), where C_2 follows from τ_2 inserted into the circuit model and the calculation proceeds until

$$\tau_2(1 + C_2) - \Gamma = \varepsilon$$

where ε is a specified allowable error limit. The technique of obtaining a one-point match described in the section, "One-Point Match and Two-Point Match Searching Procedures" can be immediately applied to the two-stage circuit by searching on τ_1 in the usual way until the desired match is obtained.

In a design simulation involving two mills in series, it is clear that the relative size of the mills, Ω, is another degree of freedom. To obtain a specific product, such as 80% less than 75 μm, there are any number of combinations of mill sizes which can achieve this product specification, ranging from a very large first-stage mill and a very small second-stage mill to a very small first-stage mill and a very large second-stage mill. The advantages of circuits with two mills in series over single mills are: (1) the classification step between the mills enables fines to be withdrawn and not overground in the second mill, thus leading to steeper size distributions at reasonable circulation ratios; (2) the grinding media can be adjusted in each mill to be more efficient for the powder being treated in each mill; and (3) the slurry rheology can be optimized for each mill. For given values of S in each mill there will be an optimum value of Ω to give a minimum of fines, and consequently energy usage, for a minimum of capital cost. Simulation for various values of Ω will define the optimum, in conjunction with power and economic calculations.

Air-Swept Ball Mills for Coal or Cement Finish Grinding

There are two types of tumbling ball mills which use dry grinding with air sweeping. The first is used for coal grinding, to prepare pulverized coal for firing into utility boiler furnaces or kilns. This type of mill is somewhat lower in capital cost than a vertical roller mill used for the same application, it has lower maintenance costs, and it can successfully prepare pulverized fuel from high ash coals or anthracites, which are abrasive in roller mills. However, it consumes a higher specific grinding energy for a normal coal (Luckie and Austin, 1980). Air is swept through the mill in sufficient quantities to carry out all the pulverized coal product. The mill has feed and discharge trunnions of large diameter to reduce the fan power required to overcome pressure drop as the gas is moved through the mill. This means that ball loadings are low, typically 25%, so that the mill has a relatively low capacity per unit volume compared to normal ball mills.

The second type of ball mill used with air sweeping is the long two (or three) compartment tube mill used for grinding cement clinker to the finished cement product. In this case, there are two product streams discharged from the mill, the airborne material and a normal flow of powder through a discharge grate.

The function of air sweeping is different in these two mills. In the coal mill, a primary purpose of the air flow is to use the mill simultaneously as a drier, since the dry grinding of wet coals is very inefficient. Thus the inlet air stream is heated to a temperature sufficient to supply latent heat of vaporization and give a cooler exit air-coal stream, with a temperature above the dewpoint of the humid air. The exit temperature must also be low enough to prevent excessive decomposition of the coal in the mill, classifier, and ductwork leading from the mill. The air-coal suspension is sometimes taken directly to the burners to the furnace (direct firing), and the velocity of the fluid through the burner must be considerably higher than the flame velocity to prevent flashback to the mill. Flame velocity increases with temperature, which again requires that the exit temperature be as low as possible consistent with being above the dewpoint.

In the cement mill, the primary function of the air stream is to cool the cement, which enters the mill hot and also picks up the mill energy released as heat. The inlet air is sometimes treated with a water spray to absorb more heat by the latent heat of vaporization.

In both types of mill, but especially the coal mill, the air sweeping has an important second effect. It is known (see Chapters 5 and 12) that the accumulation of fine material in dry grinding reduces the rates of breakage and gives direct mill inefficiency. Air sweeping, classification, and recycle keep the fines swept out of the mill and eliminates or reduces this effect (the slowing-down effect). The purpose of the mill models presented in this section is to allow for the effect of the air sweeping action on the product size distributions and mill capacities. Of course, the models must be combined in the usual way with a description of the selectivity values of the external classifier; for both systems the types of classifiers used are the adjustable mechanical air classifiers described in the section, "Mechanical Air Separators" in Chapter 13. These are set to produce the desired fineness of grind and recycle.

Air-Swept Coal Mill

It is convenient to treat the mill (Austin et al., 1983) as a series of fully mixed stages, with removal of fines in the air stream from each stage, as illustrated in Figs. 7 and 8. The rate-mass balance of a single fully mixed mill with air sweeping is "rate of removal of size i in the air stream plus rate of removal of size i in the powder flow equals rate of feed of size i plus rate of production of size i by breakage of all larger sizes minus rate of breakage of size i." Let the rate of removal of size i by air sweeping be r_i, then the total rate of removal of powder by air sweeping is $R = \sum_1^n r_i$ and Eq. 4 in Chapter 6 becomes

$$r_i + p_i(F - R) = f_i F + W \sum_{\substack{j=1 \\ i>1}}^{i-1} b_{ij} S_j w_j - S_i w_i W \qquad (26)$$

It will be assumed that the grinding action acts on the flowing powder and that the suspended powder is *not* subject to breakage action; then, $w_i = p_i$. It will also be assumed that a fraction η of the hold-up W is exposed to the air stream by tumbling, per mill revolution. For a mill

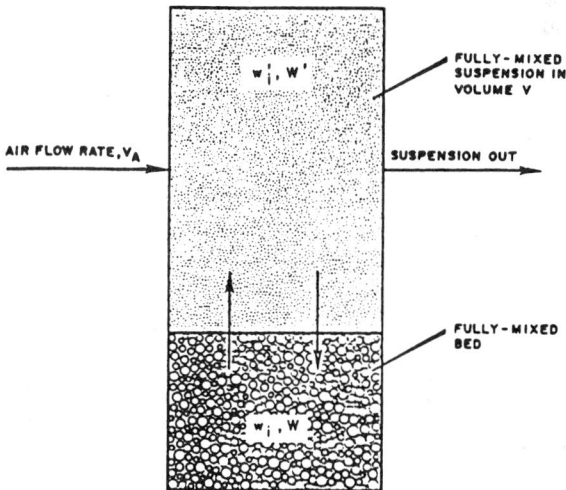

Fig. 7. Illustration of a single fully mixed stage of a grinding mill with air sweeping.

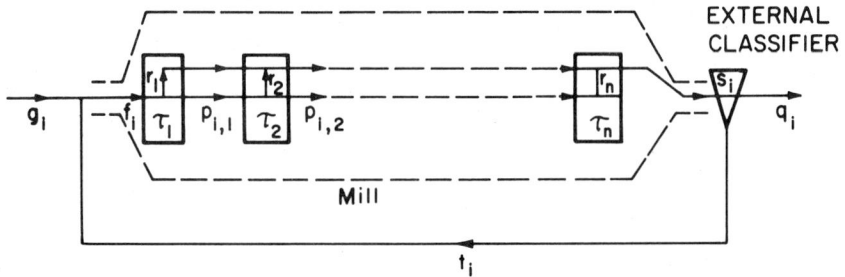

Fig. 8. Illustration of air-swept mill as a series of stages.

rotation speed ω, the amount of size i material in W exposed per unit time is $p_i \eta \omega W$. It will be assumed that the exposed material of size i has a probability c_i of being retained in the hold-up or falling back into the hold-up and hence the fraction swept up and swept out will be $1 - c_i$ or

$$r_i = p_i \eta \omega W (1 - c_i) \qquad (27)$$

This is exactly equivalent to an internal classification action. Inserting into Eq. 26 and rearranging, using $\tau = W/F$,

$$p_i = \frac{f_i + \tau \sum\limits_{j=1}^{i-1} b_{ij} S_j p_j}{(1 - \varepsilon) + S_i \tau + \eta \omega (1 - c_i) \tau}, \qquad 1 \leq i \leq n \qquad (28)$$

where

$$\varepsilon = R/F = \tau \eta \omega \sum\limits_{1}^{n} (1 - c_j) p_j \qquad (29)$$

ε is thus the fraction of the feed rate removed in the *air* stream; if the RTD of the powder flow in the mill corresponded to one fully mixed reactor, $\varepsilon = 1$. The size distribution of the airborne product is given by $p_i' R = r_i$, or

$$p_i' = \tau \eta \omega (1 - c_i) p_i / \varepsilon \qquad (30)$$

Austin et al. (1983) concluded that the tumbling ball mill for coal, in which all the coal is removed in the sweeping stream, can be approximated by a single fully mixed reactor (see Chapter 12). This is primarily because large sizes reaching the discharge end of the mill cannot flow out, and have to back-mix until they are broken small enough to be swept out.

The computational procedure for this model is given as Appendix 7. For known values of S_i, $b_{i,j}$, c_i, and $\eta \omega$, there is a unique value of τ. In practice, the air-flow rate is usually adjusted to match the coal feed rate so that a correct filling level of powder is obtained in the mill, using the sound of the mill as a guide. Since W is thus fixed, the coal feed rate follows from $F = W\tau$. Because the values of c_i vary with air-flow rates, it is necessary in the simulation to: (1) specify the air-flow rate; (2) calculate p_i, τ, p_i'; (3) calculate F for a correct value of W; and (4) repeat for different values of air-flow rate until the desired value of F is obtained by trial-and-error search or interpolation. A model for the variation of the internal classification action with particle density and air-flow rate is given in Appendix 9.

Air-Swept Cement Mill

To extend the model to the long tube mills used for cement finish grinding it is necessary to make two important modifications (Austin, Weymont, and Knobloch, 1980). First, the assumption that the mill behaves approximately as a fully mixed reactor is no longer valid because powder reaching the discharge grate can discharge from the mill. Second, it is then necessary to allow specifically for material falling back into the bed over different L/D values than that used to determine the c_i values (see the section, "A Simulation Model for an Air-Swept Ball Mill Grinding Coal" in Chapter 12). It is possible to treat the internal classification action as a continuum along the mill axis, but a sufficiently good approximation can be obtained by treating it as the series of stages shown in Fig. 9. Note that this is not the same model as Fig. 8 because of the reclassification of airborne material in the next section of the mill.

It is convenient to define arbitrarily an internal classification stage as corresponding to a single fully mixed reactor in the grinding process. If the experimentally determined residence time distribution of a mill corresponds approximately to m equal reactors in series, we will use m internal classification stages. The specific rates of breakage in each stage are $S_{i,k}$, where k denotes the kth stage, and the primary progeny fragment distribution is denoted by the symbol $b_{i,j,k}$. The simplest treatment is for a single fully mixed stage

$$F(1 + C)p_i = F(1 + C)f_i' - S_i w_i W + \sum_{\substack{j=1 \\ i>1}}^{i-1} b_{i,j} S_j W w_j \qquad (31)$$

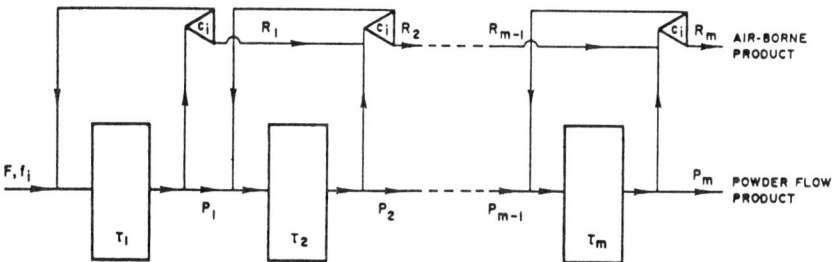

Fig. 9. Illustration of m equivalent grinding and internal classification stages in series.

where F is the external feed rate into the section, $1 + C$ is the circulating load ratio due to a fraction C falling back into the bed, and p_i is the fraction of size i material leaving the mill as powder flow. Since a fully mixed stage is considered, $p_i = w_i$. The fraction of mill feed of size i, f_i', is made up of the feed into the stage plus material falling back into the bed $(1 + C)Ff_i' = Ff_i + p_i \eta \omega W c_i$, where f_i is the fraction of size i entering in the feed. Substituting this into Eq. 31 and collecting terms in p_i gives

$$(1 + C)p_i = \frac{f_i + \tau' \sum_{\substack{j=1 \\ i>1}}^{i-1} b_{i,j} S_j p_j (1 + C)}{1 - \eta \omega \tau' c_i + S_i \tau'}$$

where $\tau' = W/[F(1 + C)]$ is the apparent mean residence time in the stage. Letting $(1 + C)p_i = p_i^*$,

$$p_i^* = \frac{f_i + \tau' \sum_{\substack{j=1 \\ i>1}}^{i-1} b_{i,j} S_j p_j^*}{1 - \eta \omega \tau' c_i + S_i \tau'}, \quad n \geq i \geq j \geq 1 \qquad (32)$$

The values of p_i^* can be sequentially calculated starting at $i = 1$, $p_1^* = f_1/(1 - \eta \omega \tau' c_1 + S_1 \tau')$, then using p_1^* in the calculation of p_2^*, etc. Then since $\sum_{i=1}^{n} p_i = 1$,

$$\sum_{i=1}^{n} p_i^* = 1 + C$$

The values of p_i follow from $p_i = p_i^*/(1 + C)$ and the normal mean residence time defined by $\tau = W/F$ is obtained.

Letting P be the flow rate of particles out of the mill as powder and R be the flow rate as airborne suspension, Fig. 10 shows that $P = (1 + C)F - \eta \omega W$ or

$$\varepsilon = (1 + C) - \eta \omega \tau$$

where ε is the fraction of the feed leaving in the *powder* stream, $\varepsilon = P/F$. Also $R = F - P$ or

$$R = F(1 - \varepsilon)$$

Finally, the size distribution of the airborne stream, p_i' can be found from $p_i' R = p_i \eta \omega W (1 - c_i)$ or

$$p_i' = p_i \eta \omega \tau (1 - c_i)/(1 - \varepsilon) \qquad (33)$$

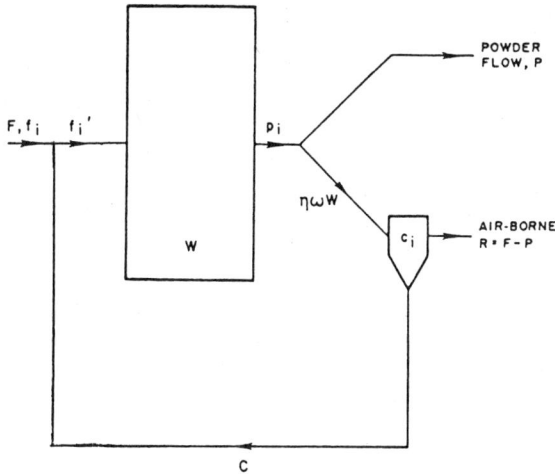

Fig. 10. Nomenclature for internal classification model for a single stage.

From a physical viewpoint, set values of η and c_i are obtained when the air-flow rate through the mill is set. Computation for a smaller value of τ' corresponds to a higher entering feed rate F, a coarser grind, less material removed in the air-sweeping stream and more removal in the powder flow. When η is zero, Eq. 32 reduces to the normal fully mixed grinding reactor equation. When $\eta\omega c_i$ becomes large compared to S_i, for the large sizes where c_i is not zero, the internal recycle is high, leading to the steepest size distribution with a minimum of fines.

The resulting equations for the size distributions and flow rates are given in Appendix 8.

The general expression for the kth stage ($k > 1$) is (see Appendix 8)

$$p_{i,k}^* = \frac{p_{i,k-1} + \left(\dfrac{1 - \varepsilon_1\varepsilon_2 \cdots \varepsilon_{k-1}}{\varepsilon_1\varepsilon_2 \cdots \varepsilon_{k-1}}\right)p_{i,k-1}'c_i + \tau_k' \displaystyle\sum_{\substack{j=1 \\ i>1}}^{i-1} b_{i,j}S_{j,k}p_{j,k}^*}{1 - \eta\omega\tau_k'c_i + S_{i,k}\tau_k'},$$

$$1 < k \leq m, \quad n \geq i \geq j \geq 1 \quad (34)$$

with $\tau_k' = (W/P_k)/(1 + C_k) = \tau_k/(1 + C_k)$. τ_k is given by $\tau_k = \tau_{k-1}/\varepsilon_{k-1}$ $= \tau_1/(\varepsilon_1\varepsilon_2 \cdots \varepsilon_{k-1})$, that is

$$\tau_k = (\tau/m)/(\varepsilon_1\varepsilon_2 \cdots \varepsilon_{k-1}) \quad (35)$$

with τ being defined by W/F, W now being the total hold-up in the mill. As before, this equation is sequentially computed $((p_i^*, k, i = 1, n), k = 1, m)$.

At each stage after the first, a trial-and-error search must be made for the value of τ_k' which gives the correct τ_k.

The size distribution of powder leaving the final stage is

$$p_{i,m} = p_{i,m}^*/(1 + C_m) = p_{i,m}^*/\sum_{j=1}^{n} p_{j,m}^* \tag{36}$$

The values of ε_k are obtained from

$$\varepsilon_k = 1 + C_k - \eta\omega\tau_k \tag{37}$$

The size distribution of the airborne dust in each stage is obtained from

$$p_{i,k}' = p_{i,k-1}'(1 - c_i)\left(\frac{1 - \varepsilon_1\varepsilon_2 \cdots \varepsilon_{k-1}}{1 - \varepsilon_1\varepsilon_2 \cdots \varepsilon_k}\right)$$

$$+ p_{i,k}(1 - c_i)\eta\omega\tau\left(\frac{1}{1 - \varepsilon_1\varepsilon_1 \cdots \varepsilon_k}\right) \tag{38}$$

The flow rate of powder out of the final stage is $P_m = F\varepsilon_1\varepsilon_2 \cdots \varepsilon_m$ or

$$P_m/W = (1/\tau)\prod_{k=1}^{m} \varepsilon_k \tag{39a}$$

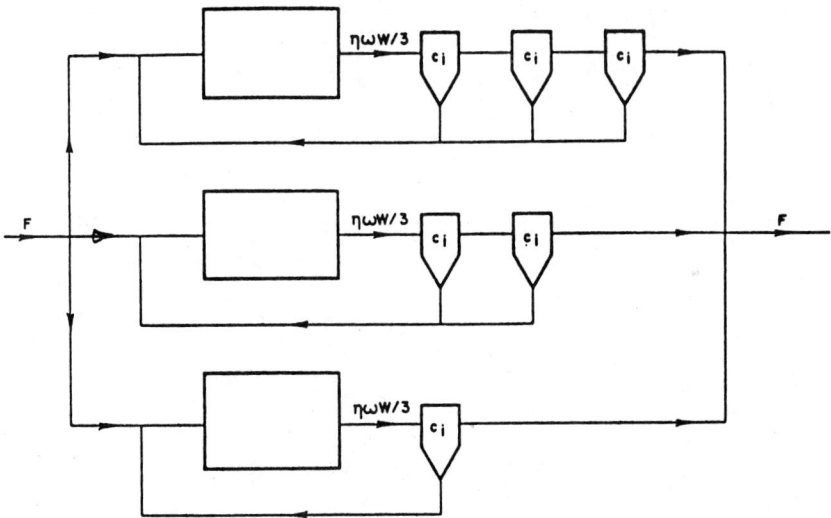

Fig. 11. Equivalent circuit for a fully mixed mill containing three stages of internal classification.

The flow rate of airborne particles out of the final stage is $R_m = R - P_m$ or

$$R_m/W = (1/\tau)\left(1 - \prod_{k-1}^{m} \varepsilon_k\right) \tag{39b}$$

The size distribution of the final airborne dust is obtained from Eq. 38 with $k = m$.

In order to convert c_i values measured for a given L/D (see Appendix 9) to another L/D, it will be assumed that the number of stages of classification are proportional to L/D. It is considered sufficiently accurate to take $0.5D$ as a single stage. The c_i values were measured on a fully mixed mill of $L = 1.5D$, so the measured c_i values can be considered to be the net result of the three mill sections shown in Fig. 11. Each section presents $\eta \omega W/3$ mass per unit time to the classification action, and since each section had the same values of w_i (fully mixed mill), the fall-back of material from the stream can be returned to any section. Thus the fall-back from section 1 is due to the following three classification actions as shown. This system is equivalent to a single fully mixed mill with a single overall internal classification action of $3\bar{c}_i = c_i + c_i[1 + (1 - c_i)] + c_i[1 + (1 - c_i) + (1 - c_i)^2]$ or

$$\bar{c}_i = c_i\left[6 + c_i^2 - 4c_i\right]/3 \tag{40}$$

The $\bar{c}_i$ values of Fig. 9 in Chapter 12 were then used to obtain c_i values using Eq. 40; the values were then fitted by the empirical classification function

$$c_i = 1\left/\left[1 + (x_i/d_{50})^{-\lambda}\right]\right. \tag{41}$$

where $\lambda = 1.54$ and $d_{50} = 320 \ \mu$m for the coal data; see Table 3.

Table 3. Estimates of Internal Classifier Parameters for an Air-Swept Coal Mill, at $\tau_A = 0.062$ Min for Each of Three Sections

Size, μm	$\bar{c}_i$	c_i
3360 × 2380	1.0	1.0
2380 × 1680	1.0	1.0
1680 × 1190	0.99	0.97
1190 × 841	0.96	0.89
841 × 595	0.92	0.80
595 × 420	0.85	0.68
420 × 297	0.76	0.56
297 × 210	0.66	0.44
210 × 149	0.54	0.34
149 × 105	0.42	0.25
105 × 75	0.30	0.17
75 × 53	0.17	0.09
< 53	0	0

Circuits for Air-Swept Mills

The algebra for closing the circuit for an air-swept cement mill is somewhat more complex than for a normal mill, because of the number of different ways in which the streams from two classifiers can be combined and recycled. This is illustrated in Fig. 12. The algebra involved for Fig. 12a is as follows.

Let P_m be the rate of flow of powder from the mill, with size distribution $p_{i,m}$; R_m is the rate of flow of airborne powder from the mill, with size distribution $p'_{i,m}$: then

$$Ff_i = Qg_i + P_m p_{i,m} s_{i,2} + R_m p'_{i,m} s_{i,1} s_{i,2}$$

The definition of overall mean residence time τ is $\tau = W/F$, where W is hold-up in the mill. Thus

$$f_i = \frac{g_i}{1 + C_e} + \tau\left(\frac{P_m}{W}\right) p_{i,m} s_{i,2} + \tau\left(\frac{R_m}{W}\right) p'_{i,m} s_{i,1} s_{i,2} \qquad (42)$$

where C_e is defined by T/Q, that is, recycle/raw feed: then $F/Q = 1 + C_e$. P_m/W, R_m/W, $p_{i,m}$ and $p'_{i,m}$ are delivered by the program computation of the mill model for a given τ. C_e is calculated from

$$1 = \frac{1}{1 + C_e} + \tau\left(\frac{P_m}{W}\right) \sum_1^n p_{i,m} s_{i,2} + \tau\left(\frac{R_m}{W}\right) \sum_1^n p'_{i,m} s_{i,1} s_{i,2} \qquad (43)$$

The final product size distribution is obtained from

$$Qq_i = P_m p_{i,m}(1 - s_{i,2}) + R_m p'_{i,m} s_{i,1}(1 - s_{i,2}) + R_m p'_{i,m}(1 - s_{i,1})]$$

that is,

$$q_i = \frac{W}{Q}\left(\frac{P_m}{W}\right) p_{i,m}(1 - s_{i,2}) + \frac{W}{Q}\left(\frac{R_m}{W}\right) p'_{i,m}[s_{i,1}(1 - s_{i,2}) + (1 - s_{i,1})]$$

$$= (1 + C_e)(\tau)\left\{\left(\frac{P_m}{W}\right) p_{i,m}(1 - s_{i,2})\right.$$

$$\left. + \left(\frac{R_m}{W}\right) p'_{i,m}[s_{i,1}(1 - s_{i,2}) + (1 - s_{i,1})]\right\} \qquad (44)$$

Note that C_e can be made zero by setting $s_i = 0$.

A more general two-stage circuit is shown in Fig. 13. In this case, some fraction ψ of the coarse product from the first milling stage is recycled and $1 - \psi$ is sent to the next milling stage, as indicated on Fig. 12. Fig. 14 shows an actual arrangement corresponding to Fig. 13, employing two mechanical separators for the powder flows and a single static separator for the combined flows of airborne material. The intermediate stage of classification allows the more efficient removal of fines, while the ability to vary ψ allows each of the two compartments to be brought to optimum filling conditions. The previous algebra is modified to

$$Ff_i = Qg_i + \psi P_m p_{i,m} s_{i,2} + \psi R_m p'_{i,m} s_{i,1} s_{i,2}$$

$$f_i = \frac{g_i}{1 + C_e} + \psi\tau\left(\frac{P_m}{W}\right)p_{i,m}s_{i,2} + \psi\tau\left(\frac{R_m}{W}\right)p'_{i,m}s_{i,1}s_{i,2} \qquad (42a)$$

$$1 = \frac{1}{1 + C_e} + \psi\tau\left(\frac{P_m}{W}\right)^n\sum_1 p_{i,m}s_{i,2} + \psi\tau\left(\frac{R_m}{W}\right)^n\sum_1 p'_{i,m}s_{i,1}s_{i,2} \qquad (43a)$$

$$q_{i,1} = (1 + C_e)(\tau)\left\{\left(\frac{P_m}{W}\right)p_{i,m}(1 - s_{i,2})\right.$$

$$\left. + \left(\frac{R_m}{W}\right)p'_{i,m}[s_{i,1}(1 - s_{i,2}) + (1 - s_{i,1})]\right\} \qquad (44a)$$

When $\psi = 1$, these reduce to Eqs. 42 to 44. The size distribution and quantity of the coarse product stream is given by

$$\left(\frac{T}{W}\right) = \left(\frac{P_m}{W}\right)^n\sum_1 p_{i,m}s_{i,2} + \left(\frac{R_m}{W}\right)^n\sum_1 p'_{i,m}s_{i,1}s_{i,2} \qquad (45)$$

$$t_{i,1} = \left[\left(\frac{P_m}{W}\right)p_{i,m}s_{i,2} + \left(\frac{R_m}{W}\right)p'_{i,m}s_{i,1}s_{i,2}\right]\Big/(T/W) \qquad (46)$$

The flow rate of fine product is now

$$\frac{Q_1}{W} = \frac{Q}{W} - (1 - \psi)\frac{T}{W} \qquad (47)$$

where

$$\frac{Q}{W} = \frac{1}{\tau(1 + C_e)}$$

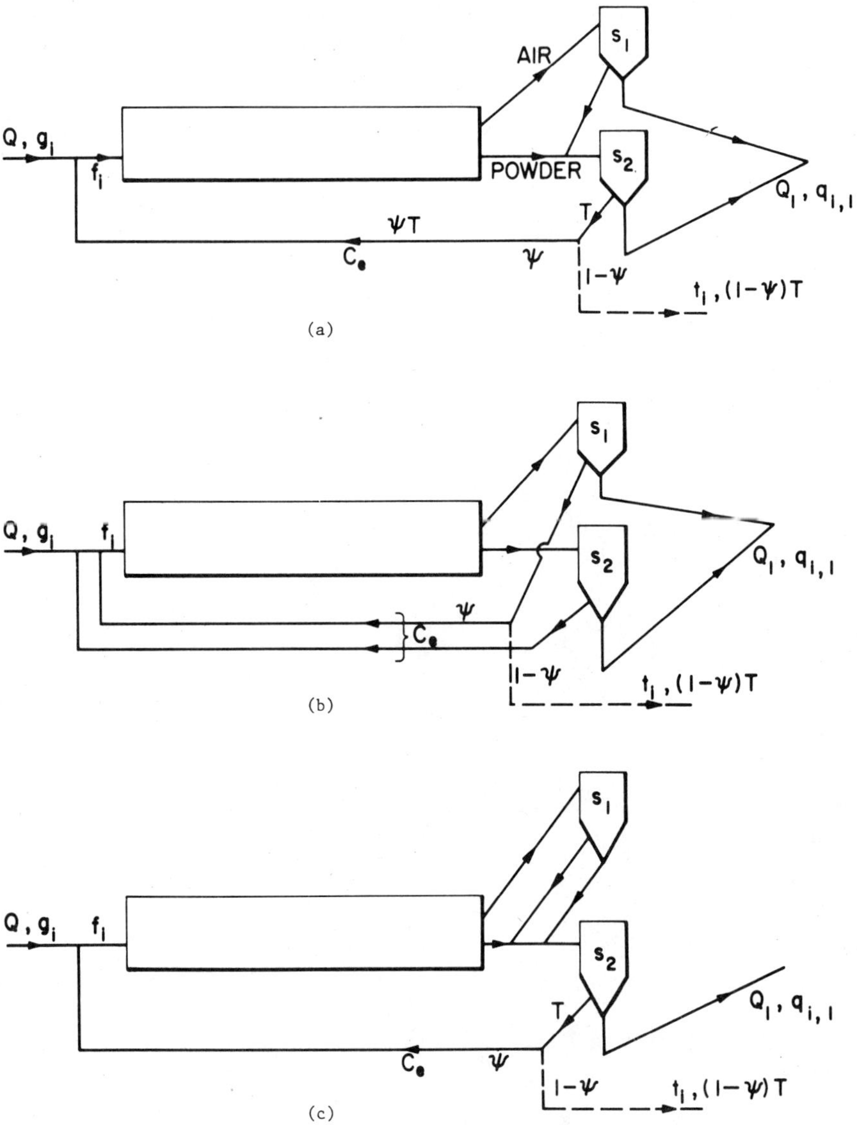

(a)

(b)

(c)

Fig. 12. Recycle possibilities for air-swept mills with classification of airborne powder and discharge powder.

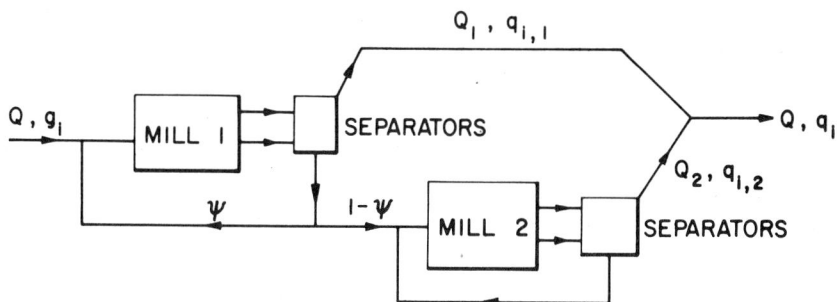

Fig. 13. More general two-stage milling circuit.

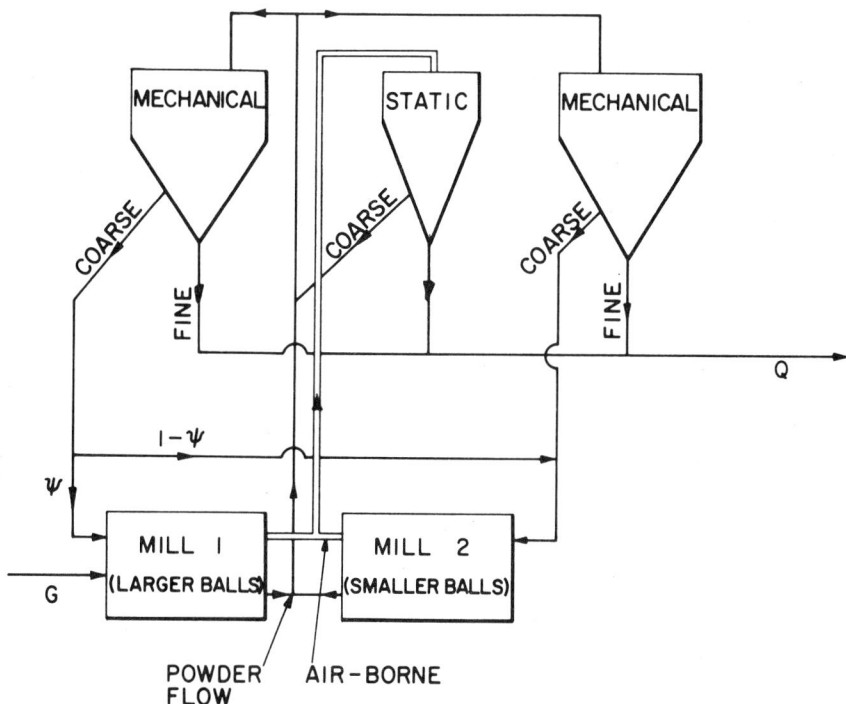

Fig. 14. Example of actual arrangement of two compartment mill feeds, corresponding to Fig. 13 (Doppelrotator, Polysius-Aerofall, Atlanta, GA).

Similarly, the algebra for the mill in Fig. 12b is

$$f_i = \frac{g_i}{1 + C_e} + \psi\tau\left(\frac{P_m}{W}\right)p_{i,m}s_{i,2} + \psi\tau\left(\frac{R_m}{W}\right)p'_{i,m}s_{i,1} \qquad (42b)$$

$$1 = \frac{1}{1 + C_e} + \psi\tau\left(\frac{P_m}{W}\right)\sum_1^n p_{i,m}s_{i,2} + \psi\tau\left(\frac{R_m}{W}\right)\sum_1^n p'_{i,m}s_{i,1} \qquad (43b)$$

$$q_i = (1 + C_e)(\tau)\left[\left(\frac{P_m}{W}\right)p_{i,m}(1 - s_{i,2}) + \left(\frac{R_m}{W}\right)p'_{i,m}(1 - s_{i,1})\right] \qquad (44b)$$

and

$$T/W = \left(\frac{P_m}{W}\right)\sum_1^n p_{i,m}s_{i,2} + \left(\frac{R_m}{W}\right)\sum_1^n p'_{i,m}s_{i,1} \qquad (45a)$$

$$t_{i,1} = \left[\left(\frac{P_m}{W}\right)p_{i,m}s_{i,2} + \left(\frac{R_m}{W}\right)p'_{i,m}s_{i,1}\right]\Big/(T/W) \qquad (46a)$$

For the mill in Fig. 12c,

$$f_i = \frac{g_i}{1 + C_e} + \psi\tau\left(\frac{P_m}{W}\right)p_{i,m}s_{i,2} + \psi\tau\left(\frac{R_m}{W}\right)p'_{i,m}s_{i,2} \qquad (42c)$$

$$1 = \frac{1}{1 + C_e} + \psi\tau\left(\frac{P_m}{W}\right)\sum_1^n p_{i,m}s_{i,2} + \psi\tau\left(\frac{R_m}{W}\right)\sum_1^n p'_{i,m}s_{i,2} \qquad (43c)$$

$$q_i = (1 + C_e)(\tau)\left[\left(\frac{P_m}{W}\right)p_{i,m}(1 - s_{i,2}) + \left(\frac{R_m}{W}\right)p'_{i,m}(1 - s_{i,2})\right] \qquad (44c)$$

etc. This can also be derived from Fig. 12a by setting $s_{i,1} = 1$.
   The computational procedure for the two-stage circuit is:
   1. Choose an arrangement as in Fig. 12a, b, or c.
   2. Choose ψ.
   3. The first mill system is computed with τ_1, giving size distributions and Q_1/W_1.
   4. The second mill system is computed with $\psi = 1.0$ and setting $g_{i,2} = t_{i,1}$. The value of τ_2 is calculated from

$$\tau_2 = (W_2/W_1)\Big/\left(\frac{Q}{W_1} - \frac{Q_1}{W_1}\right)(1 + C_{e2})$$

Thus the ratio $\Omega = W_1/W_2$ must be specified. A search is performed on τ_2 to obtain C_{e2} to satisfy this equation.
   5. The final product size distribution is

$$q_i = \left(\frac{Q_1/W_1}{Q/W_1}\right)q_{i,1} + \left(1 - \frac{Q_1/W_1}{Q/W_1}\right)q_{i,2}$$

References

Austin, L. G., Luckie, P. T., and von Seebach, H. M., 1976, "Optimization of a Cement Milling Circuit with Respect to Particle Size Distribution and Strength Development by Simulation Models," *Proceedings*, Fourth European Symposium Zerkleinern, H. Rumpf and K. Schonert, eds., Dechema Monographien 79, Nr. 1576-1588, Verlag Chemie, Weinheim, pp. 519–537.

Austin, L. G., Weymont, N. P., and Knobloch, O., 1980, "The Simulation of Air-Swept Cement Mills, Part I. The Simulation Model," European Symposium on Particle Technology, B, K. Schonert, W. Gregor, and F. Hofmann, eds., Amsterdam, pp. 640–655.

Austin, L. G., Bell, D., and Rogers, R. S. C., 1983, "Incorporation of the Slowing-Down Effect into a Mill Model," First International Meeting of Fine Particle Society, Hawaii, submitted to the *Journal of the Fine Particle Society*.

Austin, L. G., et al., 1984, "A Simulation Model for an Air-Swept Ball Mill Grinding Coal," submitted to *Powder Technology*.

Durando, A., Randolph, A., and Nuttal, H., 1970, "A Population Balance Derived Continuous Model for Prediction of Particle-Size Distributions as a Function of Feed Rate and Feed Particle-Size Distribution in a Grinding Mill," Ninth International Symposium on Decision Making in Mineral Industries, Paper 54.

Gardner, R. P. and Sukarijnajtee, K., 1973, "A Combined Tracer and Back-Calculation Method for Determining Particulate Breakage Functions in Ball Milling System," *Powder Technology*, Vol. 7, pp. 169–179.

Gardner, R. P. and Verghese, K., 1975, "A Model with Closed Form Analytical Solution for Steady-State Closed Circuit Comminution Processes," *Powder Technology*, Vol. 11, pp. 87–88.

Herbst, J. A., Grandy, G. A., and Mika, T. S., 1971, "On the Development and Use of Lumped Parameter Models for Continuous Open and Closed-Circuit Systems," *Trans. IMM*, Vol. 80, pp. C193–C198.

Herbst, J. A., Mika, T. S., and Rajamani, K., 1976, "A Comparison of Distributed and Lumped Parameter Models for Open Circuit Grinding," *Proceedings*, Fourth European Symposium Zerkleinern, H. Rumpf and K. Schonert, eds., Dechema Monographien 79, Nr. 1576–1588, pp. 467–487.

Kelsall, D. F. and Stewart, P. S. B., 1971, "A Critical Review of Implications of Models of Grinding and Flotation," Australian IMM Symposium on Automatic Control in Mineral Processing Plants, pp. 213–232.

Luckie, P. T., 1976, "The Hybrid Model for Steady-State Mill Simulation," *Powder Technology*, Vol. 13, pp. 289–290.

Luckie, P. T. and Austin, L. G., 1972, "A Review Introduction to the Solution of the Grinding Equations by Digital Computation," *Minerals Science and Engineering*, Vol. 4, pp. 24–51.

Luckie, P. T. and Austin, L. G., 1980, *Coal Grinding Technology: A Manual for Process Engineers*, NTIS, Springfield, VA.

Whiten, J. and Lynch, A., 1967, "Time Dependent Equation for Comminution Machines," ACS Symposium on Characteristics of Dispersion Systems in Chemical Engineering, pp. 11–12.

18

Economics of Grinding Processes

Introduction

The importance of the grinding process to the engineering economy of the world has been implicit throughout this book. In this chapter, an introduction to the economic aspects of grinding will be given, both from the capital purchase and the operating cost viewpoints. In terms of total dollars spent in the US, size reduction is certainly one of the most capital-intensive engineering operations in industrial usage, exceeding, for example, common engineering processes such as solids mixing, crystallization, filtration, thickening, flotation, solvent extraction, and pelletizing.

Three specific types of economic analyses representing industrial practice in the United States will be discussed in subsequent sections. The first section deals primarily with grinding economics from the viewpoint of an operating plant, with emphasis on how grinding costs are collected and their approximate 1980 values. In the next section, some cost estimation data and procedures for new equipment selection will be discussed. Finally, a summary is given of the need for a *system engineering* economic analysis of grinding coupled with related processes such as flotation, pelletizing, agglomeration, thickening, mining, and metal refining. Such an overall simultaneous analysis of all the interrelated processes in a given plant (or series of plants) is useful in identifying *true* economic costs and profits, as well as in designing optimal engineering control strategies, establishing plant production schedules, and deciding on appropriate measures of intermediate product quality control. Most importantly, such an overall system engineering analysis, the *macro-economic analysis*, is necessary to maximize total profit of a multiprocess engineering operation.

Plant Grinding Economics

The availability of economic data on existing grinding operations is limited both in quantity and consistency. Part of this lack of data is due to the proprietary nature of such information, but often the problem is further complicated by the different manner in which industrial firms collect and allocate costs and profits. Thus apparently similar grinding operations involving the same equipment, same ore type, etc., can show very different

473

Table 1. Estimated Steel and Power Costs of US Grinding (1980)

	Copper	Average	Taconite	Average
Ball/rod usage	0.2–1.5 lb/st ore	1.0	0.5–3.0 lb/st ore	1.5
Liner usage	0.05–0.3 lb/st ore	0.15	0.1–0.5 lb/st ore	0.20
Cost of balls/rods	20–80¢/lb	30	20–90¢/lb	34
Cost of liner	60¢/lb		60¢/lb	
Energy usage	3–25 kWh/st	10	5–40 kWh/st	16
Energy cost	2.0–9.0¢/kWh	4.5	2.0–8.5¢/kWh	4.0
Average Total Cost Per st of Ore				
Ball/rod	1 lb × 30¢/lb	= 30¢	1.5 lb × 34¢/lb	= 51¢
Liner	0.15 lb × 60¢/lb	= 9¢	0.2 lb × 60¢/lb	= 12¢
Energy	10 kWh/st × 4.5¢/kWh	= 45¢	16 kWh/st × 4.0¢/kWh	= 64¢
		$\overline{84¢/st}$		$\overline{\$1.27/st}$

Metric equivalents: st × 0.907 = t; lb × 0.453 = kg.

economic costs depending on the accounting *definitions* or *policies* used. Consequently, it is difficult to make direct economic comparisons.

An analysis of operating cost data from a number of published and unpublished sources, primarily from the copper sulfide and taconite processing industries, shows that the data from the different sources varied greatly. Tables 1 and 2 represent the averages of this data after logical uniformity corrections were made to put cost allocations on a common basis. The absolute numbers given may or may not be close to the actual accounting values used or quoted in a given plant because of the variance in cost accounting procedures. However, the relative comparison of the costs listed does represent a sufficiently accurate cost picture for our purposes.

Table 1 specifically identifies two major cost items in grinding operations using 1980 data: media consumption during grinding (balls, liners, rods, etc.) and mill energy usage in kWh/st ore ground. As indicated, the ranges of usages and costs are quite large and some data exists that is outside of the ranges shown. With regard to media consumption by abrasion and/or corrosion, a great deal of work has been done to isolate the costs and to test higher priced but longer lasting grinding media and liners made of special alloy materials (Nass, 1974). It has been estimated that half a billion pounds of steel are so consumed annually in the US (Remark and Wick, 1976). Also, Hoey, Dingley, and Lui (1975) and Lui and Hoey (1976) have done extensive work both to identify and to test corrosion inhibiting chemicals in mills. The results and usefulness of this work are strongly dependent on the particular ore (grindability, size distribution of feed and product, etc.) and grinding operation (pH, water quality, chemical reagents, etc.) under consideration.

The average amount of energy required per ton of ore processed and the cost of that energy can also vary greatly depending on ore type and particle size of feed and product plus the geographic location. It is, however, probable that the cost per energy unit will escalate rapidly in the future in all locations. Thus, there will be increasing economic pressure to make grinding energy usage more efficient by better equipment design (larger

Table 2. Estimated Total Cost Analysis of a Typical US Plant for Grinding Only (1980)

	Percentage of Total Cost	
	Copper	Taconite
Direct		
Power/steel	41	41
Maintenance/labor	23	20
Indirect	9	7
Fixed		
Taxes/insurance	5	5
Depreciation	22	27
	100%	100%
Average Operational Cost	$2.05/st ore	$3.10/st ore

Table 3. Estimated Variation of US Grinding Cost with Size of Plant (1980)

	Capacity	Average Operational Cost, $/st ore
Copper	25,000 stpd	2.52
	50,000 stpd	2.17
	100,000 stpd	1.89
Taconite	5,000,000 stpy	2.98
	10,000,000 stpy	2.69

Metric equivalent: st × 0.907 = t.

equipment having lower cost per unit of product) and by the use of grinding additives as discussed in Chapter 15.

Table 2 gives a typical total cost analysis in 1980 for grinding in the copper and taconite industries. It can be seen that the power and steel costs discussed earlier represent 41% of the respective total operational cost. Depreciation (cost of capital) along with maintenance/labor costs are also major cost components. The effect of economies of increased plant size (economies of scale), see Table 3, on the average cost per st of ore processed is an important variable. This concept will be discussed further in the later sections of this chapter. The cost analyses quoted include the equipment

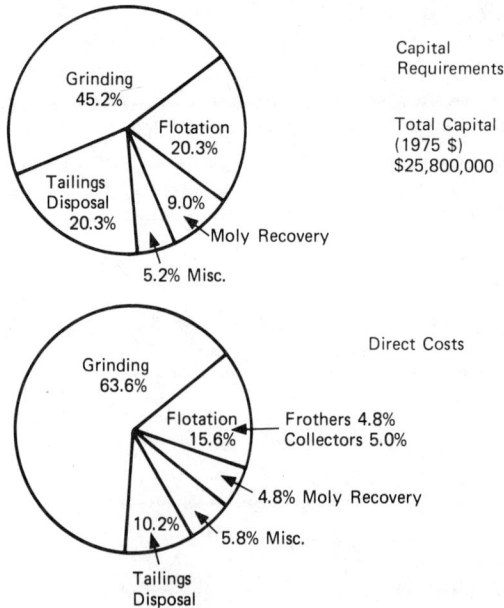

Fig. 1. Total capital and operating cost allocations for a 15 kt/d (16,500 stpd) copper-molybdenum concentrator.

normally associated with the entire grinding circuit: grinding mills, classifiers, pumps, and material handling systems.

Another perspective of the economic role of grinding in the processing of ores is illustrated in Fig. 1. Here the relative importance of grinding is compared to other engineering operations for a small but typical 15.0 kt/d (16,500 stpd) copper-molybdenum concentrator, in 1975 dollars. In terms of capital requirements, the grinding portion represents 45.2% of the total concentrator capital of $25,800,000. In terms of direct operating costs, the grinding portion is 63.6%. These percentages for a given concentrator will differ somewhat depending on the overall concentrator size, the specific equipment utilized, and the nature of the ore processed. However, the cost dominance of grinding is generally observed in most particulate processing plants. For taconite ore and cement processing, the relative importance of grinding will diminish from their portions illustrated in Fig. 1 because of the related costly operations of pelletizing and/or cement clinker firing.

The commonly accepted figures for the ore-processing capabilities of the US copper/molybdenum and taconite industries in 1980 were 318 Mt (350 million st) and 181 Mt (200 million st), respectively. If the average operational costs of Table 2 are applied assuming utilization of all the throughput capability, the costs involved in grinding in 1980 in the US in these two industries alone were in the regions of $710 million and $620 million, respectively.

New Grinding Equipment Cost Estimation

The procedures for cost estimations of new equipment in the process industries are well known and documented (Hoskins and Green, 1977; Grant, Ireson, and Leavenworth, 1976; Mular, 1982). Cost estimation is based on the availability of capital cost information of similar equipment purchased in the near past. Thus a number of historical data correlations have been developed relating equipment purchasing cost as a function of the maximum throughput of the particular type of machine under consideration. The most common correlation is the power law form

$$\text{Cost} = a(\text{an appropriate physical parameter})^b \qquad (1)$$

where a and b are curve-fitting constants. If b has a value less than 1, the piece of equipment exhibits "increasing returns to scale," that is, as the size of the equipment increases, the cost per unit of throughput decreases. This is normally the case for process equipment. A value of b greater than 1 gives increasing cost per unit of throughput, which is sometimes the case for equipment like crushers and pollution control devices.

The sources of such correlations are varied and include equipment manufacturers themselves, industrial firms who purchase large amounts of equipment, engineering contractors or consultants, and firms who specifically collect such data for use by cost estimators in the form of cost indices (Mular, 1982). Cost indices are ratios used to estimate current prices of equipment from obsolete prices once an appropriate size has been selected.

Table 4 gives a sample of the Marshall and Swift index for mining equipment.

Table 5 lists a few appropriate cost-size correlations collected by Mular (1982) for equipment commonly found in grinding circuits. To use such data in the current year, one carries out all the calculations on the proposed pieces of equipment and then applies the appropriate cost index ratio to update the information. If a cost index is not available, the data can be updated by applying yearly inflation rate factors.

The actual procedure of laying out equipment flowsheets, adding appropriate cost data, estimating work capital requirements from fixed capital, estimating safety factors, listing engineering and construction costs, construction planning, and anticipating other factors with contingency cost funds, is difficult and requires a great deal of experience. The reader is referred to appropriate references for examples of the total cost estimation procedure (Hoskins and Green, 1977; Grant, Ireson, and Leavenworth, 1976; Mular, 1982). An excellent discussion with respect to a mineral process is given by Mular (1982). The accuracy of industrial experience in cost estimation in the process industries of recent years has varied greatly because of long delivery times of equipment, nonavailability of certain

Table 4. Marshall and Swift Cost Indices*

Year	Equipment (US average) (1926 = 100)				
	All Industries	**Mining and Milling**			
1960	237.7	240.6			
1965	244.9	245.3			
1970	303.3	302.6			
1971	321.3	321.1			
1972	332.1	331.8			
1973	344.1	342.9			
1974	398.4	394.3			
1975	444.3	451.2			
1976	472.1	482.9			
1977[†]	505.4	506.0	514.9	526.6	535.8
1978	545.3	543.7	558.6	574.1	582.4
1979	599.4	599.3	613.4	625.7	638.4
1980	659.6	660.1	673.6	684.4	715.9
1981	721.3	725.1	747.0	761.2	767.9
1982	745.6	773.8	783.1	787.1	791.2
1983	—	792.3	799.0	802.0	

$$\text{Present cost} = \frac{\text{Index value for current year}}{\text{Index value for former year}} \times \text{former cost}$$

*The authors acknowledge Mr. Smith of Marshall and Swift for his permission to print the data of this table.
[†]Data for the years 1977–1983 are reported by quarter.

Table 5. Examples of Grinding Equipment Cost-Size Correlations

Cost (in Canadian \$) = a(physical parameter)^b, for Marshall and Swift Cost Index (M and M = 800)

Type of Equipment	Physical Parameter	Range of Parameter	a	b
Autogenous mill*	horsepower	3,000–10,000	4195	0.755
Ball mill (without liners)*†	diam, ft	3–20	3875.3	2.111
Pebble mill*†	diam, ft	3–7.2	12165.1	1.462
Pebble mill*†	diam, ft	7.2–20	2624.9	2.239
Rod mill (without liners)*‡	diam, ft	3–15	4364.8	2.105
Wet cyclone	diam, in.	1–13.46	414.46	0.7582
Wet cyclone	diam, in.	13.46–50	72.256	1.430
Dry cyclone (air)	diam, ft	3–12.28	631.75	1.779
Dry cyclone (air)	diam, ft	12.28–24	116.59	2.493
Vibrating screen (single)	width^2 × length, ft^3	11–1536	2141.3	0.4069
Vibrating screen (double)	width^2 × length, ft^3	11–1536	2280.4	0.4256
Vibrating screen (triple)	width^2 × length, ft^3	54–1536	1862.4	0.4908
Stationary screen (stainless steel)	width^2 × length, ft^3	10–63.19	2822.7	0.2135
Stationary screen (stainless steel)	width^2 × length, ft^3	63.19–162	1374.5	0.3870
Synchronous motor (240 rpm)	horsepower	350–5000	1629.9	0.6141
Synchronous motor (700 rpm)	horsepower	350–5000	400	0.7350
Vertical sump pump (rubber)	US gal per min	160–1356.75	818.2	0.3616
	US gal per min	1356.75–3600	53.06	0.7409

Source: Mular, 1982.

*Excludes motor starter and charge where applicable.

†Must be taken times (mill length/mill diam).

‡Must be taken times (mill length/mill diam) × 0.8.

Metric equivalents: horsepower × 0.745 = kW; ft × 0.305 = m; in. × 25.4 = mm; ft^3 × 0.0283 = m^3; gal per min × 3.785 = L/min.

Note: One US dollar = 1.19 Canadian dollars in Jan. 1984.

equipment within desired time schedules, limited availability and high cost of capital, and the overall problems created by high rates of inflation.

An interesting footnote to cost estimation is the assumption that information must be available on previous equipment of the same type before even an approximate estimate of new plant costs can be given. It is this situation which limits or discourages new process developments and major new equipment modifications. The more capital-intensive the process, as is grinding, the higher the technical and financial risk of being wrong, hence the greater the tendency to rely on previous concepts and experiences. This factor has strongly influenced the historical technical developments of grinding as a large-scale process.

Macro Economic Analyses Involving Grinding

In a series of papers, Klimpel and others have defined two levels of economic engineering analysis, macro and micro (Klimpel and Klein, 1966; Klein and Klimpel, 1967; Klimpel, 1969, 1973; Blau and Klimpel, 1972; Klimpel and Blau, 1975). They also suggested goals and quantitative representations of each type of problem with appropriate mathematical solution techniques. The micro engineering economic problem deals with the actual sizing and/or optimal operation of a single specific engineering operation such as grinding, flotation, or pelletizing. Considerable effort has been extended in both academic and industrial circles to handle such micro economic problems.

The macro problem is concerned with the total plant layout and involves the optimal economic operation of a number of simultaneous or sequential engineering operations. Macro economic analysis involves coupling simple steady-state mathematical models of the specific engineering operations involved, and includes capital and operating costs, market-product needs by contract or open market pricing, raw material availability, and appropriate mass balance relationships. The mathematical solution techniques are primarily based on techniques of mathematical optimization such as linear and separable programming (Klimpel, 1973; Blau and Klimpel, 1972; Klimpel and Blau, 1975).

The usual method for operating a macro economic system in the process industries is to divide it into several distinct parts and then assign separate management to each part. This management then operates (optimizes) each part as though it were a distinctly separate identity with little or no economic coordination with the other parts. Thus the profit of the macro system is then a function of the sum of relatively independently operated and *optimized* parts. In the mineral processing macro system these parts typically are: the mine, the concentrator, the smelter, and final metal processing. Recent work in the processing industries, especially in chemical and petroleum processing, has shown that the economic optimization of the macro system as a whole can lead to sizable increases in overall profit. In addition, true economic engineering bottlenecks within or between the parts can be identified along with the identification of realistic operating plans, process control strategies, and accounting allocations of cost.

The idea of the optimization evaluation is to select the values of a large set of decision or controllable variables (denoted by capital letters in the following examples) which must always satisfy some physical relationships and yet optimize a preselected economic criterion containing the variable values.

Klimpel and Klein (1966) were among the first to illustrate this approach on the evaluation of a newly proposed mining and processing complex containing grinding as the major component. The study, as reported, was a simplification of an actual project where the equipment cost correlation form discussed in the previous section, Cost $= a(\text{size})^b$, was extensively used. As the work represented a feasibility analysis of a plant not yet built, the mathematical relationships involved were kept extremely simple due to data uncertainties. This is in contrast to the macro economic analysis of an existing plant operation which will be discussed later. The investment study, useful for illustrating the methodology involved, is now summarized.

The study concerned the evaluation of a new processing technique for the treatment of an abundant but hard-to-process natural resource, shale, (at a rate of A stph) and its subsequent conversion to a variety of marketable products. As can be seen from the flowsheet of Fig. 2, the primary plant K_0 (consisting of grinding, heat treatment, and product separation) yields three streams: light hydrocarbons (Q), sulfur containing compounds (R), and oil (E). One stream is readily marketable (E), while (Q) has a limited market for sale *as is* or may be converted to another product (P) with unlimited markets in a special plant K_1. Finally, the third stream (R), containing the sulfur compounds, may also be converted to two alternate products (Z) and (Y) in plants K_2 and K_3.

Some of the products are visualized to enter two markets: foreign ($Z1$, $Y1$) and domestic ($Z2$, $Y2$) with different revenues per unit. One product in

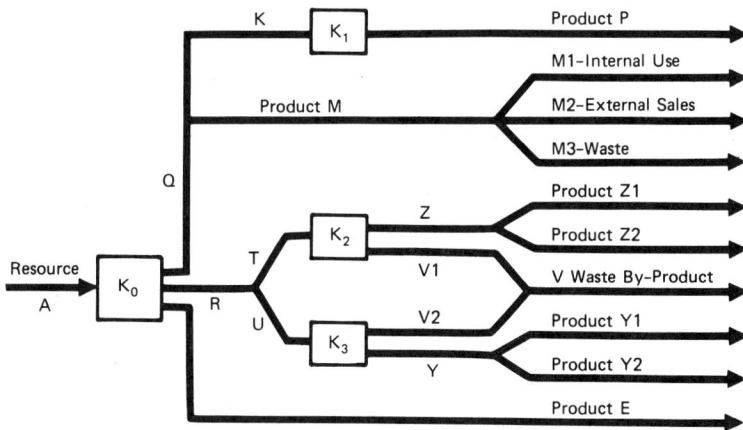

Fig. 2. The mining and processing complex examined for investment feasibility.

the domestic market ($Y2$) exhibits strongly falling revenues with increasing sales. One waste product (V) requires costly waste disposal facilities with costs increasing strongly with increasing amounts of waste.

Typical major questions asked were:

1. Should the project be undertaken?
2. If undertaken, what should be the rate of shale processing (e.g., what type and amount of grinding capacity is really required)?
3. Which supplementary conversion plants should be built and of what size?
4. What is the economic and material flow effects of variable shale quality on the overall project configuration (e.g., what is the effect of having a harder shale to grind or the need to vary the fineness of grind of a given shale for extraction purposes)?

A very important part of this type of optimization formulation is the representation of capital and operating costs by suitable mathematical formulae. This subject is covered in some depth, including a detailed process application, in a paper by Klein and Klimpel (1967). In essence, the required fixed capital (e.g., the cost of a ball mill) plus working capital for a plant K_i of capacity S_i was found to follow the function

$$(\text{total capital})_i = a_i + b_i S_i + c_i S_i^{d_i}$$

while the annual operating cost (e.g., the cost of energy) could be approximated by

$$(\text{annual cost})_i = e_i + f_i S_i + g_i S_i^2 + h_i S_i^{d_i}$$

where b, c, f, g, and h are calculated constants for a given equipment layout and a and e are vanishing fixed expenditures, $a_i = e_i = 0$ when $S_i = 0$ and for any $S_i > 0$ the full values of a_i and e_i are in force.

A variety of economic criteria, such as minimization of cost and maximization of profit, may be used to represent the decision (objective) function. The criterion of maximization of *net present value* was used in this study. In short, *net present value* is the sum of the annual discounted cash flows over the life of the project (typically 10 to 20 years). The net present value is the estimate of profit over the total life of the project. Cash flow in any given year is defined for this case as

$$\text{cash flow} = \text{net (after tax) profit} + \text{depreciation} - \text{investment}$$

where profit equals revenues minus total operating costs.

The initial year fixed capital function for plant K_0 consists of an additive series of power function terms as discussed in the previous section, with one term being used for each major piece of equipment involved. Therefore, in this simplified example involving the processing of a variable tonnage of shale rock (A),

$$\text{total capital cost of plant } K_0 = a_o + a_g \left(\text{kW to grind } A\right)^{b_g}$$

$$+ a_h \left(\text{Btu's to heat-treat } A\right)^{b_h}$$

$$+ a_s \left(\text{equipment area to separate } A\right)^{b_s}$$

where a_o for plant K_0 is assumed to be zero. The appropriate coefficients a and b are taken directly from cost indices tables as described previously. In the plant K_0, the values of the various b's were similar (≈ 0.9) and after the a's were adjusted to put the physical parameters (kW, Btu's, area) on a common basis (tons per hour of A), the relationship between the a's was $a_g \approx 4a_h \approx 10a_s$ indicating the importance of grinding. Annual costs were also described in a similar manner, with the cost indices used being based on long-term company experience.

When all of these power functions are added up, appropriate depreciation calculations performed (Grant, Ireson, and Leavenworth, 1976), revenues from products estimated, and appropriate discount factors applied (Klein and Klimpel, 1967), the decision function given in Table 6 was formulated. For plants P, Z, Y, p_o, z_o, y_o are the vanishing fixed expenditures. The portion of this function relating to the total cost of plant K_0 is equal to $-16000A^{0.9} - 1950A + 0.17A^2$. The revenue or income terms are recognizable by the terms associated with the products, e.g., $840M1$, $100E$. The values of the coefficients in the objective functions will vary, of course, with different types of equipment being considered, price changes per unit of product sold, and other factors.

The remaining portion of setting up this feasibility analysis was the development of mass balance and product yield relationships which any set of values of the variables must satisfy. Such relationships (constraint set), as shown in Table 7, consisted of simple, linearized, average (typical) steady-state yields. For example, the coefficient 5.59 in the equation $Q = 5.59A$ simply related the amount of light hydrocarbon produced in lb per hr when some particular stph of shale is ground to a specified degree of fineness, heat treated, and separated with the specific types of equipment represented in

Table 6. The Decision (Objective) Function of the Feasibility Analysis

$$
\begin{aligned}
f_{max} = \ & -16000A^{0.9} - 1950A + 0.17A^2 \\
& +840M1 + 672M2 + 100E \\
& -24000P^{0.8} + 7000P - 3000000(p_o) \\
& -38000Z^{0.85} - 12000Z + 1.5Z^2 \\
& +51000Z1 + 26000Z2 - 1000000(z_o) \\
& -12000Y^{0.87} - 26000Y + 54000Y1 \\
& +47500Y2 - 63.3(Y2)^2 - 2000000(y_o) \\
& -0.75V^{1.05} - 0.33V
\end{aligned}
$$

Table 7. Mass Balance Relationships for the Feasibility Analysis

Stream	Units		
A	stph		
Q	lb per hr	$Q = 5.59A$	
K	lb per hr	$K = Q - M$	
M	lb per hr	$M = Q - K$	
$M1$	lb per hr	$M1 = M - M2 - M3$	1200 lb per hr
$M2$	lb per hr	$M2 = M - M1 - M3$	3000 lb per hr
$M3$	lb per hr	$M3 = M - M1 - M2$	
P	stpd	$P = 0.03K$	
R	stpd	$R = 0.555A$	
T	stpd	$T = R - U$	
U	stpd	$U = R - T$	
Z	stpd	$Z = 0.855T$	
$Z1$	stpd	$Z1 = Z - Z2$	300 stpd
$Z2$	stpd	$Z2 = Z - Z1$	
V	lb per day	$V = V1 + V2$	
$V1$	lb per day	$V1 = 4010T$	
$V2$	lb per day	$V2 = 1340U$	
Y	stpd	$Y = 2.62U$	
$Y1$	stpd	$Y1 = Y - Y2$	250 stpd
$Y2$	stpd	$Y2 = Y - Y1$	300 stpd
E	lb per month	$E = 41.62A$	

Note: All streams are greater than or equal to zero.
Metric equivalents: st $\times$ 0.907 = t; lb $\times$ 0.453 = kg.

the objective function. Obviously, a variety of operating conditions is possible in any real-life plant, so separate optimization evaluations are run.

For example, if a harder rock is encountered and the throughput of the ball mill held constant, the ground material is coarser. A two percent decrease in the amount passing 75 μm (200 mesh) changes the Q coefficient from 5.59 to 5.16, which means that less potential product Q (and thus revenue) is produced for the same cost to process the same amount of shale, A. To get the coefficient 5.16 back to the neighborhood of 5.59 thus requires either a drop in throughput of shale A with longer grind time or a larger plant, with certainly an increase in operating costs in either case. Evaluating a different type of grinding configuration would lead to different constants in the objective function plus a different $Q = f(A)$ relationship. Such data was estimated from laboratory and pilot-plant studies. From these studies a number of interesting relationships were found; for example, when a given shale was ground finer [5% more passing 75 μm (200 mesh)] the distributions of recoverable products from plant K_0 changed dramatically (the coefficients of Q, R, and E from a unit of A changed by 20%). The economic implications of this are, of course, very important. All the remaining constraints in Table 7 were developed in a similar manner. Also listed are some upper market limits for products to be sold. For example, the value of $Z1$ must be chosen so that $Z1 \leq 300$ stpd.

Thus a given evaluation or optimization consisted of optimally selecting the set of variable values which maximized the net present value objective function and yet satisfied the current constraint set of relationships. Table 8 gives a solution set for the information of Tables 6 and 7. The plant capacity of plant K_1 is denoted by P and the optimal set gives a value of P of zero: thus plant K_1 was not economically justifiable. As mentioned earlier, many optimizations were run to test the economic viability of a wide variety of equipment types, operating conditions, and shale rock changes. *The ability of such an approach to realistically test the type, amount, and method of operation of the grinding capability required under a variety of circumstances had major economic planning and macro design implications.*

This philosophy of capital evaluation is very likely to grow in the future due to the rapidly escalating cost of capital. An interesting footnote to this particular study was that the grinding capability originally suggested by engineering firms was underdesigned by almost 30% in terms of operating the *overall complex* for a *maximum economic return*. Also, sensitivity of the overall success of the project to the revenues (selling prices) of certain of the products was identified and eventually led to abandonment of the project.

An example of a macro engineering analysis carried out on a large existing engineering complex has been reported by Klimpel and Blau (1975).

Table 8. Typical Problem Solution

Stream	Units	Value
A	stph	871
Q	lb per hr	4,869
K	lb per hr	0
M	lb per hr	4,869
M1	lb per hr	1,200
M2	lb per hr	3,000
M3	lb per hr	669
P	stpd	0
R	stpd	484
T	stpd	351
U	stpd	133
Z	stpd	300
Z1	stpd	300
Z2	stpd	0
V	lb per day	1,585,000
V1	lb per day	1,407,000
V2	lb per day	178,000
Y	stpd	347
Y1	stpd	250
Y2	stpd	97
E	lb per month	36,251

Net present value = $5,581,000

Metric equivalents: st × 0.907 = t; lb × 0.453 = kg.

Over six distinct managerial identities and fourteen major engineering operations were involved, including size reduction, solids mixing, and crystallization. Fig. 3 shows the highly interrelated flowsheet that was analyzed. The approach used was similar to that previously discussed except that the amount of data and the number of physical constraints involved were much greater (over 1,000 decision variables and 600 constraint equations). The quantification of physical relationships (mass balances, the effect of kinetics on production distributions, etc.) is much more difficult and time-consuming with existing plants as many of the equations involved are nonlinear in form. The systems engineer must have an accurate and detailed analysis of the process involved; otherwise an optimization model can be built which shows as its best solution an operating scheme that gives lower profit than the currently running plant—indeed an embarrassing situation!

   Plant 5 in Fig. 3 contains a size reduction operation. The representation of the product size resulting from the particular grinding equipment in place was the transfer function approach described in the section, "Continuous Milling and Residence Time Distribution," of Chapter 6 (Eq. 3 in Chapter 6),

$$p_i = \sum_{j=1}^{i} d_{ij}(\tau)f_j, \quad n \le i \le 1$$

The d_{ij} values were determined by curve-fitting pairs of p_i and f_j data taken

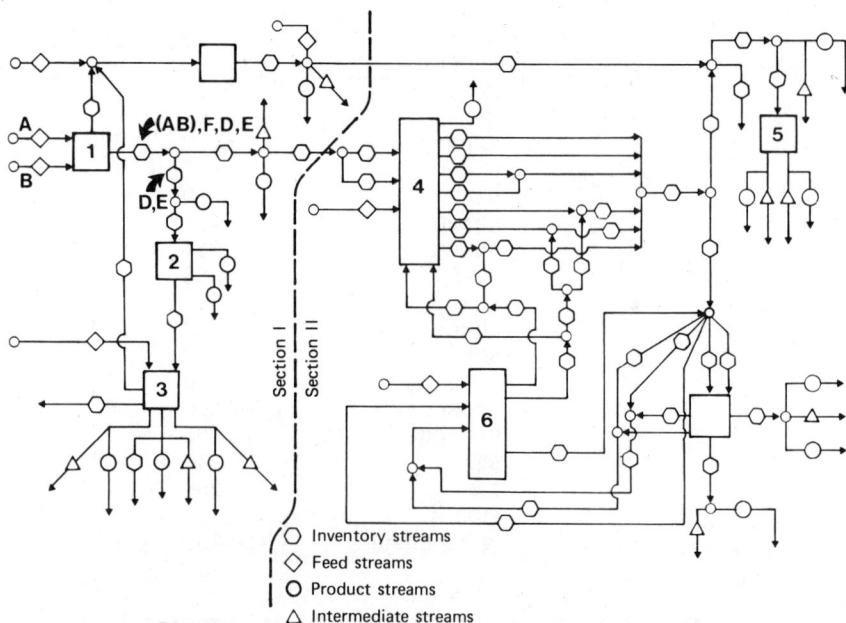

Fig. 3. Macro process model of an existing plant complex.

from existing plant grinding studies (Klimpel and Blau, 1975). Thus within the overall macro optimization, the particular portion of the constraint set dealing with the flow of products through plant 5 contained a complete set of equations of the form of Eq. 3 in Chapter 6. This set takes any current set of f_j decision variables values that may arise during the optimization and generates the appropriate p_i values. These p_i values are then used in later constraints and in the objective function for identifying product qualities and prices. The linear form of this approach makes it especially amenable to large-scale optimization techniques such as linear programming.

The optimization evaluation of this macro process helped to identify a number of questions relative to the grinding sections in plant 5:

1. Was the current grinding capability sufficient for a number of different operating conditions?

2. Was it economically beneficial to alter the f_j values coming into the grinding section so as to get a more desirable (higher values) p_i distribution?

3. How would several distinctly different types of grinding equipment (with different sets of d_{ij} values) affect the overall operation and its economic return?

It is interesting to note that the entire optimization analysis, with the consequent updating of management/operating plans, led to a 35% increase in the overall profit generated by the complex. To accomplish this, several parts of the overall complex had to change their operating criteria rather significantly. One extra piece of size reduction equipment was put into place as a result of this study.

The reader is referred to the original publication (Klimpel and Blau, 1975) for detailed information on how such macro optimization models can be formulated and solved with available digital computer techniques.

References

Blau, G. and Klimpel, R. R., 1972, Chapter II-7, "The Process Industries," *Handbook of Operations Research*, Vol. 2, J. Moder and S. Elmaghroby, eds., Van Nostrand Reinhold, New York, pp. 579–614.

Grant, E., Ireson, W., and Leavenworth, R., 1976, *Principles of Engineering Economy*, 6th ed., Wiley & Sons, New York.

Hoey, G., Dingley, W., and Lui, A., 1975, "Inhibitors Help to Reduce Ball Loss," *Canadian Chemical Processing*, pp. 36–41.

Hoskins, J. and Green, W., 1977, *Mineral Industry Costs*, 2nd ed., Northwest Mining Association, Spokane, WA.

Klein, M. and Klimpel, R. R., 1967, "Application and Linearly Constrained Nonlinear Optimization to Plant Location and Sizing," *Industrial Engineering*, Vol. 18, pp. 90–95.

Klimpel, R. R., 1969, "Some Recent Advances in the Use of Mathematical Optimization in the Mineral Industries," *Minerals Science and Engineering*, Vol. 1, pp. 15–23.

Klimpel, R. R., 1973, "Operations Research: Decision-Making Tool," *Chemical Engineering*, Vol. 80, I, pp. 103–108; II, pp. 87–94.

Klimpel, R. R. and Klein, M., 1966, "An Application of Nonlinear Optimization Using the Gradient Projection Method," *Proceedings of the Symposium on Operations Research in the Mineral Industries*, Vol. 1, University Park, PA, pp. F-1–16.

Klimpel, R. R. and Blau, G., 1975, "The Role of Optimization Theory in Process Plant Design," Symposium on Mathematical Optimization in Engineering Design, Operations Research Society of America, National Meeting, Chicago.

Lui, A. and Hoey, G., 1976, "Corrosion Inhibitors for the Reduction of Wear in Iron Ore Grinding," *Materials Performance*, Vol. 15, pp. 13–16.

Mular, A., 1982, "Mineral Processing Equipment Costs and Preliminary Capital Cost Estimations," Special Volume 25, Canadian Institute of Mining and Metallurgy, Montreal.

Nass, D., 1974, "Steel Grinding Media Used in the United States and Canada," *Materials for the Mining Industry*, Vail, CO, Climax Molybdenum Co., pp. 173–188.

Remark, J. and Wick, O., 1976, "Corrosion Control in Ball and Rod Mills," Paper 121, International Symposium on Corrosion, Houston, TX.

19

Application of the Simple Analytical Solutions of the Grinding Equations

Usefulness of Simple Solutions Which Approximate the Real Solution

In this section we will explore simple analytical solutions to the grinding equations and some applications to circuits. A simple analytical solution to a problem has considerable advantages over a computer solution since it can be hand-calculated, and general deductions can be made from the form of the solution. Unfortunately, the simple solution to the batch grinding equation is a rather poor approximation to reality, so it is not valid for exact simulation. However, it can still be very useful for the following reasons. First, the accuracy of a programmed numerical solution can be checked against the exact analytical solution by performing the computation on artificial data which satisfies the requirements for the exact solution to apply. Eqs. 7, 9, and 15 in Chapter 8 show that to compare a computed solution using $S_1 B$ values which fit the compensation condition to the algebraic simple solutions requires that $S(x_i) = S_{i-1}$, $B_{i,j} = S_{i-1}/S_j$. Second, the trend of results often can be easily deduced from the simple solutions by inspection, without time-consuming calculations. Analysis of the solutions gives valuable insight and *feel* for the behavior of the system. Third, as a grinding system becomes more complex, with more components, it is usually found that the effect of an approximation in one component is diminished in the overall effect on the system, so that precise simulation of any particular component is not essential. Fourth, there is a class of control problems involving rather small deviations from an average condition; since the deviation is small, the error caused by an approximation for the model of the deviation will also cause only small errors, and the approximation may be sufficiently good for the correct control action to be taken.

It must be recognized that the solutions given in this chapter cannot apply when overfilling or slowing-down effects are present. The capacities would have to be corrected by appropriate K_o or κ factors to get realistic capacity comparisons.

Starting Postulates and Equations

In principle, any empirical function which approximately describes the product weight-size distribution as a function of grinding time and feed size can be used for approximate circuit solutions. However, we will consider only the two solutions developed previously in Chapter 8:

Batch / plug flow grinding:

$$1 - P(x, t) = [1 - P(x, 0)]\exp[-S(x)t] \tag{1}$$

or

$$1 - P(x) = [1 - F(x)]\exp[-S(x)\tau] \tag{1a}$$

Fully mixed continuous grinding:

$$P(x) = \frac{F(x) + \tau S(x)}{1 + \tau S(x)} \tag{2}$$

It will be remembered that these arise from the stipulation

$$S(y)B(x, y) = \text{function } x \text{ only}$$

$$= S(x)$$

Therefore, it follows that

$$S(x)/S(y) = B(x, y), \quad 0 \le x \le y$$

and since $B(x, y)$ must, to be physically real, decrease continuously from unity at $x = y$ to zero at $x = 0$ [that is, $B(y, y) = 1$; $B(0, y) = 0$; $B(x, y) \le 1$; $\partial B(x, y)/\partial x \ge 0$], then $S(x)/S(y) \le 1$ and $(\partial S(x)/\partial x)/S(y) > 0$. Thus $S(x)/S(x_{max})$ must continually decrease as x decreases, and intermediate maxima or minima in $S(x)$ are not permitted. As mentioned previously, one condition which fits the criteria is $S(x) = ax^{\alpha}$, $B(x, y) = (x/y)^{\alpha}$.

The mass balance equations are as follows (see Fig. 1). Let $G(x)$ be the weight fraction less than size x in the raw feed; similarly, $F(x)$ is fraction

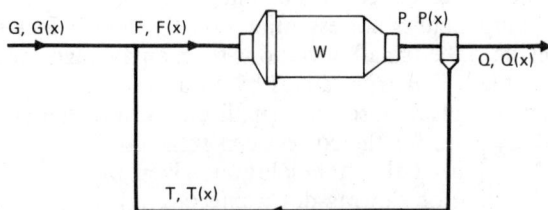

Fig. 1. Symbolism for simple circuit with size distributions as continuous functions.

less than x in the mill feed, $P(x)$ fraction less than x in the mill product, $T(x)$ fraction less than x in the tailings, $Q(x)$ fraction less than x in circuit product. Then

rate of material of size interval x to $x + dx$ sent to tailings = $Ps(x)dP(x)$

rate of material of size x to $x + dx$ sent to circuit product = $P[1 - s(x)]dP(x)$

where $s(x)$ is the continuous classifier function defined by $s(x) =$ fraction of classifier feed of size x to $x + dx$ sent to tailings. At steady state $F = P$, $G = Q$ and

$$Q(x) = (F/Q)\int_0^x [1 - s(y)]\, dP(y)$$

and hence,

$$Q(x) = (1 + C)\int_0^x [1 - s(y)]\, dP(y) \tag{3}$$

Similarly,

$$T(x) = \left(1 + \frac{1}{C}\right)\int_0^x s(y)\, dP(y) \tag{4}$$

$$F(x) = [G(x) + CT(x)]/(1 + C)$$

$$= [G(x)/(1 + C)] + \int_0^x s(y)\, dP(y) \tag{5}$$

and

$$C = \int_0^{x_{max}} s(x)\, dP(x) \Big/ \int_0^{x_{max}} [1 - s(x)]\, dP(x) \tag{6}$$

For an ideally efficient classifier (see Chapter 13) set at size μ

$$s(x) = 0, \quad x < \mu$$

$$= 1, \quad x > \mu \tag{7}$$

Then, all material above size μ is sent to tailings and no material below size μ goes to tailings so $PP(\mu) = Q$ and hence,

$$P(\mu) = \frac{1}{1 + C} \tag{8}$$

From Eq. 3,

$$Q(x) = (1 + C)P(x), \quad x \le \mu \tag{9}$$

The general technique is to combine a particular mill model equation with the mass balance and classifier equations, with algebraic rearrangement (if possible) to give explicit forms for $Q(x)$, Q/W, etc. There are, of course, a large number of combinations of mill models and arrangements which can be treated in this way. It is a relatively simple matter to perform the algebra on any desired circuit once the approach has been mastered, so no attempt will be made here to develop solutions for all possible circuits of interest. However, the simple circuit of Fig. 1 will be treated in some detail since it forms a part of all more complex circuits. There are three major mill models which are treated: (1) the batch/plug flow model, which is appropriate for long mills such as tube mills; (2) the fully mixed model, which is appropriate for short mills and mills with rapid mixing action; and (3) a mill model for conditions between (1) and (2), which can be approximately treated as m fully mixed reactors in series.

Solutions for Some Special Cases: Ideal Size Classification

CASE 1: Plug Flow
For these conditions the mill operator equation is Eq. 1a

$$1 - P(x) = [1 - F(x)]\exp[-S(x)\tau]$$

Eq. 5 becomes

$$F(x) = [G(x)/(1 + C)] + \int_{\mu}^{x} dP(x), \quad x \geq \mu$$

$$= \frac{G(x)}{1 + C} + P(x) - P(\mu)$$

and

$$F(x) = G(x)/(1 + C), \quad 0 \leq x \leq \mu$$

This mass balance is used to substitute for $F(x)$ in Eq. 1a, and the terms in $P(x)$ collected. Then from Eq. 9

$$Q(x) = (1 + C)\{1 - [1 - (G(x)/(1 + C))]\exp[-S(x)\tau]\},$$

$$0 \leq x \leq \mu$$

The relative output of the circuit, Q/W, is obtained from $\tau = W/F$, $Q/F = 1/(1 + C)$, giving,

$$Q/W = 1/\tau(1 + C) \tag{10}$$

τ and C are not independent and the relation between them is obtained from $P(x)$ at $x = \mu$ and Eq. 8

$$P(\mu) = \frac{(1 + C) - [1 + C - G(\mu)]\exp[-S(\mu)\tau]}{1 + C} = \frac{1}{1 + C}$$

Hence

$$\tau = \frac{1}{S(\mu)} \ln\left[\frac{1 + C - G(\mu)}{C}\right] \tag{11}$$

or

$$C = [1 - G(\mu)]\exp[-S(\mu)\tau]/(1 - \exp[-S(\mu)\tau]) \tag{11a}$$

$$Q(x) = (1 + C)\left\{1 - \left(\frac{1 + C - G(x)}{1 + C}\right)\left(\frac{1 + C - G(\mu)}{C}\right)^{-S(x)/S(\mu)}\right\},$$

$$0 \le x \le \mu \tag{12}$$

and

$$Q/W = S(\mu)\bigg/\left\{(1 + C)\ln\left[\frac{1 + C - G(\mu)}{C}\right]\right\} \tag{13}$$

Thus, when $G(x)$, $S(x)$, μ, and τ are set, then C, $Q(x)$, and Q/W follow. Physically, a small value of τ means that the mill size (proportional to W) is small compared to the feed rate Q, so that Q/W is large; however, τ is also interlinked to C via the classifier setting μ, so that lower values of τ also arise from high recirculation rates. When the classifier is set to less than the minimum size in the raw feed, $G(x)$ and $G(\mu)$ are zero for $x \le \mu$, which is an added simplification. Expressions for $P(x)$, $T(x)$, etc., are readily derived. Examples of use of the results will be given in the next section.

CASE 2: Fully Mixed
For these conditions, the mill operator equation is

$$P(x) = [F(x) + \tau S(x)]/[1 + \tau S(x)] \tag{2}$$

For ideal classification, at μ, following the same steps as in Case 1,

$$\tau = [1 - G(\mu)]/CS(\mu) \tag{14}$$

or

$$C = \frac{1 - G(\mu)}{\tau S(\mu)} \tag{14a}$$

Also

$$Q(x) = \frac{CG(x) + (1 + C)[1 - G(\mu)]S(x)/S(\mu)}{C + [1 - G(\mu)]S(x)/S(\mu)},$$

$$0 \le x \le \mu \quad (15)$$

and

$$Q/W = S(\mu)[C/(1 + C)]/[1 - G(\mu)] \quad (16)$$

Thus, when $G(x)$, $S(x)$, μ, and τ are set, then C, $Q(x)$, and Q/W follow; or when $G(x)$, $S(x)$, μ, and C are set, τ, $Q(x)$, and Q/W follow.

CASE 3: Residence Time Distribution Corresponding to a Series of Fully Mixed Reactors

Let the residence time distribution in the mill correspond to that from n fully mixed reactors, each of residence time $\tau_1, \tau_2, \ldots \tau_i, \ldots \tau_n$, where $\Sigma_i \tau_i = \tau$, the overall residence time. The product from the first section is the feed for the second, and so on. When $S(y)B(x, y) = S(x)$, the mill operator equation for each section is

$$P_1(x) = [F(x) + \tau_1 S(x)]/[1 + \tau_1 S(x)]$$

$$P_2(x) = [P_1(x) + \tau_2 S(x)]/[1 + \tau_2 S(x)]$$

etc., to

$$P_n(x) = [P_{n-1}(x) + \tau_n S(x)]/[1 + \tau_n S(x)]$$

These give

$$P_n(x) = \frac{F(x) + S(x)[\tau_1 + k_1 \tau_2 + k_1 k_2 \tau_3 + \cdots (k_1 k_2 \ldots k_{n-1})\tau_n]}{k_1 k_2 \ldots k_n}$$

where

$$k_i = 1 + \tau_i S(x)$$

Then, as before, $F(x) = G(x)/(1 + C)$, $x \le \mu$, and

$$P_n(x)$$

$$= \frac{[G(x)/(1 + C)] + S(x)[\tau_1 + k_1 \tau_2 + \cdots (k_1 k_2 \ldots k_{n-1})\tau_n]}{k_1 k_2 \ldots k_n}$$

$$(17)$$

$$Q(x) = (1 + C)P_n(x), \quad 0 \le x \le \mu$$

$$1/(1 + C) = P_n(\mu)$$

These equations are readily calculated for given values of τ and $S(x)$, but are not convenient for obtaining a simple relation between τ and C. However, a simpler form is obtained when the residence time distribution corresponds to *equal* reactors, giving $\tau_1 = \tau_2 = \tau_i = \tau_n = \tau/n$. Then $k_i = 1 + (\tau/n)S(x) = k$, say, and

$$P_n(x) = [1 + (\tau/n)S(x)]^{-n}$$

$$\times \left[F(x) + S(x)(\tau/n)(1 + k + k^2 + \cdots k^{n-1}) \right]$$

Using $1 + k + k^2 + \cdots k^{n-1} = (k^n - 1)/(k - 1)$,

$$P_n(x) = [1 + (\tau/n)S(x)]^{-n} \{ F(x) + [1 + (\tau/n)S(x)]^n - 1 \}$$

and, as before,

$$\tau = [n/S(\mu)] \left[\left(\frac{1 + C - G(\mu)}{C} \right)^{1/n} - 1 \right] \tag{18}$$

$$C = [1 - G(\mu)] / ([1 + (\tau/n)S(\mu)]^n - 1) \tag{18a}$$

$$Q(x) =$$

$$\frac{(1 + C) \left\{ \dfrac{G(x)}{1 + C} + \left[1 + \dfrac{S(x)}{S(\mu)} \left[\left(\dfrac{1 + C - G(\mu)}{C} \right)^{1/n} - 1 \right] \right]^n - 1 \right\}}{\left\{ 1 + \dfrac{S(x)}{S(\mu)} \left[\left(\dfrac{1 + C - G(\mu)}{C} \right)^{1/n} - 1 \right] \right\}^n},$$

$$0 \le x \le \mu \tag{19}$$

$$Q/W = [S(\mu)/n] \Big/ \left\{ (1 + C) \left[\left(\frac{1 + C - G(\mu)}{C} \right)^{1/n} - 1 \right] \right\} \tag{20}$$

It is readily shown that these equations reduced to those for Case 1 (plug flow) as $n \to \infty$, and to Case 2 for $n = 1$.

Examples of Use of the Simple Equations

CASE 1: Plug Flow

First let us consider the effect of varying the circulation ratio by increasing the feed rate, holding the ideal classifier setting fixed at a specified value of μ. The size distribution of the product from the circuit is given by Eq. 12,

$$Q(x) = (1 + C) \left\{ 1 - \left(1 - \frac{G(x)}{1 + C} \right) \left(\frac{1 + C - G(\mu)}{C} \right)^{-S(x)/S(\mu)} \right\},$$

$$0 \le x \le \mu \tag{12}$$

At very long residence times, the product is ground very fine and is principally less than the classifier size, so $C \to 0$ as $\tau \to \infty$, and $Q(x)$ is 1 except for small values of x, where

$$Q(x) = 1 - [1 - F(x)]\exp[-S(x)\tau], \quad \tau \text{ large}$$

The other limit is for $C \to$ large. Then, using $\lim(1 + x)^n = 1 + nx$, as $C \to$ large,

$$Q(x) = \left[\frac{S(x)}{S(\mu)}\right][1 - G(\mu)]\left(\frac{1 + C}{C}\right) + G(x)$$

or

$$\lim_{C \to \infty} Q(x) = \left[\frac{S(x)}{S(\mu)}\right][1 - G(\mu)] + G(x) \tag{21}$$

This is the coarsest (steepest) size distribution which can be obtained for the particular classifier setting. Fig. 2 shows some typical results. The limiting value of Q/W is obtained from Eq. 13,

$$\lim_{C \to \infty} Q/W = S(\mu)/[1 - G(\mu)] \tag{22}$$

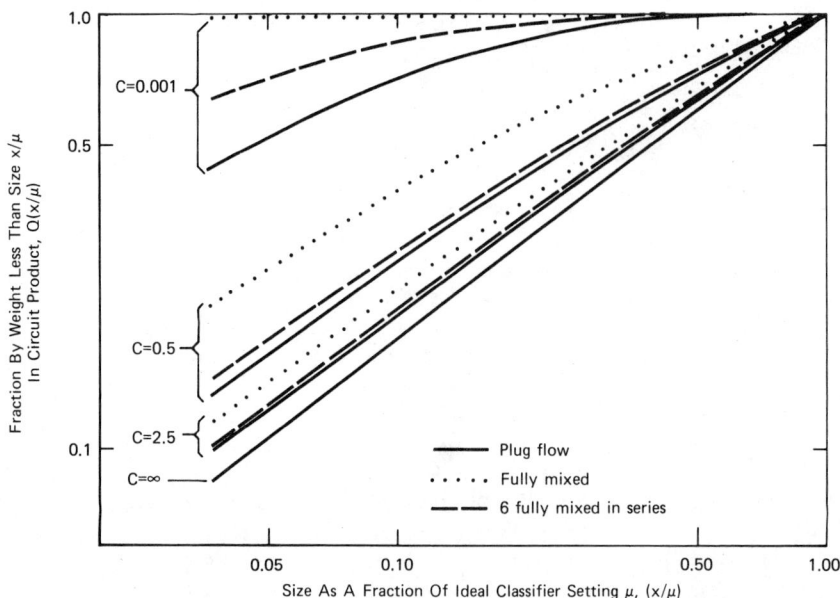

Fig. 2. Size distributions for the simple analytical solutions of simple circuit ideal classification at size μ and $G(\mu) = 0$ (no overfilling or slowing-down effects).

The physical reason for this equation is as follows (see "The Ultimate Limiting Case" in Chapter 6). At the limit of high circulating load all material less than size μ is rapidly presented to the classifier and is rapidly removed from the circuit. Material larger than size μ must break to less than size μ in order to leave the circuit. The rate of breakage is $S(\mu)W$ and this must match the feed of $Q[1 - G(\mu)]$.

Note that the figures derived from these simple solutions apply only in the absence of overfilling or slowing-down effects (see "Allowance for Variation in Filling Level" in Chapter 6). Fig. 3 shows values of $(Q/W)S(\mu)$ vs. C calculated from Eq. 13. Note that reducing the value of μ, that is, resetting the classifier setting to a lower value, gives the same curves shown in Fig. 3, but the values of Q/W are reduced in ratio to the reduction of $S(\mu)$ if $G(\mu)$ is small. Of course, the size distributions are finer for the lower μ setting.

Let us now consider the problem of running a mill to give a product which meets a specified product criterion; say ψ fraction by weight less than size x_p (a one-point criterion). Size distributions to pass through the point ψ, x_p can be obtained with a complete range of C values, but at the same time μ must be altered to produce the different values of C. At open circuit ($\mu \geq x_{max}$ of raw feed),

$$\psi = 1 - \left[1 - G(x_p)\right]\exp\left[-S(x_p)\tau_0\right] \tag{23}$$

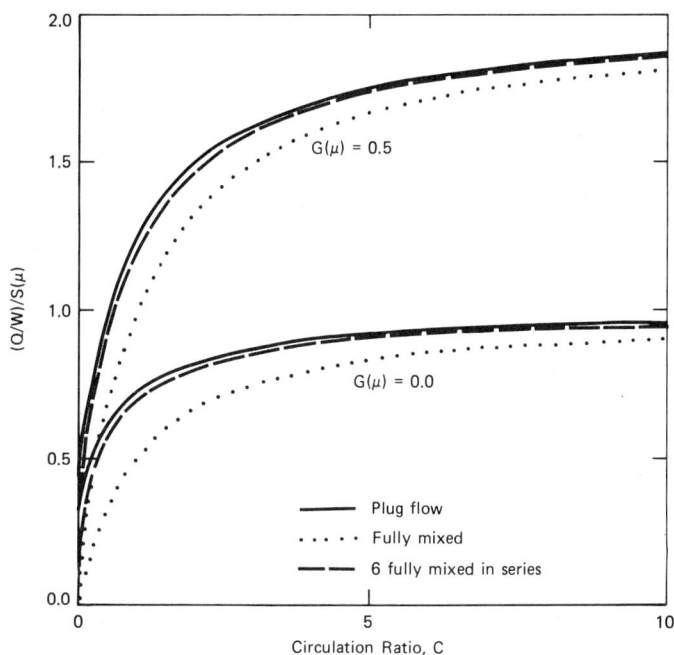

Fig. 3. Relative rate of production as a function of circulation ratio at a fixed setting μ of ideal classifier (no overfilling or slowing-down effects).

where τ_0 is the unique value of τ which satisfies the equation. Note that this obviously can only be satisfied if $G(x_p) \leq \psi$, that is, the raw feed must be coarser than the desired product. On the other hand, as C becomes large, Eq. 21 gives

$$\lim_{C \to \infty} S(\mu) = \frac{S(x_p)[1 - G(\mu)]}{\psi - G(x_p)} \tag{24}$$

The value of μ_{lim} corresponding to $S(\mu)_{\text{lim}}$ is the *lowest* classifier setting which enables the criterion to be met, and gives $C \to \infty$ (for ideal classification, of course). Between these limits, μ is related to C via

$$S(\mu) = S(x_p) \frac{\ln\left[\dfrac{1 + C - G(\mu)}{C}\right]}{\ln\left[\dfrac{1 + C - G(x_p)}{1 + C - \psi}\right]} \tag{25}$$

The size distribution at any C is obtained from Eq. 12, with $S(\mu)$ replaced from Eq. 25,

$$Q(x) = (1 + C)\left\{1 - \left[\frac{1 + C - G(x)}{1 + C}\right]\left[\frac{1 + C - G(x_p)}{1 + C - \psi}\right]^{-S(x)/S(x_p)}\right\},$$

$$0 \leq x \leq \mu \tag{26}$$

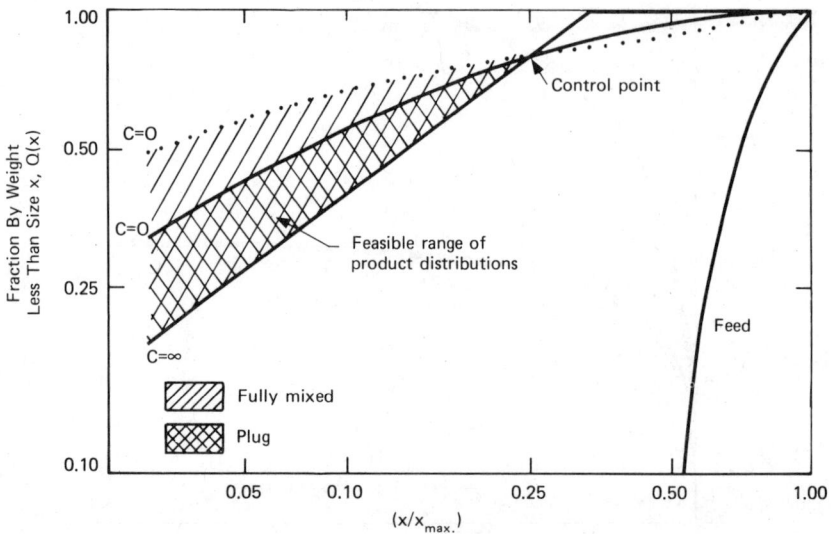

Fig. 4. Size distributions for the analytical solutions of the simple circuit; ideal classifier setting varied to give 80% through $x / x_{\text{max}} = 0.25$ (one-point fit); $S(x) = ax$ (no overfilling or slowing-down effects).

The coarsest (steepest) size distribution, at $C \to \infty$, is

$$\lim_{C \to \infty} Q(x) = Q(x) = G(x) + \left[\psi - G(x_p)\right]\left[\frac{S(x)}{S(x_p)}\right], \quad 0 \le x \le \mu$$

(26a)

Eqs. 13 and 25 give

$$Q/W = S(x_p)\Big/(1 + C)\ln\left[\frac{1 + C - G(x_p)}{1 + C - \psi}\right] \tag{27}$$

and the maximum value as $C \to \infty$,

$$\lim Q/W = S(x_p)\Big/\left[\psi - G(x_p)\right] \tag{27a}$$

These are useful expressions for estimating the effect of changes of ψ and $G(x_p)$ on output. Figs. 4, 5, and 6 show typical results.

Finally, let us consider operation of the mill to a two-point product criterion, ψ_1 at x_1, ψ_2 at x_2. The preceding paragraph demonstrated that if a control point ψ_1 at x_1 is specified, *a limited range of product size distribution results*; we have referred to this previously as the permitted band of size distributions (see "The Behavior of Different Designs on Mill Circuits" in Chapter 7). If the point ψ_2 does not lie within this band (see shaded portion

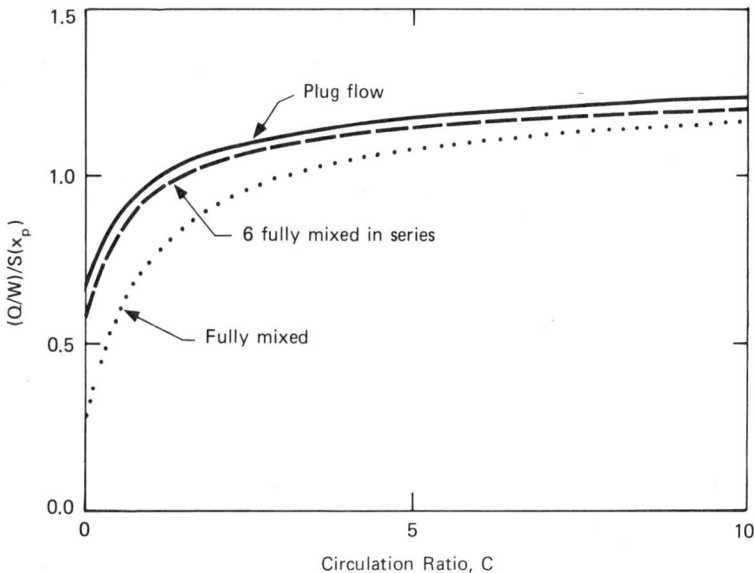

Fig. 5. Relative output rates for the analytical solution of the simple circuit; one-point fit (see Fig. 4).

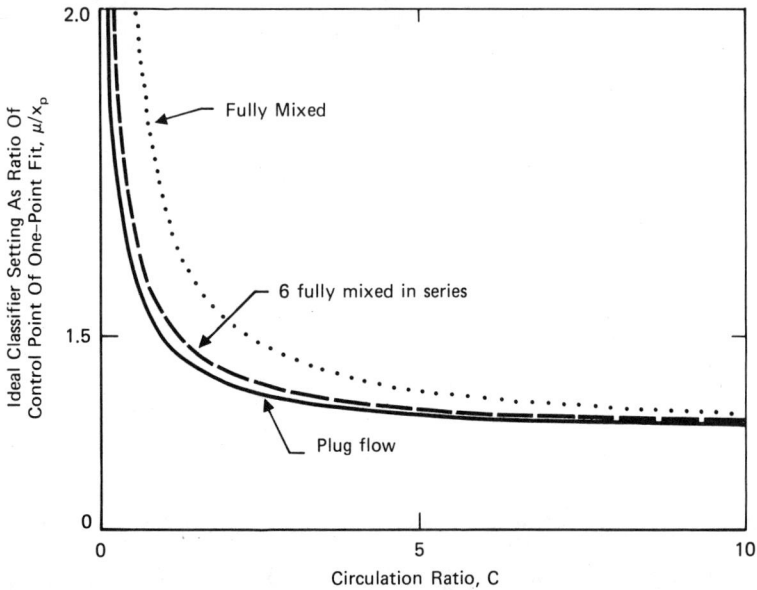

Fig. 6. Variation of ideal classifier setting μ required to get one-point fit as a function of circulating load (see Figs. 4 and 5).

of Fig. 4), it is not possible to satisfy a further equality. For a set value of x_2, the upper feasible bound of $Q(x_2)$ is obtained from:

$$\overline{Q}(x_2) = 1 - [1 - G(x_2)] \exp[-S(x_2)\tau_0]$$

where, from Eq. 23

$$\psi_1 = Q(x_1) = 1 - [1 - G(x_1)] \exp[-S(x_1)\tau_0]$$

Then

$$\overline{Q}(x_2) = 1 - [1 - G(x_2)] \left[\frac{1 - G(x_1)}{1 - \psi_1} \right]^{-S(x_2)/S(x_1)} \tag{28}$$

The lower bound is obtained from Eq. 24,

$$S(\mu) = \frac{S(x_2)[1 - G(\mu)]}{\underline{Q}(x_2) - G(x_2)} = \frac{S(x_1)[1 - G(\mu)]}{\psi_1 - G(x_1)}$$

or

$$\underline{Q}(x_2) = G(x_2) + \left[\frac{S(x_2)}{S(x_1)} \right] [\psi_1 - G(x_1)] \tag{29}$$

If $\overline{Q}(x_2) \geq \psi_2 \geq \underline{Q}(x_2)$, then an equality can be applied in Eq. 12 for both values of ψ, giving

$$S(x_2)/S(x_1) = \ln\left[\frac{1 + C' - G(x_2)}{1 + C' - \psi_2}\right] \bigg/ \ln\left[\frac{1 + C' - G(x_1)}{1 + C' - \psi_1}\right] \quad (30)$$

where C' is the *only* value of circulation ratio which will produce the exact two-point fit. Unfortunately, C' is not readily extracted from this equation, which requires trial-and-error solution for $1 + C'$. The value of output rate for the two-point fit is

$$Q'/W = S(x_1)\bigg/(1 + C')\ln\left[\frac{1 + C' - G(x_1)}{1 + C' - \psi_1}\right] \quad (31)$$

CASE 2: Fully Mixed
In the same way as above, the following equations are developed for this case

$$Q(x) = [G(x) + \tau S(x)]/[1 + \tau S(x)], \quad \tau \text{ large}$$

while for $C \rightarrow \infty$

$$Q(x) = \frac{S(x)}{S(\mu)}[1 - G(\mu)] + G(x) \quad (32)$$

and

$$Q/W = S(\mu)/[1 - G(\mu)] \quad (33)$$

Note that the limiting values for $C \rightarrow \infty$ are the same as for plug flow. This is expected for the same physical reason discussed on page 497.

For the one-point criterion,

$$S(\mu) = \frac{S(x_p)[1 - G(\mu)](1 + C - \psi)}{[\psi - G(x_p)]C}$$

Also, then,

$$Q(x) = \frac{G(x) + (1 + C)\left[\dfrac{\psi - G(x_p)}{1 + C - \psi}\right]\left[\dfrac{S(x)}{S(x_p)}\right]}{1 + \left[\dfrac{\psi - G(x_p)}{1 + C - \psi}\right]\left[\dfrac{S(x)}{S(x_p)}\right]} \quad (34)$$

$$Q/W = S(x_p)\frac{(1 + C - \psi)}{(1 + C)[\psi - G(x_p)]} \quad (35)$$

The limiting values for $C \to \infty$ are

$$Q(x) = G(x) + \left[\psi - G(x_p)\right]S(x)/S(x_p)$$

$$Q/W = S(x_p)/\left[\psi - G(x_p)\right]$$

Again these limits are the same as those for plug flow, Eqs. 26a and 27a. Figs. 2, 3, 4, 5, and 6 show typical examples.

For the two-point criterion, the upper bound is

$$\bar{Q}(x_2) = \frac{G(x_2) + \left[\dfrac{\psi_1 - G(x_1)}{1 - \psi_1}\right]\left[\dfrac{S(x_2)}{S(x_1)}\right]}{1 + \left[\dfrac{\psi_1 - G(x_1)}{1 - \psi_1}\right]\left[\dfrac{S(x_2)}{S(x_1)}\right]} \tag{36}$$

while the lower bound is

$$\underline{Q}(x_2) = G(x_2) + \left[\frac{S(x_2)}{S(x_1)}\right]\left[\psi_1 - G(x_1)\right] \tag{37}$$

If $\bar{Q}(x_2) \geq \psi_2 \geq \underline{Q}(x_2)$, then

$$S(x_2)/S(x_1) = \frac{\left[\psi_2 - G(x_2)\right]\left(1 + C' - \psi_1\right)}{\left[\psi_1 - G(x_1)\right]\left(1 + C' - \psi_2\right)} \tag{38}$$

and hence,

$$1 + C' = (\psi_1 - J\psi_2)/(1 - J)$$

where

$$J = \frac{S(x_2)\left[\psi_1 - G(x_1)\right]}{S(x_1)\left[\psi_2 - G(x_2)\right]}$$

Expressions for μ, $Q(x)$, etc., are readily derived, in particular

$$Q'/W = S(x_1)(\psi_1 - \psi_2)J/\{\left[\psi_1 - G(x_1)\right]\left[\psi_1 - J\psi_2\right]\} \tag{39}$$

CASE 3: Equal Fully Mixed Reactors in Series
Again, as above, the following equations are developed for this case. For $C \to 0$

$$Q(x) = \{G(x) + \left[1 + (\tau/n)S(x)\right]^n - 1\}/\left[1 + (\tau/n)S(x)\right]^n, \quad \tau \text{ large}$$

while for $C \rightarrow \infty$

$$Q(x) = \frac{S(x)}{S(\mu)}[1 - G(\mu)] + G(x) \tag{40}$$

and

$$Q/W = S(\mu)/[1 - G(\mu)] \tag{41}$$

For the one-point criterion

$$S(\mu) = \frac{\{[(1 + C - G(\mu))/C]^{1/n} - 1\} S(x_p)}{\{[(1 + C - G(x_p))/(1 + C - \psi)]^{1/n} - 1\}} \tag{42}$$

Also

$$Q(x) = \frac{G(x) + (1 + C)[(1 + J)^n - 1]}{(1 + J)^n}$$

where

$$J = \frac{S(x)}{S(x_p)} \left\{ \left[\frac{1 + C - G(x_p)}{1 + C - \psi} \right]^{1/n} - 1 \right\}$$

and

$$Q/W = [S(x_p)/n] \Big/ (1 + C) \left\{ \left[\frac{1 + C - G(x_p)}{1 + C - \psi} \right]^{1/n} - 1 \right\} \tag{43}$$

The limiting forms as $C \rightarrow \infty$ are the same as for plug or fully mixed flow. Figs. 2 to 6 show typical examples.

For the two-point criterion, the upper bound is

$$\overline{Q}(x_2) = \frac{G(x_2) + [(1 + J)^n - 1]}{(1 + J)^n} \tag{44}$$

while the lower bound is

$$\underline{Q}(x_2) = G(x_2) + \frac{S(x_2)}{S(x_1)}[\psi_1 - G(x_1)] \tag{45}$$

If $Q(x_2) \geq \psi_2 \geq \underline{Q}(x_2)$, then

$$\frac{S(x_2)}{S(x_1)} = \frac{\{1 - [(1 + C' - G(x_2))/(1 + C' - \psi_2)]^{1/n}\}}{\{1 - [(1 + C' - G(x_1))/(1 + C' - \psi_1)]^{1/n}\}} \tag{46}$$

The rest of the discussion is as for the plug and fully mixed cases.

Comparison of Results of Two-Point Fits

Fig. 7 shows a comparison of results for plug, fully mixed, and $n = 6$ mills producing a two-point fitted distribution. It is seen that although the classifier settings, circulation ratios, and residence times are different for the three models, *the values of size distribution and Q/W are virtually identical for all three.* Some very important conclusions can be drawn. For ideal classification, there is a limited range of two-point fitted size distributions which can be prepared, defined by $\underline{Q}(x_2)$ to $\overline{Q}(x_2)$. Within this range, there is an area common to all the mill models. Choosing a two-point fit in this common area, the different mill models give virtually identical rates and size distributions although the classifier settings and residence times are obviously different between the cases. If a satisfactory mill product is defined by this two-point fit, it can be obtained with equal efficiency from any of the mill models, irrespective of the residence time distribution of the mill. This conclusion is not valid for conditions where overfilling of the mill occurs (see "Allowance for Variation in Filling Level" in Chapter 6), because higher flow rates (smaller τ) through the mill would cause more overfilling.

The range of two-point fit which can be produced becomes wider as the system goes from plug to fully mixed flow. Note that the equations giving Q/W are *not* identities between the plug flow and fully mixed cases. The value of Q/W is almost constant for the two-point fit because of the criterion that $S(y)B(x, y) = S(x)$. The conditions in which a two-point fit gives virtually identical results for all cases between plug and fully mixed can be seen from Table 1, where R is defined by:

$$R = (Q/W)_{\text{plug}}/(Q/W)_{\text{fully mixed}}$$

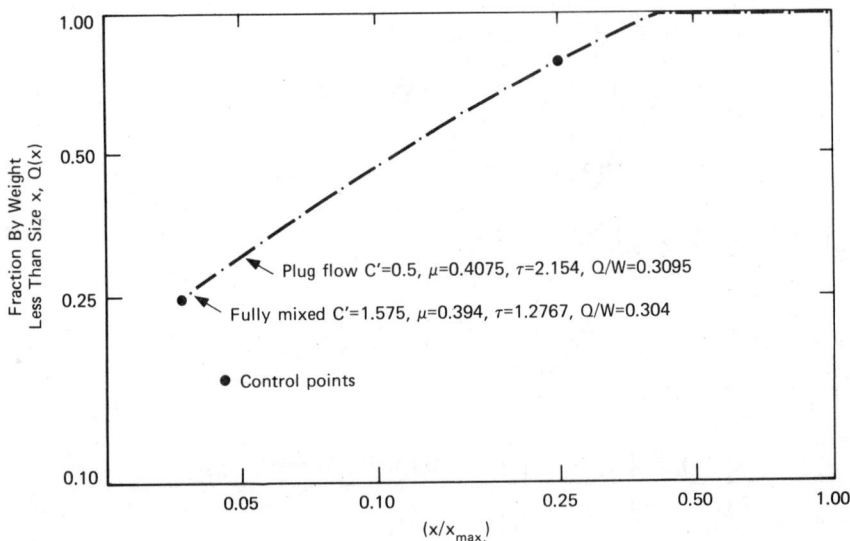

Fig. 7. Results of two-point fit, ideal classification.

From Eq. 30, for plug flow,

$$\psi_2 = K - \frac{K - G(x_2)}{K - G(x_1)} (K - \psi_1)^r$$

where

$$K = 1 + C'$$

$$r = S(x_2)/S(x_1)$$

Combining with Eqs. 31 and 38, for $G(x_1) = G(x_2) = 0$,

$$R = \left(\frac{1-r}{r}\right) \frac{\psi_1 K \left[1 - \left(1 - \frac{\psi_1}{K}\right)^r\right]}{\left\{\psi_1 - K\left[1 - \left(1 - \frac{\psi_1}{K}\right)^r\right]\right\} K \ln\left(\frac{K}{K - \psi_1}\right)}$$

This was used to calculate the results for Table 1 for a range of r, K, ψ_1 values. Note that the limits of ψ_2 for plug flow are

$$\lim K \to \text{large}, \quad \psi_2 = \frac{S(x_2)}{S(x_1)} \psi_1$$

$$\lim_{\substack{K \to 1 \\ C' = 0}}, \quad \psi_2 = 1 - (1 - \psi_1)^{S(x_2)/S(x_1)}$$

A low value of r means that x_2, x_1 are far apart, so the two points fitted are widely spaced on the size distribution. It would be expected that the wider the spacing the closer the value of R to 1, and this is indeed shown in Table 1. Increased value of C' also brings R closer to 1 since a plug flow model at high circulation rates tends to a fully mixed model.

Extension to Partially Ideal Size Classification

Even for the simple mill equations in this section, it is not possible to combine them with real classifier equations to get analytical circuit solutions, except for very special simple forms of the classifier equations (see below). It is clear that the general treatment involving computation (see Chapter 17) can be applied immediately to the simple mill equations with any classifier model of interest. It is thus always possible to *compute* the effect of real classification as compared to the ideal classification treated above. However, it is also possible to again obtain closed analytical solutions for non-ideal classification of a special kind. Considering Fig. 8, this represents what might be termed *partially ideal classification*: a set fraction a of all sizes below μ is sent to tailings (coarse return), while a set fraction b of

Table 1. Relative Output Rates for a Plug Flow Mill vs. a Fully Mixed Mill, for Simple Analytical Mill Model and Recycle by Ideal Classification: Two-Point Fit

Control Point ψ_1	$1 + C' = K$	$r = S(x_2)/S(x_1) =$	R = Relative Output, Plug to Fully Mixed								
			0.1	0.2	0.3	0.4	0.5	0.6	0.7	0.8	0.9
0.9	1.0		1.042	1.087	1.134	1.184	1.236	1.291	1.347	1.406	1.466
	1.5		1.007	1.014	1.021	1.028	1.035	1.043	1.050	1.057	1.065
	2.0		1.003	1.006	1.009	1.012	1.015	1.018	1.021	1.024	1.027
	2.5		1.002	1.003	1.005	1.007	1.008	1.010	1.012	1.013	1.015
	3.0		1.001	1.002	1.003	1.004	1.005	1.006	1.007	1.009	1.010
	3.5		1.001	1.001	1.002	1.003	1.004	1.004	1.005	1.006	1.007
	4.0		1.001	1.001	1.002	1.002	1.003	1.003	1.004	1.004	1.005
	5.0		1.000	1.001	1.001	1.001	1.002	1.002	1.002	1.003	1.003
	10.0		1.000	1.000	1.000	1.000	1.000	1.000	1.001	1.001	1.001
		$r = S(x_2)/S(x_1) =$	0.1	0.2	0.3	0.4	0.5	0.6	0.7	0.8	0.9
0.8	1.0		1.021	1.043	1.065	1.088	1.111	1.135	1.160	1.185	1.210
	1.5		1.005	1.010	1.015	1.019	1.024	1.029	1.034	1.039	1.044
	2.0		1.002	1.004	1.007	1.009	1.011	1.013	1.015	1.018	1.020
	2.5		1.001	1.002	1.003	1.005	1.006	1.007	1.009	1.010	1.011
	3.0		1.001	1.002	1.002	1.003	1.004	1.005	1.006	1.006	1.007
	3.5		1.001	1.001	1.002	1.002	1.003	1.003	1.004	1.004	1.005
	4.0		1.000	1.001	1.001	1.002	1.002	1.002	1.003	1.003	1.004
	5.0		1.000	1.001	1.001	1.001	1.001	1.002	1.002	1.002	1.002
	10.0		1.000	1.000	1.000	1.000	1.000	1.000	1.000	1.000	1.000
		$r = S(x_2)/S(x_1) =$	0.1	0.2	0.3	0.4	0.5	0.6	0.7	0.8	0.9
	1.0		1.012	1.024	1.036	1.049	1.062	1.074	1.087	1.100	1.114
	1.5		1.003	1.007	1.010	1.013	1.017	1.020	1.023	1.027	1.030
	2.0		1.002	1.003	1.005	1.006	1.008	1.009	1.011	1.012	1.014
	2.5		1.001	1.002	1.003	1.004	1.005	1.005	1.006	1.007	1.008

0.7

	$r = S(x_2)/S(x_1) =$								
	0.1	0.2	0.3	0.4	0.5	0.6	0.7	0.8	0.9
3.0	1.001	1.001	1.002	1.002	1.003	1.004	1.004	1.005	1.005
3.5	1.000	1.001	1.001	1.002	1.002	1.002	1.003	1.003	1.004
4.0	1.000	1.001	1.001	1.001	1.002	1.002	1.002	1.002	1.003
5.0	1.000	1.000	1.001	1.001	1.001	1.001	1.001	1.002	1.002
10.0	1.000	1.000	1.000	1.000	1.000	1.000	1.000	1.000	1.000

0.6

	$r = S(x_2)/S(x_1) =$								
	0.1	0.2	0.3	0.4	0.5	0.6	0.7	0.8	0.9
1.0	1.007	1.014	1.021	1.028	1.035	1.043	1.050	1.057	1.065
1.5	1.002	1.004	1.007	1.009	1.011	1.013	1.015	1.018	1.020
2.0	1.001	1.002	1.003	1.004	1.005	1.006	1.007	1.009	1.010
2.5	1.001	1.001	1.002	1.003	1.003	1.004	1.004	1.005	1.006
3.0	1.000	1.001	1.001	1.002	1.002	1.002	1.003	1.003	1.004
3.5	1.000	1.001	1.001	1.001	1.001	1.002	1.002	1.002	1.003
4.0	1.000	1.000	1.001	1.001	1.001	1.001	1.002	1.002	1.002
5.0	1.000	1.000	1.000	1.001	1.001	1.001	1.001	1.001	1.001
10.0	1.000	1.000	1.000	1.000	1.000	1.000	1.000	1.000	1.000

0.5

	$r = S(x_2)/S(x_1) =$								
	0.1	0.2	0.3	0.4	0.5	0.6	0.7	0.8	0.9
1.0	1.004	1.008	1.012	1.016	1.020	1.024	1.028	1.032	1.037
1.5	1.001	1.003	1.004	1.005	1.007	1.008	1.010	1.011	1.012
2.0	1.001	1.001	1.002	1.003	1.003	1.004	1.005	1.006	1.006
2.5	1.000	1.001	1.001	1.002	1.002	1.002	1.003	1.003	1.004
3.0	1.000	1.000	1.001	1.001	1.001	1.002	1.002	1.002	1.002
3.5	1.000	1.000	1.000	1.001	1.001	1.001	1.001	1.002	1.002
4.0	1.000	1.000	1.000	1.001	1.001	1.001	1.001	1.001	1.001
5.0	1.000	1.000	1.000	1.000	1.000	1.001	1.001	1.001	1.001
10.0	1.000	1.000	1.000	1.000	1.000	1.000	1.000	1.000	1.000

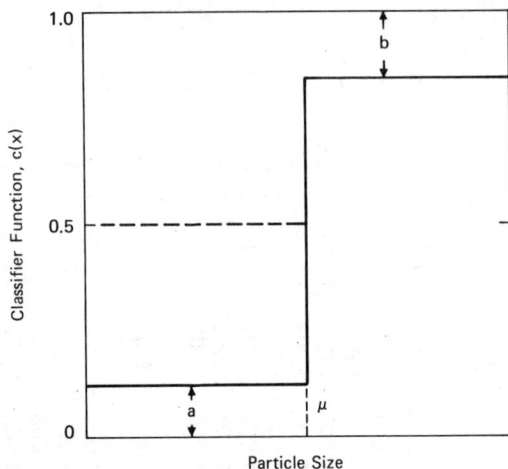

Fig. 8. Illustration of partially ideal classification.

all sizes above μ is sent to fines (circuit product). Mathematically,

$$s(x) = a, \qquad\qquad x < \mu$$
$$s(x) = a \text{ to } 1 - b, \qquad x = \mu \qquad\qquad (47)$$
$$s(x) = 1 - b, \qquad\qquad x > \mu$$

where $a + b \leq 1$. Although not sufficiently accurate for precise circuit simulation, use of this model enables the trends of the effects of non-ideal classification to be simply determined.

From Eq. 6, the equivalent of Eq. 8 using the above classifier function is

$$P(\mu) = \frac{1 - b(1 + C)}{(1 + C)(1 - a - b)} \qquad\qquad (48)$$

Note that this becomes indeterminate when $a = 1 - b$, since μ is then no longer defined. The combined feed to the mill is

$$F(x) = \frac{G(x)}{1 + C} + aP(x), \qquad\qquad x \leq \mu$$

$$F(x) = \frac{G(x)}{1 + C} + (1 - b)[P(x) - P(\mu)], \quad x \geq \mu \qquad (49)$$

These are used to replace $F(x)$ in the mill model equations. The circuit product is, from Eq. 3

$$Q(x) = (1 + C)(1 - a)P(x), \quad x \leq \mu$$

and

$$Q(x) = (1 + C)\{(1 - a)P(\mu) + b[P(x) - P(\mu)]\}, \quad x > \mu \qquad (50)$$

Considering the case of plug flow, using the mill Eq. 1a

$$1 - P(x) = [1 - F(x)]\exp[-S(x)\tau]$$

Substituting for $F(x)$ and collecting terms

$$P(x) = \left[1 - \left(1 - \frac{G(x)}{1 + C}\right)e^{-S(x)\tau}\right] \Big/ [1 - ae^{-S(x)\tau}], \quad x \leq \mu$$

Hence, from Eq. 48,

$$\tau = [1/S(\mu)]\ln\left\{\frac{[1 + C - G(\mu)](1 - a - b) - a[1 - b(1 + C)]}{C - a(1 + C)}\right\}$$

$$\bullet \qquad (51)$$

where a and b are positive and $a + b < 1$. Then, in the same way as before,

$$Q(x) = \frac{(1 + C)(1 - a)\left[1 - \left(1 - \frac{G(x)}{1 + C}\right)K^{-S(x)/S(\mu)}\right]}{1 - aK^{-S(x)/S(\mu)}}, \quad x \leq \mu$$

$$Q(x) = (1 + C)\left\{(1 - a - b)P(\mu) + b\right. \qquad (52)$$

$$\left.\left[\frac{1 - \left[\frac{G(x)}{1 + C} - (1 - b)P(\mu)\right]K^{-S(x)/S(\mu)}}{1 - (1 - b)K^{-S(x)/S(\mu)}}\right]\right\}, \quad x \geq \mu$$

where

$$K = \frac{[1 + C - G(\mu)](1 - a - b) - a[1 - b(1 + C)]}{C - a(1 + C)}$$

and $P(\mu)$ is given by Eq. 48. Also,

$$Q/W = S(\mu)/(1 + C)$$

$$\times \ln\left[\frac{[1 + C - G(\mu)](1 - a - b) - a[1 - b(1 + C)]}{C - a(1 + C)}\right]$$

(53)

The first point of interest is that $Q(\mu)$ is, from Eqs. 50 and 48,

$$Q(\mu) = (1 - a)[1 - b(1 + C)]/(1 - a - b)$$ (54)

whereas $Q(\mu)$ for ideal classification is 1. If b is zero, $Q(\mu) = 1$, which is correct since b has the physical meaning of the fraction of material above μ in size which is sent to product (fines). Second, a minimum value of C exists, corresponding to long residence time and zero Q/W, because when the mill product is ground all less than μ, *a fraction is still recirculated*, hence,

$$C_{min} = a/(1 - a)$$ (55)

Third, the presence of b introduces a finite maximum circulation ratio even when τ tends to zero at high flow rates. From Eq. 54, τ is zero when the log term is log[1], and

$$C_{max} = \frac{[1 - G(\mu)][(1 - a - b) + ab]}{b(1 - a)}$$ (56)

This goes to the required results of infinity when $b = 0$, and $a/(1 - a)$ when $a = 1 - b$ (see below). From Eqs. 54, 55, and 56, the limiting values of $Q(\mu)$ are 1 at C_{min}, and $G(\mu)$ at C_{max} ($b \neq 0$), which are the logically required results. In contrast to the case where $b = 0$, the size distribution of the coarsest possible product, at C_{max}, is $G(x)$ when b is finite. Since C is finite at $\tau = 0$, Eq. 53 gives $Q/W = \infty$ at this point.

Of particular interest is the case where $b = 0$ but a is finite. This gives a value of K in Eq. 52 of

$$K = \frac{[1 + C - G(\mu)] - \dfrac{a}{1 - a}}{C - \dfrac{a}{1 - a}}$$ (57)

$$= \frac{1 + (C - C_{min}) - G(\mu)}{C - C_{min}}$$ (57a)

The limiting values of K correspond to $C = C_{min}$ at τ large, $C = \infty$ at τ zero, that is, K ranges from ∞ to 1. This corresponds to $Q(x) = 1$ as one limit and

$$Q(x) = G(x) + \left[\frac{S(x)}{S(\mu)}\right][1 - G(\mu)]\left(\frac{1 + C}{C - C_{min}}\right) \qquad (58)$$

as the limit for C large. Comparing with Eqs. 21 and 22, it is seen that the coarsest size distribution obtained is the same with and without the presence of a, when $C \gg C_{min}$. The relative output rate is, from Eq. 53 and Eq. 57,

$$Q/W = S(\mu)\Big/(1 + C)\ln\left[\frac{1 + (C - C_{min}) - G(\mu)}{C - C_{min}}\right] \qquad (59)$$

and

$$\lim_{C \to large} Q/W = S(\mu)\Big/[1 - G(\mu)]\left(\frac{1 + C}{C - C_{min}}\right) \qquad (60)$$

The general effect of a is to decrease the value of Q/W at a given value of C, but increase the fineness of the circuit product.

The one-point match gives

$$Q/W = \frac{S(x_p)}{1 + C}\Big/\ln\left[\frac{(1 - a)[1 + C - G(x_p)] - a\psi}{(1 + C)(1 - a) - \psi}\right] \qquad (61)$$

Fig. 9 shows the result for $\psi = 0.8$, $G(x_p) = 0$, and a range of a values. It is seen that closing the circuit with a low value of C actually reduced the output rate, with the effect being more pronounced as a is larger. The physical reason is that the recycle of fines caused by the bypass leads to less volume of larger sizes in the mill, consequently lower breakage and lower

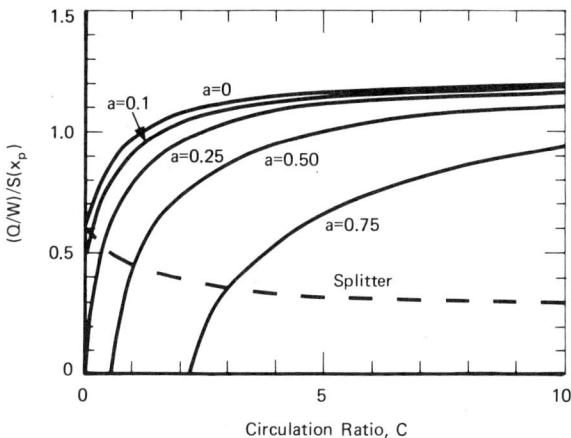

Fig. 9. Influence of bypass with partially ideal classification on relative output rates for the analytical solution of the simple circuit ($G(x_p) = 0$, $\psi = 0.8$).

output. As C increases, this effect is counteracted by the removal of fine material by the classification action. The results are readily extended to fully mixed mill models and to the ψ_1, ψ_2 criteria.

Ideally Inefficient Size Classification

This can be defined by $a = 1 - b$, that is, the classifier acts purely as a sample splitter and does not give any change in particle size distribution from feed to product to tailings. A value of μ is now not defined, although a is still a variable. In fact, $s(x) = a$ for all sizes giving

$$C = a/(1 - a) \tag{62}$$

so that C can be varied only with variation in a. The circuit is now that of Fig. 10. Then, for plug flow,

$$Q(x) = P(x) = \frac{(1 + C) - [1 + C - G(x)]\exp[-S(x)\tau]}{1 + C - C\exp[-S(x)\tau]} \tag{63}$$

For any given value of a, C is fixed but a whole range of size distributions exist depending on τ, from very fine for $\tau \to$ large to $G(x)$ for $\tau = 0$. It is interesting to consider the effect of operating a mill at a set rate Q/W but varying C via a. Fig. 11 shows a typical result. As a increases, the mill is producing a coarser product at the same rate, so it becomes more inefficient. The physical reason is the same as that discussed in the previous section: the recycle replaces some of the mill contents with slow-breaking smaller material.

A degree of freedom is removed by specifying τ at x_p which fixes τ at a given C,

$$\left.\begin{aligned} \tau &= \frac{1}{S(x_p)} \ln\left[\frac{C(1 - \psi) + (1 - G(x_p))}{(1 + C)(1 - \psi)}\right] \\ Q(x) &= \frac{(1 + C) - [1 + C - G(x)]J^{-S(x)/S(x_p)}}{1 + C - CJ^{-S(x)/S(x_p)}} \\ J &= \frac{C(1 - \psi) + (1 - G(x_p))}{(1 + C)(1 - \psi)} \end{aligned}\right\} \tag{64}$$

where

$$Q/W = S(x_p)/(1 + C)\ln\left[\frac{C(1 - \psi) + (1 - G(x_p))}{(1 + C)(1 - \psi)}\right] \tag{65}$$

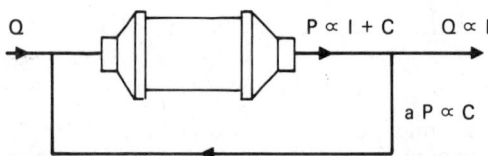

Fig. 10. Circuit for recycle without size classification.

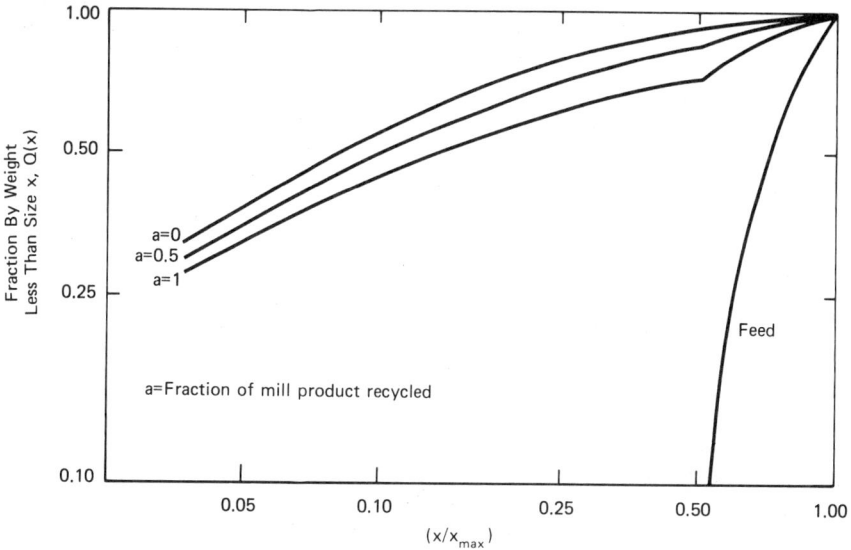

Fig. 11. Effect of recycle by a sample splitter (see Fig. 10) on size distribution from plug flow grinding, at a set output rate Q / W.

with limits of

$$\lim_{C \to 0} Q(x) = 1 - [1 - G(x)] \left[\frac{1 - G(x_p)}{1 - \psi} \right]^{-S(x)/S(x_p)} \tag{64a}$$

$$\lim_{C \to \infty} Q(x) = \frac{G(x) + \left[\dfrac{\psi - G(x_p)}{1 - \psi} \right] \dfrac{S(x)}{S(x_p)}}{1 + \left[\dfrac{\psi - G(x_p)}{1 - \psi} \right] \dfrac{S(x)}{S(x_p)}} \tag{64b}$$

$$\lim_{C \to 0} Q/W = S(x_p) / \ln \left[\frac{1 - G(x_p)}{1 - \psi} \right] \tag{65a}$$

$$\lim_{C \to \text{large}} Q/W = S(x_p) \left(\frac{C}{1 + C} \right) \left[\frac{1 - \psi}{\psi - G(x_p)} \right] \tag{65b}$$

The dashed line in Fig. 9 corresponds to the classifier acting as a sample splitter. It will be noted that in this case ($\psi = 0.8$, $G(x_p) = 0$) $C = 0$ gives the higher rate of output for a given ψ, x_p criterion. Thus higher values of C *reduce* the output, for the one-point criterion. In addition, the *maximum* rate, at $C = 0$, is the same as the *minimum* rate for ideal classification, also at $C = 0$, (Eq. 27) and thus, of course, is less than the rate at $C \geq 0$ for ideal classification.

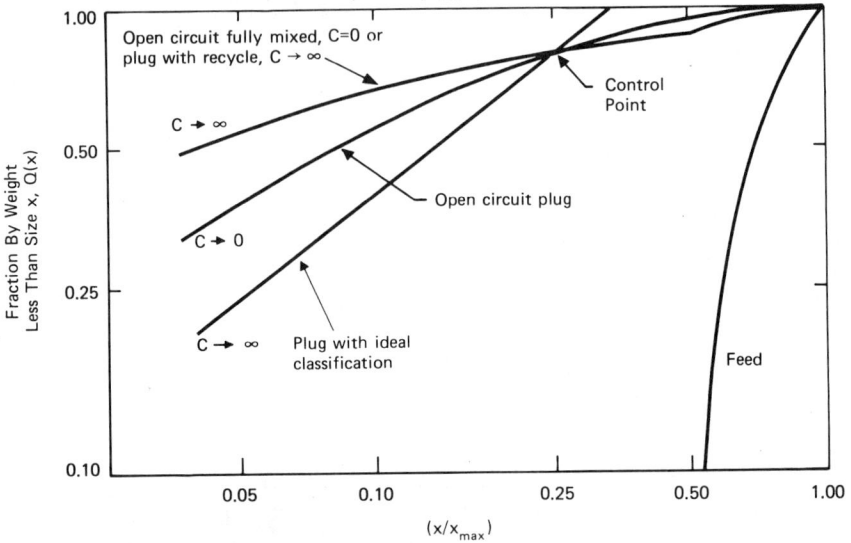

Fig. 12. Effect of recycle by a sample splitter on size distribution controlled to pass through the point 80% less than x / x_{max} equal 0.25.

A more realistic criterion for comparing rates is the two-point fit. Applied to this circuit, Eqs. 64a and 64b show that

$$\underline{Q}(x_2) = 1 - [1 - G(x_2)]\left[\frac{1 - G(x_1)}{1 - \psi_1}\right]^{-S(x_2)/S(x_1)}$$

$$\overline{Q}(x_2) = \frac{G(x_2) + \left[\dfrac{\psi_1 - G(x_1)}{1 - \psi_1}\right]\dfrac{S(x_2)}{S(x_1)}}{1 + \left[\dfrac{\psi_1 - G(x_1)}{1 - \psi_1}\right]\dfrac{S(x_2)}{S(x_1)}}$$

By comparing this to Eqs. 28 and 36, it is seen that the lower bound $\underline{Q}$ (at $C = 0$) is the *upper* bound (at $C = 0$) for plug flow, while the upper bound $\overline{Q}$ (at $C \to \infty$) is the upper bound (at $C \to 0$) for fully mixed with ideal classification. This is illustrated in Fig. 12. Thus the use of a fraction of recycle without size classification enables a plug flow mill to obtain two-point fits in the range obtainable without this recycle by a fully mixed mill.

Energy of Grinding for Different Circuits

More or less complex algebraic equations exist for the energy per ton of material ground for all of the simple analytical solutions discussed in this chapter. However, discussion will be limited to the algebraically simple cases of ideal classification. For design purposes it can be assumed that mills are run at a set W value within the optimum region. Let the energy input

Table 2. Values of Factor K

Case	K	Example K, $\psi = 0.8$, $G(x_p) \approx 0$, $C = 2.5$
Open cycle, batch or plug flow	$2.3\log\left[\dfrac{1 - G(x_p)}{1 - \psi}\right]$	1.61
Open cycle, fully mixed continuous	$\dfrac{\psi - G(x_p)}{1 - \psi}$	4
Open cycle, n fully mixed continuous	$n\left\{\left[\dfrac{1 - G(x_p)}{1 - \psi}\right]^{1/n} - 1\right\}$	$n = 2$,　2.48 $n = 3$,　2.13
Closed cycle, plug flow	$(1 + C)2.3\log\left[\dfrac{1 + C - G(x_p)}{1 + C - \psi}\right]$	0.91
Closed cycle, fully mixed	$(1 + C)\left[\dfrac{\psi - G(x_p)}{1 + C - \psi}\right]$	1.04
Closed cycle, n fully mixed	$n(1 + C)\left\{\left[\dfrac{1 + C - G(x_p)}{1 + C - \psi}\right]^{1/n} - 1\right\}$	$n = 2$,　0.98 $n = 3$,　0.94(5)

rate for a given mill be m_p. The energy per unit mass of circuit product is $E = m_p/Q$. Directly, then, the expressions for E are

$$\text{Plug flow: } E = \left[m_p/WS(\mu) \right] (1 + C)\ln\left[\frac{1 + C - G(\mu)}{C} \right]$$

$$\text{Fully mixed: } E = \left[m_p/WS(\mu) \right] \left[(1 + C)/C \right] \left[1 - G(\mu) \right]$$

$$n \text{ fully mixed: } E = \left[m_p/WS(\mu) \right] (n)(1 + C)\left[\left(\frac{1 + C - G(\mu)}{C} \right)^{1/n} - 1 \right]$$

Consider the effect of C at a set value of μ. In all cases, E decreases as C increases, which is logical because the product size distributions are becoming coarser. At a set value of C, an increase in the classifier setting μ decreases E because $S(\mu)$ increases; again, this corresponds to a coarser product size. Considering any one of the equations, it is concluded that at a given μ and C the energy per ton is proportional to $m_p/WS(\mu)$. For a given diameter mill at set conditions, m_p/W will be almost constant, giving E proportional to $1/S(\mu)$ that is, E inversely proportional to grindability. Eq. 18, for example,

$$Q(x) = (1 + C)\left\{ 1 - \left(\frac{1 + C - G(x)}{1 + C} \right)\left(\frac{1 - C - G(\mu)}{C} \right)^{-S(x)/S(\mu)} \right\}$$

shows that if two materials are compared which have the same form of $S(x)$ differing by a constant factor, $S(x)' = kS(x)$, the size distributions at a given μ and C will be identical; thus $E \propto 1/k$ for a given feed going to a given product under identical milling and circuit conditions. However, it is clear that if k varies with x, the comparative energy per ton between the two materials depends on the value of μ.

These expressions are of no use at $C = 0$, open circuit, because the size distribution is not then uniquely determined. In addition, the size distributions for the above cases at constant μ and C are different between the different mill models, so that they cannot be directly compared. However, if a product is defined by a ψ, x_p criterion, all the expressions for E are of the form

$$E = \left[m_p/WS(x_p) \right] K \tag{66}$$

where the values of K are shown in Table 2. It can be seen that the energy to grind unit mass of a given feed to a given percent less than a specified size (ψ criterion) is not independent of the circuit conditions. At a given value of circulation ratio C, closed cycle, plug flow requires the least energy; however, the overall product is finer for all other conditions. The relative values of E of different materials ground at set conditions will vary with x_p unless $S(x)' = kS(x)$.

Appendix

Appendix 1: Analysis of Batch Grinding by Simulation

It is informative to investigate the variations of batch size distributions predicted by solving the batch grinding equations, for comparison with laboratory results.

To illustrate this process it is sufficient to consider a number of cases where the value of $S_{16 \times 20}$ is arbitrarily taken as 1 min^{-1}. The first case considered is the breakage of a feed of 1.18 mm $\times$ 850 μm (16 $\times$ 20 mesh) material (which is small enough to give $Q_i = 1$ for all sizes), for a constant value of α but varying B values. The normalized B values used are shown in Fig. 1.1; they were selected to give different values of the slope γ yet keep $B_{3,1}$ the same for each set, which makes the fraction of size 1 broken into size 2 (that is, $b_{2,1}$) the same for each set. There are four major conclusions from the results shown in Fig. 1.2. First, the slopes of the lines (the Schuhmann slope or *distribution coefficient* α_s) are at first nearly constant with time. Second, they are the same as the γ value from the primary progeny fragment distribution. Third, the slopes for $\alpha = 1$, $\gamma = 1$ tend to decrease somewhat at longer grind times. Fourth, the 80%-passing size is virtually *identical* for each set up to the maximum grinding time used. Thus the value of the Bond coefficient C_B in Eq. 7 in Chapter 2 would be the same for each material, and sizing of a mill using the Bond equation would give the same answer, even though the rest of the size distributions at lower than 80%-passing sizes are radically different between the data sets.

The second case considered is the grinding of 1.18 mm $\times$ 850 μm (16 $\times$ 20 mesh) with $S_1 = 1$ min^{-1}, but keeping the B values constant and varying the α value. Fig. 1.3 shows the results. It is now clear that the Schuhmann slopes start at the γ value but tend to change in the direction of the α values as grinding proceeds. Thus if $\alpha < \gamma$, the slopes tend to decrease, and if $\alpha > \gamma$, to increase. However, for $\alpha = \gamma$ the slopes tend to decrease with degree of grinding and this acts to counterbalance the increase for $\alpha > \gamma$. Thus, in Fig. 1.3 the sets for $\gamma < \alpha$ show nearly constant slope. Fig. 1.3c shows one case extended to the subsieve size region. It shows that the Schuhmann slopes based on the sieve region [> 38 μm (400 mesh)] are lower than the true Schuhmann slopes, for long grind times. The very fine sizes always give a Schuhmann slope of γ.

Fig. 1.3a shows that lower values of α give smaller 80%-passing sizes as grinding proceeds. This is because lower values of α give higher rates of breakage of the smaller sizes. It is generally concluded that the *shape* of the product size distribution is dominated by the γ of the B values, with minor

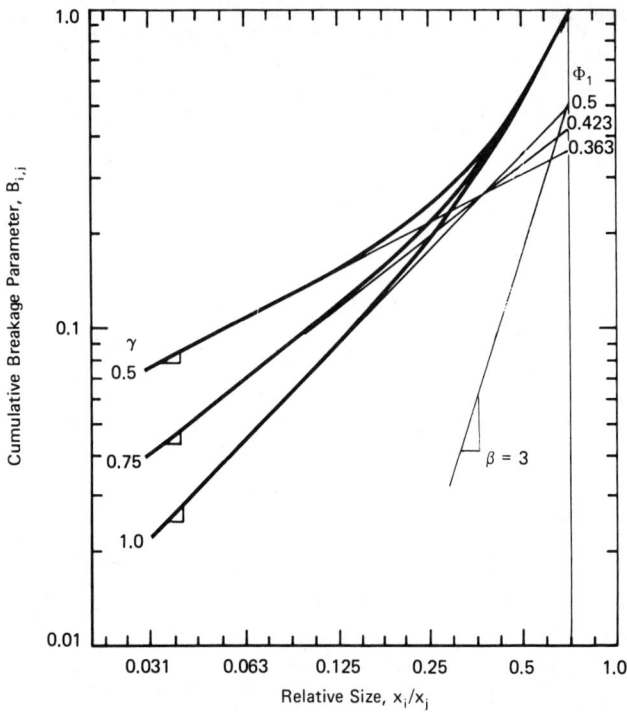

Fig. 1.1. The B values used in the batch mill simulation: chosen to give $b_{2,1} = 0.46$.

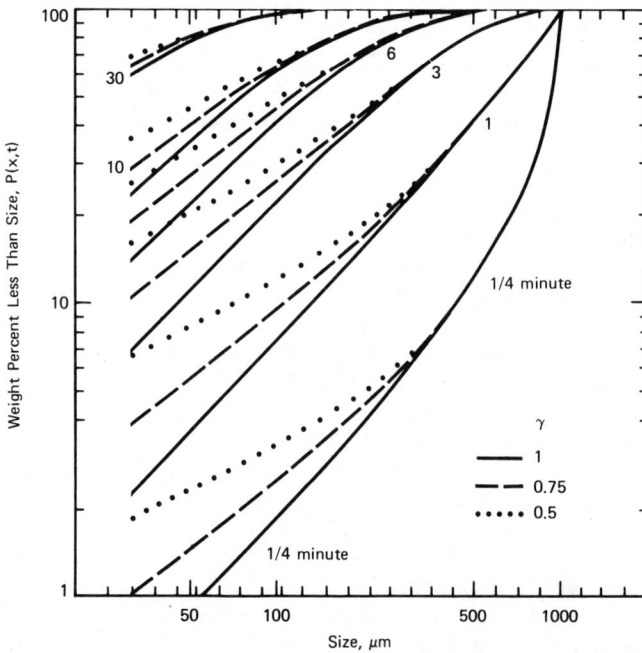

Fig. 1.2. Simulated size distribution from batch grinding 1.18 mm × 850 μm (16 × 20 mesh) feed: $\alpha = 1$, $S_1 = 1$ min^{-1}, various B values from Fig. 1.1.

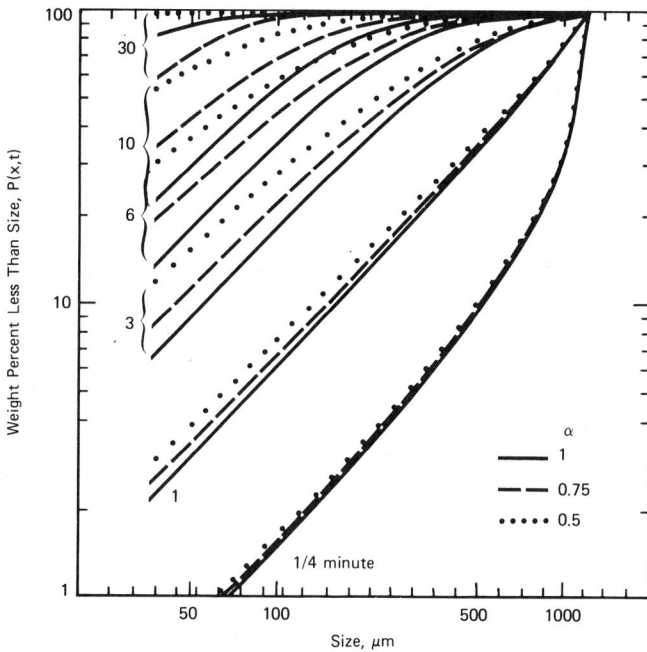

Fig. 1.3a. Simulated size distributions from batch grinding 1.18 mm $\times$ 850 μm (16 $\times$ 20 mesh) feed with constant B values, $\gamma = 1$, $S_1 = 1$ min^{-1}, three different α values. $\alpha = 1$, 0.75, 0.5; $\alpha \leq \gamma$.

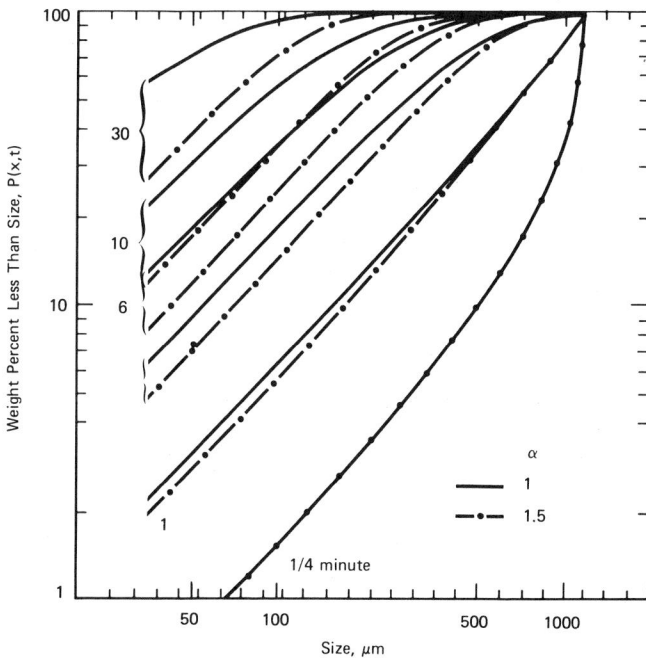

Fig. 1.3b. Simulated size distributions from batch grinding 1.18 mm $\times$ 850 μm (16 $\times$ 20 mesh) feed with constant B values, $\gamma = 1$, $S_1 = 1$ min^{-1}, two different α values, $\alpha = 1$, 1.5; $\alpha \geq \gamma$.

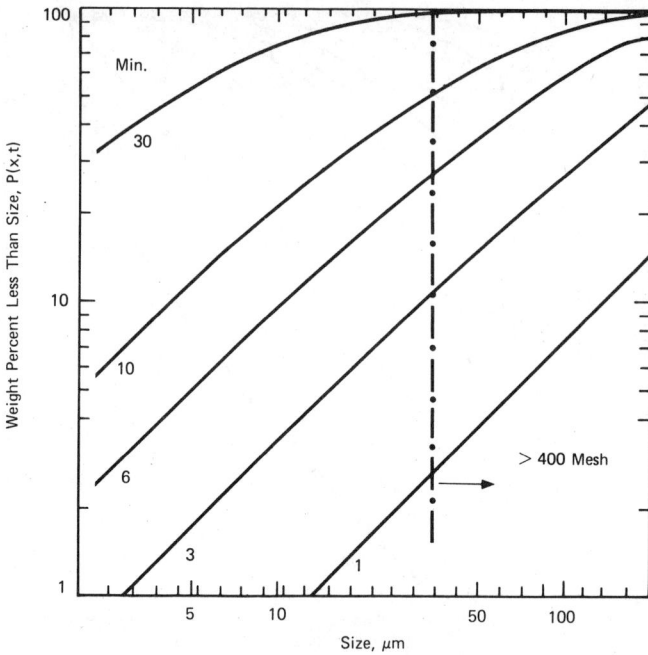

Fig. 1.3c. Simulated size distributions from batch grinding 1.18 mm × 850 μm (16 × 20 mesh) feed with constant B values, $\gamma = 1$, $S_1 = 1$ min^{-1}, three different α values. Extension to subsieve sizes, $\alpha = 0.5$.

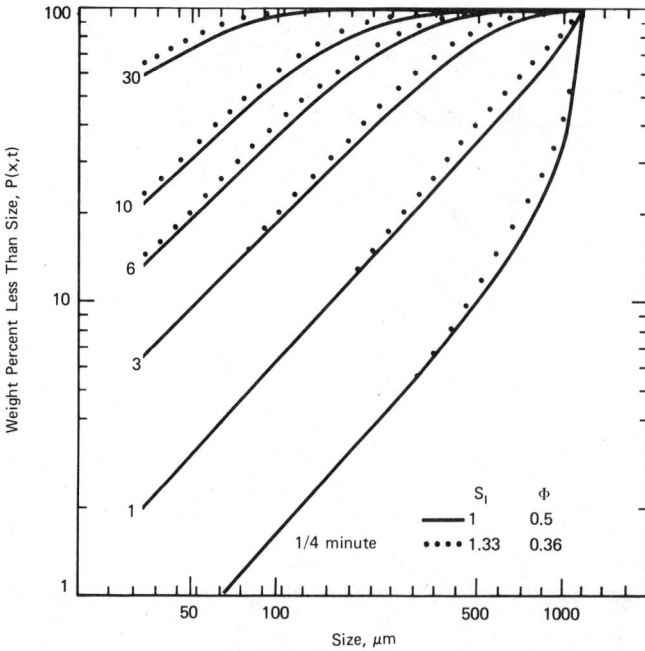

Fig. 1.4. Simulated size distributions from batch grinding 1.18 mm × 850 μm (16 × 20 mesh) feed for constant α, γ, β values, $S_1 = 1$ min^{-1} and 1.33 min^{-1}, Φ changed to give $S_1 b_{2,1}$ constant.

influence of α, but the *degree of grinding* defined by the 80%-passing size is dominated by α and, of course, S_1. The value of Φ also affects the degree of grinding via the value of $b_{2,1}$: if Φ is low it means that most of the breakage from one size goes into the next lower size interval, which represents a small degree of grinding and vice versa. Fig. 1.4 compares results where the value of S_1 and Φ were adjusted to give $S_1 b_{2,1}$ constant for the two cases, other factors held constant. It shows that the size distributions are then quite similar, so that a variation in S_1 can be partially compensated for by a change in Φ and vice versa. Thus, high S_1, low α, and high Φ all give rapid grinding.

The third case, Fig. 1.5, illustrates the effect of the shape of the feed size distribution, other factors being held constant. The steep feed size distribution, of Schuhmann slope $\alpha_s = 1.5$, rapidly turns to a slope of γ for the product size distributions. However, as the feed size distribution becomes more spread out (lower α_s), the slopes of the product size distributions still tend to γ but never reach it, so that the value of α_s lies between γ and the slope of the feed. Again, it will be noted that the 80%-passing sizes are not widely different even though the size distributions and the amount of fines are radically different for the different feeds.

Case 4 occurs when the feed size is too large for normal breakage, and the specific rate of breakage for the top sizes is to the right of the maximum.

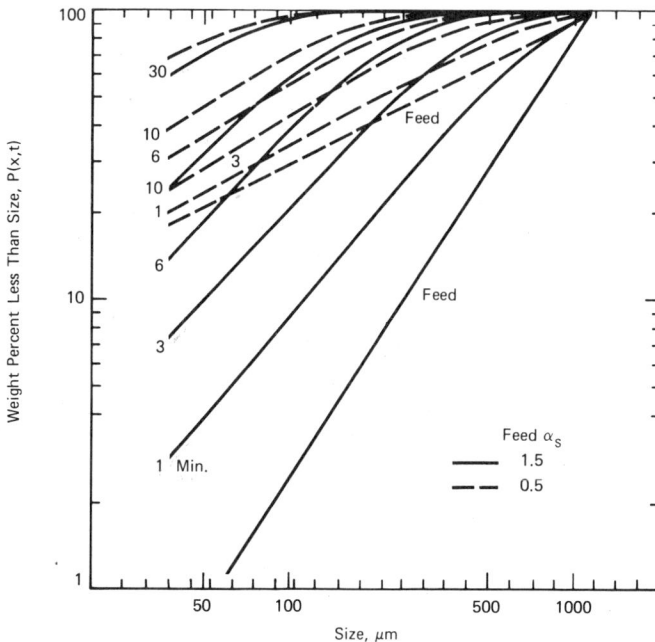

Fig. 1.5. Simulated size distribution from batch grinding of distributed feeds of 1.18 mm (16 mesh) top size: $S_1 = 1$ min^{-1}, $\alpha = 1$ and constant B values, $\gamma = 1$, $\beta = 3$, $\Phi = 0.5$.

If the normalized B values are the same for large and small sizes ($\gamma_L = \gamma$), the result is shown in Fig. 1.6 as the solid curves. Comparison with Fig. 1.2 shows that the slower rates of breakage of large material give a characteristic *plateau region* in the size distribution, because the intermediate sizes are breaking rapidly and disappear as rapidly as they are formed. It is sometimes found that the shape of the progeny fragment B values for the larger sizes is shallower than for normal breakage, that is, γ_L for the large sizes is less than γ_S for the smaller sizes. The effect of this is illustrated in Fig. 1.6, in comparison to constant γ values. Again, it does not affect the 80%-passing sizes, but it does make a significant difference to the size distribution. The difference decreases as grinding proceeds, because the influence of the abnormal large feed sizes becomes less as the product becomes finer.

Finally, it can be shown that the shapes of the Schuhmann plots of product size distributions are identical for different top feed sizes, if the particle sizes are well to the left of the maximum in S, and if the progeny fragment distributions are normalized. Then $b_{i,j} = b_{i-j}$, $S_i = ax_i^\alpha$, and $S_i/S_1 = (x_i/x_1)^\alpha = R^{(i-1)\alpha}$ where R is the geometric sieve ratio, normally $1/\sqrt{2}$. The batch grinding equation then becomes

$$\frac{dw_i}{d(S_1 t)} = -R^{(i-1)\alpha} w_i + \sum_{\substack{j=1 \\ i>1}}^{i-1} b_{i-j} R^{(j-1)\alpha} w_j$$

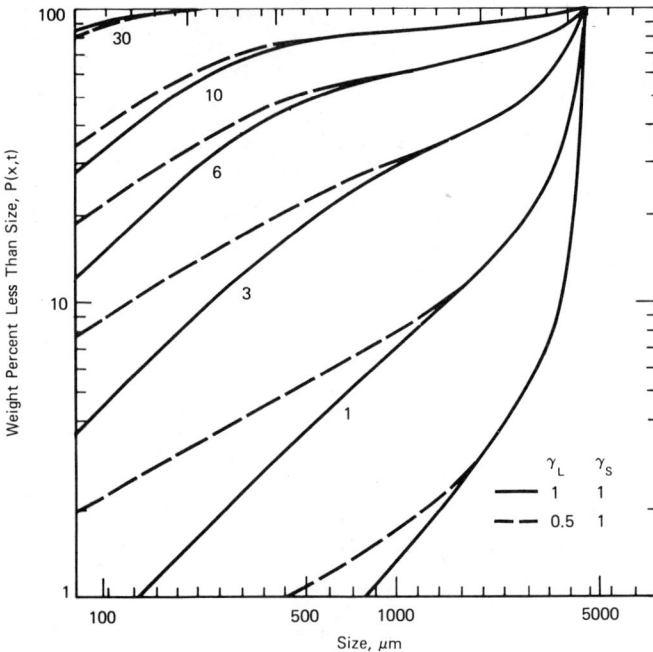

Fig. 1.6. Simulated size distributions from batch grinding of 4.75 × 3.35 mm (4 × 6 mesh) feed; $\alpha = 1$, $S_{16 \times 20} = 1$ min^{-1}; γ_L applies for the top four sizes ($\gamma_L = 0.5$, $\Phi = 0.36$; $\gamma_L = 1$, $\Phi = 0.5$).

Clearly, for given b_{i-j} and α this equation will give identical computations for any starting feed vector $w_i(0)$ and a given set of $S_i\tau$ values, irrespective of whether the top size $j = 1$ is 1.18 mm × 850 μm (16 × 20 mesh) or 425 μm × 250 μm (40 × 60 mesh) or 150 μm × 106 μm (100 × 140 mesh), etc. The computation delivers values of the cumulative amounts below interval i, where i is numbered down from the top size as 1. Plotted on the Schuhmann plot, that is, on log-log scales, the curves would be identical but shifted left to right to start from the particular top size used. Or, they would be absolutely identical if plotted in the dimensionless form of P vs. x_i/x_1. Of course, the time scale to get to a certain dimensionless size distribution would depend on S_1.

Appendix 2: Alyavdin-Harris Relation (See Chapter 2)

In addition to the relation of Eq. 7 in Chapter 2, Bond also claimed that batch grinding data included the relation (see Eq. 9a in Chapter 8)

$$1 - P(x, t) = [1 - P(x, 0)] \exp[-c(x)t] \qquad (2.1)$$

where $c(x)$ is defined by the equation; that is, $c(x) = 2.3 \log[(1 - P(x, 0))/(1 - P(x, t))]/t$. For small values of $P(x, t)$, $(1 - P(x, 0))/(1 - P(x, t))$ approaches 1 because $0 < P(x, 0) < P(x, t)$, and therefore $c(x)t$ must be small. Then this equation reduces to Eq. 15 in Chapter 2, because $\exp[-c(x)t] \approx 1 - c(x)t$ for small values of $c(x)t$, $1 - P(x, t) = 1 - c(x)t - P(x, 0) + P(x, 0)c(x)t$ and, for $P(x, 0)c(x)t \to$ small,

$$P(x, t) - P(x, 0) = c(x)t$$

However, Eq. 2.1 has limited applicability since it does not fit the data as well as the simple linear form of Eq. 15 in Chapter 2.

A very similar but more elaborate treatment has been suggested by Alyavdin (1938), with extensions by Harris (1971), who suggest the empirical relation

$$1 - P(x, t) = [1 - P(x, 0)] \exp[-c(x)t^p] \qquad (2.2)$$

Austin et al. (1974) have shown that this relation can be applied with reasonable accuracy to the finer sizes resulting from batch grinding of a single size of feed [that is, $P(x, 0) = 0$], with the exponent p being greater than 1 but less than 1.5, but that it does not work for a feed size distribution. The values $c(x)$ and p can be determined from a Weibull plot (see Fig. 3 in Chapter 8) since

$$\log \left[2.3 \log \left(\frac{1}{1 - P(x, t)} \right) \right] = p \log t + \log c(x)$$

and the set of $c(x_2)$, $c(x_3)$, etc. values obtained if desired.

Clearly, the values of $c(x)$ must depend on the size of the feed interval. To go from a *natural* feed size distribution as represented by a curve at t_1 on Fig. 1 in Chapter 2 with an amount $F(x^*)$ less than a specified size x^*, to some desired amount $\psi(x^*)$ less than size x^* occurring at t_2 will require a grind time t given by $t = t_2 - t_1$ where

$$1 - \psi(x^*) = \exp[-c(x^*)t_2^p]$$
$$1 - F(x^*) = \exp[-c(x^*)t_1^p]$$

hence,

$$tc(x^*)^{1/p} = \left[\ln \left(\frac{1}{1 - \psi(x^*)} \right) \right]^{1/p} - \left[\ln \left(\frac{1}{1 - F(x^*)} \right) \right]^{1/p}$$

As before, then, the specific energy is given by $E_A = m_p t/W$ and

$$E_A = C_A \left(\left[\ln \left(\frac{1}{1 - \psi(x^*)} \right) \right]^{1/p} - \left[\ln \left(\frac{1}{1 - F(x^*)} \right) \right]^{1/p} \right)$$

where

$$C_A = \left(m_p/W \right) \left(1/c(x^*) \right)^{1/p}$$

References

Alyavdin, V. V., 1938, "Breakage Process in Tumbling Mills," in *Proceedings: Topics on Breakage in Cement Industry*, Gyprocement, USSR.

Austin, L. G., et al., 1974, "Extension of the Empirical Alyavdin Equation for Representing Batch Grinding Data," *International Journal of Mineral Processing*, Vol. 1, pp. 107–123.

Harris, C. C., 1971, "The Alyavdin-Weibull Chart in Batch Comminution Kinetics," *Trans. IMM*, Vol. 80, pp. C42–44.

Appendix 3: The Mathematical Treatment of the Batch Grinding Equation in Size-Continuous, Time-Continuous Form

To present the results in a general manner, the equation is put into dimensionless form, using $x' = x/x_1$, $y' = y/x_1$, $t' = S_1 t$, where x_1 is an arbitrarily chosen size. For a normalized B function, $B(x, y) = B(x/y) = B(x/x_1/y/x_1) = B(x'/y')$. Eq. 5 in Chapter 4 becomes

$$\partial P(x', t')/\partial(S_1 t) = \int_{y'=x'}^{x_{max}/x_1} (S(y')/S_1) B(x'/y') d_{y'} P(y', t)$$

that is,

$$\partial P(x', t')/\partial t' = \int_{y'=x'}^{x_{max}/x_1} S'(y') B(x'/y') d_{y'} P(y', t') \qquad (3.1)$$

where

$$S'(y') = S(y')/S_1$$

The analytical solution (Filippov, 1961; King, 1972) to Eq. 3.1 with $S(x) = ax^\alpha$ and $B(x/y) = (x/y)^\beta$ and with boundary conditions

$$P(x', 0) = 0, \quad 0 \le x' \le 1$$

$$P(1, t') = 1 - \exp(-t'), \quad t' \ge 0$$

is

$$P(x', t') = t' e^{-t'} \int_0^{(x')^\beta} F\left[1 + \beta/\alpha, 2, (1 - \phi^{\alpha/\beta}) t'\right] d\phi \qquad (3.2)$$

where $F(a, b, z)$ is the congruent hypergeometric function. Note that the boundary condition is: no sizes initially present below x_1, all the initial amount in size x_1 to $x_1 + dx$. The equation is readily generalized to a complete feed size distribution using

$$P(\bar{x}', \bar{t}') = \int_0^{\bar{x}'} P(x', t) \frac{\partial P(x', 0)}{\partial x'} dx' \qquad (3.3)$$

However, we will limit the discussion to the case of Eq. 3.2. The principal interest of the analytical solution is the simpler approximate forms which result under certain conditions. For example, when t' becomes large, z becomes large in $F(a, b, z)$ and

$$F(a, b, z) \simeq \frac{\Gamma(b)}{\Gamma(a)} z^{a-b} e^z \qquad (3.4)$$

Then

$$P(x', t') \simeq t'e^{-t'}\frac{\Gamma(2)}{\Gamma(1 + \beta/\alpha)} \int_0^{(x')^\beta} \left[(1 - \phi^{\alpha/\beta})t\right]^{(\beta/\alpha)-1}$$

$$\times \exp\left[(1 - \phi^{\alpha/\beta})t'\right] d\phi$$

$$= (t')^{\beta/\alpha}\frac{\Gamma(2)}{\Gamma(1 + \beta/\alpha)} \int_0^{(x')^\beta} \left[(1 - \phi^{\alpha/\beta})t\right]^{(\beta/\alpha)-1}$$

$$\times \exp\left[-\phi^{\alpha/\beta}t'\right] d\phi \tag{3.5}$$

This expression can be immediately checked for the special case of $\alpha = \beta$, giving

$$P(x', t') \simeq t'\left[\frac{\exp(-t'_\theta)}{-t'}\right]\Bigg|_0^{(x')^\alpha}$$

$$= 1 - \exp\left[-t'(x')^\alpha\right]$$

or

$$1 - P(x', t') = \exp\left[-t'(x')^\alpha\right]$$

which is the expected result, the normalized Rosin-Rammler equation (see Chapter 8). Eq. 3.5 can be put in the form,

$$P(x', t') \simeq (t')^{\beta/\alpha}\frac{\Gamma(2)}{\Gamma(1 + \beta/\alpha)} \int \frac{(1 - \phi^{\alpha/\beta})^{(\beta/\alpha)-1}}{\phi^{(\alpha/\beta)-1}} \exp(-\phi^{\alpha/\beta}t') d(\phi^{\alpha/\beta})$$

When α/β is reasonably near 1, the term involving $(1 - \phi^{\alpha/\beta})^{(\beta/\alpha)-1}$ can be neglected, giving

$$P(x', t') \simeq (t')^{\beta/\alpha}\frac{\Gamma(2)}{\Gamma(1 + \beta/\alpha)}(\beta/\alpha) \int \frac{(\phi^{\alpha/\beta}t')^{(\beta/\alpha)-1}}{t'(t')^{(\beta/\alpha)-1}}$$

$$\times \exp(-\phi^{\alpha/\beta}t') d(\phi^{\alpha/\beta}t')$$

$$\simeq \frac{\Gamma(2)}{\Gamma(1 + \alpha/\beta)}\left(\frac{\beta}{\alpha}\right)\gamma(\beta/\alpha, (x')^\alpha t')$$

where γ is the incomplete gamma function. Since $\Gamma(z + 1) = z\Gamma(z)$, and

$\Gamma(2) = 1.0$,

$$P(x', t') \simeq \gamma\big(\beta/\alpha, (x')^{\alpha}t'\big)/\Gamma(\beta/\alpha) \tag{3.6}$$

or

$$P(x', t') \simeq \Pi(\chi^2/v)$$

where

$$\chi^2 = 2(x')^{\alpha}t'$$

$$v = 2\beta/\alpha$$

$\Pi(\chi^2/v)$ is the chi-squared probability function. Note that this is the limiting distribution for long times.

Austin, Luckie, and Klimpel (1972) have shown that when α is not too different from β, these limiting size distributions can be approximately fitted by a Rosin-Rammler equation

$$\left. \begin{aligned} R(x', t') &\approx \exp(x'/x_o')^{\beta^{0.35}\alpha^{0.65}} \\ x_o' &\approx (\beta/\alpha t')^{1/\alpha} \end{aligned} \right\} \tag{3.7}$$

At short times, of course, the slope is β, since

$$P(x', t') \approx B(x', 1)\big[1 - \exp(-t')\big] \tag{3.8}$$

and $B(x', 1) = (x')^{\beta}$.

References

Austin, L. G., Luckie, P. T., and Klimpel, R. R., 1972, "Solutions of the Batch Grinding Equation Leading to Rosin-Rammler Distributions," *Trans. SME-AIME*, Vol. 252, pp. 87–94.

Filippov, A. F., 1961, "On the Sizes of Particles Which Undergo Splitting," *Theory of Probability and Its Applications*, Vol. 6, English translation, USSR, pp. 275–294.

King, R. P., 1972, "An Analytical Solution to the Batch Comminution Equation," *Journal of the South African Institute of Mining and Metallurgy*, Vol. 73, pp. 127–131.

Appendix 4: A Procedure for Deconvolution of RTD Data

Eq. 6 in Chapter 14 is

$$c(t) - c_o(t) = \frac{C}{1 + C} \int_{\theta=0}^{t} \phi(t - \theta) c_R(\theta) \, d\theta$$

We have to extract either $\phi(t)$ or $c_o(t)$ from this expression, where by definition $\phi(t) = c_o(t)/\int_0^\infty c_o(t) \, dt = c_o(t)/M_o$, $M_o = \int_0^\infty c_o(t) \, dt$. In practice this is sometimes simpler than it might appear at first sight because there is often a substantial time period t_d before significant amounts of tracer appear in the feed back to the mill, and thus $c_R(\theta)$ is negligible for short values of time θ. The apparent delay time included two effects. First, in many mills there will be negligible amounts of tracer leaving the mill until some time after the tracer input, because of slow mixing processes in the mill. Second, there is a finite time for material leaving the mill to be transferred back to the mill feed via the classifier. For example, in cement mills, the product from the mill is raised in bucket elevators, sent through an air separator, and the oversize material sent back to the mill feed on an air slide; the total time involved in this second delay is about 60 seconds. The first type of delay means that even where $c_R(\theta)$ does become significant, the value of $\phi(t - \theta)$ in Eq. 6 in Chapter 14 will be small. The second type of delay means that the concentration of recycled tracer at the mill feed will correspond to the concentration leaving the mill at an earlier time, that is, it will be low in the initial stages. Thus

$$c_o(t) = c(t), \quad 0 \leq t \leq t_d$$

These values of $c_o(t)$ can then be used with a first estimate of M_o to calculate $\phi(t)$ over this time range, and more values of $c_o(t)$ for the next time range calculated from Eq. 6 in Chapter 14 using the estimated $\phi(t - \theta)$ function. Note that if sampling for tracer concentration in the recycle is carried out at or near the classifier exit, the real sampling times must be corrected to the times at which this recycled material gets back to the mill feed, using an estimate of the second delay time.

The value of M_o can be estimated from $c(t)$, $c_R(t)$, and C as follows. The total quantity of tracer entering the mill, including recycle, is $C\int_0^\infty c_R(t) \, dt + M_o$. This must equal $(1 + C) \int_0^\infty c(t) \, dt$, thus

$$M_o = (1 + C) \int_0^\infty c(t) \, dt - C \int_0^\infty c_R(t) \, dt \tag{4.1}$$

The procedure can be illustrated by putting the results into a finite-difference form suitable for calculation, using short time intervals of Δt, as follows. Let the effective delay time t_d correspond to N_d time intervals, that is $N_d = t_d/\Delta t$. Then Eq. 4.1 is

$$c_o(J) = c(J), \quad 0 < J \leq N_d \tag{4.2}$$

where $t = J\Delta t$, J being an integer; the first few time intervals will often give

$c(t) = 0$. Then

$$\phi(J) = c_o(J)/M_o, \quad 0 < J \le N_d \qquad (4.3)$$

The next step is to apply these values in Eq. 6 in Chapter 14, in the form,

$$c_o(J) = c(J) - \left(\frac{C}{1+C}\right) \sum_{K=N_d}^{J} \phi(J - K)c_R(K)\Delta t, \quad N_d < J \le 2N_d$$

$$c_o(J) = c(J) - \left[\frac{C\Delta t}{(1+C)M_o}\right] \sum_{K=N_d}^{J} c_o(J - K)c_R(K), \quad N_d < J \le 2N_d$$

$$(4.4)$$

The values of c and c_R are the experimental values, so the only unknown is c_o. Similarly, this set of $c_o(J)$ values is used for the next group of values

$$c_o(J) = c(J) - \left[\frac{C\Delta t}{(1+C)M_o}\right] \sum_{K=N_d}^{J} c_o(J - K)c_R(K), \quad 2N_d < J \le 3N_d$$

$$(4.5)$$

and so on.

The calculations are easily done with a hand calculator if Δt is not taken too small. If the estimate of M_o is correct, the values of $c_o(J)$ will come down to zero at higher values of J. If it is not correct, the values will become negative if M_o is too small or may start to increase if M_o is too large. The computation is stopped when either of these possibilities occur, and the values of $c_o(J)$ are used to give a new estimate of M_o using

$$M_o = \sum_{1}^{J = \text{large}} c_o(J)\Delta t \qquad (4.6)$$

and so on. If the computation was stopped because $c_o(J) > c_o(J - 1)$, we use a straight line extrapolation through the values of $c_o(J - 1)$ and $c_o(J - 2)$ to estimate the area under the rest of the curve.

This simple procedure which is based on $c_o(t) \simeq c(t)$ at short times works better when the RTD is closer to the plug flow limit, and when the transport delay time is long. If the system is operating near to the fully mixed condition with a short second (transport) delay time, the above procedure may not work. It is then necessary to estimate the $\phi(t)$ and M_o values and converge on to final values by repeated trial-and-error calculations using a search procedure. It is clearly then convenient to have a suitable functional form for $\phi(t)$.

Appendix 5: Solution of the Convective-Mixing Equation to Give the RTD for Fully Correct Boundary Conditions

The following method of solution was suggested to us by Vaillant (1970). The solution is achieved by using the identity

$$\phi^*(l^*, t^*) = \Lambda \exp\left[(2l^* - t^*)/4D^*\right] \tag{5.1}$$

which transforms the equation into

$$\frac{\partial \Lambda}{\partial t^*} = D^* \frac{\partial^2 \Lambda}{\partial (l^*)^2} \tag{5.2}$$

with boundary conditions

$$\frac{\partial \Lambda}{\partial l^*} = \frac{\Lambda}{2D^*} \quad \text{at } l^* = 0$$

$$\frac{\partial \Lambda}{\partial l^*} = \frac{\Lambda}{2D^*} \quad \text{at } l^* = 1$$

and initial conditions

$$\Lambda = \delta e^{-l^*/(2D^*)} \quad \text{at } t^* = 0, \quad l^* > 0$$

$$= \delta \quad \text{at } t^* = 0, \quad l^* = 0$$

The solution of Eq. 5.2 from Carslaw and Jaegar (1959) is

$$\Lambda|_{t^*} = \sum_{i=1}^{\infty} z_i(l^*) z_i(0) e^{-D^* \beta_i^2 t^*} \tag{5.3}$$

where

$$z_i(l^*) = \frac{\sqrt{2}\left[\beta_i \cos(\beta_i l^*) + \sin(\beta_i l^*)/(2D^*)\right]}{\left\{\beta_i^2 + \left[1/(2D^*)\right]^2 + (1/D^*)\right\}^{1/2}}$$

and β_i is the ith positive root of

$$\tan \beta_i = \frac{4D^*\beta_i}{\left[4(D^*\beta_i)^2 - 1\right]}$$

Therefore:

$$\phi^*(l^*, t^*) = \sum_{i=1}^{\infty} \frac{2\beta_i}{\beta_i^2 + [1/(2D^*)]^2 + (1/D^*)}$$

$$\times \{\beta_i \cos(\beta_i l^*) + \sin(\beta_i l^*)/(2D^*)\}$$

$$\times \exp\left\{\frac{2l^* - t^*[4(D^*\beta_i)^2 + 1]}{4D^*}\right\} \qquad (5.4)$$

Unfortunately, this gives unstable computation for small values of D^* (large Peclet Number) because the error in the first negative term in the series of alternating terms is greater than the correct sum, if normal computer accuracy is used (64 Bit Floating Point). Consequently we have also used (Trimarchi, 1978) the asymptotic solution suggested to us by Mika (1973), which applies for small values of D^*,

$$\phi^*(t^*) = \left\{\frac{1}{4}\left(\frac{1}{D^*\pi}\right)^{1/2}\left(\frac{1}{t^*}\right)^{3/2}(1 + t^*) - \left(\frac{D^*}{\pi}\right)^{1/2}\right.$$

$$\times \left[\left(\frac{1}{2t^{*1/2}(1 + t^*)} - \frac{t^{*1/2}}{(1 + t^*)^2}\right)\psi(\eta)\right.$$

$$\left.+ \frac{1 - (t^*)^2}{4D^*(1 + t^*)}\left(\frac{1}{t^*}\right)^{3/2}\psi(\eta) + \frac{t^{*1/2}}{(1 + t^*)}\left(\frac{\partial\psi(\eta)}{\partial t^*}\right)\right]\right\}$$

$$\times \exp\left[-\frac{(1 - t^*)^2}{4D^*t^*}\right] \qquad (5.5)$$

where $\eta = t^*/(1 + t^*)$ and

$$\psi(\eta) = \sum_{k=0}^{\infty} (2\eta(\eta - 1)D^*)^k \gamma_k(\eta)$$

$$\gamma_k(\eta) = \left(\frac{1}{2k + 1} - 6\eta + 4(k + 1)\eta^2\right)\prod_{i=0}^{k}(2i + 1)$$

$$\frac{\partial\psi(\eta)}{dt^*} = \frac{1}{(1 + t^*)^2}\frac{\partial\psi(\eta)}{d\eta}$$

$$= \frac{1}{(1 + t^*)^2}\sum_{k=0}^{\infty}k(2\eta(\eta - 1)D^*)^{k-1}2D^*(1 - 2\eta)\gamma_k(\eta)$$

$$+ (2\eta(1 - \eta)D^*)^k(-6 + 8(k + 1)\eta)\prod_{i=0}^{k}(2i + 1)$$

The upper limit in the summations is chosen to terminate the computation when additional terms become negligible, since the series will start to expand if too many terms are taken.

The two solutions give identical overlapping answers for $D*t*$ near 0.025; the asymptotic solution is good for lower values of $D*t*$ and the complete solution for higher values.

Also,

$$\int_0^\infty \phi*(l*, t*)\, dt* = \sum_{i=1}^\infty \left(\frac{2\beta_i}{\beta_i^2 + [1/(2D*)]^2 + (1/D*)} \right)$$

$$\times \left\{ \beta_i \cos(\beta_i l*) + \sin(\beta_i l*)/(2D*) \right\}$$

$$\times \frac{e^{[l*/(2D*)]}}{\left[\dfrac{4(D*\beta_i)^2 + 1}{4D*} \right]} \tag{5.6}$$

was found to equal 1 for the $D*$ range of interest at each $l*$.

References

Carslaw, H. J. and Jaegar, J. C., 1959, *Conduction of Heat in Solids*, 2nd ed., Oxford Clarendon Press, pp. 114, 360.

Mika, T., 1973, Xerox Corp., private communication.

Trimarchi, T. J., 1978, "Mathematical Treatments of Some Problems in Fuel Science: Residence Time Distribution in Mills, Non-First-Order Grinding Effects, and Calcination," Ph.D. Thesis, The Pennsylvania State University, University Park, PA.

Vaillant, A., 1970, Automated Process Surveys, private communication.

Appendix 6: The Rogers-Gardner RTD Model

The basic stage of the model (see Fig. 6.1) consists of two interconnected perfectly mixed tanks (a and b) and one plug flow tank (c). This basic stage may repeat itself an integer number of times. The nomenclature for the model is given as follows:

$$n = \text{total number of stages}$$

$$F_a/n = \text{fractional holdup in tank } a$$

$$F_b/n = \text{fractional holdup in tank } b$$

$$F_c/n = \text{fractional holdup in tank } c$$

$$k = \text{fractional flow between tanks } a \text{ and } b$$

$$\tau = \text{overall mean residence time}$$

The RTD for n stages in the model can be written as

$$\phi(t) = H(u) \sum_{l=1}^{n} u^{n-l} [A_{1l}\exp(-m_1 u) + A_{2l}\exp(-m_2 u)] \quad (6.1)$$

where u is $t - F_c\tau$ and $H(u)$ is the unit (positive) step function. The m_1 and m_2 are

$$\genfrac{}{}{0pt}{}{(+)m_1}{(-)m_2} = (n/\tau)\left[\frac{F_a k + F_b(1+k)}{2F_a F_b}\right]\left[1 \pm \sqrt{1 - \frac{4kF_a F_b}{[F_a k + F_b(1+k)]^2}}\right]$$

$$(6.2)$$

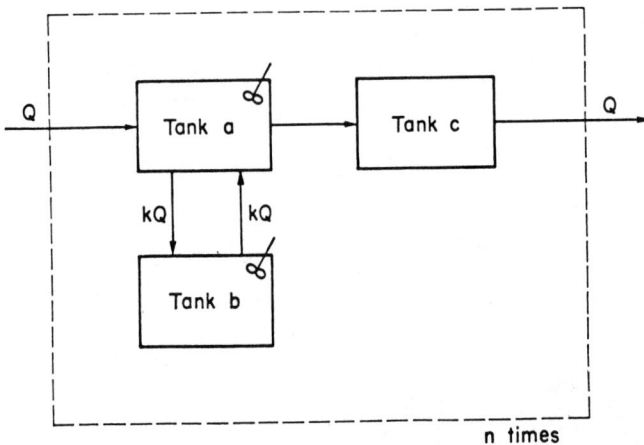

Fig. 6.1. The basic stage of the Rogers-Gardner model.

534

The A_{1l} and A_{2l} are given by $A_{jl} = (n/F_a\tau)^n \psi_{l(m_j)}/\Gamma(n - l + 1)\Gamma(l)$ where

$$\psi_{l(m_j)} = \left[\frac{m_j + nk/F_b\tau}{m_j - m_i} \right]^n$$

$$\times \sum_{r=1}^{l} \frac{\Gamma(n + r - 1)b_{lr}}{\Gamma(n - l + r + 1)(m_j + nk/F_b\tau)^{l-r}(m_i - m_j)^{r-1}},$$

$$m_j \neq m_i$$

and the b_{lr} are

$$b_{lr} = \frac{\Gamma(l)}{\Gamma(l - r + 1)\Gamma(r)}, \quad 1 \leq r \leq l$$

Appendix 7: Computational Procedure for Air-Swept Coal Mill

As discussed previously, the internal sweeping action can be treated as an apparent classification action. However, the programs used previously for mill circuits choose a value of mean residence time $\tau = F'/W$ and calculate C' and, hence, F/W. For the air-swept system, the values of c_i and η depend on the airflow so it is more convenient to set the airflow, which then defines τ, as follows.

The computation is performed as follows. Eq. 28 in Chapter 17 can be put as

$$\gamma_i = \frac{f_i + \sum\limits_{\substack{j=1 \\ i>1}}^{i-1} b_{ij}\gamma_j}{1 + \dfrac{\eta\omega(1 - c_i)}{S_i}}, \quad n > i > 1 \tag{7.1}$$

where $\gamma_i = w_i S_i \tau$. The γ values are calculated sequentially starting at $i = 1$, for known values of b_{ij}, S_i, c_i, and $\eta\omega$. (For example, the value of c_i will probably be 1 for size 1, so $\gamma_1 = f_1$.) Then, since $\sum_i w_i = 1$,

$$\tau = \sum_{j=1}^{n-1} (\gamma_j/S_j) + w_n\tau \tag{7.2}$$

Note that $\gamma_n = 0$ because $S_n = 0$, but $w_n\tau$ is not zero. It is given by

$$w_n\tau = \frac{f_n + \sum\limits_{j=1}^{n-1} B_{nj}\gamma_j}{\eta\omega(1 - \bar{c}_n)}$$

where $\bar{c}_n$ is the mean value of c for the sink interval (fortunately, $\bar{c}_n = 0$ in most cases). Thus there is a unique value of τ. The values of $w_i(= p_i)$ follow from $w_i = \gamma_i/S_i\tau$ and Eq. 30 in Chapter 17 gives the product size distribution.

When the mill circuit is closed via an external classifier it can be assumed that the external classification action can be described by a set of classifier numbers s_i. The makeup feed size distribution is g_i and the actual feed size distribution to the mill is related to the makeup recycle by $f_i F = g_i Q + t_i T$ where $F = Q + T$, Q being the makeup feed rate and T the recycle rate. Because the circulating load is defined as $C = T/Q$, then $1 + C = F/Q$. Since s_i is defined as the fraction of Fp_i' which is recycled, that is, $t_i T = Fs_i p_i'$, $f_i(1 + C) = g_i + (1 + C)s_i p_i'$. Substituting into Eq. 7.1

to eliminate f_i gives

$$\gamma_i^* = \frac{g_i + \sum\limits_{j=1}^{i-1} b_{ij}\gamma_j^*}{1 + \dfrac{\eta\omega(1 - s_i)(1 - c_i)}{S_i}}, \quad 1 \leq i < n \tag{7.3}$$

where $\gamma_i^* = (1 + C)w_i S_i \tau = w_i S_i \tau^*$. As before, the γ^* values are computed sequentially and

$$\tau^* = \sum_{j=1}^{n-1} \left(\gamma_j^*/S_j\right) + \frac{g_n + \sum\limits_{j=1}^{n-1} B_{nj}\gamma_j^*}{\eta\omega(1 - \bar{s}_n)(1 - \bar{c}_n)} \tag{7.4}$$

Then $w_i = \gamma_i^*/S_i\tau^*$, and $p_i^* = \tau^*\eta\omega(1 - c_i)w_i$ from Eq. 30 in Chapter 17 [where $p_i^* = p_i'(1 + C)$], and

$$1 + C = \sum_{j=1}^{n} p_j^* = \tau^*\eta\omega \sum_{j=1}^{n} w_j(1 - c_j) \tag{7.5}$$

Knowing C, the values of p_i' follow. The size distribution of the circuit product is given by

$$q_i = (1 + C)p_i'(1 - s_i) \tag{7.6}$$

The circuit capacity Q is W/τ^*.

It is useful to note that the closed circuit behaves like the open circuit model but with the value of $1 - c_i$ replaced with an effective value of $(1 - c_i)(1 - s_i)$. The computation program for closed circuit can thus be used for open circuit by setting $s_i = 0$ and $g_i = f_i$.

Appendix 8: Model and Computational Procedure for
Air-Swept Cement Mill

It is convenient to define arbitrarily an internal classification stage as corresponding to a single fully mixed reactor in the grinding process. If the experimentally determined residence time distribution of a mill corresponds approximately to three equal reactors in series, we will use three internal classification stages. The specific rates of breakage in each stage are $S_{i,k}$, where k denotes the kth stage. It is assumed that the primary progeny fragment distribution is the same in each stage, denoted by the symbol $b_{i,j}$, where $b_{i,j}$ is the weight fraction of material of size j which on primary breakage ends up in size interval i, $i \geq j$. The simplest treatment is for a single fully mixed stage (see Fig. 10 in Chapter 17),

$$F(1 + C)p_i = F(1 + C)f_i' - S_i w_i W + \sum_{\substack{j=1 \\ i>1}}^{i-1} b_{i,j} S_j W w_j \qquad (8.1)$$

where F is the external feed rate into the section, $1 + C$ is the circulating load ratio due to a fraction C falling back into the bed, and w_i is the fraction of size i material leaving the mill as powder flow.

Since a fully mixed stage is considered, $p_i = w_i$. The fraction of mill feed as size i, f_i', is made up of the feed into the stage plus material falling back into the bed $(1 + C)Ff_i' = Ff_i + p_i \eta \omega W c_i$, where f_i is the fraction of size i entering in the feed. Substituting this into Eq. 8.1 and collecting terms in p_i gives

$$(1 + C)p_i = \frac{f_i + \tau' \sum_{\substack{j=1 \\ i>1}}^{i-1} b_{i,j} S_j p_j (1 + C)}{1 - \eta \omega \tau' c_i + S_i \tau'}$$

where $\tau' = W/(F(1 + C))$ is the apparent mean residence time in the stage. Letting $(1 + C)p_i = p_i^*$

$$p_i^* = \frac{f_i + \tau' \sum_{\substack{j=1 \\ i>1}}^{i-1} b_{i,j} S_j p_j^*}{1 - \eta \omega \tau' c_i + S_i \tau'}, \qquad n \geq i \geq j \geq 1 \qquad (8.2)$$

The values of p_i^* can be sequentially calculated starting at $i = 1$, $p_1^* = f_1/(1 - \eta \omega \tau' c_1 + S_1 \tau')$, then using p_1^* in the calculation of p_2^*, etc.

The values of f_i, $b_{i,j}$, S_j, η, ω, and c_i must be set for the milling conditions and material used, and the computation is performed for a specific value of τ'. However, it is readily shown that the values of τ' must be bounded by 0 and $1/\eta \omega$, so that $\eta \omega \tau'$ is never greater than 1. Then since

$\sum_{i=1}^{n} p_i = 1,$

$$\sum_{i=1}^{n} p_i^* = 1 + C \qquad (8.3)$$

The values of p_i follow from $p_i = p_i^*/(1 + C)$ and the normal mean residence time defined by $\tau = W/F$ is obtained.

Letting P be the flow rate of particles out of the mill as powder and R be the flow rate as airborne suspension, Fig. 10 in Chapter 17 shows that $P = (1 + C)F - \eta \omega W$ or

$$\varepsilon = (1 + C) - \eta \omega \tau \qquad (8.4)$$

where ε is the fraction of the feed leaving in the powder stream, $\varepsilon = P/F$. Also $R = F - P$ or

$$R = F(1 - \varepsilon) \qquad (8.5)$$

Finally, the size distribution of the airborne stream, p_i' can be found from $p_i'R = p_i \eta \omega W(1 - c_i)$ or

$$p_i' = p_i \eta \omega \tau (1 - c_i)/(1 - \varepsilon) \qquad (8.6)$$

The second stage is a more general form (see Fig. 8.1). The fully mixed grinding equation is now

$$p_{i,2}(1 + C_2)P_1 = P_1(1 + C_2)f_{i,2}' - S_{i,2}p_{i,2}W_2 + \sum_{\substack{j=1 \\ i>1}}^{i-1} b_{i,j}S_{j,2}p_{j,2}W_2$$

However, f' is now substituted from $P_1(1 + C_2)f_{i,2}' + \eta \omega W_2 c_i p_{i,2} + R_1 p_{i,1}' c_i$ giving

$$p_{i,2}^* = \cfrac{p_{i,1} + \cfrac{(1 - \varepsilon_1)}{\varepsilon_1}p_{i,1}'c_i + \tau_2'\sum_{\substack{j=1 \\ i>1}}^{i-1} b_{i,j}S_{j,2}p_{j,2}^*}{1 - \eta \omega \tau_2'c_i + S_{i,2}\tau_2'}, \qquad n \geq i \geq j \geq 1 \quad (8.7)$$

where $\tau_2' = (W_2/P_1(1 + C_2))$. This equation is sequentially computed as before. However, τ_2' is not independent of τ_1', since $\tau_1 = W_1/F$ and $\tau_2 = W_2/P_1 = \tau_1/(1 - \varepsilon_1)$ as $W_1 = W_2$ for the assumptions of the model. A trial-and-error search must be made for the value of τ_2' which gives the correct value of τ_2. Thus specifying τ_1' specifies the whole subsequent calculation.

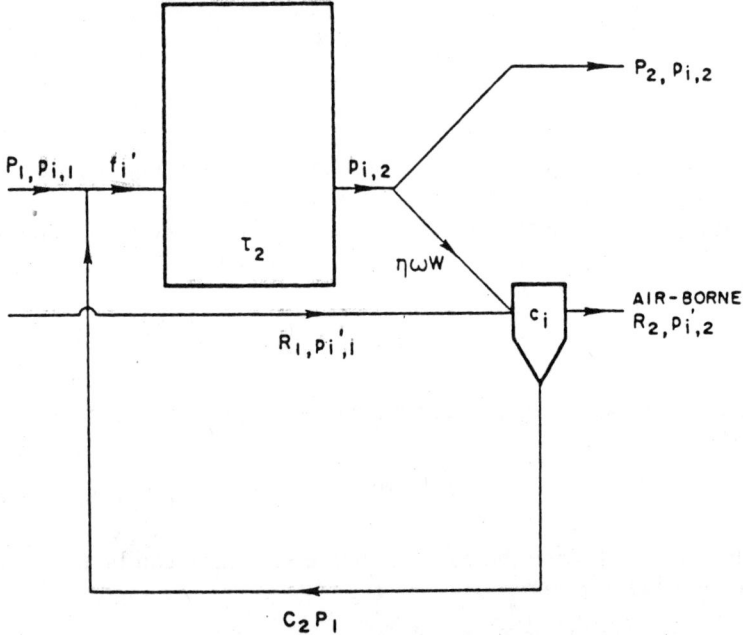

Fig. 8.1. The equivalent circuit for the second stage.

As before, $P_2 = (1 + C_2)P_1 - \eta\omega W$ and $R_2 = F - P_2$. Letting ε_2 be the fraction of P_1 which leaves as powder, $\varepsilon_2 = P_2/P_1$ and

$$\varepsilon_2 = (1 + C_2) - \eta\omega\tau_2 \qquad (8.4a)$$

$$R_2 = F - P_2 = F(1 - \varepsilon_1\varepsilon_2) \qquad (8.5a)$$

The size distribution of the airborne particles is obtained from $p'_{i,2}R_2 = R_1 p'_{i,1}(1 - c_i) + \eta\omega W_2(1 - c_i)p_{i,2}$ or

$$p'_{i,2} = p'_{i,1}(1 - c_i)\left(\frac{1 - \varepsilon_1}{1 - \varepsilon_1\varepsilon_2}\right) + p_{i,2}\eta\omega(1 - c_i)\tau_1\left(\frac{1}{1 - \varepsilon_1\varepsilon_2}\right) \qquad (8.6a)$$

The general expression for the kth stage $(k > 1)$ is clearly

$$p^*_{i,k} = \frac{p_{i,k-1} + \left(\dfrac{1 - \varepsilon_1\varepsilon_2 \cdots \varepsilon_{k-1}}{\varepsilon_1\varepsilon_2 \cdots \varepsilon_{k-1}}\right)p'_{i,k-1}c_i + \tau'_k \displaystyle\sum_{\substack{j=1 \\ i>1}}^{i-1} b_{i,j}S_{j,k}p^*_{j,k}}{1 - \eta\omega\tau'_k c_i + S_{i,k}\tau'_k},$$

$$1 < k \le n, \quad n \ge i \ge j \ge 1 \qquad (8.8)$$

with $\tau_k' = (W/P_k)/(1 + C_k) = \tau_k/(1 + C_k)$. τ_k is given by $\tau_k = \tau_{k-1}/\varepsilon_{k-1}$ $= \tau_1/(\varepsilon_1\varepsilon_2 \cdots \varepsilon_{k-1})$, that is,

$$\tau_k = (\tau/m)/(\varepsilon_1\varepsilon_2 \cdots \varepsilon_{k-1}) \tag{8.9}$$

with τ being defined by W/F, W now being the total hold-up in the mill. As before, this equation is sequentially computed for $k = 1$, $i = 1, 2, \cdots n$; $k = 2$, $i = 1, 2, \cdots n$; etc. At each stage after the first, the program conducts a trial-and-error search for the value of τ_k' which gives the correct τ_k.

The size distribution of powder leaving the final stage is

$$p_{i,m} = p_{i,m}^*/(1 + C_m) = p_{i,m}^*/\sum_{j=1}^{n} p_{j,m}^* \tag{8.10}$$

The values of ε_k are obtained from

$$\varepsilon_k = 1 + C_k - \eta\omega\tau_k \tag{8.11}$$

The size distribution of the airborne dust is obtained from (see Eqs. 8.6 and 8.6a),

$$p_{i,k}' = p_{i,k-1}'(1 - c_i)\left(\frac{1 - \varepsilon_1\varepsilon_2 \cdots \varepsilon_{k-1}}{1 - \varepsilon_1\varepsilon_2 \cdots \varepsilon_k}\right)$$

$$+ p_{i,k}(1 - c_i)\eta\omega\tau\left(\frac{1}{1 - \varepsilon_1\varepsilon_1 \ldots \varepsilon_k}\right) \tag{8.12}$$

The flow rate of powder out of the final stage is $P_m = F\varepsilon_1\varepsilon_2 \ldots \varepsilon_m$ or

$$P_m/W = (1/\tau) \prod_{k=1}^{m} \varepsilon_k \tag{8.13}$$

The flow rate of airborne particles out of the final stage is $R_m = R - P_m$ or

$$R_m/W = (1/\tau)\left(1 - \prod_{k=1}^{m} \varepsilon_k\right) \tag{8.14}$$

The size distribution of the airborne dust is obtained from Eq. 8.12 with $k = m$.

Appendix 9: Variation of Internal Classification With Air-Flow Rate

The concept used is that a particle of a given size (and density) entrained in the air stream will follow a falling path determined by its starting position, settling velocity, and the time for the air to reach the end of the mill. If the path does not reach the bed before the particle is swept to the end of the mill, it will leave in the air stream. When the velocity of air flow through the mill is increased, an entrained particle will follow the same path and have the same probability of leaving if its settling velocity is higher than the original particle so that its time for settling is decreased in the exact ratio that the time for the air to reach the end of the mill is decreased. Thus, as air velocity is increased, the size of particle which leaves the mill by a particular path also increases.

In quantitative form, the treatment is as follows. For spheres of coal (specific gravity 1.4) of 40 to 6000 μm (0.0016 $\times$ 0.24 in.), the variation of settling velocity v_s in air with particle radius r given by Fuchs (1964) was converted to the air viscosity and density at 66°C (150°F), corresponding to typical air temperatures in an air-swept coal mill. The result was fitted by the empirical expression

$$\log v_s = 2.352 - 1.383/r^{0.24} \tag{9.1}$$

where r is in mm, v_s in m/s. At a given air-sweeping velocity u in the mill, the experimental value of c_i represents the mean effect of all possible paths for powder of radius r_i with a flow time for air in the mill of t. At any other flow rate u' the time is $t' = tu/u'$. In order for particles of radius r' to go through the same set of paths to give the same c_i, the settling velocity must increase proportionally to $1/t$. Thus $v_s't' = v_s t$ and $v_s'/v_s = u'/u$. Using Eq. 9.1,

$$r = 1/\left[(r')^{-0.24} + 0.723 \log(u'/u)\right]^{4.17} \tag{9.2}$$

The equation is arranged in this form so that setting a value for r' enables the calculation of r for a given u'/u. The value of c for this value of r can be interpolated from an experimental c_i vs. r_i curve.

In the case given here $u \propto 0.61$ m³/s for the c_i curve of Fig. 9 in Chapter 12 (at 1290 ACFM). This was fitted by the empirical curve

$$c = 1 - \exp\left[-\left(\frac{x - 53}{219}\right)^{0.72}\right] \tag{9.3}$$

where x was in μm, that is, $x = 2r \times 10^3$. Thus for x' in the desired $\sqrt{2}$ sequence, Eq. 9.2 enables the calculation of x, and Eq. 9.3 then gives c_i.

For particles of different density ρ_s or gas of different density ρ_g or viscosity η, the following relations are used:

$$Re = 2r\rho_g v_s/\eta \tag{9.4}$$

$$Re^2 C_D = \left(\frac{32}{3}\right) r^3 \rho_s \rho_g g/\eta^2 \tag{9.5}$$

where Re is Reynold's number and C_D the drag coefficient: Eq. 9.5 applies for settling under gravity. For a given Reynold's number the value of C_D is fixed so the value of Re^2C_D is also fixed. Thus the relations between v_s, r, ρ_s, ρ_g, and η for a given Re are

$$r_1 \rho_{g1} v_{s1}/\eta_1 = r_2 \rho_{g2} v_{s2}/\eta_2$$

and

$$r_1^3 \rho_{s1} \rho_{g1}/\eta_1 = r_2^3 \rho_{s2} \rho_{g2}/\eta_2$$

hence

$$v_{s2}/v_{s1} = \left(\rho_{s2}/\rho_{s1}\right)^{1/3}\left(\rho_{g1}\eta_2/\rho_{g2}\eta_1\right)^{2/3}$$

Thus if Eq. 9.1 applies for ρ_{s1}, ρ_g, and η_1, for other conditions it becomes

$$\log v_s = (2.352 + A) - (1.383B)/r^{0.24} \tag{9.1a}$$

where

$$A = \log\left[\left(\rho_{s2}/\rho_{s1}\right)^{1/3}\left(\rho_{g1}\eta_2/\rho_{g2}\eta_1\right)^{2.3}\right]$$

$$B = \left(\rho_{s1}\rho_{g1}\eta_2/\rho_{s2}\rho_{g2}\eta_1\right)^{0.08}$$

The development then proceeds as before.

With respect to mill diameter, the same set of *geometrically similar* settling paths might be expected for the same L/D at the same gas flow velocity. The ratio of u' to u is calculated from volume flow rates V by

$$u'/u = \frac{V'/(D')^2}{V/D^2} \tag{9.6}$$

If the c relation of Eq. 9.3 is to be used, $V = 0.61$ m³/s (1290 ACFM) and $D = 0.98$ m (3.2 ft), for $L/D = 1.5$, that is, $u = 1.04$ m/s (3.4 fps).

Reference

Fuchs, N. A., 1964, *The Mechanics of Aerosols*, The Macmillan Co., New York.

Appendix 10: BIII Method for Calculating *B* Values

This method requires as input an estimate of the rate constants of breakage, although the estimate need not be very accurate. The estimate is obtained from experimental data on rates of breakage; the values of S for the various size intervals can be estimated by interpolation.

In principle, the size distribution produced at some time of grinding from a specified feed is a unique result of the S and B parameters, and knowing S, B can be back-calculated. However, the algebra of the back-computation is nontractable, except for the two top size intervals 1 and 2. For these two intervals, the fractional amounts broken to less than size i in time t of grinding are:

fraction less than size i from size 1

$$= B_{i,1}\big[w_1(0) - w_1(t)\big], \quad i > 1$$

$$= B_{i,1}\Delta w \qquad (10.1)$$

fraction less than size i from size 2

$$= B_{i,2}\int_0^t S_2 w_2(t)\, dt$$

$$= B_{i,1}\Delta_2, \quad i > 2 \qquad (10.2)$$

Now $w_2(t)$ is known from the Reid solution

$$w_2(t) = \frac{S_1 b_{2,1}}{S_2 - S_1}\left(e^{-S_1 t} - e^{-S_2 t}\right)w_1(0) + e^{-S_2 t}w_2(0) \qquad (10.3)$$

Thus

$$\Delta_2 = S_2\left[\frac{S_1 b_{2,1} w_1(0)}{S_2 - S_1}\left(\frac{1 - e^{-S_1 t}}{S_1} - \frac{1 - e^{-S_2 t}}{S_2}\right) + w_2(0)\left(\frac{1 - e^{-S_2 t}}{S_2}\right)\right]$$

and substituting for $S_1 b_{2,1} w_1(0)/(S_2 - S_1)$ from Eq. 10.3

$$\Delta_2 = S_2\left[\frac{w_2(t) - e^{-S_2 t}w_2(0)}{e^{-S_1 t} - e^{-S_2 t}}\left(\frac{1 - e^{-S_1 t}}{S_1} - \frac{1 - e^{-S_2 t}}{S_2}\right)\right.$$

$$\left. + w_2(0)\left(\frac{1 - e^{-S_2 t}}{S_2}\right)\right]$$

Since $1 - \exp(-S_1 t) = \Delta w/w_1(0)$, and $1 - \exp(-S_2 t) = 1 - (1 - \Delta w/w_1(0))^{S_2/S_1}$,

$$\Delta_2 = \frac{w_2(t) - w_2(0)(1 - A)^r}{A + (1 - A)^r - 1}\big[1 - (1 - A)^r - rA\big] + \big[1 - (1 - A)^r\big]w_2(0)$$

$$(10.4)$$

where for printing convenience

$$A = \Delta w / w_1(0) \qquad (10.5)$$

and

$$r = S_2 t / \ln[1/(1 - \Delta w / w_1(0))] \qquad (10.5a)$$

Δ_2 has the physical significance of the amount broken from the second interval, and is calculated from Eqs. 10.4, 10.5, and 10.5a.

Similar use of the Reid solution for the third, fourth, etc., intervals becomes intractable, and therefore the amount broken out of the jth interval is approximated by a simple linear mean rate:

fractional amount broken to less than size i from breakage

of j material $\simeq B_{i,j} S_j t \left[w_j(0) + w_j(t) \right] / 2, \quad i > j \geq 3$

This is an excellent approximation for short grind times where $w_j(t)$ varies almost linearly with time.

Now the fractional amount below the upper size of size 3 produced during grinding is the sum of contributions from sizes 1 and 2, and hence

$$P_3(t) - P_3(0) = B_{3,1} \Delta w + \Delta_2 .$$

or

$$B_{3,1} = \left[P_3(t) - P_3(0) - \Delta_2 \right] / \Delta w \qquad (10.6)$$

In the same way,

$$P_4(t) - P_4(0) \simeq B_{4,1} \Delta w + B_{4,2} \Delta_2 + S_3 t \left[w_3(0) + w_3(t) \right] / 2$$

Now if B is normalized, $B_{i,j} = B_{i-j+1,1}$; if it is not, this can be used as an approximation for the one size fraction method at short grind times because the value of $P_4(t)$ is dominated by $B_{4,1}$ and the error in setting $B_{4,2} = B_{3,1}$ is negligible. Then

$$B_{4,1} \simeq \frac{P_4(t) - P_4(0)}{\Delta w} - B_{3,1} \frac{\Delta_2}{\Delta w} - \frac{S_3 t \left[w_3(0) + w_3(t) \right]}{2 \Delta w}$$

where $B_{3,1}$ is known from Eq. 10.6.

In general

$$B_{i,1} \simeq \frac{P_i(t) - P_i(0)}{\Delta w} - B_{i-1,1} \frac{\Delta_2}{\Delta w} - \sum_{k=i-1}^{3} \frac{B_{i-k+1,1} S_k t \left[w_k(0) + w_k(t) \right]}{2 \Delta w}$$

$$n \geq i \geq 4 \qquad (10.7)$$

where the solution marches down from $i = 4$.

To summarize, the required data are: values of S, $w_1(0) = 1 - P_2(0)$, $\Delta w = w_1(0) - w_1(t) = P_2(t) - P_2(0)$, and values of $P_3(t), P_4(t), \ldots P_n(t)$; $P_3(0), P_4(0), \ldots P_n(0)$. Then, Δ_2 is calculated using Eqs. 10.4 and 10.5; $B_{3,1}$ is calculated from Eq. 10.6; the values of $B_{i,1}$ are calculated from Eq. 10.7 starting with $i = 4$ and working down.

INDEX

547

M